ADVANCES IN
HEMODYNAMICS AND HEMORHEOLOGY

Volume 1 • 1996

ADVANCES IN HEMODYNAMICS AND HEMORHEOLOGY

Editor: T. V. HOW
Department of Clinical Engineering
University of Liverpool

VOLUME 1 • 1996

JAI PRESS INC.

Greenwich, Connecticut *London, England*

55 Old Post Road No. 2
Greenwich, Connecticut 06836

JAI PRESS LTD.
The Courtyard
28 High Street
Hampton Hill
Middlesex TW12 1PD
England

ISBN: 1-55938-634-7

Transferred to digital printing in 2008.

CONTENTS

LIST OF CONTRIBUTORS

D. Barthes-Biesel	Université de Technologie de Compiègne
R.A. Black	Department of Clinical Engineering University of Liverpool
Michael W. Collins	Thermo-Fluids Engineering Research Centre
G.G. Couch	Institute of Biomedical Engineering University of Toronto
David N. Firmin	Magnetic Research Unit Royal Brompton Hospital, London
Belinda Ha	Department of Biomedical Engineering University of North Carolina, Chapel Hill
J. Michael Hasenkam	Department of Thoracic and Cardiovascular Surgery and Institute of Experimental Clinical Research Aarhus University Hospital, Denmark
Frank S. Henry	Thermo-Fluids Engineering Research Centre City University, London
William Henry	Department of Pediatric Cardiology University of North Carolina, Chapel Hill
T.V. How	Department of Clinical Engineering University of Liverpool
P.E. Hughes	Department of Clinical Engineering University of Liverpool
Carol Lucas	Department of Biomedical Engineering University of North Carolina, Chapel Hill

Raad H. Mohiaddin	Magnetic Research Unit Royal Brompton Hospital, London
M. Ojha	Institute of Biomedical Engineering University of Toronto
Hsing-Wen Sung	School of Chemical Engineering Georgia Institute of Technology
George B. Thurston	Department of Bio-Medical Engineering and Mechanical Engineering University of Texas, Austin
Takami Yamaguchi	School of High-Technology for Human Welfare Tokai University
Ajit Yoganathan	School of Chemical Engineering Georgia Institute of Technology

PREFACE

Over the last 30 years, considerable progress has been made in hemodynamics and hemorheology, spurred on by continual advances in cardiovascular medicine and surgery, on the one hand, and new developments in theoretical and numerical analyses and measurement techniques and instrumentation, on the other. During this period, several landmark volumes consisting of exhaustive reviews written by acknowledged experts have appeared, albeit infrequently.

The purpose of this series is to present up-to-date reviews covering a broad range of topics in this rapidly advancing field. It is anticipated that all aspects of hemodynamics and hemorheology will eventually be covered, ranging from the rheology of blood and blood vessels and the mechanics of blood flow in arteries and veins to the highly complex phenomena of microcirculation. The contributions will try to reflect the advances being made in experimental techniques and instrumentation for laboratory and clinical measurements, and in mathematical and numerical modeling. While this series should rightly focus on the scientific and engineering principles involved, special attention will also be given to the clinical significance of this important area of research.

T.V. How
Series Editor

VISCOELASTIC PROPERTIES OF BLOOD AND BLOOD ANALOGS

George B. Thurston

Advances in Hemodynamics and Hemorheology,
Volume 1, pages 1–30.

ISBN: 1-55938-634-7.

I. INTRODUCTION

The heart delivers energy to the blood with each pulse, part of which is dissipated through viscosity, another part stored in the elastic structure, and another stored in the kinetic mass of the blood. Because early studies of blood flow were carried out using steady flow only, they do not provide a complete basis for understanding pulsatile flow in the circulation (see Appendix A). Though elastic energy is stored in the fluid during the initiation of flow, its presence is hidden from observation when the flow reaches a steady state. This means that steady flow measurements reflect only the rate of energy dissipation.

The timing of the heart beat establishes the time scale for events occurring in the larger arteries. Therefore, the pressure pulse can be decomposed into sinusoidal components with the bulk of the energy in the frequency range from approximately 1 to 10 Hertz, making the viscoelastic properties of the blood in that frequency range particularly significant.

In the early theoretical work analyzing oscillatory blood flow carried out by Womersley[1] blood was treated as a Newtonian viscous fluid. Using oscillatory tube flow measurements, however, Thurston[2,3] showed that human blood exhibits pronounced elasticity in addition to viscosity, and that the theory of oscillatory flow of a viscoelastic fluid in cylindrical tubes[4] was applicable to the flow of blood. Viscoelasticity now is known to be a basic rheological property of blood, which is amenable to measurement in any geometrical confinement.

The viscoelastic properties of blood depend primarily on the elastic behavior of the red blood cells. Since their abundance at normal hematocrits leaves little space for cell motion or deformation without direct interaction with neighboring cells, the tendency for interacting cells to aggregate and undergo elastic deformation are important underlying factors in blood viscoelasticity. But aggregation and deformability alone do not provide a complete picture of events that occur in blood flow, since flow-induced cell organization is a third major factor.[5] Together, these three factors make human blood viscoelastic, nonNewtonian, thixotropic, and dilatant.[6]

With the development of cardiovascular prosthetic devices, prosthetic heart valves, pulsatile blood pumps, models of the circulatory system, and sophisticated studies of flow through the complex geometries of the circulation, the need developed to produce an analog fluid with the rheological properties of blood. The classical analog, a mixture of glycerine and water, lacks elasticity and provides only viscous and inertial effects, while a few polymer solutions and particulate suspensions provide some of the viscoelastic properties of human blood.

In this review, the spectrum of rheological properties of normal human blood are described together with the properties of selected viscoelastic analogs. The roles of hematocrit, temperature, time, plasma composition, red cell aggregation, and deformability in blood viscoelasticity are presented.

II. ORIGIN AND CHARACTER OF BLOOD VISCOELASTICITY

The red blood cell is the structural element primarily responsible for blood viscoelasticity. Calculations show that the closest packing of red blood cells without cell deformation gives an hematocrit of 58%,[7] so in the range of normal hematocrits there is little plasma space available for free cell motion without deformation. Consequently blood flow is accomplished through cell deformation and cell-to-cell sliding on thin plasma layers. The stretching of red cells and recovery of initial shape is commonly observed[8] and its significance to the circulation has prompted many studies to assess red cell deformability (e.g., ref. 9). However, the time for a deformed cell to recover has been measured by optical methods to be near 0.1 s,[10] which places it on the same scale as the time between cell-to-cell contacts at physiological flow rates and frequencies. Therefore, the kinetics of deformation of the red cell must be considered along with aggregation and organization as important to blood flow.

The viscoelasticity of blood directly reflects how the aggregation and deformability of the red blood cells affect their ability to flow. In order for flow to occur, the red cells must rearrange to provide plasma pathways on which they can slide.[5, 11] For oscillatory flow with shear strain amplitudes <1, cells can "wiggle" as their motion relative to nearest neighbors remains small. When the strain >1, however, the cells must pass from one set of neighbors to an adjacent set, and this is accomplished by cell deformation plus realignment (see Figure 1). At higher strains, the formation of cell layers becomes perfected and flow is due to these layers sliding on adjacent plasma layers. The layering is retained if the frequency of flow reversal

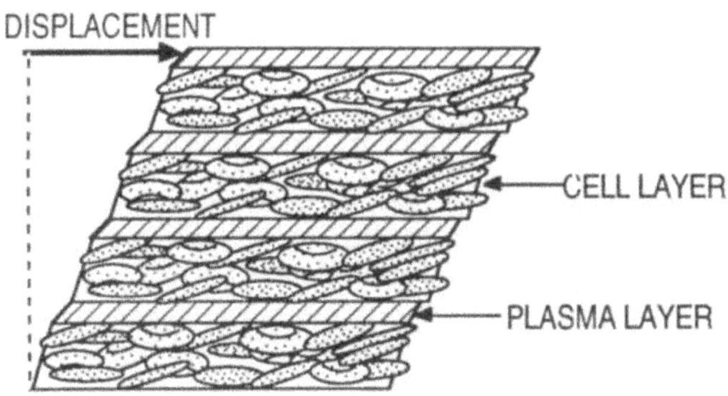

Figure 1. Formation of compacted cell layers separated by plasma layers. When the shear displacement produces greater than unit strain, the red blood cells tend to orient to form ordered layers of cells separated by plasma layers on which they slide. Higher strains and higher shear rates produce more compact cell layers and enlarged plasma layers.[5]

is so rapid that the cells do not have time to return to their quiescent state of disorganization. With normal blood, cell layering occurs at low stress and produces a "super-fluid" having reduced viscosity and elasticity. But for cells with elevated aggregation tendency or for cells that resist deformation, the stress required is elevated. Consequentially the viscosity and elasticity are increased. Because of these changes in microstructure with increasing oscillatory strain, the viscoelastic properties change with flow amplitude.

A. Basic Parameters and Methods of Measurement

Blood viscoelasticity has been measured using both oscillation and step transient flows with conventional rotational instruments.[12–15] However, the nonlinear viscoelastic properties of blood were first measured using principles of oscillatory flow of a viscoelastic fluid in a rigid, cylindrical tube of circular cross section.[2,3] The tube flow method has the distinct advantage of sensitivity as well as being a first approximation to flow in the longer vessels of the circulatory system and consequently is most often used, especially in clinical applications. The physical principle for oscillatory tube flow was first developed for a Newtonian, viscous fluid[16–18] but it was not until the theory for oscillatory flow of a viscoelastic fluid was developed[4] that nonlinear elastic properties were found in blood. The character of the velocity profile in the tube was shown to depend on a dimensionless parameter,

$$Y = a[\rho\omega/\eta]^{1/2} \tag{1}$$

where a is the tube radius, ρ is the density of the fluid, ω is the radian frequency of the oscillatory flow, and η is the magnitude of the complex coefficient of viscosity, with

$$\eta^* = \tau^*/\gamma' \tag{2}$$

$$= \eta' - i\eta'' = \eta \exp(-i\phi) \tag{3}$$

where η^* is the complex viscosity, τ^* is the complex shear stress, γ' is the shear rate, η' is the viscosity, and η'' is the elasticity. The shear stress can be given in terms of its viscous and elastic components, where

$$\tau^* = \tau' - i\tau'' \tag{4}$$

with the viscous stress $\tau' = \eta'\gamma'$ and the elastic stress $\tau'' = \eta''\gamma'$. The velocity profile approaches a piston-like movement when Y >20, but for Y<1 it is parabolic, the condition under which measurements often are performed.

To determine the rheological parameters of stress, strain, and shear rate, measurements are made of the magnitude and phase of oscillatory pressure and volume flow through a cylindrical tube of known dimensions (see Appendix B for a summary of the equations relating the rheological parameters to the pressure-flow relations.) Most of the measurements reported herein were performed with an automated instrument using the above principle (Vilastic Scientific, Inc.), where the frequency of oscillatory flow and amplitude are precisely controlled while the measurements are performed and the rheological properties are calculated.

B. Normal Blood: Stress, Strain, and Shear Rate

An evaluation of the viscoelastic properties of blood most closely related to the behavior of blood flow *in vivo* is obtained by performing measurements at a frequency near that of the normal pulse. These measurements are made over a range of amplitudes of oscillation to cover the extreme conditions ranging from near stasis to the maximum shear rates *in vivo*. Figures 2 and 3 show the result of such a set of measurements on blood with an hematocrit of 43%. Measurements were made at 2 Hz using a tube 1 mm in diameter at a temperature of 22°C, stepping from lower to higher amplitudes while holding each flow for 12s, during which time the magnitude and phase of the pressure and volume flow were determined.

At the frequency of measurement, 2 Hz, the magnitude of the strain and shear rate are related by $\gamma' = \omega\gamma = (2\,\pi \text{ frequency})\,\gamma = 12.56\,\gamma$. The effects of oscillatory flow on the blood structure change with amplitude. For strain <0.1, the quiescent structure of the blood is not modified by the flow, and the stress components are directly proportional to the strain while the viscosity and elasticity values are constant. As the strain is increased between 0.1 and 1, the cell aggregates are progressively reduced in size; when the strain >1, the cells are forced into a degree

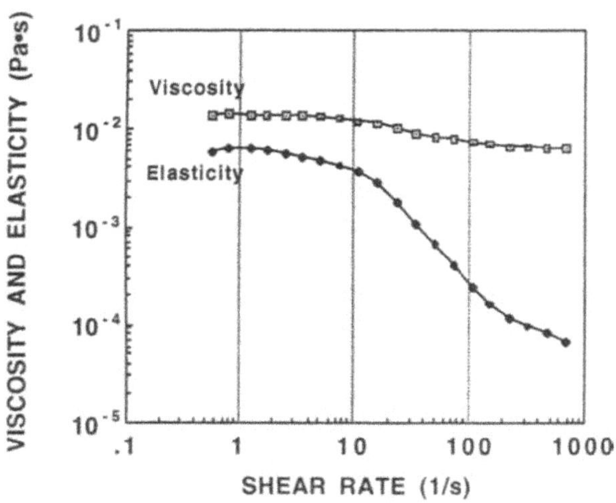

Figure 2. Viscosity and elasticity of a normal blood sample measured at 2 Hz in a 1 mm diameter tube at 22°C. The shear rates are given as rms values of the sinusoidal oscillation.

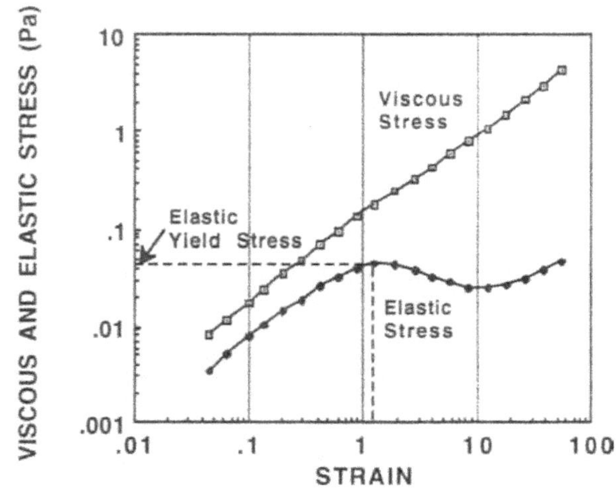

Figure 3. Viscous and elastic components of the shear stress using the same data base in Figure 2. The stress and strain are rms values.

of alignment. This marks a final yielding of the aggregated cell structures and is identified by the elastic yield stress value of 0.045 Pa at a strain of 1.27, as shown in Figure 3. As the strain is further increased in the range from 1 to 10, the cell alignment and compaction forming cell layers separated by plasma layers (shown in Figure 1) becomes essentially complete. This stable layered structure is retained at higher strains.

C. Steady, Oscillatory, and Pulsatile Flow

The fundamental difference between measured steady flow viscosity and viscoelasticity is that in steady flow the strain is increasing continuously whereas in oscillatory flow it reverses periodically. In oscillatory flow, measurements can be made sensing the response of blood to very small strains that produce minimal microstructural changes, while in steady flow the relative displacement of neighboring cell groups increases without limit and the blood structure is modified from the quiescent state. The changes in microstructure are very similar when comparing steady flow viscosity at shear rates >100 s^{-1} and oscillatory flow shear strain >10. In addition, as the frequency is reduced, the shear rate of congruence of the steady flow and oscillatory flow viscosities is lowered. Figure 4 shows measurements performed at frequencies ranging from 0.1 to 2 Hz for a blood sample with 0.46 hematocrit together with the steady flow viscosity for a 0.43 hematocrit blood. Steady flow measurements were performed in a concentric cylinder viscometer having an annular gap of 0.5 mm, while the oscillatory flow measurements were made using a 1 mm diameter cylindrical tube.

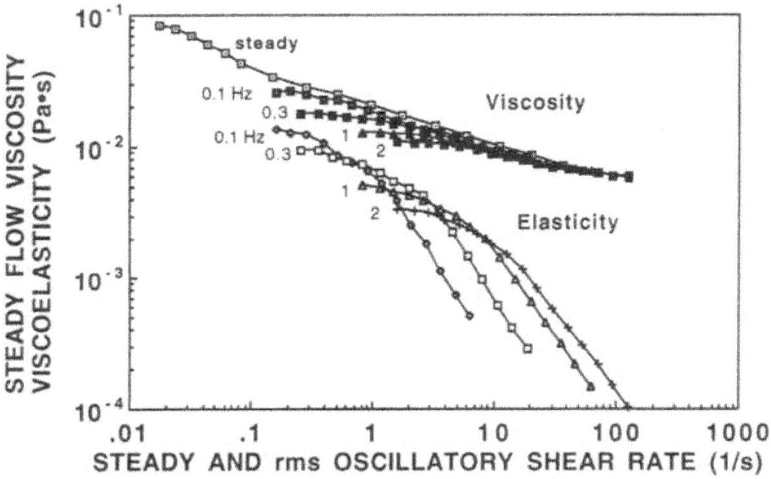

Figure 4. Steady flow viscosity for 0.43 hematocrit blood and oscillatory flow viscoelasticity for 0.46 hematocrit blood measured at frequencies of 0.1, 0.3, 1, and 2 Hz.

Pulsatile flow is produced by superimposing oscillatory flow on steady flow, providing a closer approximation to blood flow *in vivo*. Because of the nonlinear relation between viscosity and viscoelasticity versus shear rate, change in one flow component has an effect on the pressure to flow relation for the other[19] so that a high shear rate of one component puts the blood into the high shear rate condition for the second component even though that shear rate may be low.

D. Frequency and Relaxation Processes

Due to the complex structure of the red blood cell and the overall microstructure of blood, the spectrum of relaxation processes is very broad.[6] Consequently, within the range of Fourier frequency components of the pulse (~1–10 Hz), the viscoelastic properties are moderately constant. As seen in Figure 4, the changes in viscoelasticity are more strongly dependent on shear rate than frequency. If unit strain is selected, $\gamma = 1$, the changes in viscoelasticity with frequency are as shown in Figure 5, which shows the viscosity and elasticity near the yield strain together with the apparent relaxation time T_r for each frequency, where

$$T_r = \eta''/(\omega\eta'). \qquad (5)$$

It is seen that the apparent relaxation times range over several decades when excited in the frequency range of the measurements.

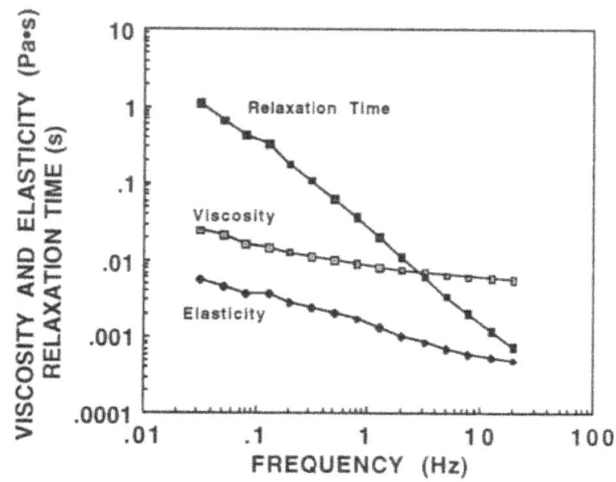

Figure 5. Frequency dependence of the viscoelasticity of normal blood at an oscillatory strain of 1 (rms). The apparent relaxation time is shown for each frequency.

III. CONTROL FACTORS FOR VISCOELASTICITY OF NORMAL BLOOD

The two factors, hematocrit of the blood and temperature at which measurements are made, are usually specified with viscoelasticity data. A third factor, the time history of the flows used during measurement, often is not given, yet all three of these factors require special attention when establishing controlled conditions of measurement.

A. Hematocrit

Because of the central role of the elastic red cell, the hematocrit H, is a strong parameter in establishing the level of viscoelasticity of blood.[3] Since the microstructure of blood depends on the rate of flow, the functional dependences on the hematocrit, $\eta'(H)$ and $\eta''(H)$, vary with the shear rate as well.[20] The shear rate dependence of the viscosity and elasticity at 2 Hz of normal cells in autogenous plasma at 22°C are shown in Figure 6, where it is seen that the character of the curves is similar for the hematocrits of 0.35 to 0.65 except in the elasticity at high shear rates. The viscosity at shear rates of 1 s^{-1} and 10 s^{-1} increases exponentially with hematocrits from 0.25 to 0.95, while the elasticity increases nearly proportional to the cube of the hematocrit.

Functional relations can be developed for normalization of measured viscosity and elasticity to a reference hematocrit. To see how the viscoelasticity changes with

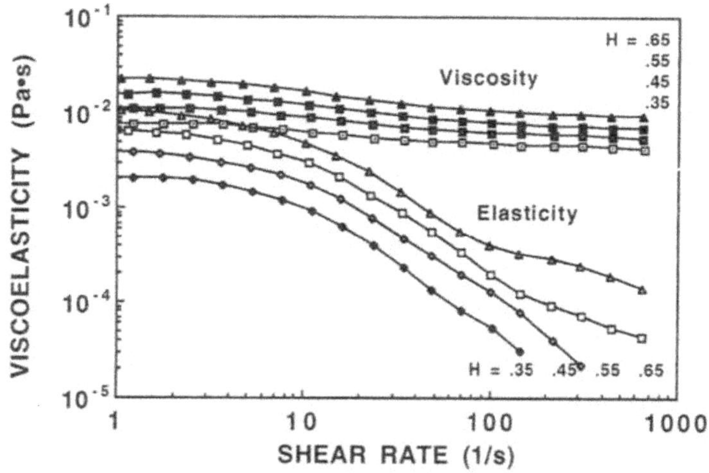

Figure 6. Shear rate dependence of viscoelasticity for hematocrits of 0.35, 0.45, 0.55, 0.65. Measurements were made at 2 Hz and 22°C.

hematocrit in the physiologic range, values of the viscosity and elasticity at a shear rate of 10 are plotted versus the hematocrit in Figure 7.

The solid lines in the figure represent the functional relations,

$$\eta' = 0.0019736 \exp(3.2361H) \tag{6}$$

and

$$\eta'' = 0.00015344 \exp(5.240H), \tag{7}$$

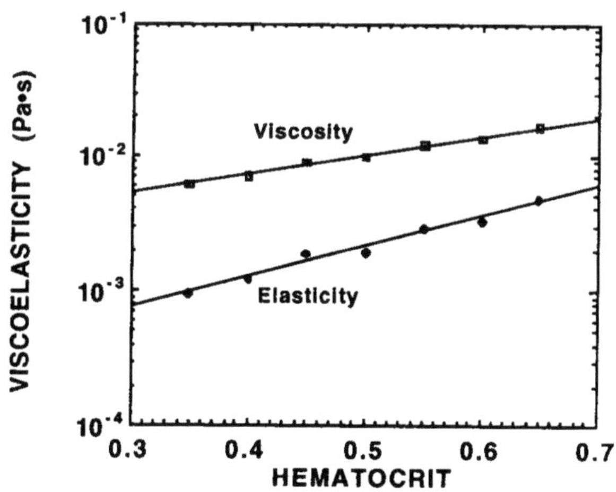

Figure 7. Viscosity and elasticity measured at 2 Hz, 22°C, and at a shear rate of 10 s^{-1} rms. The solid lines are exponential fits to the data.

where the viscoelasticity units are *Pa·s* and *H* is the hematocrit fraction. Similar expressions have been derived by statistical analysis of 233 normal donors at 2 Hz, shear rate of 10 s^{-1}, and 37°C[21] and for 113 donors differentiated by age and gender.[22] The results are in general agreement considering possible differences in the normal populations, sample handling, and measurement protocol to accommodate for thixotropic effects.

B. Temperature

The rheological properties of blood are usually measured either at 37°C or near "room" temperature between 20 and 25°C. For comparison purposes, it is important to know how the measured properties change in this temperature range. Figure 8 shows the 2 Hz shear rate dependence of viscoelasticity at 15, 21.8 and 37°C for normal blood at 0.47 hematocrit. Though the values shift with temperature the essential character of the curves is unchanged.

The rate of change of plasma viscosity, water viscosity, and blood viscoelasticity measured at 2 Hz and unit strain (shear rate = 12.56 s^{-1}) is shown in Figure 9. In the temperature range from 9 to 42°C all of the data can be fitted with exponential functions of the form $\eta = \alpha\ 10^{-\beta T}$. The solid lines shown in the figure are as follows:

$$\eta' = 0.02147 \bullet 10\text{^}(-0.01264\ T) \tag{8}$$

$$\eta'' = 0.004213 \bullet 10\text{^}(-0.01022\ T) \tag{9}$$

$$\eta_{plasma} = 0.003362 \bullet 10\text{^}(-0.01017\ T) \tag{10}$$

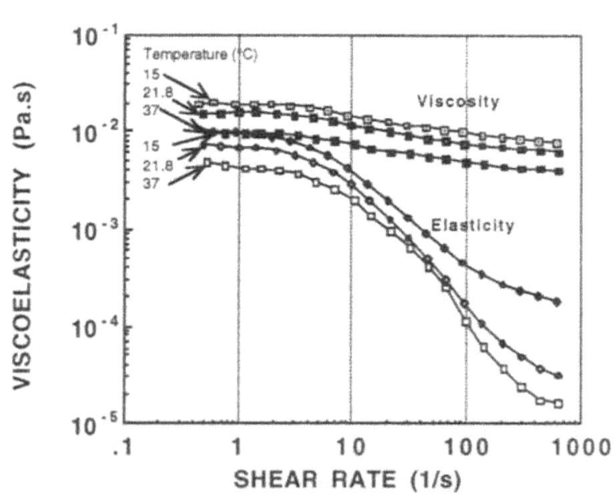

Figure 8. Shear rate dependence of viscoelasticity at three temperatures. The frequency was 2 Hz and the blood hematocrit was 0.47.

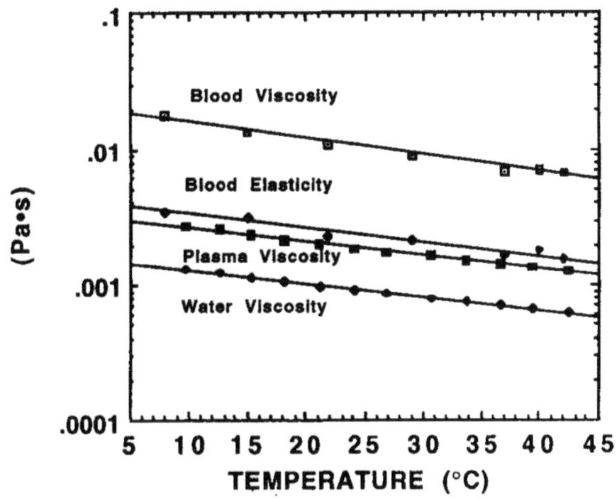

Figure 9. Change of plasma viscosity, water viscosity and blood viscoelasticity with temperature. Measurements were at 2 Hz and unit strain (shear rate = 12.56 s^{-1}). The blood is that shown in Figure 8.

$$\eta_{water} = 0.001598 \bullet 10\text{^}(-0.009867\ T) \qquad (11)$$

where T is in (°C) and η is in ($Pa{\cdot}s$). While the coefficients α of η' and η'' are specific to this blood sample, the exponents can be used for other normal blood samples to project values measured at one temperature to those expected at a reference temperature T_o. This can be done using

$$\eta(To) = \eta\ (T)\ 10^{-\beta(To-T)} \qquad (12)$$

where η *(T)* is the viscosity or elasticity at the temperature T and β is the corresponding constant in the experimentally determined exponent.

C. Time (Thixotropy)

The relatively long time required for the microstructure to stabilize following a rapid change in rate of flow makes blood thixotropy readily observable. This effect on viscosity has been studied using steady flow[23] and on viscoelasticity using oscillatory flow.[6] The time required for a flow induced microstructural change to occur increases with the size of the change in shear rate. At higher shear rates, structural change occurs more rapidly that at lower shear rates. Figure 10 shows the thixotropy of blood due to sudden changes in shear rate. The frequency was 2 Hz and the blood hematocrit was 0.46. The initial step was from the no-flow condition to a shear rate of 10 s^{-1} rms, a rate near the unit strain condition marking major structural rearrangement of cell alignment to form cell and plasma layers. The viscosity and elasticity decrease over a period of approximately 20 s before

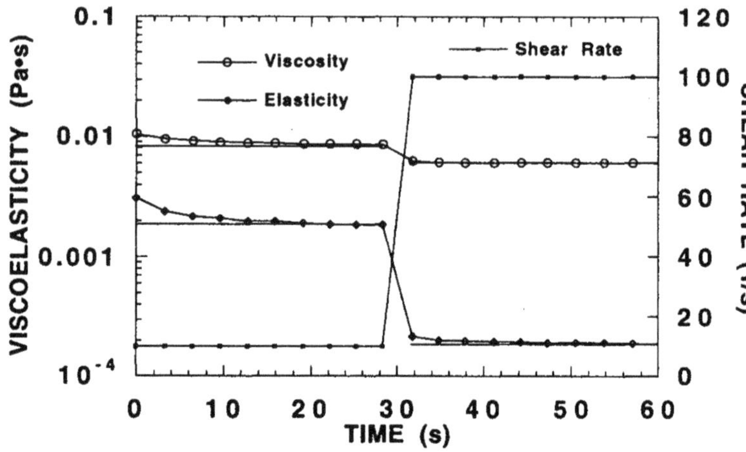

Figure 10. Time dependent change in viscoelasticity following step change in shear rate. The initial step is from 0 to 10 s^{-1} and the next step is from 10 to 100 s^{-1}. The blood hematocrit is 0.46. Measurements were at 2 Hz at 23°C.

final state is attained. Next, the shear rate steps from 10 to 100 s^{-1} and the viscosity and elasticity drop abruptly to within a few percentage points of final values. The time dependence of the viscoelasticity following increasing or decreasing shear rate steps is described by a sum of one or more exponential functions,[6]

$$\eta(t) = \eta_i + S\sum_{p=1}^{N} W_p[1 - \exp(-t/\lambda_p)] \tag{13}$$

where η_i is the initial viscosity or elasticity, W_p is a weighting function, λ_p is a characteristic time, and $S = -1$ for decreasing viscoelasticity and $S = +1$ for increasing values.

Attaining equilibrium following a decrease in shear rate is much slower than with an increase in shear rate. The structural recovery following cessation of steady flow has been thoroughly analyzed and attributed to recovery of aggregates.[24] The optical response to cessation and initiation of oscillatory flow is shown in Figure 11 where it is seen that the recovery of quiescent structure requires approximately one minute while the high shear rate structure is attained in a few seconds.[25]

When establishing a measurement protocol for clinical purposes, the role of thixotropy must be taken into consideration. After introduction of a blood sample into a measurement apparatus, a specific period of rest prior to initiation of measurements is required. The timing between steps in shear rate should be precise and reproducible. Each step may be accompanied by a period during which flow is established before the measurement is made. With such precautions, highly reproducible measurements can be obtained.

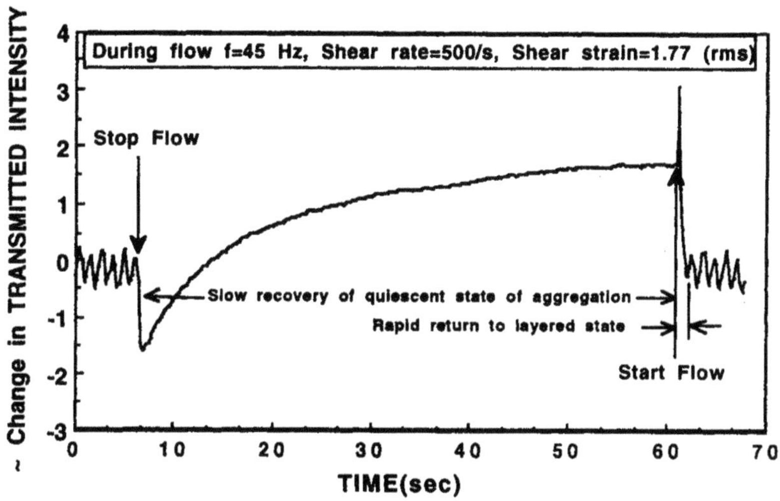

Figure 11. Changes in light transmission through a layer of blood following sudden cessation of oscillatory shear and subsequent reinitiation of flow. Approximately 60 s are required for the high shear rate layered structure of blood to return to the quiescent aggregated state. The return to the layered state requires only a few seconds.

IV. CAUSES OF CHANGE IN BLOOD VISCOELASTICITY

To understand both normal and abnormal pressure-flow relations and blood viscoelasticity, it is helpful to examine the effects of several independently identifiable properties of the red cells and of the suspending plasma. The dependence of viscoelasticity of whole blood on laboratory parameters attainable from analysis of blood and plasma has been examined by use of statistical methods. It is found that the viscosity and elasticity measured at 2 Hz and a shear rate of 10 s^{-1} are strongly dependent on hematocrit, with weaker dependence on plasma viscosity, platelets, and plasma proteins (fibrinogen, albumin, and cholesterol).[26] Other studies concern properties that can be modified or controlled, including the tendency for red cells to aggregate, deformability of the red cells, mean red cell volume, and composition of the suspending plasma.

A. Influence of Red Cell Deformability

In order to see the effects of loss of deformability (increase of cell rigidity), red cells can be modified by hardening with acetaldehyde and by changing cell volume by changing osmotic pressure. The effect of cell hardening is to increase viscosity and elasticity at higher shear rates, producing an effect called viscoelastic dilatancy.[27] When the oscillatory strain amplitude exceeds unit value, the red cells resist stable layer formation because of their limited ability to conform to neighboring

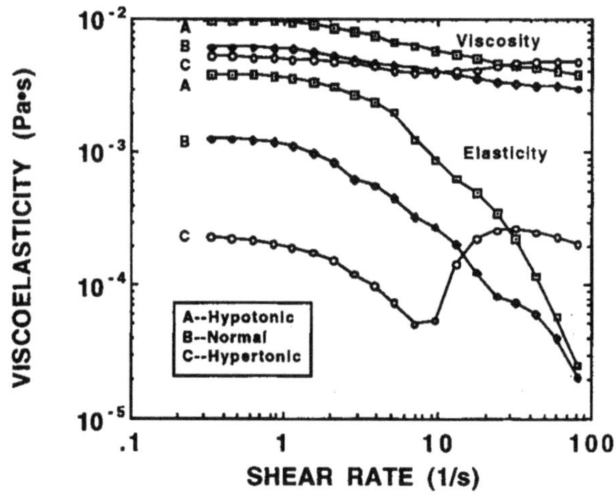

Figure 12. Effects of osmotic pressure on viscoelasticity of three cell suspensions with the same number of cells per unit volume. Case A (215 mOSM/l, H = 0.43), B (310 mOSM/l, H = 0.33), and C (946 mOSM/l, H = 0.19). Measurements were a 2 Hz and 22°C.[27]

cells and to release trapped plasma, making the cell layers unstable at higher rates of flow. Similar results have been obtained with cells hardened with glutaraldehyde.[28] The effects of cell hardening are also seen at low cell concentrations.[29]

With normal cells, the down turn in the elastic stress in the range of strains from 1 to 10 (see Figure 3) is indicative of the ease with which the cells conform into layers by cell deformation; reduced negative slope generally indicates some loss of cell deformability.

The effects of modified deformability are illustrated in Figures 12 and 13 which show the effects on the viscosity and elasticity versus shear rate and the viscous and elastic stress versus shear strain measured at 2 Hz. A single blood sample was modified by addition of hypotonic saline, isotonic saline, and hypertonic saline to swell the cells, leave them normal, or shrink the cells. In the figures case A (215 mOSM/l), B (310 mOSM/l), and C (946 mOSM/l) show major differences in viscoelastic properties for the same number of cells per unit volume. The swollen cells in case A show the highest viscosity and elasticity, primarily due to elevation of the effective hematocrit to 0.43, whereas for the normal cells the effective hematocrit is 0.33, and the viscosity and elasticity are lower. However, with the shrunken and hardened cells of case C, a pronounced viscoelastic dilatancy appears, with upturns in both viscosity and elasticity for shear rates above 10 s^{-1}. The level of the elastic stress at a strain of 5 compared to that at 0.2 is a clear indicator of the red cell deformability.

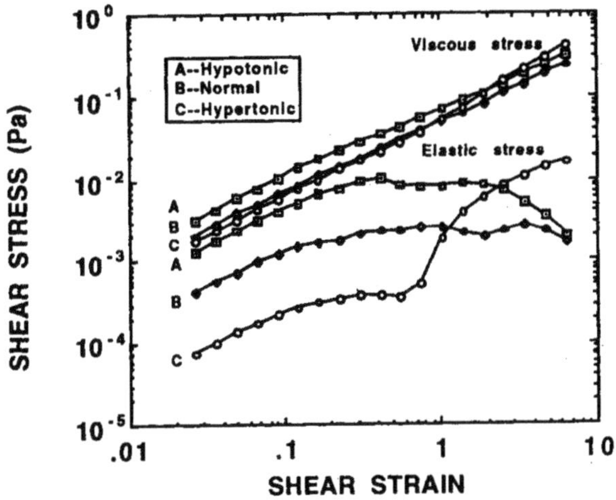

Figure 13. Viscous and elastic components of the shear stress versus shear strain for the cases shown in Figure 12.

B. Influence of Red Cell Aggregation

Red cell aggregation has been investigated extensively as it relates to sedimentation[30] optical transmission[25,31] and microscopic examination of aggregates.[32] The influence of higher molecular weight macromolecules in increasing red cell aggregation tendency is well established. Using this effect, it has been shown that the viscosity and elasticity of a cell suspension are increased by addition of 1% Dextran 150[33] and by addition of Dextran T110.[28]

Aggregation tendency can be lowered by reducing the protein concentration of the plasma, including fibrinogen that has been shown to play a significant role in aggregation.[34] The effect on viscoelasticity of diluting blood plasma is shown in Figures 14 and 15. The blood sample was divided into two parts. One part was modified by diluting the plasma 75% with isotonic saline while retaining the normal hematocrit. This gave two matched samples, one with normal aggregation tendency and one with greatly lowered aggregation tendency as well as lowered plasma viscosity. The measurements show the shape of the elasticity and the elastic stress is substantially modified while the viscous component is shifted in level together with some shape change. These changes due to plasma dilution can be explained on the basis that by reducing aggregation, the red cells have enhanced freedom of movement and consequently can align and form cell layers at lower elastic stress levels. Once the transition is completed, however, the viscoelastic properties become nearly identical, being affected only by the reduction in plasma viscosity.

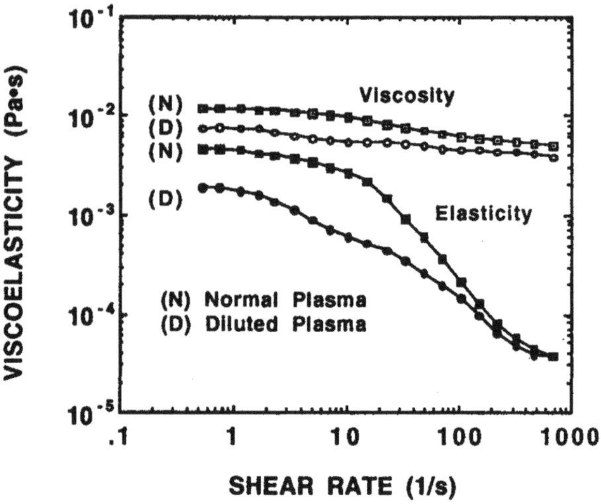

Figure 14. Viscosity and elasticity versus shear rate for 0.44 hematocrit cells in normal plasma [*N*] and for the same cells suspended in the plasma diluted 75% with isotonic saline[*D*]. Measurements were at 2 Hz and 22°C.

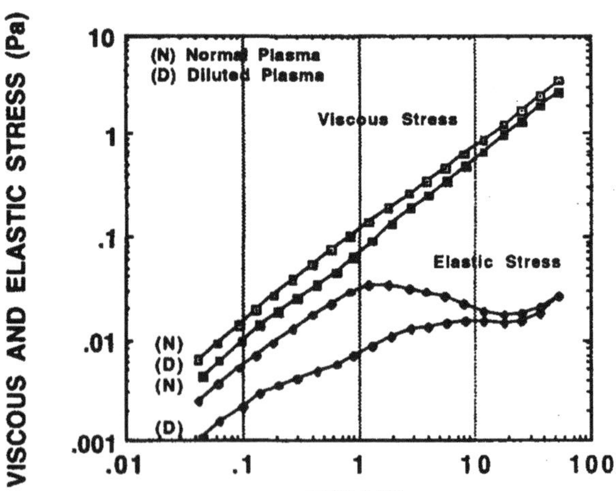

Figure 15. Viscous and elastic stress versus shear strain for the cell suspension of Figure 14.

C. Influence of Suspension Medium

In laboratory tests, the natural plasma in which the red cells are suspended is usually modified by addition of an anticoagulant such as ethylenediaminotetracetic acid or sodium citrate. These anticoagulants have little effect on the blood viscoelasticity (other than dilution of plasma) when measured within a few hours of drawing the blood.

In clinical procedures requiring hemodilution, blood volume expanders may have major effects on viscoelasticity. To demonstrate this, three equal portions of a normal blood sample of 0.46 hematrocrit were diluted 50% with autogenous plasma, with lactated Ringer's injection solution (Travenol Laboratories), and Dextran 40 (Cutter Laboratories, 10% Dextran 40 plus 5% dextrose), reducing the hematocrit to 0.31. The viscosities of the suspending media were 0.00179 Pa·s (plasma), 0.00127 Pa·s (Ringer's plus plasma), and 0.00368 Pa·s (Dextran plus plasma). The resulting viscoelastic properties measured at 2 Hz and 22°C are shown in Figures 16 and 17.

The figures show that the dilution media have distinctly different effects on viscoelasticity. Comparing values at low shear rates and strains for which the red cell organization is near the quiescent state, the viscosity and elasticity are in the same order as the viscosity of the modified plasmas, though the levels are not directly proportional to plasma viscosity. This order is also true for the viscosity at high shear rates and strains, but the elasticity and the elastic stress differ in character.

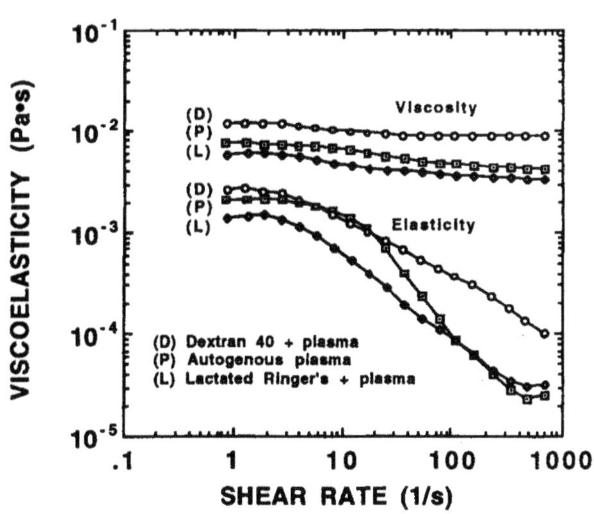

Figure 16. Shear rate dependence of viscoelasticity for normal 0.46 hematocrit blood diluted to 0.31 hematocrit by addition of (D) Dextran 40, (P) autogenous plasma, and (L) lactated Ringer's solution. Measurements are at 2 Hz and 22°C.

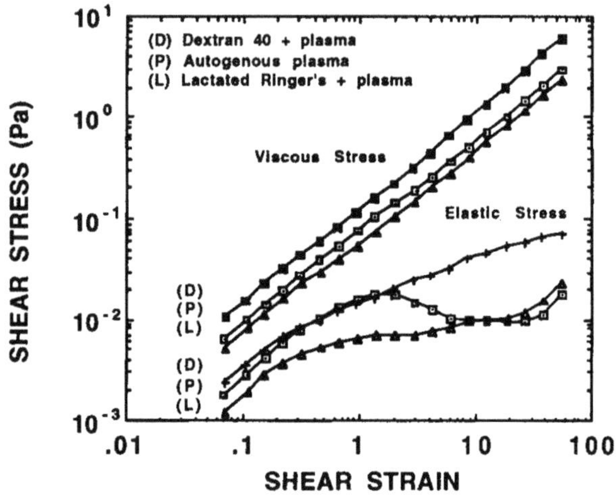

Figure 17. Viscous and elastic components of the shear stress versus shear strain for the three cases shown in Figure 16.

The plasma sample appears normal in that the elastic stress undergoes a pronounced transition in the strain range from 1 to 10, and the lactated Ringer's shows a similar transition but with a lower elastic stress below unit strain. The Dextran case does not show this transition, which could be related to the close match between the mechanical properties of the red cells and the viscosity of the surrounding medium.[29]

V. CLINICAL APPLICATIONS OF BLOOD VISCOELASTICITY

The viscoelasticity of blood measured at a few well defined flow conditions often is used for routine clinical evaluation. Measurements are usually made at 2 Hz because this is close to the period of the normal pulse. Most of the measurements reported are for flow in a cylindrical tube approximately 1 mm in diameter, which is large compared to the size of the red blood cell yet small enough that simple equations for tube flow are applicable (see Appendix B). An assessment of the blood under one or more of three oscillatory flow conditions provides the principal characterization information:

1. In the fully aggregated quiescent state ($\gamma' \approx 1\ s^{-1}$ or $\gamma \approx 0.1$).
2. Near the transition point for yield stress ($\gamma' \approx 10\ s^{-1}$ or $\gamma \approx 1$).
3. In the superfluid state ($\gamma' \approx 50\ s^{-1}$ or $\gamma \approx 5$).

These three conditions represent distinct states of the microstructure. Measurements made at a shear rate near 1 s^{-1} or a strain near 0.1 (rms values) give values of the viscosity and elasticity near the quiescent, fully aggregated state. The major transition in blood structure occurs near unit strain, which at 2 Hz is a shear rate of 12.56 s^{-1}, while the point of maximum stress with normal blood usually occurs at a strain between 1 and 2. Assessment of the properties near this transition, where the cells must reorganize in order to facilitate movement of one cell past another, is often done at a shear rate of 10 s^{-1}, slightly below unit strain. Finally, the degree of perfection of the transition to the superfluid state can be obtained from measurement at shear rates above approximately 50 s^{-1}, or at a shear strain of 5 or above. These higher flow conditions reflect the ability of the cells to compress into layers while releasing trapped plasma on which to slide. Loss of deformability reduces the efficiency of layer formation, and is reflected in elevated elasticity and elastic stress as well as in elevated viscosity.

Normal values for the viscosity, elasticity, and shear stress for the several flow conditions vary with hematocrit, measurement temperature, and demographics of the normal population. The plots shown in Figures 2 and 3 are typical of normal values measured at 2 Hz and 22°C, with 0.43 hematocrit blood. Statistically determined values measured at a shear rate of 10 s^{-1} for larger populations have received the most attention.[20–22] In many clinical studies, the selected control is a population of normals. Some researchers find it expedient to adjust the hematocrit prior to measurement to a desired reference value, such as 0.45 hematocrit, while others prefer to measure at natural hematocrits. The measured values can be projected to a reference condition using methods previously described (for both hematocrit and temperature), recognizing that the projections may be strictly valid only for normals.

Abnormal viscoelasticity can be attributed to laboratory parameters[26] such as the plasma viscosity and concentration of specific proteins, or to red cell aggregation and deformability. Aside from such attribution, Witte and Anadere have proposed[35] that among atherosclerotic risk factors, hyperviscoelasticity is an individual pathological finding, independent of other known pathogenic factors, and should be treated directly.

Pathological bloods tested show elevated viscoelasticity, with the elasticity exhibiting the greater deviation from normality. Values of viscosity and elasticity have been correlated with several disease conditions: diabetic microangiopathy,[36] acute necrotizing pancreatitis,[37] cardiovascular disease, peripheral vascular disease, stroke,[38,39] diabetes and myeloma,[40] hyperthyroid activity,[41] sickle cell anemia,[42] cardiovascular disease and atheriosclerosis,[35] cerebrovascular disease, myocardial infarction, coronary heart disease,[21] atheriosclerotic lesions,[43] apoplexy,[44] and peripheral vascular disease.[45]

Other correlations have been made with smoking,[45,46] age,[46] hypervolaemic hemodilution,[47] and blood leeching.[48] Physiological function of increased oxygen uptake has been linked to an elevated viscoelasticity.[49]

Therapeutic aspects of Ginko biloba flavone glucosides (Tebonin®) have been correlated with blood hyperviscoelasticity in arterial circulatory disturbances[35] and stroke,[50] and the local anesthetic procaine hydrochloride has been shown to elevate viscoelasticity,[51] as is the case with oral contraceptives.[46]

VI. ANALOGS FOR HUMAN BLOOD

In many applications the natural qualities of blood present special problems:

- Composition differs from donor to donor, precluding a fixed standard.
- Properties change with time because of metabolic processes.
- Opacity restricts optical studies of flow.
- It can carry disease, presenting a hazard to workers.
- It is a two-phase, unstable system subject to sedimentation.
- It can clot, blocking its fluid-like flow.
- The volume obtainable from a single donor is limited, restricting high volume applications.

Consequently, in special applications it is desirable to use blood analogs that are free of complicating qualities. In addition the stress-strain-shear rate relation for blood varies with the timing of flow, rate of flow, and duration of flow because it is viscoelastic, nonNewtonian, thixotropic, and dilatant. A safe, stable, transparent, low cost alternative having well established rheological properties matching those of human blood has not been found. Rather available analogs mimic blood only in selected ways.

Applications for analogs include research into the relation between the rheology of blood and how flow occurs through complex structures.[52] Analogs are also needed for testing wear and failure of heart valve prostheses, testing the qualities of blood pumps, for use in modeling the circulatory system and in pulse duplication systems, and as a reference fluid for instrument calibration. The most commonly used analog has been an aqueous glycerol solution, which at 45% glycerine gives a Newtonian viscosity of 3.97 mPa·s at 25°C, about twice the viscosity of blood plasma, but in the neighborhood of that of normal blood at high steady shear rates. Glycerol is low cost and transparent, but exhibits none of the viscoelastic flow properties of human blood.

Since the discovery of the viscoelasticity of blood, much effort has been devoted to finding a suitable analog that possesses this property. Ideally, a suspension of ellipsoidal particles with oblate shape, having elastic properties similar to red blood cells and some aggregation tendency would be a good candidate. In this direction, 1 μm diameter polystyrene spheres suspended in water with addition of Dextran 70 and calcium chloride to induce aggregation[53] produced some limited range of viscoelastic properties. Suspensions of biconcave, disc-shaped particles in Dextran 70 also have be used to produce a viscoelastic analog.[52]

Shear rate dependent viscoelastic character very similar to blood has been found using solutions of water soluble polymers. These are usable for modeling blood flow in uniform channels with cross sections which are large compared to the granularity of blood. These solutions may not be suitable models for flow through irregular channels because they show measurable normal forces[54] while blood does not.[55] Also, no information is available on expected differences in extensional properties of blood compared with polymers. Differences may be due to the layered character of the flow induced microstructure in blood as compared to more rapidly responding entanglements with polymer chains.

Many polymer-solvent mixtures are possible that are nontoxic, stable, low cost, and optically transparent. Aqueous-based systems are attractive from the standpoint of handling and cleanup. In general, the higher molecular weight polymers are required in order to exhibit the spectrum of relaxation times characteristic of blood.[6] Hydrolyzed polyacrylamide has been the basis for several analogs,[52,54,56] and the biopolymer Xanthan gum has been well characterized as a blood analog.[54,56,57] Polyacrylamide solutions are sensitive to ionic environment and when dissolved in distilled water, a maximum of viscoelasticity is obtained. While the polyacrylamide solutions are transparent, Xanthan gum solutions are slightly opaque and insensitive to ionic environment. By the addition of glycerine and adjustment of Xanthan gum concentration, these analogs have been produced for blood over a wide range of hematocrits.[57]

Figures 18 and 19 show an example of the viscoelastic properties of a hydrolyzed polyacraylamide (Pfizer Flopaam 2000) analog for normal blood. The molecular weight of the polymer was approximately 12 million Daltons, and the solutions contained 300 ppm in distilled water. The shear rate dependence of the steady flow viscosity is shown along with the 2 Hz oscillatory flow viscosity and elasticity. The

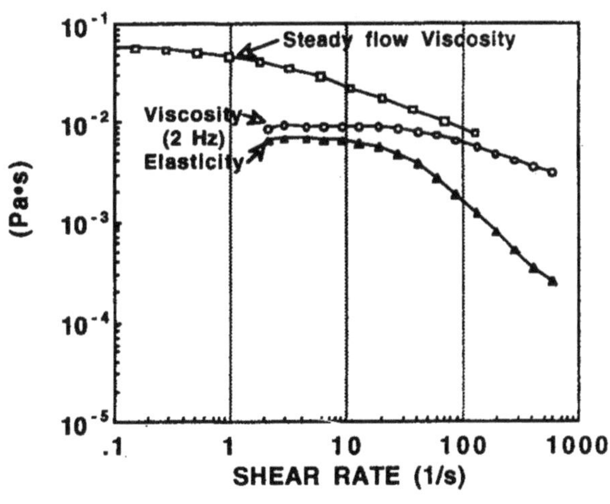

Figure 18. Viscoelasticity versus shear rate for a 12 million Dalton polyacrylamide 300ppm in distilled water. Measurements are at 2 Hz, 23°C.

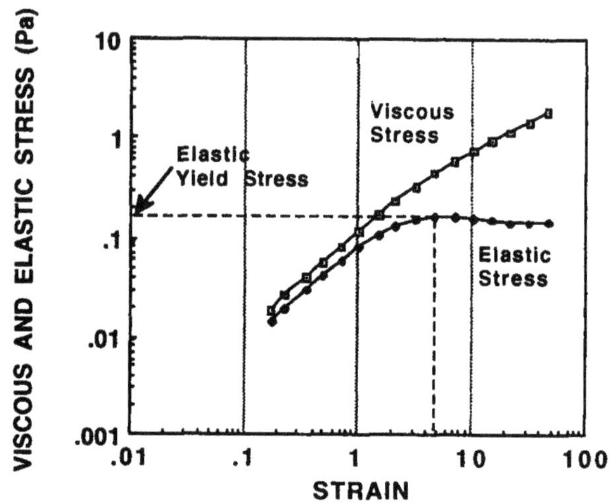

Figure 19. Shear stress versus strain for the polyacrylamide solution shown in Figure 18.

viscosity and elasticity versus shear rate in Figure 18 and the shear stress versus strain in Figure 19 can be compared with the data for a normal blood shown in Figures 2 and 3. While the character of the curves are similar, quantitative differences are evident. Compared with blood, the polymer solution viscosity is low, the elasticity is high and the yield stress is elevated.

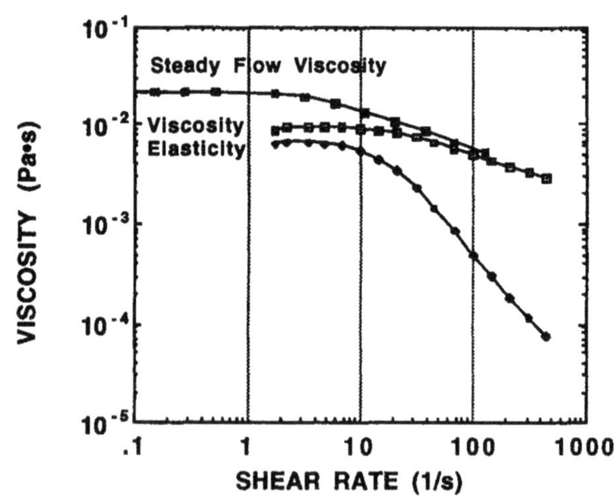

Figure 20. Steady flow viscosity and viscoelasticity versus shear rate for a Xanthan gum 500 ppm in distilled water. Viscosity and elasticity measurements are at 2 Hz, 22°C.

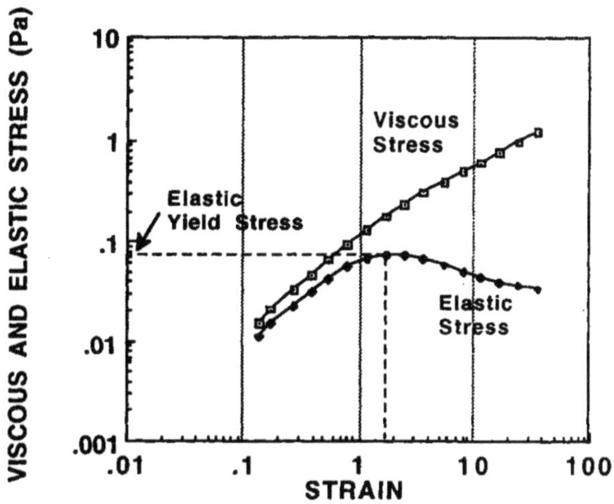

Figure 21. Shear stress versus strain for the Xanthan gum solution shown in Figure 20.

Xanthan gum is a water soluble macromolecule obtained from the bacterium *Xanthomonas campestris.* It is a relatively stiff, rod-like molecule with persistence length of 500–800 nm and molecular weight of approximately 2 million Daltons.[58] The viscoelastic properties of solutions have been measured over a wide range of flow excitation and the relaxation processes analyzed.[59] The viscoelastic properties measured at 2 Hz, 22°C for a Xanthan gum (Pfizer Biopolymer 4800P) solution are shown in Figures 20 and 21. The solution contained 500 ppm gum in distilled water. The elasticity and elastic yield stress are very similar to that of the blood shown in Figures 2 and 3, but the viscosity is below that for blood. By use of glycerol solutions as solvent, an excellent balance between viscosity and elasticity can be achieved with Xanthan gum[57] as well as with suspensions of chopped gel fragment.[56]

VII. CONCLUSION

Compared with research on the nonNewtonian viscosity of human blood, work on viscoelasticity is relatively new. Much work remains in order to integrate knowledge of the origin and applications of blood viscoelasticity to flow *in vivo* where vessel geometry is complex and vessel sizes range from large where bulk blood properties may be used, to small where the granularity of blood must be considered. But the basic concept prevails: since blood flow *in vivo* is time varying, the relation between the time varying pressure gradient and volume flow must be attributed in part to the viscoelastic properties of blood. Therefore, when the circulation fails, abnormal blood viscoelasticity must be considered as a possible contributing factor.

APPENDIX A: DEFINITIONS

The red blood cell is an elastic entity that dominates the way blood flows, both in large vessels and in the microcirculation. Consequently the viscoelastic properties of blood are evident throughout the circulation. This review is concerned with the bulk properties of blood, that is, those displayed in confined spaces that are large compared to the dimensions of the red blood cell (RBC). In such large spaces, the blood can be treated as a continuum from the perspective of definitions of the fundamental parameters, shear stress, shear strain, and shear rate (time rate of shear strain).

Consider a small cubical volume of blood. Under the action of forces that produce shear stress, the shape shifts to a parallelogram. Figure 1 shows such a cube at rest and immediately following application of force. The change in shape of the cube has two components, one is elastic deformation E and the other is slippage S. The elastic deformation is accompanied by storage of elastic energy within the structure of the blood, while the slippage is associated with a continuous input of viscous energy. When the force is removed, the deformed cube undergoes a partial recovery of shape as the elastic energy is recovered; the shape change due to slippage is permanent. Thus, in steady flow, the displacement component S continues to increase, and measurements of the non-time-varying force and velocity provide no information about the elastic energy component. But in any time varying flow, the elastic energy component also varies with time, and may be either increasing or decreasing, while the viscous energy is always increasing. Consequently, the relation between the time varying force and velocity reflects both the elastic and viscous properties of the blood.

Figure A1 can be used for defining the following fundamental parameters:

$$\text{Shear Stress} = \tau = F / A \tag{A1}$$

$$\text{Shear Strain} = \gamma = D / H \tag{A2}$$

$$\text{Shear Rate} = \gamma' = V / H. \tag{A3}$$

Sinusoidal time varying flow provides a basis for clear differentiation of the elastic and viscous properties of blood, and for understanding the role of viscoelasticity in more complex time varying flow, such as pulsatile flow.

Figure A2 shows the cubical volume of fluid in oscillatory shear with accompanying sinusoidal functions descriptive of the shear rate and the shear stress.

The sinusoidal time variations in τ and γ' are as shown in the Figure A2. The phase angle $\phi = 0$ if the fluid is purely viscous, $\phi = 90°$ if it is purely elastic, and ϕ is between 0° and 90° if it is viscoelastic. The size and phase relation between the stress, strain, and shear rate are described using complex numbers,

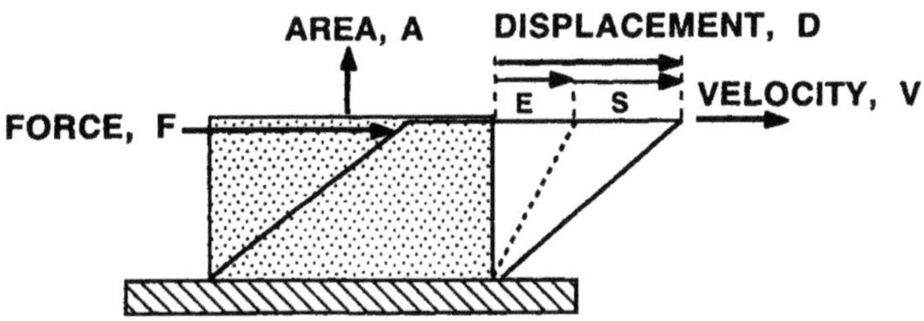

Figure A1. Diagram showing a small cubical volume of height *H* in shear. The displacement *D* due to deformation is composed of two parts: an elastic part *E* and a sliding part *S*. With a constant force *F*, *E* remains constant but *S* continues to increase. When the force is removed, *E* diminishes to zero while *S* remains unchanged.

$$\text{Shear Stress:} \quad \tau = \tau^* e^{i\omega t} \ \text{ with } \tau^* = \tau_m e^{-i\phi} \tag{A4}$$

$$\text{Shear Strain:} \quad \gamma = \gamma^* e^{i\omega t} \ \text{ with } \gamma^* = \gamma_m e^{-i\pi/2} \tag{A5}$$

$$\text{Shear Rate:} \quad \gamma' = \gamma'^* e^{i\omega t} \ \text{ with } \gamma'^* = \gamma_m' e^{-i0} \tag{A6}$$

where the radian frequency $\omega = 2\pi f$, f is the frequency in Hertz, and the phase of the shear rate is arbitrarily selected as 0. The complex shear stress can be written as

$$\tau^* = \tau' - i\tau''$$

where

$$\tau' = \tau_m \cos\phi \text{ and } \tau'' = \tau_m \sin\phi \tag{A7}$$

where τ' is the viscous stress and τ'' is elastic stress. Using these terms the complex coefficient of viscosity is defined:

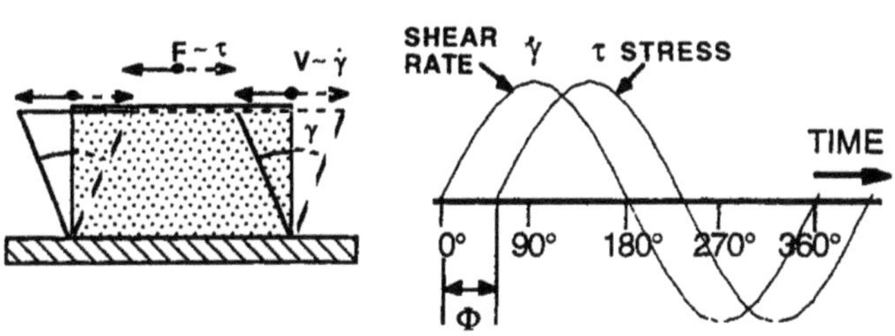

Figure A2. Sinusoidal deformation of a cubical volume of fluid. The sinusoidal time varying shear rate and shear stress differ in phase by the angle Φ as shown.

$$\eta^* = \tau^*/\gamma'^*$$
$$= (\tau'/\gamma') - i\,(\tau''/\gamma')$$
$$= \eta' - i\,\eta'' \qquad \text{(A8)}$$

where η' is the viscosity and η'' is the elasticity.

The instantaneous viscous energy loss and elastic energy storage vary with time at twice the frequency of the flow. When averaged over an integral number of cycles, the power dissipated per unit volume is

$$(\text{Power/Volume}) = \eta'\,\gamma'^2 = \tau'\,\gamma' \qquad \text{(A9)}$$

and the elastic energy stored builds to a maximum value before recovery, which is given by

$$(\text{Maximum elastic energy/Volume}) = \eta''\,\gamma'^2/\omega = \tau''\,\gamma. \qquad \text{(A10)}$$

Thus, the viscous stress τ' is the rate of energy dissipation per unit volume, per unit shear rate, and the elastic stress τ'' is the maximum energy stored during the cycle per unit volume, per unit strain.

APPENDIX B: OSCILLATORY TUBE FLOW THEORY

Oscillatory flow of a viscoelastic liquid in a rigid cylindrical tube of circular cross section has been analyzed to obtain the relations between the pressure gradient and the volume flow (time rate of volume displacement). The relation between the pressure and flow for a tube of length L is described by the impedance Z,

$$Z = P^*/U^* = R + iX \qquad \text{(B1)}$$

where P^* is the complex pressure difference across the length of tube, U^* is the complex volume flow, R is the resistance, X is the reactance of the tube, and $i = (-1)^{1/2}$. Direct measurements of the magnitudes of the pressure P and volume flow U together with the phase difference $\Theta = (\Theta_P - \Theta_U)$ can be used to determine the resistance and reactance

$$R = (P\cos\Theta/U) \text{ and } X = (P\sin\Theta/U). \qquad \text{(B2)}$$

The impedance can also be calculated from a knowledge of tube dimensions together with the density and viscoelastic properties of the liquid in the tube. It is assumed that the liquid is uniform and continuous throughout the tube, a fulfilled condition as long as the granularity of the liquid is small compared to the tube diameter. (See ref. 5 for discussion of the effects of cell layering in small and large diameter tubes.) The radius of the tube is a, the density of the liquid is ρ, and the viscoelastic modulus of the liquid is η^*. If the radian frequency of the oscillation is ω, then impedance is a function of the dimensionless parameter Y,[4] where

$$Y = a\,[(\rho\omega)\,/\,(|\eta^*|)]^{1/2} \tag{B3}$$

and an angle $\sigma = (\phi/2) - (\pi/4)$ where ϕ is the angle of the complex coefficient of viscosity, $\phi = \tan^{-1}(\eta''\,/\,\eta')$. Then the complex impedance is

$$Z = [(i\rho\omega L)/(\pi a^2)]\{1 - [2J_1(Y\exp(i\sigma)]/[Y\exp(i\sigma)J_0(Y\exp(i\sigma)]\}^{-1} \tag{B4}$$

where J_0 and J_1 are zero and first order Bessel's functions of complex argument. Graphical forms are available to facilitate calculation of the resistance and reactance components of Z.[4,60,61] The theoretical impedances have been verified for wide ranges of the parameter Y by experimental measurements using blood[61] and other fluids.[62]

The velocity profile in the tube varies in character with Y, and approaches piston-like behavior for large values of Y while becoming parabolic for $Y < 1$.[4] The parabolic profile is applicable to tube diameters < 2 mm when filled with a fluid similar to normal blood (viscoelasticity magnitude = 0.11 poise and density = 1.04 g cm^{-3}) at 1.5 Hz. The equation for the impedance is greatly simplified for the parabolic profile, in which case $Y < 1$ and the resistance and reactance become

$$R_0 = (8\eta' L)/(\pi a^4) \tag{B5}$$

$$X_0 = [(4\rho\omega L)/(3\pi a^2)] - [(8\eta'' L)/(\pi a^4)]. \tag{B6}$$

Also, under this same condition, the magnitude of the shear rate at the tube wall is

$$\gamma' = (4\,U\,/\,\pi\,a^3)\,|[1 + i\,(\rho\omega a^2)\,/\,(24\,\eta^*)]| \tag{B7}$$

and the shear stress at the tube wall can be calculated using

$$\tau_w = \tau' - i\,\tau'' = (\eta' - i\,\eta'')\,\gamma', \tag{B8}$$

where the viscous stress $\tau' = \eta'\,\gamma'$ and the elastic stress $\tau'' = \eta''\,\gamma'$. From measurements of pressure and volume flow under the condition $Y < 1$, the viscoelastic properties of the fluid can be calculated using equations (B5) and (B6). The shear rate magnitude is given by equation (B7), the shear strain magnitude is $\gamma = \gamma'\,/\,\omega$, and the two components of the shear stress is given by equation (B8).

REFERENCES

1. Womersley, J.R. Method for the calculation of velocity, rate of flow and viscous drag in arteries when the pressure gradient is known. *Amer. J. Physiol.* **1955**, *127*, 553–563.
2. Thurston, G.B. The viscoelasticity of blood and plasma during coagulation in circular tubes. *Proceedings of the Sixth Conference of the European Society for Microcirculation, Aalborg.* S. Karger Basel, Aalborg, Denmark, 1971, pp. 12–15.
3. Thurston, G.B. Viscoelasticity of human blood. *Biophysical J.* **1972**, *12*, 1205–1217.
4. Thurston, G.B. Theory of oscillation of a viscoelastic fluid in a circular tube. *J. Acoust. Soc. Amer.* **1960**, *32*, 210–213.

5. Thurston, G.B. Plasma release-cell layering theory for blood flow. *Biorheology* **1989**, *26*, 199–214.
6. Thurston, G.B. Rheological parameters for the viscosity, viscoelasticity, and thixotropy of blood. *Biorheology* **1979**, *16*, 149–162.
7. Burton, A.C. *Physiology and Biophysics of the Circulation.* Year Book Medical Publishers Inc., Chicago, 1965, p. 53.
8. Fischer, T.; Schmid-Schonbein, H. Tank tread motion of red cell membranes in viscometric flow: Behavior of intracellular and extracellular markers. *Blood Cells* **1977**, *3*, 351–365.
9. Dormandy, J., Ed. *Proceedings of the Second Workshop held in London (1982).* Martinus Nijhoff Publishers, The Hague, 1983.
10. Kage, H.S.; Engelhardt, H.; Sackmann, E. A precision method to measure average viscoelastic parameters of erythrocyte populations. *Biorheology* **1990**, *27*, 67–68.
11. Thurston, G.B. Light transmission through blood in oscillatory flow. *Biorheology* **1990**, *27*, 685–700.
12. Lessner, A.; Zahavi, F.; Silberberg, A.; Frei, E.H.; Dreyfus, F. The viscoelastic properties of whole blood. In: *Theoretical and Clinical Hemorheology* (Hartert, H.; Copley, A.L., Eds.). Springer–Verlag, Berlin, Heidelberg and New York, 1971, pp. 194–205.
13. Chien, S.; King, R.G.; Skalak, R.; Usami, S.; Copley, A.L. Viscoelastic properties of human blood and red cell suspensions. *Biorheology* **1975**, *12*, 341–346.
14. Bureau, M.; Healey, J.C.; Bourgoin, D.; Joly, M. Etude experimentale *in vitro* du comportement rheologique du sange en regime transitoire a faible vitesse de cisaillement. *Rheol. Acta* **1978**, *17*, 612–625.
15. Stoltz, J.F.; Gaillard, S.; Lucius, M. A study of the viscoelastic properties of blood in transient flow. *J. Biomechanics* **1980**, *13*, 341–334.
16. Crandall, I.B. *Theory of vibrating systems and sound.* D. Van Nostrand Co. Inc., New York, 1926, pp. 229–241.
17. Lambossy, P. Oscillations forcees d'un liquide incompressible et visqueux dans un tube rigide et horizontal. Calcul de la force de frottment. *Helv. Phys. Acta* **1952**, *25*, 371–386.
18. Thurston, G.B. Periodic fluid flow through circular tubes. *J. Acoust. Soc. Amer.* **1952**, *24*, 653–656.
19. Thurston, G.B. Elastic effects in pulsatile blood flow. *Microvascular Research* **1975**, *9*, 145–157.
20. Thurston, G.B. Effects of hematocrit on blood viscoelasticity and in establishing normal values. *Biorheology* **1978**, *15*, 239–249.
21. Walitza, E.; Anadere, I.; Chmiel, H.; Witte, S. Evaluation of viscoelasticity measurements of human blood. *Biorheology* **1988**, *25*, 209–217.
22. Kasser, U.; Heimburg, P. Quality control in blood viscoelastometry. *Clinical Hemorheology* **1988**, *8*, 93–103.
23. Huang, C.R.; Siskovic, M.; Robertson, R.W.; Fabisiak, W.; Smitherberg, E.J.; Copley, A.L. Quantitative characterization of thixotropy of whole human blood. *Biorheology* **1975**, *12*, 279–282.
24. Schmid Schonbein, H.; Kline, K.A.; Heinich, L.; Volger, E.; Fischer, T. Microrheology and light transmission of blood. III. The velocity of red cell aggregate formation. *Pflugers Arch.* **1975**, *354*, 299–317.
25. Gaspar-Rosas, A.; Thurston, G.B. Erythrocyte aggregate rheology by transmitted and reflected light. *Biorheology* **1988**, *25*, 471–487.
26. Kasser, U.; Heimburg, P.; Walitza, E. Viscoelasticity of whole blood and its dependence on laboratory parameters. *Clinical Hemorheology* **1989**, *9*, 307–312.
27. Thurston, G.B. Erythrocyte rigidity as a factor in blood rheology: Viscoelastic dilatancy. *J. Rheology* **1979**, *23*, 703–719.
28. Ribitsch, V. Determination of red blood cell suspensions rheological properties in oscillatory flow. *Clinical Hemorheology* **1989**, *9*, 313–317.
29. More, R.B.; Thurston, G.B. Intrinsic viscoelasticity of blood cell suspensions: Effects of erythrocyte deformability. *Biorheology* **1978**, *24*, 297–309.

30. Thorsen, G.; Hint, H. Aggregation, sedimentation, and intravascular sludging of erythrocytes. *Acta Chir. Scand. Suppl.* **1950**, *154*, 1–50.
31. Klose, H.J.; Volger, E.; Brechtelsbauer, H.; Schmid-Schonbein, H. Microrheology and light transmission of blood. I. The photometric effects of red cell aggregation and red cell orientation. *Pflugers Arch.* **1972**, *333*, 126–139.
32. Brooks, D.E.; Greig, R.G.; Janzen, J. Mechanisms of erythrocyte aggregation. *Erythrocyte Mechanics and Blood Flow.* Alan R. Liss, Inc., New York, 1980, pp. 119–140.
33. Thurston, G.B. Blood viscoelasticity and relaxation processes: Influence of aggregation tendency. In: *Hemo-Rheology and Diseases* (Stoltz, J.F.; Drouin, P., Eds.). Doin Editeurs, Paris, 1980, pp. 51–66.
34. Chien, S.; Usami, S.; Dellenback, R.J.; Gregersen, M.I. Shear-dependent interaction of plasma proteins with erythrocytes in blood rheology. *Amer. J. Physiology* **1970**, *219*, 143–153.
35. Witte, S.; Anadere, I. Modifications of viscoelastic properties during cardiovascular diseases. *Clinical Hemrheology* **1989**, *9*, 831–837.
36. Isogai, Y.; Ikemoto, S.; Kuchiba, K.; Ogawa, J.; Yokose, T. Abnormal blood viscoelasticity in diabetic microangiopathy. *Clinical Hemorheology* **1991**, *11*, 175–182.
37. Chen, H.-Q.; Yan, L.-N.; Li, L.; Zhong, G.-H.; Wu, H.-G. Alteration in viscoelasticity of whole blood from dogs with experimental acute necrotizing pancreatitis. *Clinical Hemorheology* **1991**, *11*, 25–33.
38. Chmiel, H.; Anadere, I.; Walitza, E. The determination of blood viscoelasticity in clinical hemorheology. *Clinical Hemorheology* **1990**, *10*, 363–374.
39. Chmiel, H.; Anadere, I.; Walitza, E. The determination of blood viscoelasticity in clinical hemorheology. In: *Hemorheologie et Aggregation Erythrocytaire,* Vol. 3 (Stoltz, J.F.; Donner, M.; Copley, A.L., Eds.). Editions Medicales Internationales, Cachan, France, 1991, pp. 196–206.
40. Ikemoto, S.; Tanaka, T.; Yamamoto, J.; Kuchiba, K.; Akiyama, M. ; Maeda, T.; Yokose, T.; Isogai, Y. Blood viscoelasticity in clinical medicine. In: *Hemorheologie et Aggregation Erythrocytaire,* Vol. 3 (Stoltz, J.F.; Donner, M.; Copley, A.L., Eds.). Editions Medicales Internationales, Cachan, 1991, pp. 189–195.
41. Koltringer, P.; Langsteger, W.; Lind, P.; Eber, O. Erythrocyte aggregation in hyperthyroid patients: measurement under fluid conditions as elasticity of blood at low shear-rates. In: *Hemorheologie et Aggregation Erythrocytaire,* Vol. 3 (Stoltz, J.F.; Donner, M.; Copley, A.L., Eds.). Editions Medicales Internationales, Cachan, 1991, pp. 261–264.
42. Drasler, W.J.; Smith II, C.M.; Keller, K.H. Viscoelastic properties of the oxygenated sickle erythrocyte membrane. *Biorheology* **1989**, *26*, 935–949.
43. Hell, K.M.D.; Balzereit, A.; Diebold, U.; Bruhn, H.D. Importance of blood viscoelasticity in arteriosclerosis. *Proceedings of the 29th Annual Meeting, International College of Angiology.* Montreux, Switzerland, 1987.
44. Chmiel, H.; Anadere, I.; Walitza, E.; Witte, S. The measurement of density and its significance in blood rheology. *Biorheology* **1983**, *20*, 685–696.
45. Anadere, I.; Chmiel, H.; Hess, H.; Thurston, G.B. Clinical blood rheology. *Biorheology* **1979**, *16*, 171–178.
46. Kollegger, H.; Oder, W.; Zeiler, K.; Baumgartner, Ch.; Lind, C.; Oder, B.; Sagmeister, Ch.; Deecke, L. Viscoelasticity of whole blood as a function of age, gender, cigarette smoking and intake of oral contraceptives. *Clinical Hemorheology* **1990**, *10*, 499–503.
47. Langsteger, W.; Koltringer, P.; Pierer, G.; Eber, O. The effect of rapid hypervolaemic hemodilution on blood viscoelasticity: Course observation over 5 days. *Clinical Hemorheology* **1992**, *12*, 93–100.
48. Chmiel, H.; Anadere, I.; Moser, K. Hemorheological changes under blood leeching. *Clinical Hemorheology* **1989**, *9*, 569–576.
49. Wolf, H.R.D.; Witte, S. Blood rheology and oxygen uptake. *Clinical Hemorheology* **1990**, *10*, 393–399.

50. Anadere, I.; Chmiel, H.; Witte, S. Hemorheological findings in patients with completed stroke and the influence of a ginkgo biloba extract. *Clinical Hemorheology* **1985**, *5*, 411–420.
51. Sowemimo-Coker, S.O.; Yardin, G.; Meiselman, H.J. Effect of procaine hydrochloride on the aggregation behavior and suspension viscoelasticity of human red blood cells. *Biorheology* **1989**, *26*, 951–972.
52. Liepsch, D.; Thurston, G.; Lee, M. Studies of fluids simulating blood-like rheological properties and applications in models of arterial branches. *Biorheology* **1991**, *28*, 39–52.
53. Fukada, E.; Seaman, G.V.F.; Liepsch, D.; Lee, M.; Friis-Baastad, L. Blood modeling using polystyrene microspheres. *Biorheology* **1989**, *26*, 401–413.
54. Mann, D.E.; Tarbell, J.M. Flow of non-Newtonian blood analog fluids in rigid curved and straight artery models. *Biorheology* **1990**, *27*, 711–733.
55. Copley, A.L.; King, R.G.; Chien, S.; Usami, S.; Skalak, R.; Huang, C.R. Microscopic observations of viscoelasticity of human blood in steady and oscillatory flow. *Biorheology* **1975**, *12*, 257–263.
56. Thurston, G.B. Rheological analogs for human blood in large vessels. In: *2nd International Symposium on Biofluid Mechanics and Biorheology* (Liepsch, D., Ed.). Springer-Verlag, Berlin, 1990, pp. 517–526.
57. Brookshier, K.A.; Tarbell, J.M. Evaluation of a transparent blood analog fluid: Aqueous xanthan gum/glycerin. *Biorheology* **1993**, *30*, 107–116.
58. Carriere, C.J.; Amis, E.J.; Schrag, J.L.; Ferry, J.D. Dilute-solution dynamic viscoelastic properties of xanthan polysaccharide. *J. Rheology* **1993**, *37*, 469–478.
59. Thurston, G.B.; Pope, G.A. Shear rate dependence of the viscoelasticity of polymer solutions. II Xanthan gum. *J. Non-Newtonian Fluid Mechanics* **1981**, *9*, 69–78.
60. Thurston, G.B. Elastic effects in pulsatile blood flow. *Microvasc. Res.* **1975**, *9*, 145–157.
61. Thurston, G.B. The effects of frequency of oscillatory flow on the impedance of rigid, blood-filled tubes. *Biorheology* **1976**, *13*, 191–199.
62. Thurston, G.B. Measurement of the acoustic impedance of a viscoelastic fluid in a circular tube. *J. Acoust. Soc. Amer.* **1961**, *33*, 1091–1095.

MICRORHEOLOGICAL MODELS OF RED BLOOD CELL MECHANICS

D. Barthes-Biesel

Advances in Hemodynamics and Hemorheology,
Volume 1, pages 31–65.

ISBN: 1-55938-634-7.

I. INTRODUCTION

Human blood is a concentrated suspension of different cells in plasma. As such it exhibits complex rheological properties that depend strongly on microscale mechanics, that is, on the suspended cell's motion and deformation, on the type of flow, and on the diameter of the vessel. Consequently, the approach to blood rheology will not be the same depending on whether one is interested in circulation at the capillary level or in large vessels. The suspending phase, the plasma, is itself a complex fluid, since it is a solution of various macromolecules. However, for low shear rates, plasma may be treated as a Newtonian liquid with a viscosity varying between 1.75 mPa s at 23°C and 1.5 mPa s at 37°C. The volume concentration of cells, the hematocrit, varies between 40 and 45%. The suspended particles consist of red blood cells (RBC), white blood cells (WBC), and platelets. The most numerous are by far the RBCs with a normal concentration of about 5×10^6 cells/mm^3, which represents about 99% of the total cell volume. White blood cells concentration is of order 7,000 cells/mm^3. Although these cells have a volume about twice that of RBCs, their impact on whole blood rheology is negligible. From the mechanical point of view, their role is important essentially at the microcirculation level, where they will need a time longer than RBCs to deform enough to flow in the smallest capillary vessels. Platelets are very small, flat cells with their larger dimension not exceeding 2 μm. Their normal concentration is about 300,000 cells/mm^3, but the volume they occupy is negligible. Altogether, they do not influence blood rheology, but of course, they play an important role in coagulation processes.

It thus follows that whole blood rheological properties are essentially determined by the mechanical behavior of RBCs, their tendency to form aggregates, and their overall deformability. Similarly, at the microcirculation level, the RBC ability to deform plays an essential role since the cells have to squeeze into capillary vessels that have a diameter smaller than theirs.

This chapter thus focuses on the mechanical properties of RBCs. Different experimental means of measuring it will be presented together with the pertinent mechanical models that allow a physical interpretation of the experimental data.

II. RED BLOOD CELLS' PROPERTIES

In their normal state, RBCs are small biconcave disks, filled with a hemoglobin solution and surrounded by a very thin membrane. Due to this composition, these cells deform readily and this has important consequences. At the macrocirculation level, the high deformability of RBCs ensures that whole blood viscosity is low, of order 4 to 5 mPa s at high shear rates, even though the volume concentration is large. Indeed a suspension of rigid spheres at the same hematocrit, would have a viscosity of order 30 mPa s. In fact, any reduction in the deformability of RBCs would lead to an increase in whole blood viscosity and to a corresponding increase of the heart energy expenditure to pump blood in the body. Similarly, at the microcirculation level, the RBCs have to squeeze through capillary vessels, which have a diameter smaller than the cell's. This can only be achieved if the cell takes a parachute or slipper shape.

It thus follows that RBC deformability is an essential property that governs blood circulation. This has been recognized for a long time and many experimental techniques have been proposed over the years to measure "deformability." However, a close inspection of the structure of a RBC, indicates that "deformability" is the consequence of an interplay of complex interrelated physical properties of the cell, such as geometry, fluidity of the hemoglobin solution, and mechanics of the surrounding membrane. These different factors can be measured independently and each is presently well understood.

Red blood cells are very sensitive to their physico-chemical environment, such as pH or osmolality. Unless otherwise specified, the results presented here pertain to normal physiological conditions with a pH of 7.4 and an osmolality of 300 mOsm/l.

A. Geometry

The detailed geometry of an RBC can be obtained by means of microscopic observations[1] with a precision of ±0.02 μm. The normal RBC, also known as erythrocyte, is an axisymmetric biconcave disk. The equation of a meridian curve can be approximated by the following equation in a z-r plane, where Oz is the revolution axis:

$$z = 1.96\sqrt{1-\rho^2}\,(C_0 + C_1\rho^2 + C_2\,\rho^4), \tag{1}$$

where

$$\rho = r/3.91,$$

$$C_0 = 0.207161,\ C_1 = 2.002558,\ C_2 = -1.122762. \tag{2}$$

The coefficients C_0, C_1, and C_2 have been obtained from measurements made on 50 normal cells. The corresponding profile is represented in Figure 1. The cell diameter is roughly 8 μm, its volume 94.1 μm^3, and surface area 134.1 μm^2. It is interesting to note that in this configuration, the erythrocyte has an excess surface area as compared to a spherical cell with the same volume. This is measured by the so-called sphericity index K, that represents the ratio of the surface of the cell to that of the sphere having the same volume as the cell. For the erythrocyte of Figure 1, the sphericity index is then:

$$K = 1.33.$$

When suspended in a hypotonic medium, the cell swells progressively until it reaches a spherical shape ($K = 1$) for an osmolality of about 140 to 130 mOsm/l.

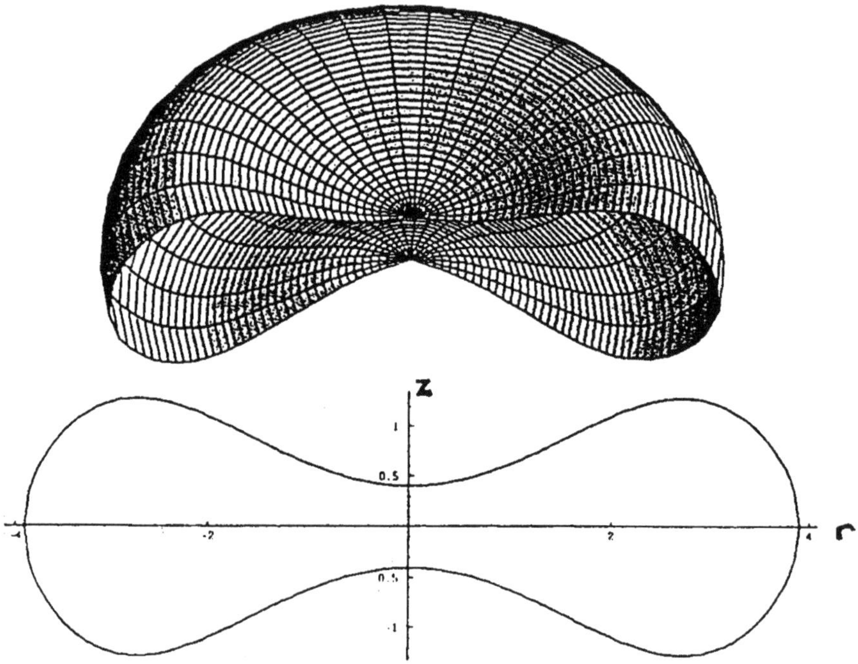

Figure 1. Normal red blood cell shape, computed from equations (1) and (2).

During the initial swelling stages, the surface of the cell remains constant, whereas, during the final phase, a slight increase of surface area of about 7% is observed before hemolysis, that is, bursting of the membrane, occurs.

B. Internal Medium

The internal hemoglobin solution may be analyzed by hemolyzing the cell and removing the membrane by centrifugation. The hemoglobin cellular concentration is then easily deduced from the values of the hematocrit and of the hemoglobin concentration in the lyzed cells solution. Its normal value is 32.5 ± 2.5 g/dl. In order to infer the viscosity of the internal medium of the RBC, a rheological study of hemoglobin solutions with varying concentrations must then be performed. Cokelet and Meiselman,[2] Rogauch,[3] and Chien[4] have conducted a systematic study of the rheology of hemoglobin solutions. They find a Newtonian behavior in a shear rate range between 0.5 and 200 s^{-1}. However, the viscosity is strongly concentration dependent for hemoglobin concentrations in excess of 20 g/dl: it increases from 6 to 8 mPas at 37°C when the hemoglobin concentration varies from 30 to 32 g/dl (Figure 2). Furthermore, Rogauch[3] showed that the viscosity of hemoglobin solutions is also temperature dependent. For example, at the normal concentration

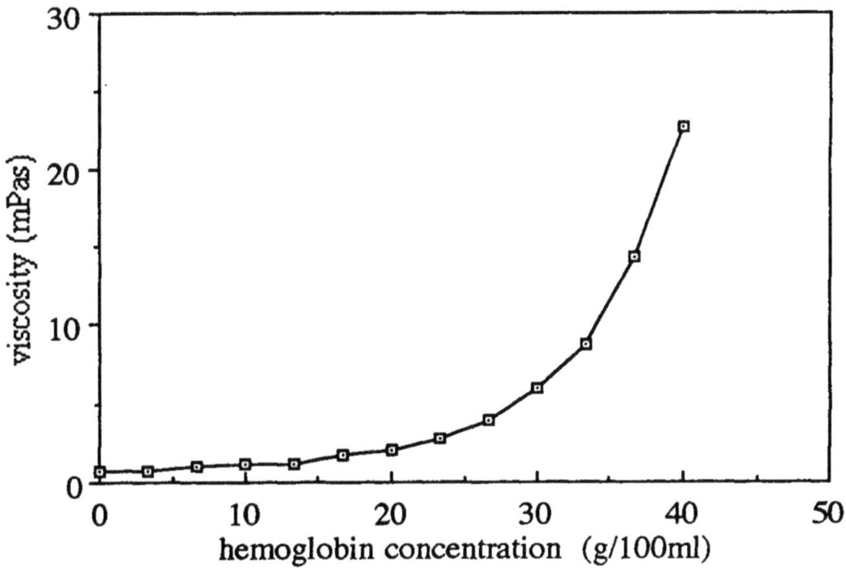

Figure 2. Viscosity of hemoglobin solutions at 37 °C. The viscosity increases steeply with concentrations larger than 25 g/100 ml (from Refs. 2,3,4).

of 32 g/dl, when the temperature drops from 37 to 20°C, the hemoglobin solution viscosity is almost doubled. This is an important fact to be kept in mind when *in vitro* measurements on RBCs are conducted at a temperature lower than 37°C, as is often the case for experimental simplicity reasons.

In pathological cases, the hemoglobin may be altered, and the internal viscosity is correspondingly modified. In particular, sickle cell anemia is linked to a structural change of the hemoglobin. When fully oxygenated, sickle cell hemoglobin is Newtonian and has the same viscosity as normal hemoglobin. However, when the oxygen saturation is decreased, then sickle cell hemoglobin becomes non-Newtonian, with a very large low shear viscosity whereas normal hemoglobin remains Newtonian.[4]

C. Red Blood Cell Membrane Mechanics

Experimental Evidence

A RBC membrane consists of a lipid bilayer lined by a protein network that confers some elasticity to the system. The thickness of the bilayer is of order 10^{-8} m, so that the membrane can be treated as a two-dimensional elastic solid. The mechanical properties can be measured by means of a micropipette with an internal diameter of 1 to 2 μm. A small part of the membrane is sucked into the pipette (Figure 3), and the height of aspiration is then measured as a function of the applied

Figure 3. Suction of a normal red blood cell in a micropipette. (Picture obtained by C. Bucherer, Unité de Biorhéolgie, and Hôpital Pitié-Salpêtrière.)

depression. Such measurements indicate that the membrane is easily shearable, but strongly resists any increase in the local surface area.[5,6,7] The bending resistance is small.[8] Consequently, bending effects should be taken into account only when they are the main elastic forces acting (e.g., rest shape of the RBC) or when the local radius of curvature of the membrane becomes small, of the same order of magnitude as the thickness. Relaxation experiments[11] after suction in a micropipette, clearly demonstrate a measurable surface viscosity. More details on membrane mechanics can be found in the two extensive reviews of Hochmuth and Waugh[12] and of Hochmuth and Berk.[13]

Membrane Constitutive Law

The membrane can be modeled as a two-dimensional shell with negligible thickness, subjected to large deformations. Bending effects are first neglected, and only shear deformations are considered. Then the normal direction to the surface is invariant during deformation. Stresses are replaced by tensions, that is, forces per unit length. In order to simplify the presentation, it is convenient to consider in a tangent plane of the shell, a set of principal axes of deformation and stress, denoted OX_1, OX_2 (resp. Ox_1, Ox_2) before (resp. after) deformation. Correspondingly, a membrane element with initial dimensions dX_1, dX_2 is deformed into an element dx_1, dx_2 (Figure 4).

The principal stretch ratios are then

$$\lambda_1 = dx_1/dX_1 \quad \text{and} \quad \lambda_2 = dx_2/dX_2. \tag{3}$$

The corresponding principal Cauchy tensions per unit length of deformed membrane are denoted T_{11} and T_{22}. In the following, only the expression for T_{11} is given. The corresponding one for T_{22} can be easily obtained after a permutation on the indices 1 and 2. Skalak et al.[6] proposed a constitutive law for the RBC membrane that takes into account the shearing ability and the large resistance to surface area

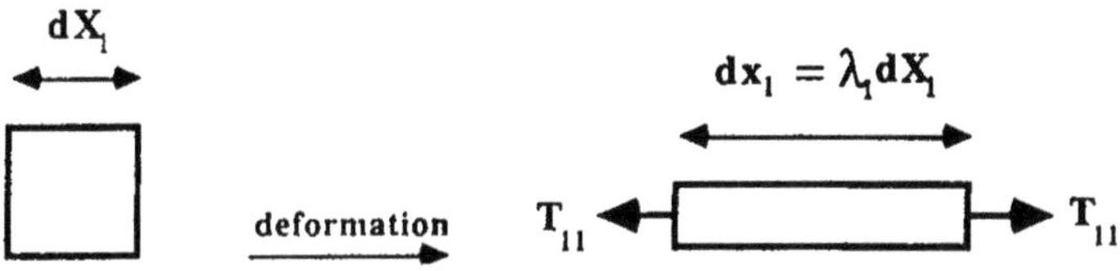

Figure 4. Deformation of a membrane element in the x_1 direction.

changes. Two surface elastic moduli are thus introduced: E_s corresponding to shear deformations and A_s corresponding to area changes and such that $A_s/E_s >> 1$ (A_s is of order $10^5\ E_s$). The elastic tensions in the membrane may be written as:

$$T_{11} = E_s(\lambda_1^4 - \lambda_1^2)\lambda_2 + T_0, \tag{4}$$

where the first term results from shear deformations, whereas the second term, which is isotropic, corresponds to changes in the local surface area, and can then be written as:

$$T_0 = \frac{A_s\lambda_1\lambda_2}{2}(\lambda_1^2\lambda_2^2 - 1), \tag{5}$$

if small area changes are allowed. For interpretation of experimental measurements, it is usually enough to use a simplified version of this law, proposed by Evans,[7] where the shear behavior is simply quadratic in λ_1 and where the membrane is assumed to be exactly area incompressible:

$$T_{11} = E_s(\lambda_1^2 - 1) + T_0. \tag{6}$$

Then T_0 plays the role of an isotropic pressure that must be determined from the additional constraint:

$$\lambda_1\lambda_2 = 1, \tag{7}$$

that expresses the invariance of the local surface area. Measurements with a micropipette yield the following value of E_s[11]:

$$E_s = 4 \text{ to } 5 \times 10^{-3}\ \text{mN m}^{-1}.$$

The surface viscosity is usually modelled by means of a linear viscoelastic law where the viscous contribution T_{11}^{v} is simply added to the elastic terms (4) or (6). Neglecting the expansion viscosity, the corresponding stress is then:

$$T_{11}^{v} = 2\,\mu_s \frac{\partial\lambda_1/\partial t}{\lambda_1}, \tag{8}$$

where $\partial\lambda_1/\partial t$ denotes a time derivative, and where μ_s is the surface viscosity, measured by Chien et al.[11]

$$\mu_s = 1 \text{ to } 5 \times 10^{-4} \text{ mN s m}^{-1}.$$

The relative dispersion of values of μ_s is due to the fact that the RBC membrane viscosity seems to be somewhat shear thinning.

If bending effects are included, then in addition to tensions, elastic bending moments appear, with principal components M_{11} and M_{22} per unit length of deformed membrane. Zarda et al.[14] suggest the following constitutive equation:

$$M_{11} = \frac{B_s}{\lambda_2}(\kappa_{11} + \frac{1}{2}\kappa_{22}), \qquad (9)$$

with a similar expression for M_{22}, and where κ_{11} and κ_{22} are the two principal curvatures of the deformed surface. The surface bending modulus has been measured by Evans[8]:

$$B_s = 1.8 \times 10^{-16}\, mN\, m.$$

The red blood cell membrane has other complex mechanical properties. In particular it can exhibit a plastic behavior and form tethers when the applied forces exceed a given threshold. Furthermore, the membrane mechanics have been presented here in a simplifed form expressed in a set of principal axes of deformation and stress. The formulation is obviously more complicated in a general set of axes and the reader is then referred to any classical text on shell mechanics.[9,10]

In conclusion, the RBC is a very deformable particle. This property is of fundamental physiological importance and it is thus useful to be able to detect any alterations. Red blood cell deformability results from three different intrinsic factors. One of them is geometry, since it is comparatively easier to modify the shape of a flaccid disk than that of a sphere. The other is the low viscosity of the internal liquid and the final one is the low resistance of the membrane to shear deformations. These physical parameters may be measured individually as explained above. However, such measurements are difficult to perform, often require micromanipulations and must be repeated many times in order to yield a mean value of these parameters over a cell population.

III. MOTION OF A RED BLOOD CELL IN SIMPLE SHEAR FLOW

It is of interest to study the motion of a red blood cell as a whole, when it is suspended in another liquid subjected to flow. This situation corresponds to the physiological reality and thus leads to a better understanding of *in vivo* behavior of the cell.

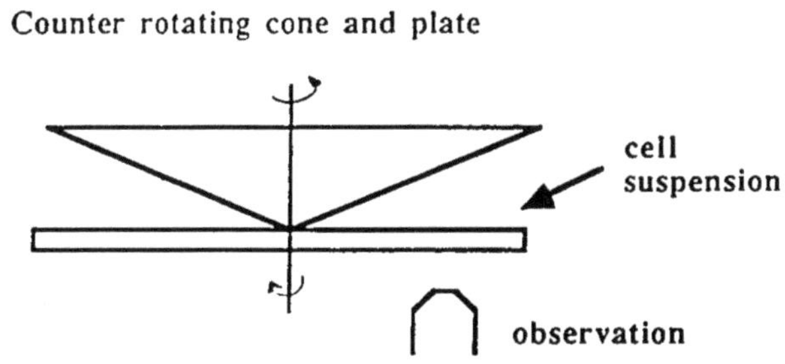

Figure 5. Schematics of a rheoscope. The plate is made of glass, and the cells that are in the stationary velocity cone, are observed by means of an inverted microscope.

A. Experimental Evidence

A very interesting device, that has provided considerable insight regarding RBC mechanics, consists in observing the motion and deformation of one cell freely suspended in a simple shear flow. This is achieved in a rheoscope, first proposed by Schmid-Schonbein and Wells.[15] The rheoscope is a glass counter-rotating cone-and-plate viscometer, in the gap of which is placed a dilute red blood cell suspension (Figure 5). The deformation of the cells which have their center of mass in the zero velocity cone and which are thus stationary, is observed and measured by means of an inverted microscope. It is then found, that under the influence of the viscous forces due to the flow, all cells take similar prolate ellipsoidal shapes oriented with respect to the streamlines. The deformation D of the cell in the shear plane, is defined as:

$$D = (L - B)/(L + B),$$

where L and B denote respectively the length and the breadth of the deformed particle.

The experiments show that deformation increases with the external shear stress and also depends on the viscosity ratio λ between the hemoglobin solution and the suspending liquid (Figure 6). This last fact was established by Pfafferott et al.16 who used suspending media of different viscosities and measured young and old RBCs, respectively, obtained from the top or bottom 10% fraction of density separated cells. The bottom 10% cells are more viscous than the top 10% ones. It is found that, when the viscosity ratio λ is less than about 2, the deformation at a given shear stress level, decreases when λ increases. Furthermore, the cells show no tendency toward breakup for stresses less than 50 Pa, but rather exhibit a limiting maximum deformation.

By attaching microspheres to the membrane and by suspending some in the internal hemoglobin solution, it is possible to demonstrate that both the membrane and the internal liquid have a rotational motion, called "tank-treading".[17,18] The

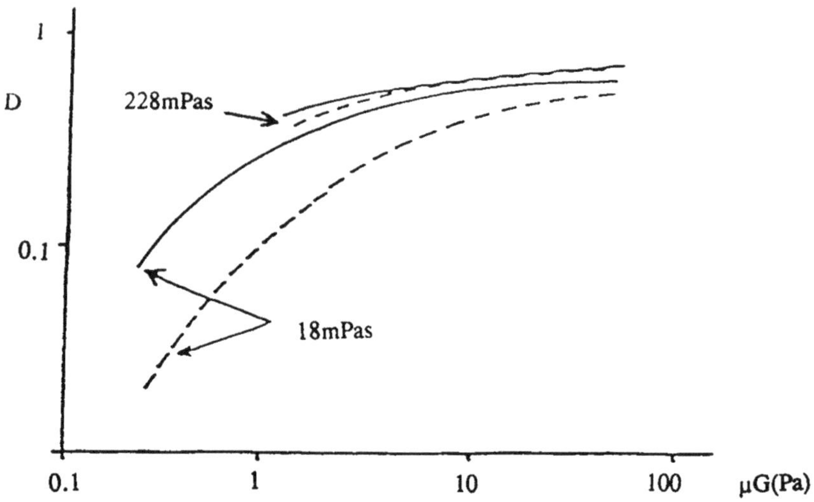

Figure 6. Deformation of density separated RBCs in two Dextran solutions with viscosity 18 and 228 mPa s. —— top 10% fraction, - - - bottom 10% fraction (From Ref. 16).

rotation frequency depends on the shear rate (and thus on the cell deformation) and weakly on λ. However, when λ is greater than about 2, the RBC flips and tumbles in the flow like a flexible disc. It should be noted that similar devices have been recently developed,[19,20,21] to study artificial capsules consisting of a liquid drop enclosed by a thin polymeric membrane. As compared to the rheoscope, the new experimental setups have the advantage of allowing observation of the particles in two orthogonal directions and thus lead to an evaluation of three-dimensional effects, which are found to be significant at least for these particles.

B. Equations of Motion of a Cell in Shear Flow

The interpretation of rheoscope data is impossible without a good mathematical model of the motion and deformation of a cell freely suspended in a shear flow. There has been some work over the years to develop such a model. The generally agreed upon problem statement is now briefly outlined, before the available solutions are presented.

A cell, with characteristic dimension a (e.g., the radius of the isovolume sphere) is freely suspended in a Newtonian incompressible liquid of viscosity μ, subjected to a shear flow with shear intensity G. The internal hemoglobin solution has a viscosity denoted $\lambda\mu$. It is surrounded by an infinitely thin membrane with mechanical properties as described in Section II. Buoyancy effects are ignored. The

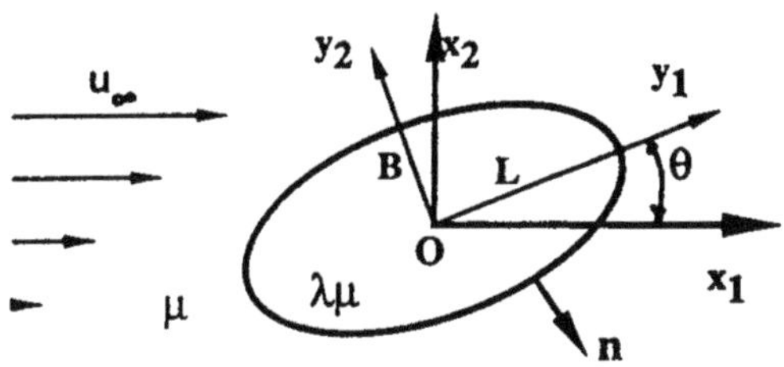

Figure 7. Schematics of a cell freely suspended in a linear shear flow.

equations of motion are written with respect to a reference frame $(O, x_1, x_2, x_3 3)$ centered on the particle center of mass O and moving with it (Figure 7). The external liquid is subjected far from the particle to the undisturbed flow field:

$$\mathbf{u}_\infty = [\mathbf{E}(t) + \mathbf{\Omega}(t)]\cdot\mathbf{x}, \tag{10}$$

where $\mathbf{E}$ and $\mathbf{\Omega}$ denote respectively the rate of strain and the vorticity tensors which depend only on time. This last assumption does not really limit the validity of the model, since in many cases, the spatial scale of the velocity variations is much larger than the particle size, and the flow field may then be linearized. For example, equation (10) would be a good approximation of flow field around a cell suspended in a Poiseuille flow in a tube of radius R_T, provided that $R_T/a >> 1$ and that the cell is not too near the tube wall. In the gap of a Couette or of a cone-and-plate viscometer, however, equation (10) applies exactly, and the only non zero components of $\mathbf{E}$ and $\mathbf{\Omega}$ are:

$$E_{12} = E_{21} = \Omega_{12} = -\Omega_{21} = G/2. \tag{11}$$

Under the influence of the viscous stresses, the cell deforms and the equation of its interface is given by:

$$r = (x_1^2 + x_2^2 + x_3^2)^{1/2} = f(x_1, x_2, x_3). \tag{12}$$

The function f is itself unknown and must be determined as part of the problem solution. In absence of inertia effects, the internal and external velocity and stress fields, respectively $\mathbf{u}^*$, $\mathbf{u}$, $\boldsymbol{\sigma}^*$, and $\boldsymbol{\sigma}$, are governed by the Stokes equations:

$$\nabla\cdot\mathbf{u} = 0,\ \nabla\cdot\boldsymbol{\sigma} = 0, \text{ for } r \geq f; \tag{13}$$

$$\nabla\cdot\mathbf{u}^* = 0,\ \nabla\cdot\boldsymbol{\sigma}^* = 0, \text{ for } r \leq f. \tag{14}$$

For Newtonian liquids, the stresses are given by Newton's law:

$$\boldsymbol{\sigma} = -p\mathbf{I} + \mu(\nabla\mathbf{u} + {}^T(\nabla\mathbf{u}), \tag{15}$$

$$\boldsymbol{\sigma}^* = -p^*\mathbf{I} + \lambda\mu(\boldsymbol{\nabla}\mathbf{u}^* + {}^T\boldsymbol{\nabla}\mathbf{u}^*), \tag{16}$$

where p (resp. p^*) denotes the external (resp. internal) pressure and **I** the identity matrix. The associated boundary conditions are:

1. no flow disturbance far from the cell:

$$\mathbf{u} \to \mathbf{u}_\infty \quad \text{as} \quad r \to \infty, \tag{17}$$

2. continuity of velocities at the interface:

$$\mathbf{u}^* = \mathbf{u} = \mathbf{u}_s = \partial\mathbf{x}/\partial t \quad \text{at} \quad r = f, \tag{18}$$

   where $\mathbf{u}_s$ represents the velocity of the membrane, as measured by the time derivative of the position **x** of the interface material points.

3. dynamic equilibrium of the interface:

$$(\boldsymbol{\sigma} - \boldsymbol{\sigma}^*) \cdot \mathbf{n} = \mathbf{q} \quad \text{at} \quad r = f, \tag{19}$$

   where **n** denotes the outer unit normal vector to the deformed interface. Condition (19) states that the viscous load exerted by the flow of the internal and external liquids is balanced by the force ($-\mathbf{q}$) exerted by the membrane on the internal and external liquids because of its deformation. In order to close the formulation of the problem there remains to relate **q** to the surface deformation. This is achieved by means of the shell equilibrium equations, which take the following form, when bending forces are ignored:

$$\mathbf{q} + \boldsymbol{\nabla}_s \cdot \mathbf{T} = 0, \tag{20a}$$

   where $\boldsymbol{\nabla}_s$ represents a surface gradient defined as:

$$\boldsymbol{\nabla}_s = (\mathbf{I} - \mathbf{n} \cdot {}^\mathrm{T}\mathbf{n}) \cdot \boldsymbol{\nabla}.$$

The components of the surface Cauchy stress tensor **T**, are related to the deformation of the membrane by means of one of the constitutive equations (4)–(8). The inclusion of bending effects complicates considerably equation (20a) which then becomes:

$$\mathbf{q} + \boldsymbol{\nabla}_s \cdot \mathbf{T} + \boldsymbol{\nabla}_s \cdot [-\boldsymbol{\kappa} \cdot \mathbf{M} + \mathbf{n} \cdot {}^T(\boldsymbol{\nabla}_s \cdot \mathrm{M}) \cdot (\mathbf{I} - \mathbf{n} \cdot {}^\mathrm{T}\mathbf{n})] = 0, \tag{20b}$$

where $\boldsymbol{\kappa}$ and **M** are respectively the curvature and bending moment tensors.

A dimensional analysis shows that the solution to the problem described by Equations (10)–(20) depends on two important dimensionless parameters: the viscosity ratio λ and the Capillary number C, which represents the ratio of viscous to elastic forces and which is defined as:

$$C = \mu\, Ga/E_s.$$

Obviously, the solution may depend on other parameters linked to the constitutive behavior of the cell membrane, such as:

A_s/E_s, the ratio of compression to shear moduli.
B_s/a^2E_s, the ratio of bending to shear forces
$\beta = \mu_s G/E_s$.

This last parameter β appears for a viscoelastic membrane and represents the ratio between the externally imposed shear time, 1/G, and the intrinsic response time of the interface.

Altogether, the problem described by Equations (10)–(20a) and (20b), represents a complicated case of continuum mechanics, involving fluid-structure interactions and a free surface, since the deformed position of the cell membrane is unknown. As a consequence, it will be amenable to an analytical solution only in some particular cases. There are presently two types of available solutions for the motion of a RBC in simple shear flow, obtained either for initially spherical cells or for cells which have an a priori given ellipsoidal deformed shape.

C. Spherical Capsule Models

The cell, also termed a capsule, is assumed to be initially spherical and to be subjected to moderate deformations. The membrane bending resistance is ignored. Cell deformation will be limited either if the viscosity ratio is very large ($\lambda >> 1$) or if the viscous to elastic forces ratio is very small ($C << 1$). Then, the problem described by Equations (10)–(20a) is solved exactly by means of a perturbation method, which consists in expanding all quantities in terms of a small parameter that measures cell deformation, and in obtaining successive approximations to the shape. The interest of this approach lies in the fact that all mechanical aspects of cell deformation are exactly taken into account. Even though the initial geometry (a sphere) is fundamentally different from the discocyte, such models give information as to the important physical processes that govern cell motion in shear flow. Furthermore, they allow to investigate the effect of different interfacial properties on the deformation of a capsule.[22]

When deformation is limited by the smallness of C, Barthes-Biesel and Rallison[23] show that, to first order in C, the deformed surface of the membrane may be characterized by one symmetric and traceless second-order tensor **J**, which is a function of time only. The deformed profile equation (12) becomes:

$$r = (\mathbf{x}\cdot\mathbf{x})^{1/2} = a\{1 + C\,{}^T\mathbf{x}\cdot\frac{\mathbf{J}}{a^2}\cdot\mathbf{x} + O(C^2)\}. \tag{21}$$

The surface and volume variations are then of order C^2 and thus negligible. The solution of equations (10)–(20a), for the case of an area incompressible membrane characterized by equations (6) and (7), leads to the following time evolution equation for **J**:

$$C\left\{\frac{\partial \mathbf{J}}{\partial t}+\mathbf{J}\cdot\mathbf{\Omega}-\mathbf{\Omega}\cdot\mathbf{J}\right\}=G\left\{\frac{60\,\mathbf{E}}{G(23\lambda+32)}-\frac{8\mathbf{J}}{(23\lambda+32)}+O(C)\right\}. \tag{22}$$

For a viscoelastic area incompressible membrane, characterized by equations (6), (7), and (8), the time evolution equation for **J** has been obtained by Barthes-Biesel and Sgaier[24]:

$$\beta\left\{\frac{\partial \mathbf{J}}{\partial t}+\mathbf{J}\cdot\mathbf{\Omega}-\mathbf{\Omega}\cdot\mathbf{J}\right\}=\frac{15(\beta+1)\mathbf{E}}{4}-\frac{G\mathbf{J}}{2}. \tag{23}$$

These models predict that in simple shear, the capsule deforms into an ellipsoid since to first order in C, equation (21) may also be written in the following form:

$$r^2=a^2\left\{1+2C\,{}^T\mathbf{x}\cdot\frac{\mathbf{J}}{a^2}\cdot\mathbf{x}+O(C^2)\right\}.$$

which is the equation of an ellipsoid. Furthermore, the tank-treading motion of the membrane is retrieved through the product of **J** with the vorticity tensor $\mathbf{\Omega}$.

It is simple to solve equations (22) or (23) for the steady simple shear flow given by equation (11). In the case of a purely elastic area incompressible membrane, the steady state solution leads to

$$J_{12}=J_{21}=15/4,$$

all other components being zero. The capsule is thus oriented at 45° with respect to stream lines (i.e., axis Ox_1) and its deformation is a linear function of Capillary number:

$$D=15C/4+O(C^2). \tag{24}$$

However, for a viscoelastic membrane (equation (23)), the angle θ of the ellipsoid longest major axis with the streamlines decreases with β, that is, with shear rate:

$$\theta=\frac{1}{2}\tan^{-1}\frac{1}{2\beta}. \tag{25a}$$

Furthermore, as shown on Figure 8, the deformation increases with shear rate and reaches a high shear limiting value:

$$D=\frac{15}{8}C\frac{1}{(\beta^2+\frac{1}{4})^{1/2}}. \tag{25b}$$

The predictions for a capsule surrounded by a viscoelastic membrane are in much better qualitative agreement with experimental observations of RBCs than those

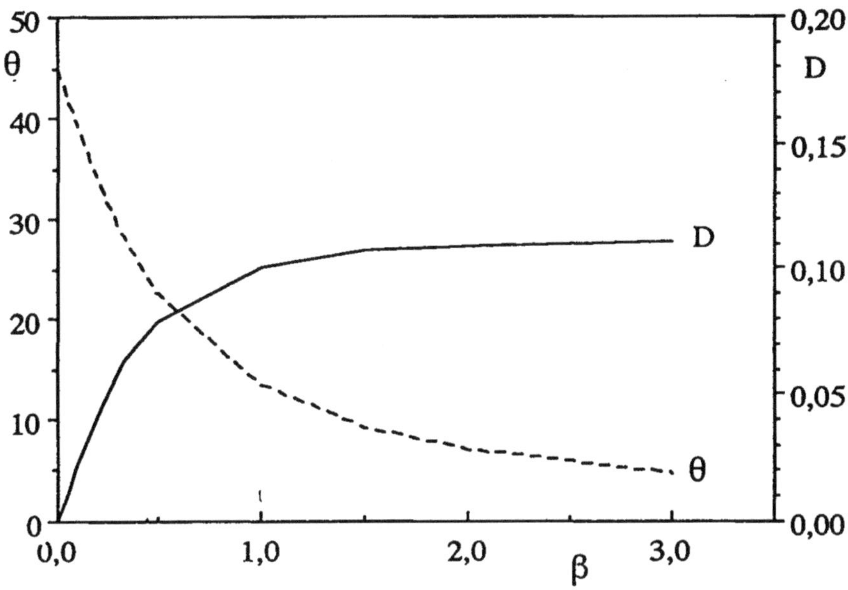

Figure 8. Deformation and orientation of an initially spherical capsule with a viscoelastic membrane. Computed from (25a) and (25b).

obtained with a purely elastic membrane. This might indicate that membrane viscosity plays an important role in the tank-treading motion. This is not surprising, since although the overall cell profile is steady, the membrane material points undergo time varying deformations due to the convective rotational motion of the membrane.

It should be pointed out that the spherical capsule model yields good predictions for liquid droplet deformation[25] or for artificial capsules[19,20] even when the deformation is not small. The agreement between experiments and theory indicates that the main physical phenomena are properly taken into account by the asymptotic model. This model is not applicable for the interpretation of rheoscopic data on RBCs because the aspect ratio of the cell that tank-tread in a rheoscope, is fairly large[16] and thus falls well out of the range of validity of the spherical capsule model. However, it gives quite good predictions regarding the relaxation time τ_r of a deformed cell after cessation of flow. Indeed, equations (22) and (23) lead to the following values of τ_r:

$$\tau_r = \frac{(23\lambda + 32)}{8}\frac{\mu a}{E_s} \quad \text{for a purely elastic membrane,} \tag{26}$$

$$\tau_r = \frac{2\mu_s}{E_s} \quad \text{for a viscoelastic membrane.} \tag{27}$$

The experimental values of τ_r obtained by Sutera et al.[30] are of order 0.1s for RBCs suspended in a 35 mPa s Dextran. With $E_s = 5 \times 10^{-3}$ mN s^{-1} and $a = 2.8$ μm, the value predicted by equation (26) is $\tau_r = 0.09s$. For a viscoelastic membrane with a surface viscosity varying from 1 to 5×10^{-4} mN s m^{-1}, equation (27) gives values of the relaxation time varying from 0.2 to 1 s. These values are consistent with those measured by Sutera. Furthermore, the fact that they are all of the same order of magnitude indicates that in the relaxation process of a RBC, the membrane and internal viscosities play equally important roles.

D. Ellipsoidal Capsule Models

Another approach to the problem defined by equations (10)–(20b) was proposed by Keller and Skalak,[26] who a priori assume that the deformed shape of the cell is an ellipsoid, with principal semi diameters denoted L and B in the shear plane, and oriented by an angle θ with respect to stream lines (Figure 7). The principal axes of the ellipsoid are denoted Oy_1 and Oy_2, with corresponding unit vectors denoted $\mathbf{e}_1'$ and $\mathbf{e}_2'$. A simple membrane velocity field is then assumed in accordance with experimental observations of tank-treading cells:

$$\mathbf{u}_s = F(L/B\, y_2\mathbf{e}_1' - B/L\, y_1\mathbf{e}_2'), \tag{28}$$

where F is the rotational frequency of the membrane. The external flow Stokes equations are solved subject to boundary conditions (17) and (18). This is achieved by means of an adaptation of Jeffery's solution obtained for the flow around a solid ellipsoid. Then the velocity distribution given by equation (28) is linearly extended in the interior of the cell, and the corresponding rate of energy dissipation is computed. The problem is closed by assuming that the membrane does not dissipate energy and that the energy brought by the external liquid is all dissipated by the internal motion. The model predicts two types of behavior for the cell: an unsteady flipping motion or a steady orientation with a tank-treading membrane. The transition from the first behavior to the second one occurs when the viscosity ratio λ is decreased and/or when the cell elongation L/B is increased. This is in good qualitative agreement with rheoscope observations. However, it should be noted that in this model, the cell geometry and the viscosity ratio both are considered as two independent parameters in spite of the fact that they are linked by the mechanics of the system. This is a consequence of the fact that the membrane mechanics are not treated rigorously.

The velocity field (28) does not respect the local area incompressibility. Secomb and Skalak[27] have suggested two modified versions of equation (28) that do take into account area incompressibility, but that are obviously much more complicated. However, the predictions of these improved models are qualitatively the same as those obtained from equation (28), so that it is often quite sufficient to use equation (28) for practical purposes.

This is tank-treading ellipsoid model is used by Sutera et al., in a series of papers,[18,28,29,30] to interpret rheoscope data in terms of intrinsic cell membrane properties. After the internal and external velocity fields are determined as indicated above, it is simple to compute the stress distribution in the fluids from equations (15) and (16). The load **q** on the membrane is given by equation (19), and the elastic tensions can be obtained directly from equation (20a) since the system is statically determinate. A linear viscoelastic behavior for the membrane, as given by equations (6) and (8), is then assumed. This makes it possible to compute the membrane shear elastic modulus and surface viscosity. Sutera et al.[30] find a shear stiffening behavior for the membrane with average values of E_s varying from 3.3 to 8.6 10^{-3} mN m^{-1} when the shear rate increases from 28.6 to 285.7 s^{-1}. This interesting effect cannot be corroborated by micropipette experiments which are essentially static, and which involve much higher extension rates than those produced in a rheoscope. They also find that the internal pressure in the cell is different from the external pressure, owing to the membrane deformation. The excess internal pressure increases with the level of deformation.

An average value of the membrane viscosity can be determined by stating that the rate of energy supplied by the external fluid is balanced by the sum of the rates of energy dissipated in the membrane and in the internal liquid. This procedure yields surface averaged values of the membrane viscosity of order 0.6 to 1×10^{-4} mNs m^{-1}, with a tendency toward shear thinning. This evaluation has been refined by Sutera et al.,[30] who are able to determine the local value of μ_s. They find that μ_s varies from 0.18 to 0.1×10^{-4} mN s m^{-1} as the shear rate increases from 28.6 to 285.7 s^{-1}. These values are substantially less than those obtained from micropipette experiments or from those averaged over the surface, which correspond to a constant viscosity membrane that would produce the same rate of energy consumption.

The ellipsoidal tank-treading capsule model thus allows an interpretation of rheoscope data. However, the model predictions regarding the shear dependency of the membrane elastic modules and the exact value of the surface viscosity have yet to be validated by independent measurements.

IV. MICRORHEOLOGY OF A DILUTE SUSPENSION OF RED BLOOD CELLS

The direct observation of cells in a rheoscope is a sophisticated experimental technique that is reserved for research purposes. Besides, many cells must be measured before a mean value of their physical properties can be obtained. Conversely bulk experimental measurements of a cell suspension are usually fairly easy to perform and directly provide an average over a population. As such they are good candidates for routine investigations. However, in order to interpret correctly such bulk measurements in terms of the intrinsic physical properties of the suspended cells, a microrheological model is needed.

A. Experimental Evidence

Whole blood rheological properties result from a series of complicated phenomena that involve dissociation of aggregates, deformation of cells, and hydrodynamic interactions between cells. In order to assess cell deformability from rheological measurements, Drochon et al.[31] suggested use of a *dilute* suspension of red blood cells in a viscous suspending medium where no aggregating molecules are present. The dilution hypothesis eliminates hydrodynamic interactions between the cells, and thus any observed non-Newtonian property of the suspension can be attributed to cell deformation. Correspondingly, they conducted a series of experiments on dilute suspensions of normal cells in Dextran solutions, where they measured their apparent viscosity μ_a as a function of shear rate. They used a Couette viscometer, where the flow field is a simple shear flow, analogous to the one that exists in a rheoscope. Rather than μ_a it is more convenient to introduce an intrinsic viscosity $[\mu]$, that eliminates the dependency on hematocrit :

$$[\mu] = (\mu_a - \mu)/\mu\phi \tag{29}$$

Figure 9 shows the intrinsic viscosity of two suspensions of normal red blood cells suspended in Dextran solutions with viscosities equal to 14.3 mPa s and 8.4 mPa s. The suspension exhibits a definite shear thinning effect that may be attributed to the deformation and orientation of the cells in shear flow, as observed in a rheoscope. The total viscosity drop between the high and low shear limits

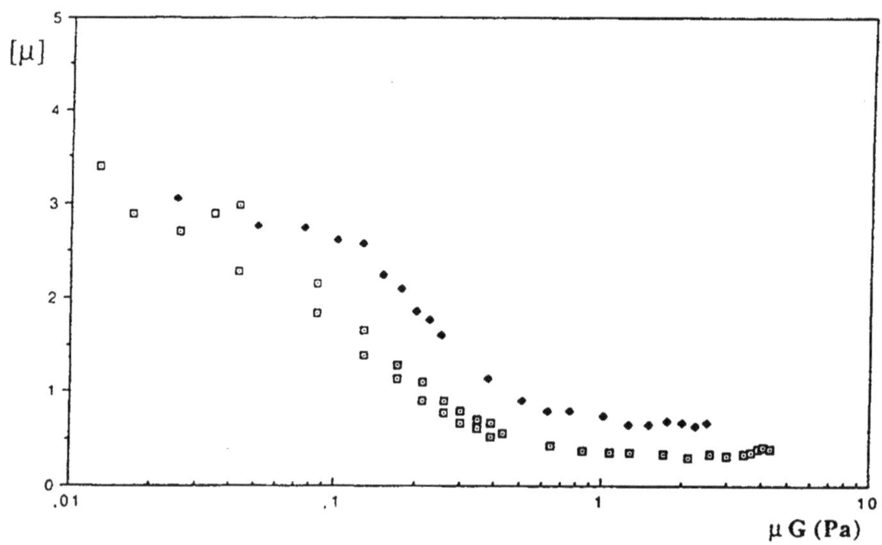

Figure 9. Intrinsic viscosity of a suspension of normal RBCs in two Dextran solutions: ▣ 14.3 mPa s, ♦ 8.4 mPa s. The two suspensions exhibit a shear thinning effect. From Drochon, private communication.

depends on the suspending fluid viscosity and increases with it. It is also found that intrinsic viscosity curves obtained in the same Dextran solution for different hematocrits, superimpose within the experimental errors.[31] This phenomenon occurs for hematocrits up to 12%, thus indicating that interactions effects between cells are small, even though a volume concentration of 12% is already fairly high from the point of view of suspension mechanics. This, in itself, is an interesting result, since the higher the hematocrit, the better the experimental precision. A volume concentration of order 10% is thus recommended by Drochon et al.

The discriminating power of the method can assessed with normal cells treated with diamide, following the method used by Fischer et al.[32] and by Drochon et al.[33] Diamide is a bifunctional SH-reagent which cross-links membrane SH-groups and thus decreases the membrane deformability. Indeed Fisher et al. have observed diamide treated cells in a rheoscope and found that, under equal applied shear stresses, the diamide cells are less elongated than the normal ones, but still tank-tread, which indicates that the internal hemoglobin solution is still fluid. The effectiveness of the diamide treatment can be assessed by means of micropipette measurements from which, it appears that with the diamide concentrations used here (0.1 mMol), the membrane shear elastic modulus is roughly multiplied by a factor 2.5 ± 0.9 as compared to the normal value.[33] Figure 10 shows the intrinsic viscosity curves of two suspensions of red blood cells that are normal or diamide treated, respectively. The viscosity decrease for the diamide rigidified cells occurs for larger shear stresses than for normal ones. As a matter of fact, the suspension containing the diamide-treated cells has not yet reached the high shear limit, even for the maximum stress that can be applied with the viscometer. Since the two suspensions were prepared with the same Dextran solution, the total viscosity drops should be almost the same, but this cannot be exactly verified because of the apparatus limitations.

Those experimental results indicate clearly that the rheological properties of a dilute suspension of RBCs depend in a complex way not only on the membrane mechanics, but also on the viscosity ratio between the internal hemoglobin solution and the suspending medium.

B. Rheology of a Dilute Suspension of Spherical Capsules

The bulk properties of a dilute suspension of identical capsules are defined as averages of the corresponding local quantities over a representative volume containing many particles, but small with respect to the length scale of the flow.[34] When the cells are identical and hydrodynamic interaction effects are ignored, the particle contribution to the bulk stress $\mathbf{\Sigma}$ is simply identified with the local stress averaged over one representative cell and weighted by the hematocrit:

$$\mathbf{\Sigma} = -p\mathbf{I} + 2\mu\mathbf{E} + \phi\int_{S_p} [(\boldsymbol{\sigma}.\mathbf{n}).^{T}\mathbf{x} - \frac{1}{3}(^{T}\mathbf{x}.\boldsymbol{\sigma}.\mathbf{x})\mathbf{I} - \mu(\mathbf{u}.^{T}\mathbf{n}+\mathbf{n}.^{T}\mathbf{u})]\frac{3dS_p}{4\pi a^3},$$

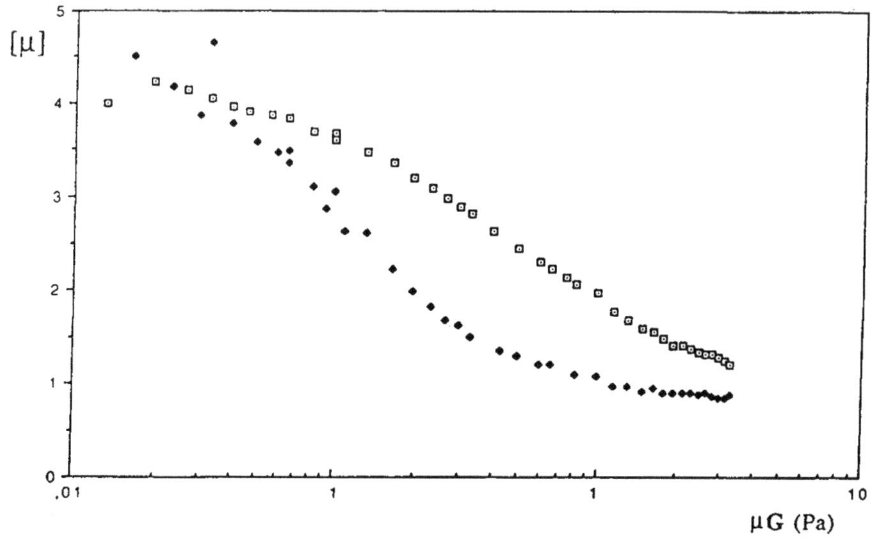

Figure 10. Intrinsic viscosity curves for normal and diamide treated cells. ♦ normal cells, ◘ diamide treated cells. The hardened cells require higher shear stresses to deform. From Drochon, private communication.

where p is the pressure, **E** the bulk shear rate, ϕ the hematocrit, and S_p the surface of the cell. The spherical capsule model can thus be extended to yield a constitutive equation that relates the bulk stress $\mathbf{\Sigma}$ to the bulk shear rate **E**.[23] In the case of a suspension of purely elastic, area incompressible, initially spherical capsules, this relation becomes:[31]

$$\mathbf{\Sigma} = -p\mathbf{I} + 2\mu\mathbf{E} + 2\mu G\phi \left\{ \frac{5(23\lambda - 16)}{2G(23\lambda + 32)}\mathbf{E} + \frac{16}{(23\lambda + 32)}\mathbf{J} + O(C) \right\} \tag{30}$$

In principle, this equation is valid for asymptotically small values of ϕ and C. Coupled to the time evolution equation (22) for **J**, equation (30) predicts a viscoelastic rheological behavior for the suspension. The deforming effect of the external flow is accounted for through the terms containing **E**, and the elastic response of the membrane through the terms in **J**. The left-hand-side of the time evolution equation (22) may be large for rapid time variations but is negligibly small for steady flow. However, since these terms account for tank-treading because of the occurrence of the corotating derivative, they are important from the physical point of view and are thus retained. The missing $O(C)$ terms in equations (30) and (22) correspond to the modification of the external flow field due to the deformation of the capsule and to the elastic contribution of the next order approximation to the deformed shape.

Drochon et al.[31] suggested using equations (22) and (30), where the missing $O(C)$ terms are ignored, as a phenomenological constitutive law to represent the rheological behavior of a suspension of red blood cells even if the Capillary number is not small. The parameters λ and C are assumed to keep their original physical significance and to represent average apparent values respectively of the characteristic viscosity ratio and of the Capillary number based on the cell dimensions and shear elastic modulus of the membrane. In the case of a simple shear flow that exists in the gap of a Couette viscometer, the components of **E** and **Ω** are given by equation (11). It is then simple to express equations (22) and (30) in this particular situation and to deduce the apparent shear viscosity μ_a of the suspension defined as:

$$\mu_a = \Sigma_{12}/G. \tag{31}$$

The intrinsic viscosity is then found to have the following expression:

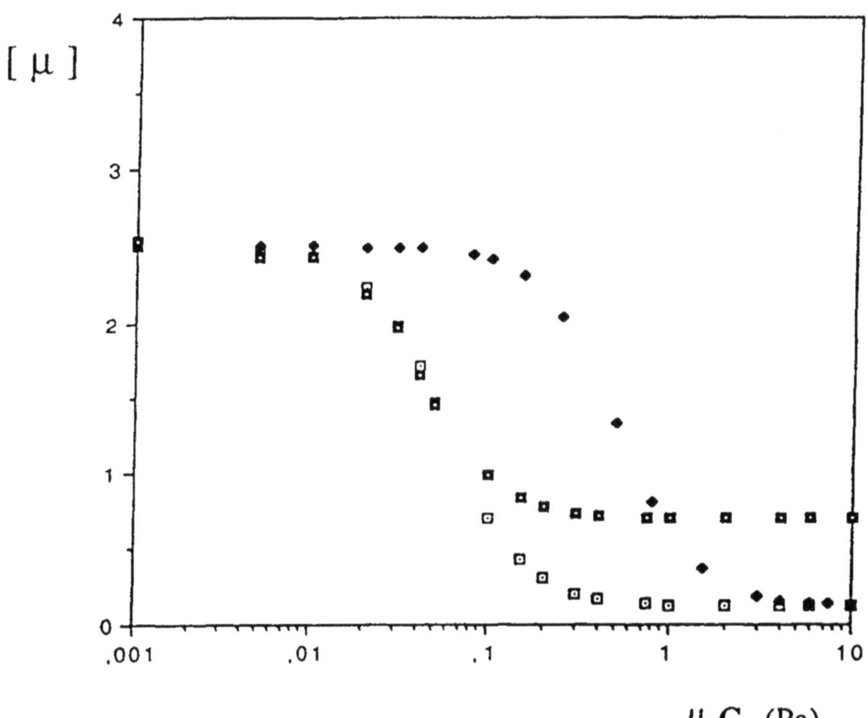

Figure 11. Theoretical predictions for the shear dependence of the intrinsic viscosity. ◘ $E_S = 1 \times 10^{-3}$ mN m^{-1}, $\lambda = 0.8$; ■ $E_S = 1 \times 10^{-3}$ mN m^{-1}, $\lambda = 1.5$; ♦ $E_S = 9 \times 10^{-3}$ mNm^{-1}, $\lambda = 0.8$. The shear thinning effect is a complicated function of both λ and E_S. Computed from equation (32).

$$[\mu] = \frac{5(23\lambda - 16)}{2(23\lambda + 32)} + \frac{7680}{(23\lambda + 32)[64 + (23\lambda + 32)^2 C^2]} . \tag{32}$$

The corresponding theoretical viscosity versus shear stress curve is shown on Figure 11. The suspension is shear thinning with a zero shear viscosity $[\mu]_0$ of 2.5 (corresponding to Einstein's hard spheres value) and a limiting high shear viscosity that depends on λ:

$$[\mu]_\infty = \frac{5(23\lambda - 16)}{2(23\lambda + 32)} .$$

The intrinsic viscosity evolution between these two limits depends upon E_s. Specifically, the model predicts that the drop in viscosity is shifted toward higher shear stresses when the membrane elastic rigidity is increased. This of course is a direct consequence of the dependency of $[\mu]$ on C rather than on just μG.

C. Interpretation of Experimental Results

It is immediately apparent that the experimental and theoretical graphs shown on Figures 9, 10, and 11, respectively, are qualitatively similar. A difference occurs for the zero-shear value of the intrinsic viscosity, which may be attributed to geometric effects. The comparison between the experimental and theoretical curves is straightforward. The parameter a that appears in the Capillary number C is taken to be the radius of the sphere which has the same volume as the RBC. The value of λ is inferred from the viscosity of the Dextran solution and from standard data on the viscosity of the hemoglobin solution (see Section IIB). Since the experiments were conducted at 25 °C, the viscosity of the hemoglobin solution was taken to be equal to 10 mPas. The high shear limit $[\mu]_\infty$ of the intrinsic viscosity is found from the experimental graphs. When the inverse of the partial viscosity drop $[\mu] - [\mu]_\infty$ is plotted as a function of $(\mu G)^2$, a linear correlation is obtained (Figure 12). The slope and the intercept with the ordinate axis are related to the mean value of E_s and to the total viscosity drop, respectively. This last value is a check of the consistency of the model, since it can be compared to the experimental viscosity drop which is directly measured from the difference between the high and low shear limits of $[\mu]$. The corresponding values of the membrane shear elastic modulus that are found in this fashion are

$E_s = 3.3 \times 10^{-3}$ mNm^{-1} for normal cells,

$E_s = 9.1 \times 10^{-3}$ mNm^{-1} for diamide treated cells.

These values are of the same order of magnitude as those measured with a micropipette. Furthermore, the diamide treated cells have a modulus that is 2.8 times that of normal cells, which also compares well with micropipette measure-

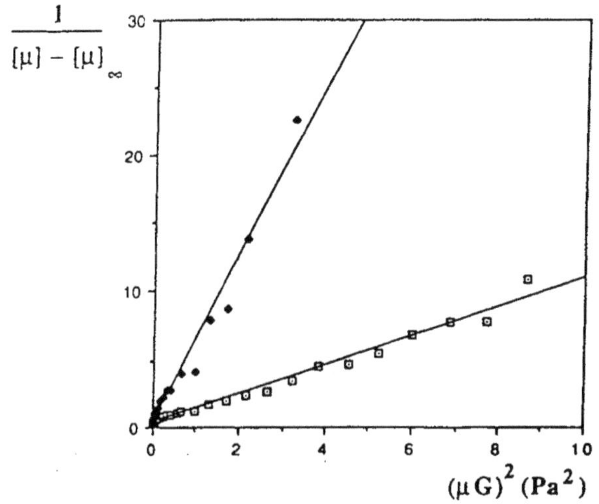

Figure 12. Correlation between the experimental data and the rheological model equation (32). The slope of the straight line allows to determine E_S. ♦ normal cells, ▣ diamide treated cells. From Drochon private communication.

ments. This indicates that the viscometric method could be used to measure directly a mean value of the elastic modulus of a population of RBCs.

The E_s values obtained by this viscometric method correlate also quite well with those computed by Sutera et al.[30] from rheoscope observations (Section IIID). In the latter case, the modulus is found to increase with shear rate. In the viscometric method, the shear dependency of the modulus cannot be assessed since it only yields a value averaged over a shear rate range from is 1 s^{-1} to 100 s^{-1}.

This technique gives a good estimation of the average shear elastic modulus of a cell population. However, it requires a very precise viscometer since the measurements are performed on dilute suspensions for which small shear thinning effects are measured. Furthermore, it cannot give detailed information on the stress dependency of the membrane elastic properties. Its main advantage remains its experimental simplicity.

V. RED BLOOD CELL FILTRATION

A class of very interesting devices that are also used to measure RBC deformability consists of filtration apparati. The common underlying principle of the different setups is to force a cell suspension through a filter with pores having a diameter smaller than the cell's and to measure simultaneously the flow rate and the pressure drop across the filter. The advantage of such measurements is linked to their experimental simplicity and also the fact that they are directly relevant to microcirculation. A review by Nash[35] gives a description and an assessment of the different techniques used over the years. The experimental protocols differ in the nature of the microsieve, the pore geometry, the suspension hematocrits (from 1 to 90%) and the flow driving force (constant pressure head or constant flow rate). With a large number of pores (typically of order of 300,000) the measured bulk quantities correspond to averages of the pore level micromechanics over all the filter pores. Other techniques use a limited number of pores (of order 20–30) and are able to follow the passage of one cell through one pore. The Cell Transit Analyser (CTA), is based on this principle. The technique consists in measuring the variation in the electrical conductivity of the filter when one cell flows through it. This apparatus can thus yield data on the total transit time of a cell in one pore, and on the relative importance of entry and exit phenomena. Another advantage of the method is that it also gives average data on a cell population. Indeed, many passages are detected and measured in the course of one experiment, and the results are then properly averaged.[36]

It is difficult to compare experimental data obtained in different devices, because the measured values depend not only on the imposed flow forces but also on the determinants of RBC deformability. In order to do so, one needs a very good understanding of the mechanics of the flow of one cell in one pore, before a link can be established between the bulk measured quantities and the cell intrinsic properties.

A. Experimental Evidence

Filtration has been shown to be able to detect modifications of the internal viscosity of the cells. Indeed, Kiesewetter et al.[37] have measured the transit time of an erythrocyte through a single pore of 5.8 μm diameter under a driving pressure of 100Pa. They find that the transit time increases with the suspending buffer osmolality: from 18.4 ± 8.6 ms for RBCs suspended in an hypotonic medium (200 mOsm/kg) to 28.0 ± 9.1 ms for normal cells (300 mOsm/kg) and 46.8 ± 11.4 ms for RBCs suspended in an hypertonic medium (480 mOsm/kg). The faster passage time for hypoosmolar swollen cells is a consequence of a decrease of the internal viscosity due to the resulting decrease in hemoglobin concentration. This effect is important enough to compensate the volume increase of the cells. Conversely, exposure to a hyperosmolar environment leads to an increase in internal viscosity and to significantly higher passage times even though the dehydrated cells are smaller. It thus appears that the viscosity ratio λ plays a major role in the flow process of a cell in a pore.

In order to assess the ability of filtration to detect membrane alterations, Kiesewetter et al.[37] have also measured the transit time for normal and diamide treated erythrocytes, suspended in a saline solution, in a 5.8 μm pore under a 100 Pa driving pressure. The transit times of the diamide treated cells (29.2 ± 9.7 ms) and of the normal cells (28.0 ± 9.1 ms) are not significantly different. Schmid-Schonbein and Gaehtgens[38] reach a similar conclusion after comparing the filtration rates for normal and diamide treated RBCs at a higher diamide concentration (1.5 mM) in a 5 μm pore, under a 98 Pa driving pressure. A decrease of the tube diameter to 3.75 μm does not change the conclusion.[38]

All the previously-mentioned experiments were conducted in a low viscosity suspending medium, that is, plasma or a saline buffer solution. Then, the flow process in the pore and the cell deformation are essentially controlled by viscous effects, and the main energy expenditure is due to cytoplasmic motion. If the viscosity of the suspending liquid is roughly matched to the internal viscosity of the cells, the flow and deformation of a cell in a pore should be controlled by the membrane elasticity. Drochon et al.[33] have performed filtration measurements on dilute (8%) suspensions of normal and diamide treated cells suspended in a Dextran solution with a viscosity of 8 ± 0.5 mPas. They use an apparatus with a Nuclepore filter with some 300,000 pores of diameter 4.7 μm and length 11 μm. They conclude that it is possible to distinguish between the normal and membrane-hardened cells. However, although the elastic modulus of the diamide treated cells is about 2.5 times that of the normal cells, the increase in transit time after the diamide treatment is only about 15%.

This result has been recently confirmed by Drochon et al.,[40] who also used a CTA and measured the entry time of cells in pores with 5 μm diameter and 15 μm length. It is found that the entry times for normal and diamide treated cells in a 8 mPas Dextran are 1.55 ± 0.05 ms and 1.78 ± 0.08 ms, respectively. Although the difference is statistically significant, it is small and difficult to measure. This indicates that filtration is not a very sensitive device to measure the properties of the cell membrane.

B. Analysis of Filtration Experiments

When a large filter is used, the bulk filtration index depends not only on the micromechanics of the motion of one cell through one pore, but also on the cell repartition among the filter pores. Skalak et al.[41,42] have proposed an analysis of the filtration process when red and white blood cells are present in the suspension. The model is derived for a constant flow rate device where the pressure drop ΔP across the filter varies with time, but it can easily be extended to constant pressure drop situations. The suspension is assumed to contain different types of cells (identified by the index k). The ratio of a cell to pore volume is denoted V_k and the relative pore hydraulic resistance when a cell is flowing through as compared to clear suspending fluid, is denoted δ_k. Expressing dynamic boundary conditions and mass balance leads to a set of coupled equations, valid for steady state:

$$\frac{\Delta P}{\Delta P_0} = \left\{ \sum_k \frac{\alpha_k}{\delta_k} + \gamma \right\}^{-1}, \tag{33}$$

$$\alpha_k = \frac{\phi_k \, \delta_k}{V_k} \left\{ 1 + \sum_k \frac{\phi_k(\delta_k - 1)}{V_k} \right\}^{-1}. \tag{34}$$

where ΔP_0 is the pressure drop across the filter when only clear fluid is flowing, γ is the proportion of pores that are not occupied by any cell, α_k is the proportion of pores occupied by species k and ϕ_k is the relative feed hematocrit of species k. This model is valid for low feed hematocrits since it is assumed that at any time there is only one cell occupying a given pore. It is also restricted to tightly fitting cells since the presence of a liquid film between the cell and the pore wall has been ignored. As it stands, it requires some knowledge of the increase in pore resistance δ_k due to the presence of a cell. This information is obtained from the experimental measurements of the passage time of a cell through a single pore.[37,38] The model can thus give good qualitative predictions of filtration measurements performed on large filters. It has also been extended to unsteady flows, and is thus able to predict filter clogging due to the presence of less deformable cells (WBCs or hardened RBCs). It is also possible to take into account a continuous distribution of the properties of any cell species.[42] However, although this analysis is extremely useful to analyze bulk filtration results obtained on filter containing many pores, it cannot provide any direct information on the intrinsic physical properties of the suspended cells.

C. Motion of a Capsule Through a Pore

The formulation of a mechanical model of the motion of a cell through a pore is a very complicated problem of continuum mechanics that has so far received little attention due to its complexity. The problem that must be addressed is the one described by equations (12)–(20), with additional conditions that arise from the presence of the channel walls and from input and output conditions. The flow domain is deliminated by solid boundaries B, corresponding to the pore, and by two entrance and exit sections, respectively S_1 and S_2, where the velocity profiles $\mathbf{u}_1(\mathbf{x})$, $\mathbf{u}_2(\mathbf{x})$ and pressures p_1, p_2 are specified (Figure 13).

The corresponding additional conditions are thus:

$$\mathbf{u}(\mathbf{x}) = \mathbf{0}, \quad \mathbf{x} \text{ on } \mathbf{B}, \tag{35}$$

$$\mathbf{u}(\mathbf{x}) = \mathbf{u}_i(\mathbf{x}), \text{ and } p(\mathbf{x}) = p_i, \quad \mathbf{x} \text{ on } S_i,\ i = 1,2. \tag{36}$$

The sections S_1 and S_2 are taken far enough from the particle, for the flow perturbation due to the presence of the cell to have died out. Depending on the flow

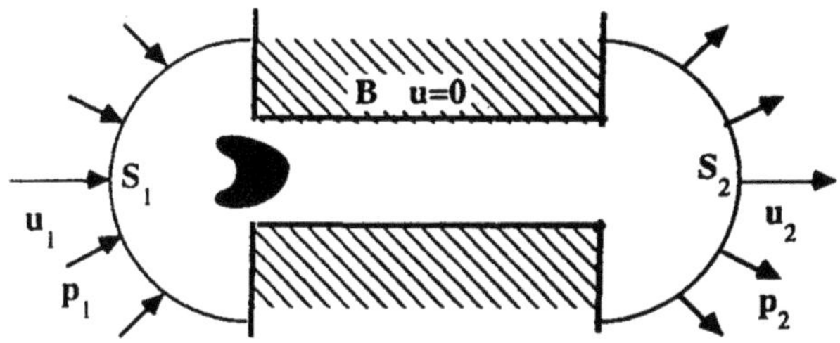

Figure 13. Definition of the flow domain for a filtration model.

configuration, either the total pressure drop Δp or the total flow rate Q across the pore is kept constant. The corresponding bulk boundary conditions are then:

- constant pressure drop:

$$p_1 - p_2 = \Delta p_0, \quad Q(t) = \frac{\Delta p_0 - \Delta p_{\text{add}}(t)}{R_{Hc}}, \tag{37}$$

- constant flow rate:

$$p_1 - p_2 = \Delta p(t) = R_{Hc} Q_0 + \Delta p_{\text{add}}(t), \quad Q(t) = Q_0, \tag{38}$$

where $\Delta p_{\text{add}}(t)$ denotes the additional pressure drop due to the presence of the cell and where R_{Hc} is the pore hydraulic resistance that depends on pore geometry.

Long Cylindrical Pores

A semi-analytical solution to the problem outlined above can be found in the particular case of a closely fitting cell.[43,44] Then, lubrication theory may be used to simplify the equations of motion of the fluids. Similarly, the case of flow channels with slowly varying cross section and closely fitting cells can also be treated,[43] provided that the hypotheses inherent to lubrication are fulfilled. Such models have direct applications to microcirculation.

In general, a numerical solution must be sought. In the case of a discoidal cell having a surface incompressible membrane with bending resistance (equations (4), (5) and (9)), flowing in a long cylindrical pore, it is found that the cell reaches a steady deformed shape, that is parachute-like.[45] Its elongation increases with the capillary number. The case of a tapered pore has also been studied with the same numerical method,[45] in order to model entrance effects in a capillary tube. As the cell enters the channel, its elongation increases until it reaches a limiting value in the cylindrical part.

Such models are essentially useful for understanding microcirculation. They are of limited use, however, for interpretation of filtration experiments, where the pores are short and the process transient.

Flow of a Capsule through a Hyperbolic Pore

In order to account for entrance/exit phenomena, Leyrat-Maurin and Barthes-Biesel[46] have considered the flow of a capsule through a hyperbolic constriction. The flow channel B is an axisymmetric hyperboloid with revolution axis Ox (Figure 14). The throat radius is denoted r_t, and the angle of the hyperbola asymptote with Ox is 45°. The cell is assumed to be also axisymmetric with its axis aligned with Ox and its center of mass is located on Ox. The problem is thus fully axisymmetric. The constriction represents a short pore or a thin filtration membrane. The advantage is that the flow field in absence of a cell, is known exactly and is very close to the entrance flow in a cylindrical pore. Due to the channel geometry, entrance and exit effects prevail and the flow is essentially transient.

The Capillary number takes a slightly different definition:

$$C = \frac{\mu Q_0}{\pi r_t^2 E_s}. \tag{39}$$

where Q_0 is the initial flow rate. An important new parameter appears, namely the size ratio between the cell and the pore:

$$R = a/r_t. \tag{40}$$

This problem is solved by means of a numerical model, based on the use of boundary integrals for the solution of the Stokes equations (12)–(15).[46] Results

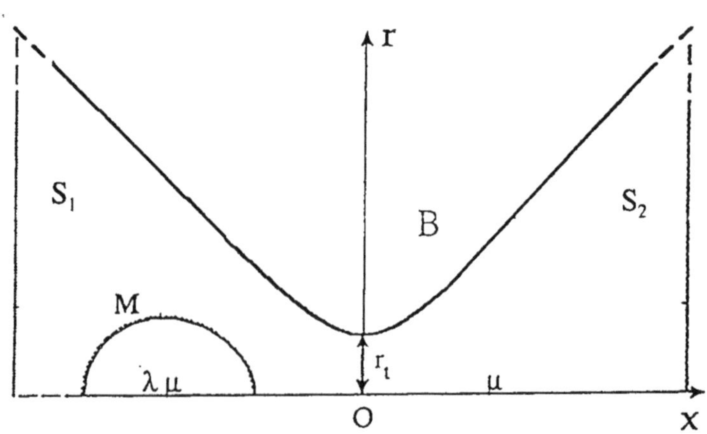

Figure 14. Schematics of a cell flowing through an hyperbolic constriction.

have so far been obtained for the case of initially spherical capsules (radius a) surrounded by a Mooney-Rivlin type membrane that is not area incompressible and filled with an internal liquid that has the same viscosity as the suspending medium ($\lambda = 1$). The membrane elastic law is given by:

$$T_{11} = \frac{E_s}{3\lambda_2\lambda_2}\left(\lambda_1^2 - \frac{1}{\lambda_1^2\lambda_2^2}\right),$$

and is thus similar to equation (6) when the product $\lambda_1\lambda_2$ is constrained to be unity.

The model can simulate the motion and deformation of a cell as it flows through the pore. The time evolution of local quantities is calculated, such as particle geometry, center of mass velocity, and elastic tensions in the membrane. At the macroscopic level, some global quantities are computed, such as the entry time of the capsule through the pore, the additional pressure drop, or the perturbation of flow rate (according to the type of flow condition chosen, respectively constant flow rate or constant pressure head). A convenient way to represent the bulk results is to compute the apparent pore hydraulic resistance defined by:

$$R_{Ha} = \Delta P(t)/Q(t).$$

The relative perturbation of R_{Ha} as a function of the position of the cell center of mass x_G along the pore axis, is shown on Figure 15 for a size ratio $R = 1.4$, a Capillary number $C = 0.1$, and either constant pressure drop or constant flow rate conditions.

Capsules that are deformable and small enough to squeeze through the pore all present the same general features. The cell motion can be roughly decomposed into three phases: a suction phase during which all of the capsule is upstream of the constriction, a transit phase during which the front of the cell has passed the throat while the rear is still downstream, and finally a recoil phase where the cell relaxes back to its initial shape. Parachute shapes very analogous to those observed in microcirculation, are predicted during the recoil phase. They are due to the deceleration of the flow after the throat and to the recoil of elastic tensions that have built up in the cell membrane. It is very interesting to note that the maximum flow perturbation, that is, the maximum energy expenditure occurs *upstream* of the throat, and is thus due to entry phenomena. These correspond roughly to the suction phase. Then, once the cell is properly deformed, it creates a smaller flow perturbation. Finally, the model indicates that, from the experimental point of view, it is more judicious to operate under constant pressure drop rather than constant flow rate conditions. Indeed, all things being equal, the flow perturbation is more accentuated, and thus easier to detect under constant pressure drop flow (Figure 15).

The model also predicts two different mechanisms of pore plugging. Figures 16a and 16b show the successive profiles of cells with size ratio $R = 1.8$ and capillary number of respectively $C = 0.05$ and 0.1, under constant pressure drop conditions. For the smaller Capillary number, the cell is blocked before being able to pass

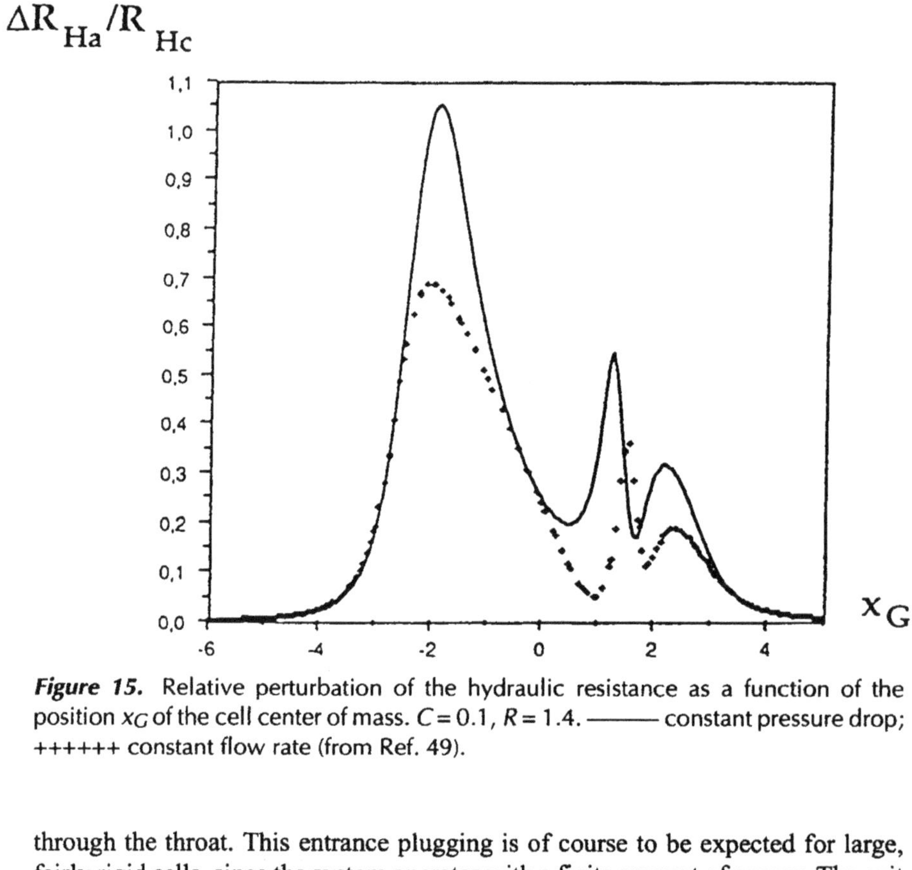

Figure 15. Relative perturbation of the hydraulic resistance as a function of the position x_G of the cell center of mass. $C = 0.1$, $R = 1.4$. ——— constant pressure drop; ++++++ constant flow rate (from Ref. 49).

through the throat. This entrance plugging is of course to be expected for large, fairly rigid cells, since the system operates with a finite amount of energy. The exit plugging is more surprising. It is due to the conjugate effects of the presence of a constriction wall and of the elastic relaxation process that tends to alleviate the membrane tensions by creating a radial expansion of the nose. However, it should be noted that exit plugging is commonly observed experimentally. Chien et al.[47] report the formation of clusters of cells on the downstream side of filtration membranes. This phenomenon is attributed to adhesive forces between cells and the filter wall. The corresponding physico-chemical processes are not accounted for by the model, which can only predict by which mechanical processes the cells can get near enough the wall for adhesive forces to become important.

The hyperbolic pore model provides enough information for allowing an analysis of filtration data in terms of the cell intrinsic properties. In particular, the relative hydraulic resistance δ_k can be derived from the curves of Figure 15 and then used in the analysis of Section VB.

The entry time t_e into the pore is defined as the duration of the suction phase. It is then possible to correlate the maximum flow rate perturbatrions and t_e to both C and R, as shown in Figure 17a and 17b. It appears that these two bulk flow

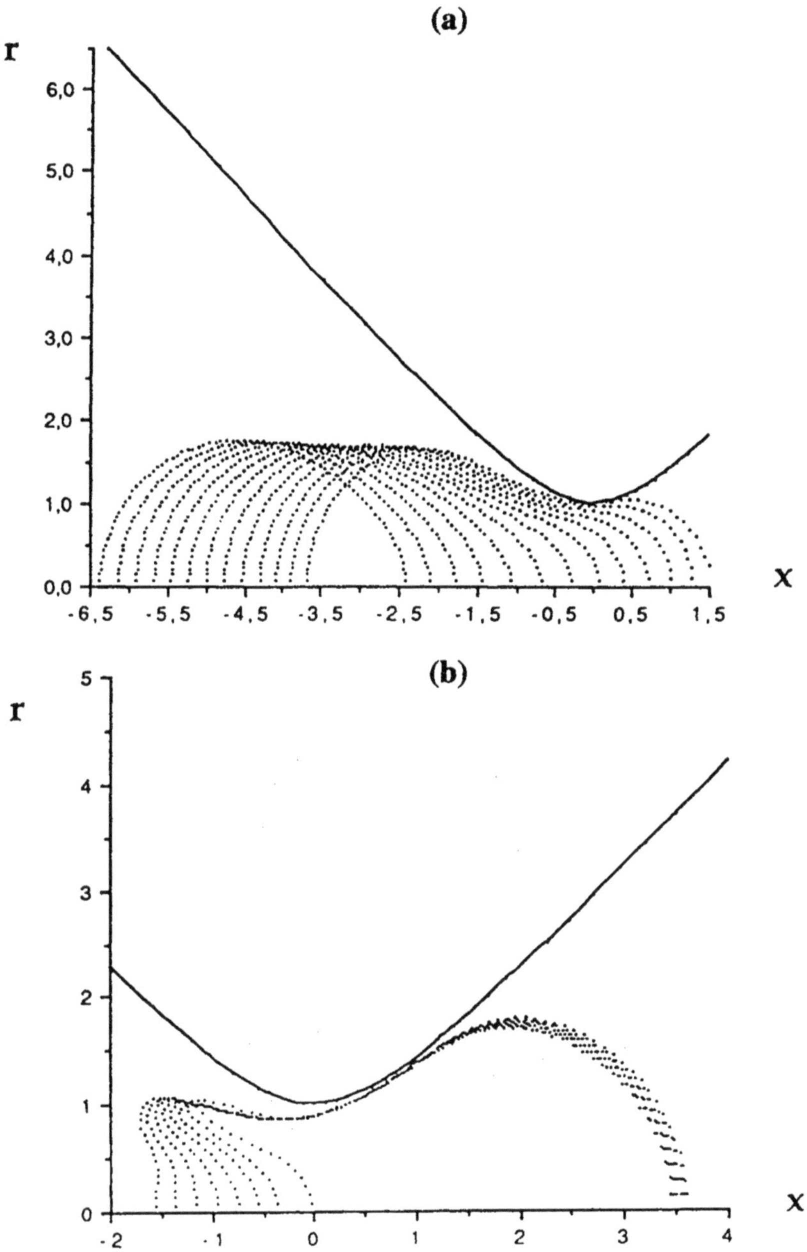

Figure 16. Successive cell profiles showing entry and exit plugging. (a) entry plugging, $C = 0.03$, $R = 1.4$. (b) exit plugging, $C = 0.1$, $R = 1.44$ (from Ref. 49).

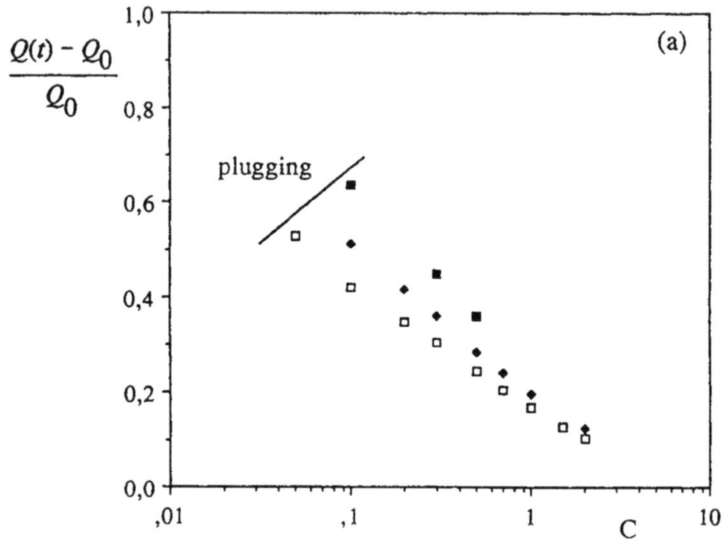

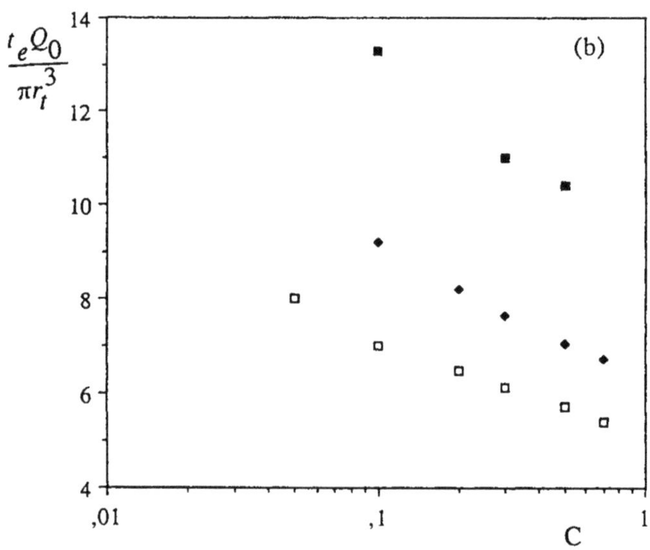

Figure 17. Correlation between (a) the relative flow rate decreases $[Q(t)-Q(0)]/Q(0)$ and (b) the entry time t_e of a cell in the pore as a function of Capillary number C and aspect ratio R. □ $R = 1.2$, ♦ $R = 1.4$, ■ R = 1.8 (from Ref. 49).

parameters are not very sensitive to capillary number. This is particularly clear for entry time where a 10-fold increase in C results in a 15–20% change in t_e. This model thus confirms the experimental evidence regarding the influence of membrane properties on filtration parameters. Indeed, filtration experiments cannot detect alterations of the membrane elastic properties unless very small pores are used. However, in such a situation interactions between the filter wall and the cells may occur, and the experiments become very difficult to perform.

VI. CONCLUSION

The red blood cells have complex properties that are both difficult to measure and to model. The membrane mechanics are now well understood in *static* conditions by means of micropipette experiments. However, in *dynamic* situations, the time and stress dependency of the membrane properties must still be investigated.

In actual *in vivo* conditions, the RBC membrane is constantly moving and deforming. As a consequence, the membrane material points are subjected to time varying stress levels, and viscoelastic phenomena become important. It is thus of considerable interest to be able to devise a good constitutive law for the cell membrane mechanics. A model is also needed of the three-dimensional motion and deformation of a cell freely suspended in a general shear flow. Such a model is missing for a good interpretation of rheoscope data where cells are subjected to a simple shear flow. It is also missing for the interpretation of filtration data or the observation of cells in capillary tubes, where the flow is complex and where the influence of the solid walls is very important.

REFERENCES

1. Evans, E.A.; Fung, Y.C. Improved measurements of the erythrocyte geometry. *Microvasc. Res.* **1972**, *4*, 335–347.
2. Cokelet, G.R.; Meiselman, H.J. Rheological comparison of haemoglobin solutions and erythrocyte supensions. *Science* **1968**, *162*, 275–277.
3. Rogauch, A. Modifications of the erythrocyte deformability alter the effect of temperature on the relative viscosity of human blood. *Biorheology* **1982**, *19*, 237–244.
4. Chien, S. Rheology of sickle cells and erythrocyte content. *Blood Cells* **1977**, *3*, 283–303.
5. Evans, E.A.; Skalak, R. *Mechanics and Thermodynamics of Biomembranes*. CRC Press, Boca Raton, 1980.
6. Skalak, R.; Tozeren, A.; Zarda, R.P.; Chien, S. Strain energy function of red blood cell membranes. *Biophys. J.* **1973**, *13*, 245–264.
7. Evans, E.A. A new material concept for the red cell membrane. *Biophys. J.* **1973**, *13*, 926–940.
8. Evans, E.A. Bending elastic modulus of red blood cell membrane derived from buckling instability in micropipette aspiration tests. *Biophys. J.* **1983**, *43*, 27–30.
9. Flugge, W. *Stresses in Shells*. Springer-Verlag, Berlin, 1973.
10. Batoz, J.L.; Dhatt, G. *Modélisation des structures par éléments finis*. Hermes, 1973.
11. Chien, S.; Sung, P.K.; Skalak, R.; Usami, S.; Tozeren, A. Theoretical and experimental studies on viscoelastic properties of erythrocyte membrane. *Biophys. J.* **1978**, *24*, 463–487.

12. Hochmuth, R.M.; Waugh, R.E. Erythrocyte membrane elasticity and viscosity. *Ann. Rev. Physiol.* **1987**, *49*, 209–219.
13. Hochmuth, R.M.; Berk, D.A. Analytical solutions for shear deformation and flow of red cell membrane. *J. Biomech. Eng.* **1984**, *106*, 2–9.
14. Zarda, P.R.; Chien, S.; Skalak, R. Elastic deformations of red blood cells. *J. Biomechanics* **1977**, *10*, 211–221.
15. Schmid-Schonbein, H.; Wells, R.E. Fluid to drop like transition of erythrocytes under shear. *Science* **1969**, *165*, 288–291.
16. Pfafferott, C.; Wenby, R.; Meiselman, H.J. Morphologic and internal viscosity aspects of RBC rheologic behavior. *Blood Cells* **1982**, *8*, 68–78.
17. Fischer, T.M.; Stohr, M.; Schmid-Schonbein, H. Red blood cell microrheology: Comparison of the behaviour of single RBC and liquid droplets in shear flow. *AIChE Symp. Ser.* **1978**, *74*, 34–45.
18. Sutera, S.P.; Tran-Son-Tay, R.; Boylan, C.W.; Williamson, J.R.; Gardner, R.A. A study of variance in measurements of tank-treading frequency in populations of normal human red cells. *Blood Cells* **1983**, *9*, 1121–1127.
19. Chang, K.S.; Olbricht, W.L. Experimental studies of the deformation of a synthetic capsule in extensionnal flow. *J. Fluid Mech.* **1993**, *250*, 587–608.
20. Chang, K.S.; Olbricht, W.L. Experimental studies of the deformation and breakup of a synthetic capsule in steady and unsteady simple shear flow. *J. Fluid Mech.* **1993**, *250*, 609–633.
21. Burger, A.; Rehage, H. From two-dimensional model networks to microcapsules. *Ang. Makromol. Chemie* **1992**, *202/203*, 31–44.
22. Barthes-Biesel, D. Role of interface properties on the motion and deformation of capsules in shear flow. *Physica A.* **1990**, *172*, 103–124.
23. Barthes-Biesel, D.; Rallison, J.M. The time dependent deformation of a capsule freely suspended in a linear shear flow. *J. Fluid Mech.* **1981**, *113*, 251–267.
24. Barthes-Biesel, D.; Sgaier, H. Role of membrane viscosity in the orientation and deformation of a capsule suspended in shear flow. *J. Fluid Mech.* **1985**, *160*, 119–135.
25. Bentley, B.J.; Leal, L.G. An experimental investigation of drop deformation and breakup in steady, two-dimensional linear flows. *J. Fluid Mech.* **1986**, *167*, 241–283.
26. Keller, S.R.; Skalak, R. Motion of a tank-treading ellipsoidal particle in a shear flow. *Quart. J. Fluid Mech.* **1982**, *120*, 27–47.
27. Secomb, T.W.; Skalak, R. Surface flow of viscoelastic membranes in viscous fluids *J. Appl. Math.* **1982**, *35, 2*, 233–247.
28. Tran-Son-Tay, R.; Sutera, S.P.; Rao, P.R. Determination of RBC membrane viscosity from rheoscopic observations of tank-treading motion. *Biophys. J.* **1984**, *46*, 65–72.
29. Tran-Son-Tay, R.; Sutera, S.P.; Zahalak, G.I.; Rao, P.R. Membrane stress and internal pressure in a RBC freely suspended in a shear flow. *Biophys. J.* **1987**, *51*, 915–924.
30. Sutera, S.P.; Pierre, P.R.; Zahalak, G.I. Deduction of intrinsic mechanical properties of the erythrocyte membrane from observations of tank-treading in the rheoscope. *Biorheology* **1989**, *26*, 177–197.
31. Drochon, A.; Barthes-Biesel, D.; Lacombe, C.; Lelievre, J.C. Determination of the red blood cell apparent membrane elastic modulus from viscometric measurements. *J. Biomech. Eng.* **1990**, *112*, 241–249.
32. Fischer, T.M.; Haest, C.W.M.; Stohr, M.; Kamp, D.; Deuticke, B. Selective alteration of erythrocyte deformability by SH-reagents. Evidence for an involvement of spectrin in membrane shear elasticity. *Biochimica Biophysica Acta* **1978**, *510*, 270–282.
33. Drochon, A.; Barthes-biesel, D.; Bucherer, C.; Lacombe, C.; Lelievre, J.C. Viscous filtration of red blood cell suspensions. *Biorheology* **1993**, *30*, 1–7.
34. Batchelor, G.K. The stress system in a supension of force-free particles. *J. Fluid Mech.* **1970**, *41*, 545–570.

35. Nash, G.B. Filterability of blood cells: Methods and clinical applications. *Biorheology* **1990**, *27*, 873–882.
36. Fischer, T.C.; Wenby, R.B.; Meiselman, H.J. Pulse shape analysis of RBC micropore flow via new software for the cell transit analyser (CTA). *Biorheology* **1992**, *29*, 185–201.
37. Kleisewetter, H.; Dauer, U.; Teitel, P.; Schmid-Schonbein, H.; Trapp, R. The single erythrocyte rigidometer (SER) as a reference for RBC deformability. *Biorheology* **1982**, *19*, 737–753.
38. Schmid-Schonbein, H.; Gaehtgens, P. What is red cell deformability? *Scand. J. Clin. Lab. Invest.* **1981**, *41*, *Suppl. 156*, 13–26.
39. Hanss, M. Erythrocyte filterability measurement by the initial flow rate method. *Biorheology* **1983**, *20*, 199–211.
40. Drochon, A.; Barthes-Biesel, D.; Lacombe, C.; Lelievre, J.C. How to detect alterations of the RBC mechanical properties from CTA filtration experiments. *Proc. 14th Congress of the Int. Soc Biomech.* Paris, July 4–8, 1993.
41. Skalak, R.; Impelluso, T.; Schmalzer, E.A.; Chien, S. Theoretical modeling of filtration of blood cell suspensions. *Biorheology* **1983**, *20*, 41–56.
42. Skalak, R.; Soslowsky, L.; Schmalzer, E.A.; Impelluso, T.; Chien, S. Theory of filtration of mixed blood suspensions. *Biorheology* **1987**, *24*, 35–52.
43. Tozeren, H.; Skalak, R. The flow of closely fitting particles in tapered tubes. *Int. J. Multiphase Flow* **1979**, *5*, 395–412.
44. Secomb, T.W.; Skalak, R.; Ozkaya, N.; Gross, J.F. Flow of axisymmetric red blood cells in narrow capillaries. *J. Fluid Mech.* **1986**, *163*, 405–423.
45. Skalak, R.; Oskaya, N.; Skalak, T.C. Biofluid mechanics. *Ann. Rev. Fluid Mech.* **1986**, *21*, 167–204.
46. Leyrat-Maurin, A.; Barthes-Biesel, D. Motion of a spherical capsule through a hyperbolic constriction. *J. Fluid Mech.* **1993**, *279*, 135–163.
47. Chien, S.; Schmid-Schönbein, G.W.; Sung, K.L.P.; Schmalzer, E.A.; Skalak, R. Viscoelastic properties of leukocytes. In: *White Cell Mechanics: Basic Science and Clinical Aspects*. Alan R. Liss, Inc., 1994, pp. 19–51.
48. Leyrat-Maurin, A.; Drochon, A.; Barthes-Biesel, D. Flow of a capsule through a constriction: Application to cell filtration. *J. Phys. III* **1993**, *3*, 1051–1056.
49. Leyrat-Maurin, A. Thèse de Doctorat. Université de Compiègne, 1993.

NUMERICAL MODELING OF BLOOD FLOW

Frank S. Henry and Michael W. Collins

Advances in Hemodynamics and Hemorheology, Volume 1, pages 67–112.

ISBN: 1-55938-634-7.

I. INTRODUCTION

The three dimensionality and unsteadiness of the flow of blood in much of the cardiovascular system makes the mathematical analysis of such flows a challenging problem. The analysis is further complicated by the compliant nature of the vessels in which the blood flows and the fact that blood is a suspension of deformable cells in a plasma. However, numerical methods for the solution of the partial differential equations governing the flow of fluids have been developed to the point where it is possible to solve for unsteady, three-dimensional flows in complex, moving geometries. This together with the advent of supercomputers has spawned a new area of fluids research now fairly universally known as Computational Fluid Dynamics (CFD). This chapter is concerned with the application of CFD to the problem of blood flow in the cardiovascular system.

Knowledge of the time-dependent pressure and shear stress in blood flow, particularly on the vessel walls, is of interest for a number of reasons. For example, there appears to be compelling evidence that hydrodynamic factors play a part in the development of atherosclerosis. However, it remains to be established precisely what factors are important. Initially, Fry[1] proposed that atheromatous lesions were more likely to be found in regions of high wall shear stress. However, Caro et al.,[2] found the opposite to be true. More recently, Giddens et al.,[3] Ku et al.,[4] and Jones et al.,[5] among others, have shown that the preferred sites for atheromatous lesions are areas of the arterial wall which experience low and fluctuating shear stress.

The detailed description of the flow in the vicinity of some abnormality in vessel geometry is another area of interest. The abnormality might be caused by disease, for example, an arterial stenosis due to atheroma; or by surgery, for example, a bypass graft anastomosis. In the case of a stenosis it is hoped that if it can be demonstrated that there is a detectable change in some flow variable distal of a mild stenosis, then this change can be measured and curative measures taken before total blockage occurs. Details of the flow through bypass grafts are also of interest because they are known to fail due to plaque build-up which is thought to be due to irregularities in the flow fields. Numerical simulation of flows through anastomoses of varying geometries could be used to design grafts with more desirable flow fields.

The flow in artificial hearts and cardiac assist devices are also areas of active interest. Simulations of flows in such devices can be used to study the influence of ventricle size, stroke volume, rates of contraction, and the orientation of inlet and outlet tubes. It is known that these devices are prone to failure due to atheroma, hemolysis, and thrombosis. Hence, as well as using numerical simulation to help design more efficient devices, it is possible to use them to search for areas of low and high shear stresses, and regions of stasis. Similar numerical searches may be made for other cardiac flow situations, such as the back flow through an incompetent mitral valve or the flow through an artificial heart valve.

The new three-dimensional, transient, codes are being matched in engineering with the application of whole-field, non-invasive, optical measurement methods, for example holographic interferometry.[6] While these methods in themselves are probably less in fine detail than state-of-the-art medical measurement techniques, for example, magnetic resonance imaging, the interface between the methods and CFD results for code validation and post-processing is of considerable interest in our current studies in hemodynamics, see for example Jones et al.[5]

This chapter will concentrate on the simulation of blood flow in the larger vessels. A brief review of the concepts of CFD will first be given, and then recent applications of these concepts to the flow through bifurcations, stenoses, grafts, and natural and artificial ventricles will be discussed. Mostly, two and three-dimensional simulations will be considered. The many one-dimensional method-of-characteristic calculations, which are primarily concerned with prediction of wave phenomena, will not be discussed. The treatment concludes with a discussion of some problems outstanding and future likely developments. The objective of this chapter is not to present an exhaustive review of the subject at hand, but to bring to a wider audience the benefits and limitations of this highly computational form of analysis. The authors recognize that we have concentrated on those aspects of CFD with which we are most familiar, namely the finite difference method, and that the following is inescapably colored by our own experience in the field. For this we apologize in advance.

II. COMPUTATIONAL FLUID DYNAMICS METHODS

A. Introduction

Computational fluid dynamics is generally known to mean the study of fluid flows by numerical simulation. That is, the fluid's velocity and pressure at discrete points within the flow are determined by solving approximate, algebraic forms of the partial differential equations governing the flow of a fluid. While the equations solved are approximations of the original set, it can be shown that, provided that certain well defined conditions are met, they become exact representations in the limit as the distance between the discrete points goes to zero. Once the velocity and pressure fields are known, stresses within the body of the flow, and forces and shear stresses applied by the fluid to the surrounding structure can be estimated. A variety of numerical methods are used, and the increasing power of personal computers and work stations means that some quite substantial problems can now be tackled without recourse to supercomputers. However, supercomputers are still required for the more complex applications.

The set of partial differential equations alluded to above comprises four equations; one describes the conservation of mass, and the remaining three describe the conservation of momentum in three mutually perpendicular directions. The latter are known collectively as the Navier-Stokes equations. The derivation of these equations may be found in a variety of sources, for instance, Collins and Ciofalo.[7] Restricting the discussion to a Newtonian fluid with constant density, ρ, and viscosity, μ, the governing equations can be given in vector notation as

Conservation of mass:

$$\nabla \cdot \mathbf{V} = 0. \tag{1}$$

Conservation of momentum:

$$\frac{\partial \mathbf{V}}{\partial t} + \mathbf{V} \cdot \nabla \mathbf{V} = \mathbf{F} - \frac{1}{\rho} \nabla p + \nu \nabla^2 \mathbf{V}. \tag{2}$$

where **V** is the velocity vector, **F** is the body force per unit mass, p is the pressure, and $\nu = \mu/\rho$ is the fluid's kinematic viscosity. As the above equations are given in vector notation, they apply to any coordinate system. The ∇ operator contains first derivatives with respect to the particular coordinates used. Hence, before approximations of these derivatives can be formulated equations (1) and (2) have to be cast in a particular system. General coordinates, and their application to the above, are discussed in subsequent parts of this section.

The body force vector, **F**, can represent the sum of several forces. In many engineering applications, the only significant body force is that due to the earth's gravitational field. However, in medical applications, other forces may be present. This is because the above equations are written in terms of an inertial frame of reference, that is, coordinates are assumed to be fixed in space. If the coordinates are in fact accelerating, additional terms will appear in **F**. There is in fact some evidence[8] that the flow in the coronary arteries, for example, is affected by the cyclic motion of the heart wall.

An important feature of the above set of equations is that while there are four equations and four unknowns (i.e., three velocity components and the pressure), there is not a distinct equation for pressure. As will be discussed later, this presented a major obstacle to early attempts to solve for complex three-dimensional flows. In the case of two-dimensional flows, the original equations can often be manipulated in such a way as to remove the pressure from the resulting equations. A further complication to the solution of Navier-Stokes equations is that they can exhibit different mathematical characteristics (i.e., parabolic, hyperbolic, and elliptic) depending on the nature of the particular problem. Hence, solution techniques developed specifically for boundary layer studies differ from those used for tidal flow in estuaries, say, and neither would be appropriate for the simulation of flow through a stenosed artery.

In many ways CFD is the numerical analogue of experimental fluid mechanics. In both cases, the raw data is found at discrete points in the flow field. Indeed the possibility of using CFD to build a numerical wind tunnel is being seriously pursued with simulated interferograms and other flow visualization records. Further, the new field of virtual reality has been applied to CFD as a way of presenting the results of complex flow simulations. Using this technique it is possible for an observer to "walk through" the data. As in the experimental case, a particular flow configuration has to be considered. This is contrary to the third option of investigation; mathematical analysis. In the latter case, if a solution can be found, it is usually applicable to a class of flows. However, unlike experimental analysis, CFD is not restricted by the physical scale of a flow situation. For example, flows ranging from those in the smallest arteries to the movements of oceans and the atmosphere can all be simulated. It is equally possible to simulate flows in hostile environments, such as combustion chambers, and nuclear reactors. This is largely because the Navier-Stokes equations are valid over laminar, transitional, turbulent, and sonic range flows.

There are two main classes of methods used to generate CFD programs or codes. Historically, the preferred approach has been the finite difference method (FDM). However, more recently, the finite element method (FEM) has been used, particularly in the field of blood flow. The FEM has its roots in solid mechanics, where it is now routinely used to perform analyses of complex structures. Before more generalized gridding techniques were developed for the FDM, it was argued that the FEM could be used to solve problems with geometries too complex for FDM analysis. However, today that difference has largely disappeared. The choice between using FDM or FEM is now largely a matter of personal preference and experience. Both methods will be briefly outlined below, but space limitation will permit no more than a cursory look at a rapidly growing body of literature on this subject. Readers wishing for further details should consult some of the many excellent books and review articles on the subject. For example, Baker,[9] or Hutton et al.[10] for FEM and Anderson et al.[11] or Patankar[12] for FDM.

It is noted that a third approach known as the spectral method has relatively recently been developed to the point where it can now be applied to problems with complex geometries. Its inherent attraction is partly because the method reflects the spectral character of turbulence, and partly because of its superior accuracy. Hussaini and Zang[13] have reviewed recent developments in this field. The original method was best suited to problems with simple geometries and periodic boundary conditions. Spectral methods have been used for meteorological simulations for many years. For these flows, the aforementioned restrictions on geometry and boundary conditions do not represent major obstacles. A crucial development in terms of applying the method to more realistic engineering problems was the development of the multi-domain technique. An example of this technique is the spectral-element method due to Patera,[14] which is well suited for time dependent laminar flows calculations, and offers the geometric flexibility of FEM with the accuracy of the spectral method. Patera's method has been applied to a variety of engineering problems, see for instance Petera,[14] and recently to biomedical type flows.[15]

Today there are numerous vendors of powerful CFD programs based on FDM (e.g., HARWELL-FLOW3D, FLUENT, PHOENICS, STAR-CD). Many of these programs started life either in government research laboratories or in universities, but are now sold commercially. Most solve the governing equations in generalized, body-fitted, coordinates, and some allow moving boundaries. A number of FEM-based codes are also available commercially (e.g., FIDAP and NEKTON). Grid generation and post-processing packages are often included, but also both pre- and post-processing, the latter increasingly referred to as data visualization, are available from companies specializing in such software (e.g., PATRAN, FEMVIEW, UNIRAS, AVS). A recent survey of a number of CFD codes has been given by Wolfe,[16] see also the introduction section of Collins and Ciofalo.[7] The power and flexibility of the CFD packages available today is such that there is now little need to develop purpose-build programs for specific problems. While considerable work can often be required to apply these programs to a specific flow problem, it is in general still more efficient to use such packages.

B. The Computational Grid

Regardless of whether the FEM or FDM is used, the first step in a simulation is to define the geometry of the problem. That is, the spatial coordinates of all solid

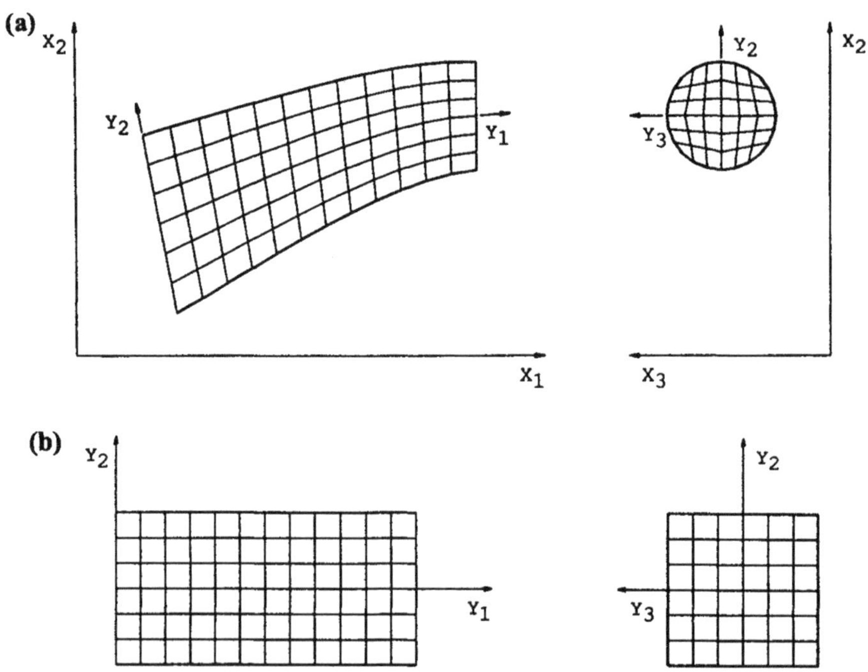

Figure 1. Sample single-block grid for an artery segment; (a) Physical domain, and (b) computational domain.

surfaces, symmetry planes, inlets, outlets, and other geometrical features of interest, have to be defined. With this accomplished, the solution domain is divided into small elements, control volumes, or cells. This subdivision is termed grid or mesh generation.

Modern CFD codes use body-fitted grids. These allow the codes to be applied to a wide range of complex geometries and greatly assist in the imposition of boundary conditions. A simple body-fitted grid for a segment of artery is shown in Figure 1. As the name implies, certain grid surfaces coincide with the solution domain boundaries. As indicated in Figure 1, each grid surface can be considered to be a surface of constant value of a particular coordinate in the general, non-orthogonal curvilinear coordinates system (y_1, y_2, y_3). As will be discussed later, the cost of the greater flexibility afforded by the use of body-fitted grids and general coordinates is an increase in complexity of the equations to be solved.

There are three basic types of computational grid in current use. The grid discussed above is of the type traditionally used in FDM. This is termed a single-block, or fully structured, grid. As shown in Figure 1, a single block grid has to transform, from physical space to computational space, into a hexahedron (a brick-like shape). This makes it difficult to solve accurately flows in some more complex geometries, for example, those in branched tubes. Another problem with single-block grids is highlighted in the cross-sectional views of the grid in Figure 1. It can be seen that the corner cells in computational space are distorted in physical space to the extent that two of their sides are very nearly at 180° to each other. This can be shown to degrade the accuracy of the solution near the wall in this area.

A recent innovation particularly useful in the case of flow through a bifurcation, is the development of the concept of multi-block grids. In a multi-block grid, each block has to transform into a hexahedron, but no such restriction is placed on the grid as a whole. A five-block grid of the model artery discussed above is given in Figure 2a. It can be seen that in this case the cells near the wall are relatively undistorted. Other examples of multi-block grids will be given in the section on applications.

The third type of grid is that traditionally used in FEM, that is, a fully unstructured grid. A typical unstructured grid for the model artery is given in Figure 2b. It can be seen that the individual finite elements are not restricted to be rectangular. It is also possible to have a mix of rectangular and triangular elements. The triangular elements shown are straight sided, but boundary-fitting, curved-sided, isoparametric elements are also available. The unstructured nature of a finite element grid means that each element requires its own local coordinates. Information relating one element to its neighbors is also required. These and other requirements of the FEM result in larger storage requirements than would be the case for a comparable finite difference solution. However, the benefit of a completely unstructured grid is that complex geometries can be dealt with more efficiently than they can with a block-structured finite difference grid.

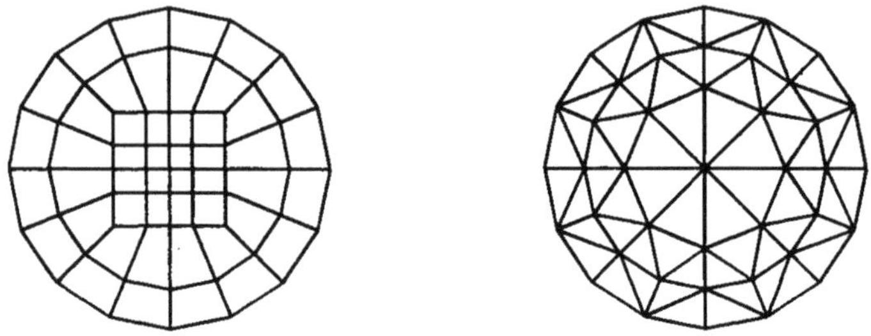

Figure 2. Sample multi-block and finite element grids; (a) multi-block, and (b) finite element.

As the geometric complexity of problems which can be numerically simulated has increased, the need for some form of automatic grid generation has similarly increased. Some of the more powerful grid generation programs have come from the FEM modeling community since it is they who have had a longer history of tackling complex geometries. However, the introduction of body-fitted grids written in curvilinear coordinates to FDM codes has made numerical grid generation more widely used in CFD as a whole. Essentially, numerical grid generation programs are based either on algebraic methods, or the solution of partial differential equations. The numerical generation of such grids has been outlined by Thompson et al.,[17] among others. Transfinite interpolation appears to be the most commonly used algebraic method, while procedures based on the solution of elliptic partial differential equations seem to be the most popular differential method.

Notwithstanding the increased use of numerical grid generation packages, defining an efficient grid remains something of an art. There are many factors to consider in defining a grid. For instance, in FDM grids, solution accuracy is degraded if the individual volumes have high aspect ratios or are severely skewed. Also, the grid should be as nearly orthogonal as possible near boundaries. An overall requirement is to use the minimum number of cells, or element for the desired accuracy, but have sufficient cells in areas of the solution domain where there are large gradients of flow variables. Attempts have been made, with some success, to generate automatic grid refinement, or adaptive grid, programs. These codes automatically re-grid the problem as the solution proceeds. Cells are moved away from areas where the gradients of important flow variables are relatively small and toward areas where these gradients are large.

C. The Finite Difference Method

While there are in fact several different procedures available to generate the finite difference equations, the finite volume method is unquestionably the one most often used, and it is this method that will be described. In this method algebraic versions of the governing partial differential equations are formed by integration over each cell of the grid. Partial differentials are approximated using values of the relevant variables at neighboring volumes. The result of this discretization process is a set of non-linear, coupled, algebraic equations. This set consists of four equations, three for velocity and one for pressure, for each control volume. While direct methods for solving such sets of equations exist, they are expensive in computer memory and time. However, of more importance is that due to the coupled, non-linear nature of the original equation each algebraic equations has to be solved repeatedly in order to update the coefficients that depend on the solution of the other equations. For these reasons, the resulting set of algebraic equations are solved using iterative methods. The process is stopped when some preset convergence criterion has been reached. To enhance convergence, block iterative schemes are usually employed, for example, line-by-line Gauss-Seidel, ADI, or Stone's method. Transient calculations are usually solved using either a two-step method, such as the Crank-Nicolson scheme, or the single step, fully implicit Backward Euler scheme. Explicit time marching schemes are usually not employed because they require excessively small time steps for the solution to remain stable.

Because early calculations were restricted to two dimensions, mainly due to the lack of computing power, they were able to use the streamfunction-vorticity formulation of the basic, Navier-Stokes, equations. In this formulation the pressure does not appear, which is its main attraction over the alternative approach of solving for the primitive variables, that is, velocity and pressure. In moving to three dimensions, the solution of the pressure field had to be tackled. A Poisson equation for pressure can be derived from equations (1) and (2), but the solution of this equation has proved difficult, in general. The difficulty is largely due to the boundary conditions on this equation being of the Neumann type, that is, the pressure gradient normal to the boundaries are required. As these gradients are found from the velocity field, in an iterative manner, mass conservation cannot easily be assured and the solution tends to drift. Roache[18] has reviewed some of the earlier attempts to solve this problem.

The problems associated with the solution of the Poisson pressure equations were circumvented in the early 1970s by the development of the pressure correction scheme known as SIMPLE (Semi-Implicit Method for Pressure-Linked Equations).[19] This algorithm and its variants, for example, SIMPLEC,[20] are arguably the most widely used solution to the problem of solving for the pressure field. These schemes do not actually solve for pressure, but rather for a correction to a guessed pressure field in combination with an iterative procedure involving the velocities. The original implementation of this scheme required the use of staggered grids in order to prevent a numerical instability known as "checkerboarding," which occurred on a single grid because the velocities were insensitive to small oscillations in the pressure solution, and vice versa.

An alternative pressure-correction scheme has been proposed by Issa[21] and is known as PISO (Pressure-Implicit with Splitting of Operators). In this algorithm, a second pressure correction equation is solved, and it was originally formulated to remove completely the need to iterate between the pressure and velocity equations.

However, this requires the individual pressure and velocity equations to be solved to tight convergence tolerances, and in practice the scheme is often implemented in an iteration loop similar to that used for SIMPLE. The method does appear to have some advantages over SIMPLE for problems where conservation of mass and momentum are particularly important.

Another area where care has to be taken is in the formulas used to approximate the convection or advection terms (the spatial acceleration terms on the left-hand side of equation (2)). Inaccurate modeling of these terms can seriously degrade the overall accuracy and stability of the solution. For all but the slowest of flows, some form of upwinding has to be used. Upwinding is the process of estimating the values of the flow variables, principally momentum, on the cell faces from values that are upstream of the particular cell face. Obviously, this means the direction of the flow in the vicinity of the cell face has to be determined, at each iteration. Some considerable effort has gone into developing accurate and stable models of the convection terms, however, it is still the case that the modeler has to make a trade-off between robustness and accuracy.

To help clarify subsequent discussion some of the more common advection models will be described here in terms of a one-dimensional example. Consider the quantity, ϕ being advected at a constant (positive) velocity u. Then, integrating the relevant advection term over the control volume shown in Figure 3 gives

$$\int u \frac{\partial \phi}{\partial x} dx = u(\phi_e - \phi_w). \tag{3}$$

As values of ϕ are only known at the center of each control volume, some prescription has to be devised to estimate ϕ_e and ϕ_w. The obvious choice of taking the average of the two nodal values on either side of the face in question can be shown to be unstable (a converged solution is difficult to achieve) unless the cell

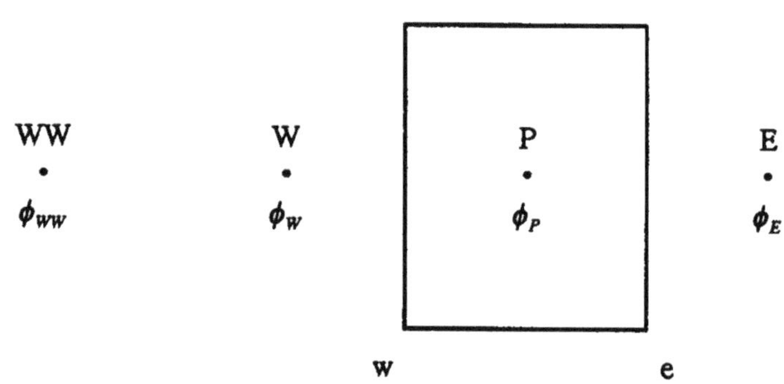

Figure 3. Typical control volume.

Peclet number ($u\Delta x/v$, where Δx is the cell width, and v is the fluid kinematic viscosity) is less than 2. An unconditionally stable but rather crude approximation is to assume that the value at the face is the same as the nodal value upwind, or upstream, of the face. The default advection model in many FDM program uses one or another of these schemes depending on whether the Peclet number is greater or less than 2. This is known as the Hybrid scheme, which can be given for the w face as:

$$\phi_w = \frac{1}{2}(\phi_W + \phi_P) \tag{4a}$$

if the mesh Peclet number is less than 2, or

$$\phi_w = \phi_W \tag{4b}$$

otherwise. A more accurate upwind scheme uses an additional nodal value one spatial step further away from the face. This is known as Higher-order Upwinding (HUW), and can be given for the w face, assuming equal spatial increments, as

$$\phi_w = \frac{3}{2}\phi_W - \frac{1}{2}\phi_{WW}. \tag{5}$$

A model that is formally more accurate than HUW is the quadratic upwinding scheme known as QUICK. For the w face, this can be given, again assuming equal spatial increments, as

$$\phi_w = \frac{3}{8}\phi_P + \frac{3}{4}\phi_W - \frac{1}{8}\phi_{WW}. \tag{6}$$

The use of a nodal value downstream of the w face (i.e., ϕ_p) makes this scheme slightly unstable.

As mentioned previously, a major advance in the flexibility of finite difference CFD programs in recent years has come from the introduction of body-fitted grids. This has allowed the method to be applied to problems with solution domains with almost arbitrary shape. However, solving for flows in arbitrary geometries requires the calculation of various combinations of what is termed the metric tensor of the grid. The metric tensor defines the transformation form a Cartesian computational space, to the curvilinear physical space. The elements of the theory of curvilinear coordinates and general tensor analysis necessary for even a cursory discussion of the solution of the governing equations in general coordinates are unfortunately beyond the scope of this introductory chapter. A brief but readable review of these concepts and their application to the Navier-Stokes equations is given in Appendix 2 of Yih.[22] However, some idea of the complexity of the problem can be achieved by considering equations (1) and (2) in a two-dimensional orthogonal coordinate set, (ξ,η). equations (1) and (2) become:

Conservation of mass:

$$\frac{\partial}{\partial \xi} h_\xi u_\xi + \frac{\partial}{\partial \eta} h_\eta u_\eta = 0. \tag{7}$$

Conservation of momentum:

$$\frac{\partial u_\xi}{\partial t} + \frac{u_\xi}{h_\xi}\frac{\partial u_\xi}{\partial \xi} + \frac{u_\eta}{h_\eta}\frac{\partial u_\xi}{\partial \eta} + \frac{u_\xi u_\eta}{h_\xi h_\eta}\frac{\partial h_\xi}{\partial \eta} - \frac{u_\eta^2}{h_\xi h_\eta}\frac{\partial h_\eta}{\partial \xi}$$

$$= F_\xi - \frac{1}{\rho}\frac{\partial p}{\rho \partial \xi} + \frac{\nu}{h_\xi h_\eta}\left[\frac{\partial}{\partial \xi}\frac{h_\eta}{h_\xi}\frac{\partial u_\xi}{\partial \xi} + \frac{\partial}{\partial \eta}\frac{h_\xi}{h_\eta}\frac{\partial u_\xi}{\partial \eta}\right] \tag{8a}$$

$$\frac{\partial u_\eta}{\partial t} + \frac{u_\xi}{h_\xi}\frac{\partial u_\eta}{\partial \xi} + \frac{u_\eta}{h_\eta}\frac{\partial u_\eta}{\partial \eta} + \frac{u_\xi u_\eta}{h_\xi h_\eta}\frac{\partial h_\eta}{\partial \xi} - \frac{u_\xi^2}{h_\xi h_\eta}\frac{\partial h_\xi}{\partial \eta}$$

$$= F_\eta - \frac{1}{\rho}\frac{\partial p}{\partial \eta} + \frac{\nu}{h_\xi h_\eta}\left[\frac{\partial}{\partial \xi}\frac{h_\eta}{h_\xi}\frac{\partial u_\eta}{\partial \xi} + \frac{\partial}{\partial \eta}\frac{h_\xi}{h_\eta}\frac{\partial u_\eta}{\partial \eta}\right] \tag{8b}$$

where

$$h_\xi^2 = \left[\frac{\partial x}{\partial \xi}\right]^2 + \left[\frac{\partial y}{\partial \xi}\right]^2,$$

$$h_\eta^2 = \left[\frac{\partial x}{\partial \eta}\right]^2 + \left[\frac{\partial y}{\partial \eta}\right]^2$$

where x and y are the two-dimensional Cartesian coordinates. For the special case of axisymmetric polar coordinates (r,z), $\xi = r$, $\eta = z$, $h_\xi = h_\eta = 1$, and $u_\xi = u_r$, $u_\eta = u_z$.

In a two-dimensional orthogonal grid, the only non-zero terms in the metric tensor are equal to the square of the scale factors h_ξ and h_η. The metric tensor is a symmetric tensor of rank two, and as such, in general has six independent elements. As shown in Figure 1b, the grid in computational space is rectangular (hexahedral in three dimensions) with equal increments. That is, the basic cell, or control volume, is a cube. The regular structure of the grid in computational space means that while the equations to be solved are more complex, the algebraic approximations used to model the individual terms can be less so. For example, equation (5) for the case of unequal spatial increments would read

$$\phi_w = \frac{2\Delta x_{WW} + \Delta x_W}{2\Delta x_{WW}}\phi_W - \frac{\Delta x_W}{2\Delta x_{WW}}\phi_{WW} \tag{9}$$

where Δx_{WW} and Δx_W are the distances between points WW and W, and between points W and P, respectively. It can be seen that equation 5 is recovered when the spatial increments are equal. The simplicity of equation 5 compared to equation 9 illustrates the advantage of spatially uniform grids. While the various derivatives (e.g., $\partial x/\partial \xi$ and $\partial y/\partial \xi$) contained in the metrics can usually be calculated once and stored, they represent a significant overhead in terms of computer memory above that which is required for a calculation in Cartesian coordinates. Voke and Collins[23] have considered various equivalent forms of the governing equations which may be more suitable for efficient numerical solution, and Burns and Wilkes[24] have discussed further strategies to increase efficiency.

An algorithm due to Rhie and Chow[25] has proved extremely useful in the context of three-dimensional, body-fitted grids. Rhie and Chow showed how to compute the required velocity components on the cell faces from the values of velocity and pressure at the cell centers without inducing checkerboarding. As was mentioned above, previous to this, the SIMPLE algorithm was implemented on a staggered grid. In the case of three-dimensional flows this meant four separate grids, one for pressure and one each for the three velocity components. If these three-dimensional grids were also to incorporate general coordinates the storage requirements quickly became prohibitive.

Numerical procedures for the solution of flows in domains with moving boundaries have recently been developed by a number of research groups. However, these calculations tend to be costly in terms of CPU time, as it is necessary to recalculated the terms involving the grid's metric tensor each time the boundary, and hence, the grid is moved. It is also necessary to include terms in the original equations to account for the grid movement so that the grid in computational space is fixed. For example, if a two-dimensional, Cartesian grid, is considered, the continuity equations would become, assuming the grid remains Cartesian while moving,

$$\frac{1}{V_c}\frac{\partial V_c}{\partial t}+\frac{\partial}{\partial x}\left(u-\frac{\partial x}{\partial t}\right)+\frac{\partial}{\partial y}\left(v-\frac{\partial y}{\partial t}\right)=0 \qquad (10)$$

where V_c is the cell volume, and $\partial x/\partial t$ and $\partial y/\partial t$ are the grid velocities in the x and y coordinate directions, respectively. It can be shown that V_c is related to the determinant of the metric tensor. Obviously modeling these terms also add to the cost of the computation. The papers by Demirdzic and Peric,[26] and Hawkins and Wilkes[27] describe in more detail the application of these concepts to the full Navier-Stokes equations. Some of the more advanced commercially available CFD codes now include this facility.

D. The Finite Element Method

Approximate versions of the governing equations are generated for each element using a variational principle, or a weighted-residual method, the Galerkin method being a popular choice for the latter. The local equations are then collected together to form a global system of algebraic equations. Solution of this system of equations requires inversion of a large matrix, either by direct or indirect methods. If the equations are non-linear this inversion is repeated until the required convergence is obtained. This process means that FEM programs are usually slower than finite difference programs, although it is argued that the FEM is inherently more accurate.

As with the FDM method, the solution of the pressure field can cause difficulty in FEM solutions, and again like FDM, early calculations using FEM circumvented this problem by using the streamfunction vorticity formulation in which pressure does not appear. One popular approach used in the solution of the primitive equations is known as the penalty function method, this was apparently first presented by Zienkiewicz.[28] In this method, mass conservation is not satisfied exactly, and the pressure is recovered, after the velocity field has been found to some pre-defined tolerance, from the mass residual. In another method known as the Lagrange multiplier method,[29] the pressure is solved, but the approximation for pressure has to be of lower degree than that used for velocity. Also, solution existence can only be assured for those velocity-pressure element nodal arrangements that satisfy the Babuska-Brezzi condition.[30,31] It is noted that the simplest nodal arrangement on a triangular element, that of one velocity node on each vertex and one pressure node at the centroid, does not satisfy the above condition. Often, the penalty and Lagrange methods are combined. Recently Perktold[32] and Hilbert[33] devised a scheme reminiscent of the pressure correction scheme used in FDM codes. This scheme uses simple hexahedral elements that are linear in velocity and constant in pressure. Perktold and his colleagues have successfully applied this method to a variety of blood flow problems, some of which are discussed in later sections, however, Perktold et al.[34] reported the need to filter the pressure to suppress checkerboarding.

The problem leading to the need to upwind the advection terms in FDM codes also has to be addressed in FEM. Various methods have been proposed to alleviate this problem, including upwinding schemes that are comparable to those used in FDM. However, other methods are available[35] including those that use elements of the methods of characteristics. The non-linearity of the advection terms also has to be addressed. Newton-Ralphson or Picard iteration methods are commonly used for this purpose. The manner in which pressure is taken into account has an effect on how the discrete equations are solved. In general, the system matrices derived from the use of the penalty method have to be solved by direct means while other approaches allow iterative solutions. Iterative methods offer savings in computer memory requirements. As with the FDM codes, Crank-Nicolson or Backward Euler time stepping may be used for transient problems. However, more general time-splitting techniques are becoming popular. Some of these have been recently reviewed by Glowinski and Pironneau.[35]

E. Turbulence

While most hemodynamic flows are laminar, there are elements of the cardiovascular system, for example, the aortic arch, in which the flow is sufficiently strong for turbulence to occur, at least for some part of the cardiac cycle. Also, turbulence may be induced due to disease, for example, in a stenosed artery. As the majority of flows of engineering significance are turbulent there has been much research into the nature of turbulence and into the mathematical description and numerical simulation of turbulent flows. Despite this body of work, turbulence remains relatively poorly understood, and even its definition tends to be somewhat subjective, particularly in pulsatile flow. Hence, a precise definition will not be given, but instead some relevant characteristics of turbulence will be listed below:

1. A fundamental feature of turbulence is that each flow variable shows random fluctuations in space and time about some mean value.
2. Transport of heat, mass, and momentum is much more vigorous than is the case for laminar flow. The corollary of this is that turbulent flows have significantly higher levels of shear stresses than do their laminar counterparts.
3. Turbulence is three dimensional, even if the mean flow is one or two dimensional. Turbulence is also rotational and displays a high degree of fluctuating vorticity. The fluctuating, three-dimensional vorticity is essential in extracting energy from the mean flow and in cascading it down to the smaller scales where it is dissipated by the fluid viscosity.
4. Turbulence is dissipative, and hence, requires a constant supply of energy. Without a source of energy the turbulence quickly decays.

Turbulence may be viewed as the motion of unsteady three-dimensional eddies whose sizes range from that of the confining duct in internal flows to that where dissipation to thermal energy extinguishes the turbulence. The large range of spatial and temporal scales occurring in turbulent flows makes direct numerical stimulation (DNS) impossible for most flows of engineering interest. This is because the computational grid has to be sufficiently fine to capture all the significant spatial scales, and the time step is similarly restricted if all the important time scales are to be resolved. For these reasons DNS has to date only been applied to turbulent flows with low Reynolds numbers and simple geometries. An approach designed to alleviate the resolution requirement of DNS is known as Large Eddy Simulation (LES). In this case, only the motion of the larger, slower turbulent eddies is resolved, and the remaining scales are modeled using semi-empirical relationships. However, LES still requires considerable computing resources.

Despite this, DNS and LES are starting to be used where fine details of flow and turbulence are essential. A recent special issue of *Aerospace America* (see for instance Moin[36]) shows the aeronautical engineering relevance of DNS and LES.

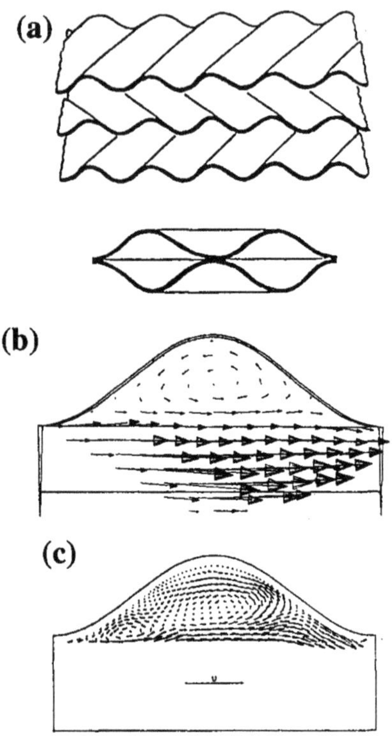

Figure 4. Flow in a cross-corrugated heat exchanger; (a) Typical cross-corrugated plate configuration, (b) predicted secondary flow using the *k*–ε model, and (c) predicted secondary flow using LES.

The authors' group has considerable experience in LES.[37–40] We have extensively studied the complex problem of flow and heat transfer in cross-corrugated heat exchangers. This has recently included DNS.[40] A schematic of the cross-corrugated plates is shown in Figure 4a. The geometric periodicity of this problem allows only a small typical cell to be considered. In the cell, two fluid streams cross, and each stream produces a secondary, swirling flow in the other. There is some conjecture as to the precise role played by this swirl in enhancing cross-flow heat transfer. Henry et al.[41] predicted the flow using a standard turbulence model (the *k*–ε model, described below). A typical prediction is given in Figure 4b. This can be compared with the LES prediction of Ciofalo et al.,[40] given in Figure 4c, for a similar flow situation. It can be seen that the two approaches predict significantly different secondary flows, and hence, they suggest somewhat different conclusions as to the importance of secondary flow in terms of enhancing cross-flow transfer.

In the following, we give a brief account of what can be considered to be the "standard" engineering approach to predicting turbulent flows. We include this because it has been applied by some to certain problems in hemodynamics. However, such simulations are of limited utility, as flows of hemodynamic interest violate many of the basic assumptions (especially continuous turbulence) on which this approach is based.

Equations (1) and (2) hold for both laminar and turbulent flows. However, for the reasons stated above, DNS of the latter is not usually attempted. Instead, each flow variable is decomposed into a (time) mean and a fluctuation about the mean. The decomposed variables are then substituted into equations (1) and (2) and the equations are themselves averaged over time. This process is usually referred to as Reynolds averaging. Obviously, for the time averaging process to be completely unambiguous, it is necessary for the flow to be steady in the mean. However, the resulting equations are routinely applied to transient flows. In this case the averaging process is assumed to be over many occurrences of the same flow. This requires the boundary conditions and initial conditions for each occurrence to be identical, which is in practice almost impossible to achieve. The time-averaged equations differ from the original set by the appearance of six additional terms known as the Reynolds stresses. These represent six independent correlations of fluctuating velocities. Their appearance means that the averaged equation set contains more unknowns than equations. In order to proceed, it is necessary to formulate a model of the Reynolds stresses.

Of the methods available to close the equations, the k–ε model of turbulence is probably the method most often used. This model uses the eddy viscosity concept to relate the Reynolds stresses to gradients of mean velocity. Using dimensional analysis and experimental evidence, it can be shown that the eddy, or turbulent, viscosity, ν_t, is given by

$$\nu_t = C_\mu k^2/\varepsilon \tag{9}$$

where C_μ is a constant, k is the kinetic energy of the turbulence, and ε is the rate of kinetic energy dissipation. Values of k and ε for each point in the flow field are estimated by solving model partial-differential transport equations for both variables.

The standard k–ε model only applies strictly to areas of the flow where the effects of molecular viscosity can be ignored, that is, in the fully turbulent parts of the flow. Close to a solid wall this cannot hold, and hence, the model equations can only be solved to some "near-wall" point. The near-wall boundary conditions are defined using equations known as wall functions. The wall functions are based on the assumption that there exists in most turbulent flows a universal velocity profile near a solid boundary. A number of what are termed low-Reynolds-number turbulence models have been proposed. These allow the solution to be continued down to the wall, but they have not been shown to be particularly universal.

More sophisticated turbulence models are available, such as the non-linear k–ε model, algebraic stress models, and full Reynolds stress models. Speziale[42] has recently reviewed many of these models. However, all such models use the same basic averaging approach, and of crucial importance to the subject at hand, assume that the flow remains turbulent at all times. The latter is actually not strictly true of some low-Reynolds number models, as they were designed to model relaminarization (i.e., transition from turbulent to laminar flow) with limited success, but they have not been successfully applied to problems of transition from laminar flow to turbulent. Those hemodynamic flows that do become turbulent do so over a relatively small part of the pulsatile cycle. Hence, any simulation procedure used to model this type of flow needs to be able to model the process of transition from laminar to turbulent and back again to laminar flow.

The stark truth is that it is not yet possible to simulate the type of transition processes that may occur in flows of hemodynamic interest. This is due in part to a lack of understanding of the processes involved and in part to a lack of computing power. The direct numerical simulation of transition to turbulence in wall-bounded flows has recently been reviewed by Kleiser and Zang.[43] They show that while great progress has been made in the last few years, direct simulation of the full transition process all the way to fully developed turbulence can only be done for a few idealized cases. It would appear that we are some years away from a prediction scheme that could simulate transition in pulsatile flow in complex geometries. However, it is not altogether certain that what is termed "turbulence" in hemodynamic flows ever becomes fully developed. It may be that the disorganized flow structures seen at some points in the cycle reflect essentially laminar characteristics. This point will be returned to in the closing section of the chapter.

F. Code Validation

It is obviously necessary to have some degree of assurance that the predicted flows are accurate representations of the physical situation. Unfortunately, no rigorous mathematical proof exists to ensure that this is so. However, certain practical steps can be taken, which come under the general heading of "Code Validation." This includes making a series of predictions for problems for which either analytical solutions or experimental data of known accuracy exist. In the context of hemodynamics, Womersley's pioneering analysis[44] is useful in this regard. Obviously, the model problems should be as close as possible to the actual problem being considered. However, the sequence of code validation should include all features of the code's potential; that is, three-dimensionality, transience, wall movement, and so forth (see Xu et al.[45] for a careful comparison of predictions with *in vivo* data).

Other elements of code validation include demonstrating that the solution is "grid independent." This is achieved by refining the grid (increasing the number of control volumes) until no significant change is seen in the solution. This is an ideal that is not always possible for large problems with complex geometries. A similar procedure should be carried out for transient problems, to ensure that the solution accuracy is not significantly degraded by too large a time step. However, defining the "best" grid and time step is not always a simple matter. It will be shown in the next section that a more accurate solution can sometimes be achieved by using an advection model of higher accuracy rather than continued grid refinement.

III. APPLICATIONS

A. The Heart and Assist Devices

The flow of blood in the heart offers a wealth of challenging problems to the would-be numerical modeler. Principle among these is the fact that it is hard to envision a realistic model of the left ventricle, say, which does not address the movement of the cardiac wall, this being the driving mechanism of the flow. It has only been in the last two or three years that the more powerful CFD codes have been able to cope with moving walls, and even these advanced codes require the wall movement to be known to a precision that is currently not available for the heart. For example, little detailed information is available on the movement of the septa, but its motion must be influenced by the flow in both left and right ventricles. The internal geometry of both ventricles is also quite complex. The ventricle walls are covered with bands of randomly orientated, protruding muscles known as the trabeculae carneae. The shape and orientation of the atrio-ventricular valves and the fibrous threads, the chordae tendineae, attached to them which are in turn attached to papillary muscles protruding from the ventricular wall must all affect the flow within the ventricle. However, accurate representation of these features would require a grid of a size that is beyond current capabilities.

Some ground breaking work in this field has been carried out by a group lead by C.S. Peskin at the Courant Institute in New York.[46,47] Their rather novel approach has been to model the cardiac wall as an immersed system of fibers. For reasons of computational efficiency they have to date restricted their analysis to rather abstract shapes and have yet to address the heart itself. A further issue to be resolved is a consequence of the central assumption of the model, namely that the fibers are immersed in the fluid. In practice, this requires the assumption that the model heart is somehow suspended in a box of fluid and that this box is surrounded by identical boxes containing identical model hearts.

The flow through both natural and prosthetic heart valves is another area where CFD can be usefully applied. However, the geometric complexity of both is such that a great deal of idealization is currently required. Also, simulation of the flow around moving valve leaflets will probably have to wait for future CFD codes. Peskin's group have, however, also made some progress in this area.[48] As in their ventricular flow calculations, the moving boundary, the valve in this case, is represented by a system of massless fibers. The movement of the fibers defines a force field which is fed into the fluid flow calculation. The fluid does not actually see the solid boundary, only its influence in terms of the defined force field. The predicted velocity field is then used to define a new wall position. This procedure is repeated for each time step. Calculations for both natural and prosthetic valves[48] have been performed using this method, but both were limited to two-dimensions and rather unrealistically low Reynolds numbers. Using more conventional means, Rief[49] simulated the flow in a St. Jude valve using the Galerkin FEM. The objective of this study was to find areas of shear stress in the flow large enough to produce hemolysis. Lei et al.[50] used a similar method to investigate the flow through a Bjork-Shiley valve. Both simulations were restricted to steady flow in two-dimensional models with the valves in fixed positions. Thus, the results can only be viewed as qualitative.

There have been a few simulations of flows in artificial hearts and cardiac assist devices. For instance, Kim et al.[51] have reported a calculation of the flow in a model of a total artificial heart. The analysis was restricted to two-dimensional, steady flows. However, both laminar and turbulent cases were studied. The equations were solved using the finite analytic method of Chen and Chen[52] on a staggered Cartesian grid. This method derives the necessary algebraic equations from local analytic solutions of the governing equations over each cell. Pressure-velocity decoupling was accomplished using one of the standard modifications to SIMPLE. Turbulence was modeled using the k–ε model. Kim et al.[51] report reasonable agreement with available experimental data, but recognize the need to extend their model to pulsatile, three-dimensional flows, and to include body-fitted moving grids to allow the modeling of the moving valve leaflets and the pumping diaphragm. The latter would seem essential as it is this that produces the flow. Also, as discussed in a preceding section, extending the calculation to pulsatile laminar/turbulent flow is not possible with the k–ε model.

The authors are co-investigators with the group, headed by Professor Salmons at the University of Liverpool, which is conducting a comprehensive research program on skeletal muscle ventricles. A skeletal muscle ventricle (SMV) is an assist device that is fashioned from one of the recipient's own muscles (see Salmons and Jarvis[53] for a review of various approaches). The program comprises three parts: *in vivo*, *in vitro*, and numerical studies. The main objective of the numerical studies is to aid in the design of an optimum ventricle shape and inlet/outlet configuration by simulating the flow within model SMVs throughout the cardiac cycle. The compliant nature of the SMV puts quite severe demands on the numerical flow code to be used. However, as has been discussed in a previous section, codes are now available that are able to model transient flow problems with moving walls.

Both two-dimensional (axisymmetric) and three-dimensional model SMVs have been considered. In both cases HARWELL-FLOW3D was used to solve the discrete versions of equations governing the flow. FLOW3D is a general purpose code for the numerical solution of laminar and turbulent flows and heat transfer. It uses the finite volume method to solve the governing equations on general, three-dimensional, non-orthogonal (body-fitted) grids. While Release 2 of FLOW3D uses a single-block grid structure, a multi-block, or block-structured, grid has been adopted in Release 3. The code offers a variety of velocity-pressure coupling schemes, equation solvers, convection models, and time marching schemes. The Rhie-Chow algorithm[25] is incorporated into the solution procedure to allow the use of non-staggered grids. Both the single and multi-block versions of FLOW3D allow the grid to move in some prescribed manner.

Some preliminary results of the investigation into the effect of inlet size and shape on the resulting flow field in the SMV are given in Figure 5. Fully-implicit, Backward Euler, time marching was employed with a time step equal to one-twenty-fourth of the cardiac cycle. The cardiac period was taken to be one second. The walls were assumed to move radially in a simple sinusoidal manner, varying 20% about the mean value. The inlet pipe was taken to be rigid. The entrance plane of the inlet pipe was defined to be a surface of constant pressure. Hence the flow is solely driven by the wall movement. Included in Figure 5 is a sketch of the multi-block grid. As can be seen, the grid used was rather sparse, but it was thought sufficient for these exploratory calculations. More refined grids will be used in future calculations.

Given in parts a, b, and c of Figure 5 are predictions of the flow in three different SMV models toward the end of the filling phase, and parts d, e, and f show the corresponding predictions for the flow in the same three models close to the end of the pumping phase. The given velocity vectors are in the plane of symmetry. The models were designed to have approximately the same volume change. The vector plots were produced using Jasper, which is FLOW3D's post-processing package. It should be noted that the internal solid lines in these and subsequent plots do not represent physical surfaces but simply depict the grid block boundaries. It can be seen that a curved inlet produces a flow with a significantly different structure from that of a straight inlet, and that reducing the diameter of the inlet pipe increases the strength of flow within the SMV. More complex inlet/outlet configurations and more physiologically realistic flow situations are to be investigated in the near future.

B. Arterial Bifurcations

The study of blood flow through arterial bifurcations has been driven to a large extent by the fact that these areas of the cardiovascular system appear to be preferred sites for atheroma. There is evidence[3,4,5] that the location of atheromatous lesions correlate to areas of the arterial wall experiencing low and fluctuating wall shear stress. There is further evidence[5] that due to the complex nature of the flow occurring in arterial bifurcations the walls in these regions experience such shear stress variations. Three-dimensional pulsatile flow simulations in realistic geometrical models of arterial bifurcations are now becoming possible with the latest CFD codes and computers. These simulations are expected to clarify the role played by hemodynamics in the formation of atheromatous plaques.

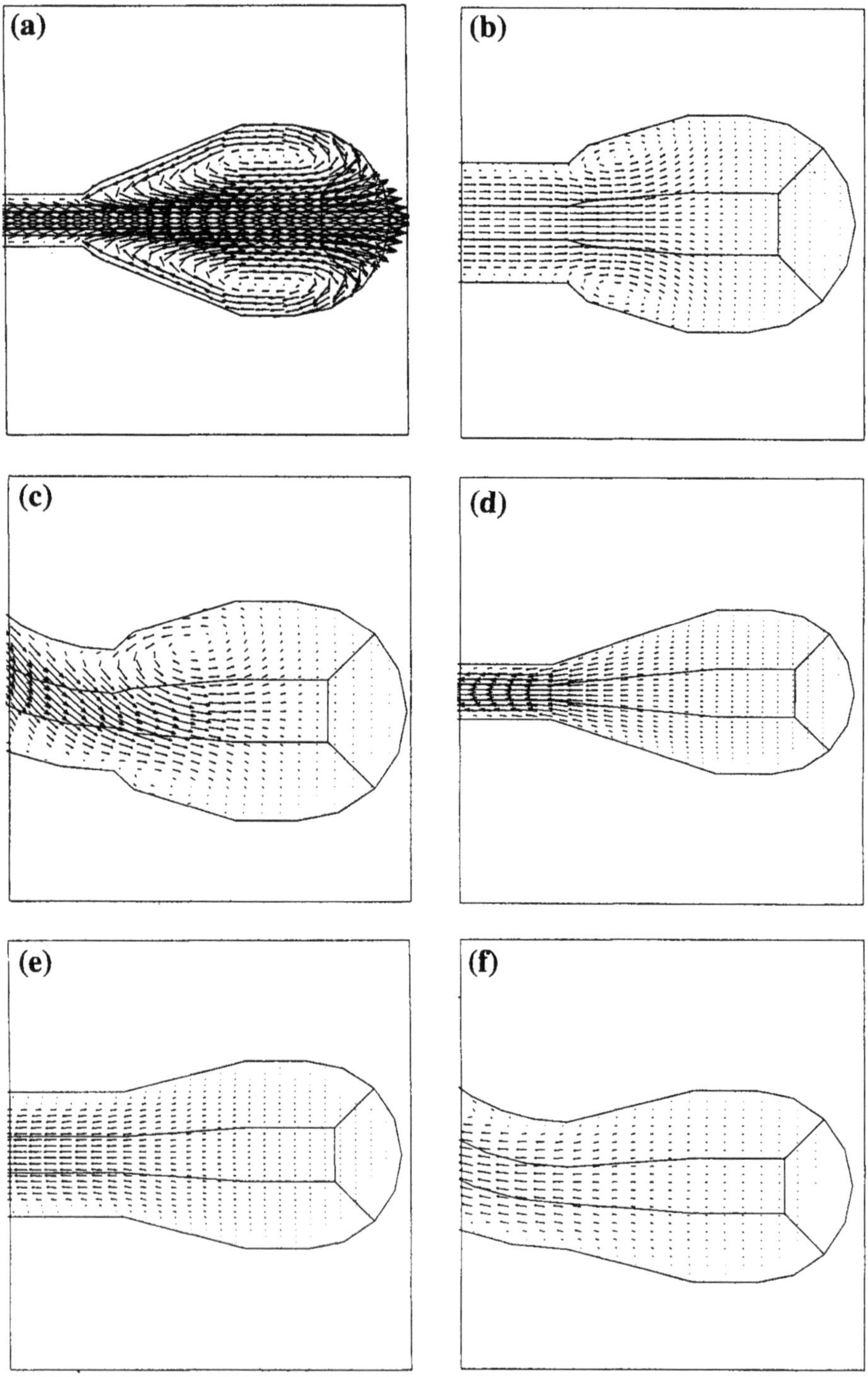

(continued)

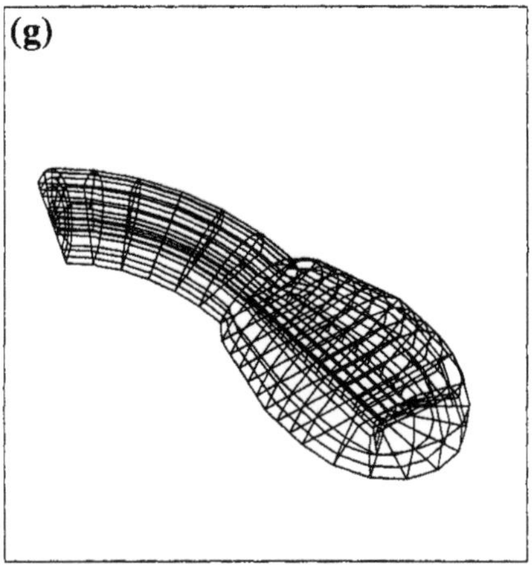

Figure 5. (Continued) Flow in model skeletal muscle ventricle; (a) filling phase—small diameter inlet tube, (b) filling phase—large diameter inlet tube, (c) filling phase—curved inlet tube, (d) pumping phase—small diameter inlet tube, (e) pumping phase—large diameter inlet tube, (f) pumping phase—curved inlet tube, and (g) typical gird.

A sketch of an idealized carotid bifurcation[54] is given in Figure 6. This figure indicates the geometrical complexity required of a numerical model for a realistic simulation of flow through a typical arterial bifurcation. The data given in Table 1 due to Forster et al.[55] indicates the significant variation in geometry from one person to another. A realistic simulation should also address the pulsatile nature of the flow. Measured velocity and pressure waveforms[45] for the canine femoral artery are given

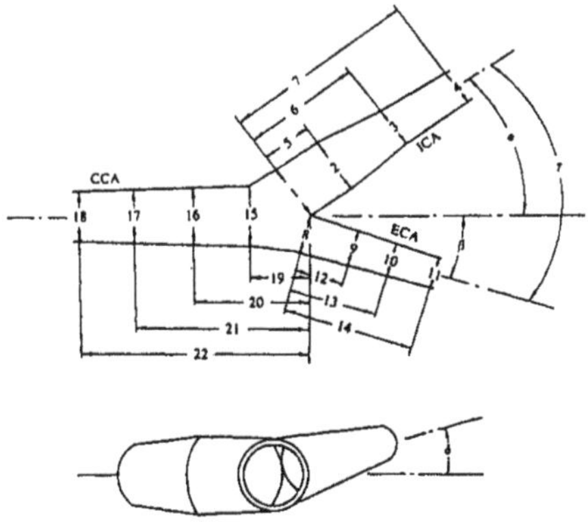

Figure 6. Generalized carotid bifurcation.

Table 1. Dimensions of Generalized Carotid Bifurcation

Angles	*Mean value (deg)*	*Standard deviation*
α	33.9	13.0
β	22.3	12.2
γ	52.6	19.5
δ	0.9	35.7
Diameters and location	*Mean value (mm)*	*Coefficient of variation (%)*
1	9.3	20.6
2	9.4	17.5
3	6.3	22.8
4	5.5	13.6
5	4.6	66.5
6	16.7	32.7
7	28.0	36.3
8	6.1	24.7
9	5.0	17.8
10	4.4	22.7
11	4.1	24.2
12	6.2	23.4
13	12.6	23.2
14	18.9	22.3
15	8.6	15.9
16	8.1	13.6
17	8.0	11.6
18	7.8	12.7
19	10.4	24.7
20	18.8	16.9
21	27.6	16.0
22	36.1	16.0

Source: Data from Forster et al.[55]

Note: See Figure 6 for definitions of angles, diameters, and locations.

in Figure 7. It can be seen that these are far from simple sinusoids. The profiles of velocity and pressure in a human artery are of similar form.

Inspection of the relevant literature reveals that there has probably been more simulations of flow through arterial bifurcations than any other part of the cardiovascular system. Initial studies were confined to two dimensions (see Xu and Collins[54] for a summary), primarily due to a lack of computing power at the time. However, since experimental measurement of model bifurcations[4,5] have shown

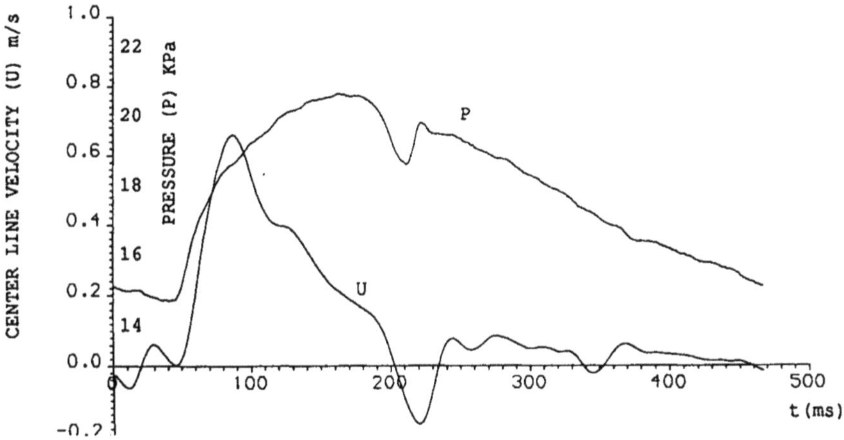

Figure 7. Typical velocity and pressure waveforms in a canine artery.

that the flow in bifurcations is highly three dimensional, these simulations were of limited validity.

Wille[56] was probably the first to present a three-dimensional simulation of the flow in a bifurcation. This FEM model was restricted to steady flow in a symmetric aortic bifurcation and assumed an unrealistically low Reynolds number of 10. Even with these limitations the method required a prohibitively long time to reach convergence. Yung et al.[57] have also considered steady three-dimensional flow in a symmetric bifurcation, but used an FDM (SIMPLE) method. They were able to achieve reasonable Reynolds numbers, and showed the importance of three-dimensional simulations by predicting secondary flows that would not occur in plane calculations. They noted qualitative agreement with experimental results, but admitted that the grid used (13×9×31) was a compromise between accuracy and available computer power. Rint et al.[58] presented a three-dimensional analysis of steady flow in a model of the carotid artery bifurcation. They used the Galerkin FEM and eliminated the need to solve for the pressure by invoking the penalty method. A typical simulation apparently took approximately 10 hours of CPU time on a mini-supercomputer, a Convex-1c-xp. The geometry used was similar to that shown in Figure 6, with δ set to zero. Their predictions were found to be in good agreement with measurements of steady flow in model bifurcations of similar geometry.

A three-dimensional calculations of unsteady flow through a 90° T-bifurcation was presented by Dinnar et al.[59] using FDM. While the scheme was found to be extremely efficient when compared to an FEM approach, the results could not be directly compared to a physiological bifurcation as the numerical model was defined to be composed of channels of rectangular cross section. In a series of papers (see for instance refs. 34,60) Perktold and his colleagues[60] have studied three-dimensional, pulsatile flow in realistic models of the human carotid artery bifurcation. The model is based on the Galerkin FEM and uses a pressure correction technique to decouple the velocity and pressure. The fully implicit backward Euler method is used to march forward in time, and the non-linear terms are linearized using Picard iteration. The model includes a realistic velocity waveform at inlet, and non-Newtonian effects have been considered.[60] These simulations have been able to predict the expected oscillating wall shear stress at the outer sinus wall, and

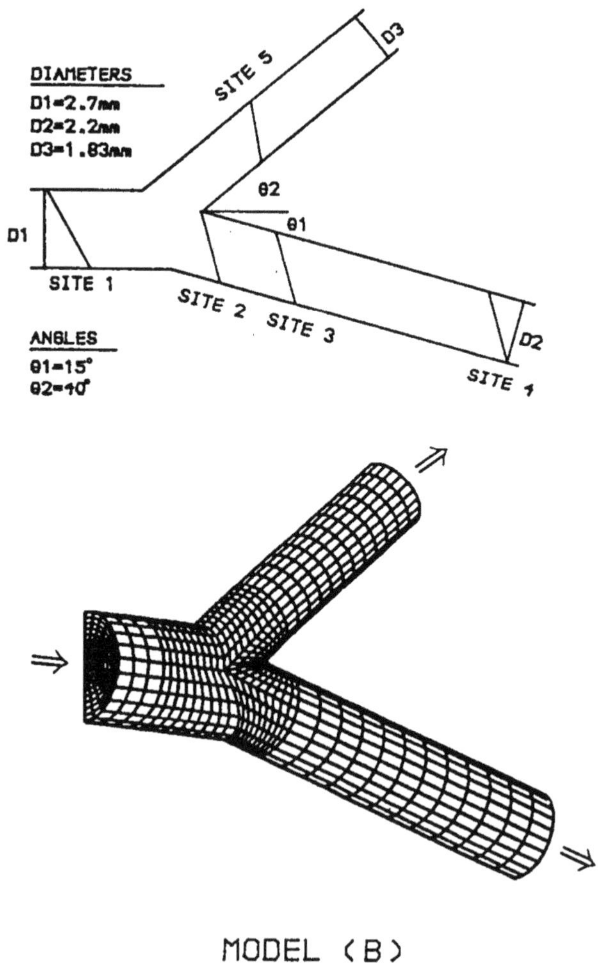

Figure 8. Model canine femoral bifurcation.

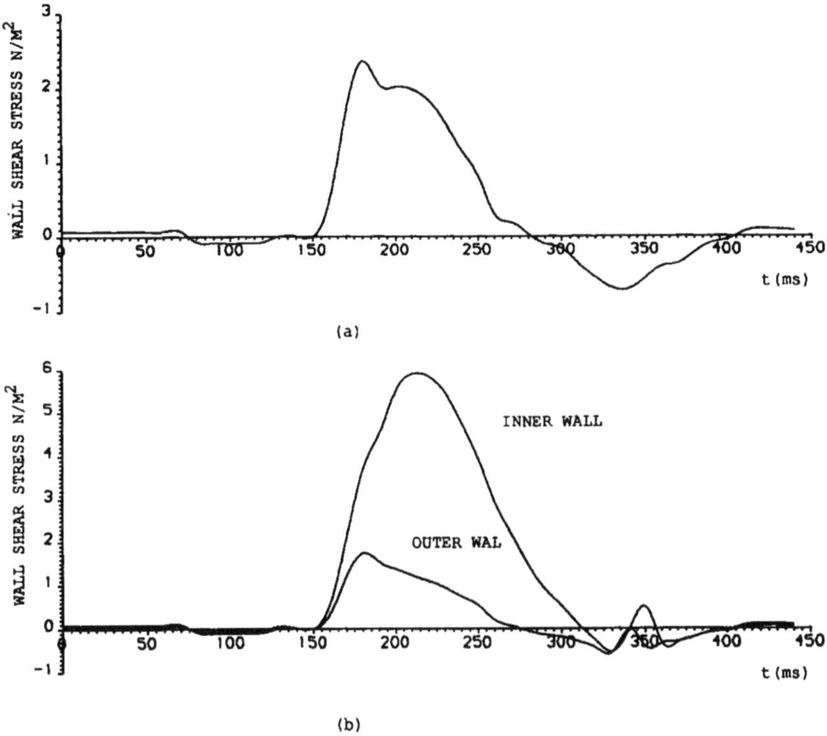

Figure 9. Predicted wall shear stress in canine femoral bifurcation; (a) One diameter above the flow divider in the parent vessel, and (b) 0.2 diameters downstream of the flow divider in the larger daughter vessel.

comparison of Newtonian and non-Newtonian simulations have revealed little differences in the basic flow characteristics.

Xu et al.[45] have predicted three-dimensional simulations of the flow through canine femoral bifurcations under physiological flow situations. Both Newtonian and non-Newtonian blood models were considered. The models' geometry and inlet velocity waveforms were based on *in vivo* measurements.[5] Discrete versions of the Navier-Stokes equations were solved using ASTEC[61] which uses the Finite Volume Method incorporating the SIMPLE algorithm on a Finite Element grid. Schematics of the one of models and corresponding grid used are given in Figure 8. Sample predictions of wall shear stress are given in Figure 9. It can be seen that the outer wall of the larger daughter vessel is predicted to experience relatively low and oscillating shear. Comparison of the results for Newtonian and non-Newtonian simulations appear to agree with the findings of Perktold et al.[60] in that the overall features of the flow appear to be little effected by the choice of rheological model although they could have local significance. The calculations were performed on grids of 5,040 elements and 6,090 nodes, with 18 elements in the radial direction. Using these grids, one pulse cycle took approximately two hours of CPU time on a Cray X-MP 28.

To the best of the authors' knowledge, only one attempt to simulate pulsatile flow through a compliant bifurcation has been published. However, in this calculation Reuderink[62] decoupled the problem by first finding the wall motion from a one-dimensional wave analysis, and then applying the resulting wall motion to a moving-wall flow model. The FEM was used for calculation of the flow and the wall movement. However, the inertia of the wall was neglected, and the wall was modeled as a thick shell, thus ignoring any variation of stress over the wall thickness. Reuderink[62] gives figures that indicate it took approximately 6.5 hours on a Cray-YMP to compute the flow field over one pulse cycle. It would appear that it took a little more than 4.25 hours to generate the wall displacements for one pulse cycle on a Convex C1 mini-super computer. The authors' group is working on the problem of flow through compliant arterial models. We have recently[63] simulated the flow through a complaint tube by solving the equations of fluid and wall motion simultaneously, and Dr. Xu of our group has very recently successfully simulated flow in a compliant T-junction using a simplified pressure/diameter relationship.

C. Grafts

Aorto-coronary bypass grafts are now routinely used to alleviate chronic ischaemic heart disease. Bypass grafts are also used to restore blood supply to other areas of the body distal of severely stenosed arteries, for example, the lower leg. However, it is an unfortunate fact that these grafts can fail due to the formation of atheromatous plaque in the anastomoses' region. This is particularly ironic as it is often atheroma that has caused the need for the surgery in the first place. While the picture is clouded by the obvious fact that the grafts fail in patients who have shown a tendency toward atheroma, it seems reasonable to assume that local flow dynamics play a role in the formation of the atheromatous lesions.

There is obviously some similarity in geometry between a bypass anastomosis and a naturally occurring arterial bifurcation. However, a fundamental difference between most arterial bifurcations and the distal anastomosis is that in the latter the flow is in the reverse direction. Apparently, the only naturally occurring case of flow in this direction is in the basilar artery[64] which is formed by the union of the two vertebral arteries. Also, in both proximal and distal anastomoses, the flow situation can be quite different to that occurring naturally. Bypasses are usually performed when the flow in an artery is either severely stenosed or totally occluded. Thus, in the distal case at least, most of the flow can be expected to pass through the graft and very little through the host artery proximal to the anastomosis site. In the proximal case, the flow division depends on the exact location of the anastomosis site. In the axillo-bifemoral bypass graft, for example, the blood supply is

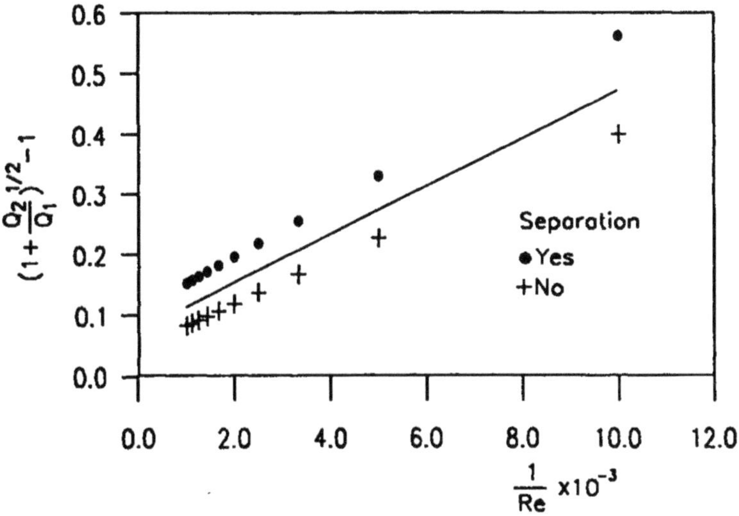

Figure 10. Numerical (symbols) and analytical (line) prediction of the occurrence of separation in a T-junction.

taken far from the site of occlusion, and in this case the flow into the graft can be expected to be less than that remaining in the host artery. However, in other situations, for example, the femoro-popliteal bypass, the proximal anastomosis may be much nearer the site of occlusion, and hence, most of the flow enters the graft.

Richardson and Christo[65] have considered the occurrence of flow separation from the wall of the parent tube opposite the daughter tube inlet in a T-junction. Using simple analytical arguments, they have shown that the ratio of flow into the daughter tube to flow continuing in the parent tube is an important parameter in determining whether the flow in the parent tube separates. Xu and Collins[66] have confirmed this finding in a numerical study of flow in a T-junction. Figure 10 shows the comparison between analytical and numerical predictions. Richardson and Christo[65] hypothesize that flow division in the circulatory system may have evolved in such a way as to suppress this separation. If this is true it is clear that a surgically constructed bifurcation will almost certainly not generate the correct division of flow.

It can be expected that a careful and detailed program of numerical simulations of flows in proximal and distal anastomosis models will help develop an understanding of the important factors governing these flows. Analysis of the predicted wall shear stress and separation sites can be used to determine the optimal anastomosis angle and graft to artery diameter ratio for a particular flow situation. Also, it will be possible to identify any areas of abnormally high shear in the body of the

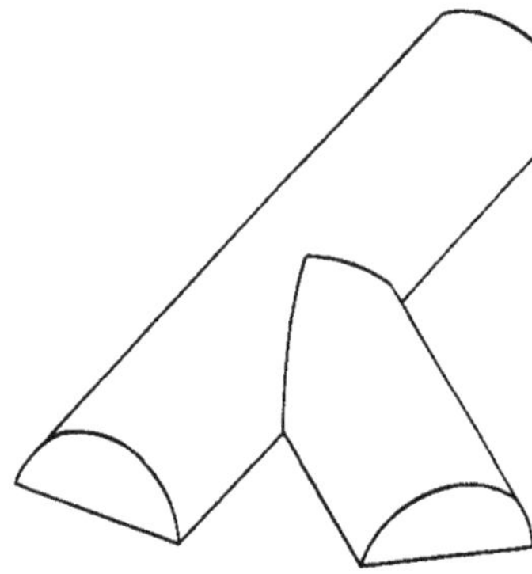

Figure 11. Model anastomosis.

flow caused by the artificiality of the flow geometry that could damage the blood itself.

Some two-dimensional simulations have been reported in the literature,[67,68] but these will not be discussed here in detail. It is clear from experimental measurements[65,69] in branched tubes that the flow in bypass anastomoses, as in bifurcations, is both complex and three dimensional. For instance, a prominent feature of the flow in plane T-junctions is a large separation region at the heel (proximal side) of the branch tube.[54] In a two-dimensional flow, separated regions have to be closed, that is, the streamlines within these regions are closed, and there can be no communication between the bulk flow and that in the separated region apart from that produced by diffusion. However, experimental investigations[65,69] of T-junctions of circular cross section have shown that separation at this site, when it occurs, is accompanied by a similar separation in the host tube, and that fluid travels from one to another in a complex, three-dimensional manner.

While experimental investigations of flow through models of three-dimensional anastomoses have been performed, the authors have been unable to find any articles on comparable numerical simulations. The authors are currently engaged in a numerical study of flow through anastomoses in collaboration with researchers at Liverpool who are conducting *in vitro* studies. Some preliminary findings, mostly for steady flow, have been reported in Henry et al.[70] A sketch of the model anastomosis considered is given in Figure 11. The flow is assumed to be symmetrical about the plane defined by the tubes' axes. This assumption reduces by one half the size of the numerical computation. The geometry of the model anastomoses and the fluid properties were set to conform to those of the *in vitro* models. In both the flow was Newtonian. The equations governing the flow were solved numerically using HARWELL-FLOW3D, Release 3.1, general details of FLOW3D having been given in a previous section.

Figure 12. Typical grid for model anastomosis.

An example of the multi-block grids used for the proximal anastomoses is given in Figure 12. The distal case is topologically similar. Both grids are composed of 15 blocks. The basic structure of these grids is a six block arrangement over the semi-circular cross section, similar to that shown in Figure 2a. The grids were defined in USRGRD, which is a user-defined routine within FLOW3D.

Initial computations of the flow in the proximal anastomosis with 100% bypass flow were performed using a fairly sparse grid (1,600 control volumes). While the overall flow field seemed to be in reasonable agreement with the experiment, closer

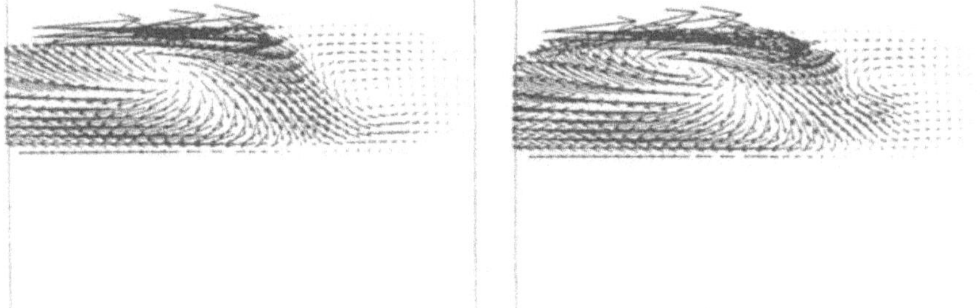

Figure 13. Predictions of the flow field in the host artery distal of the anastomosis with 100% bypass flow; (a) Higher Upwind advection model, and (b) QUICK advection model.

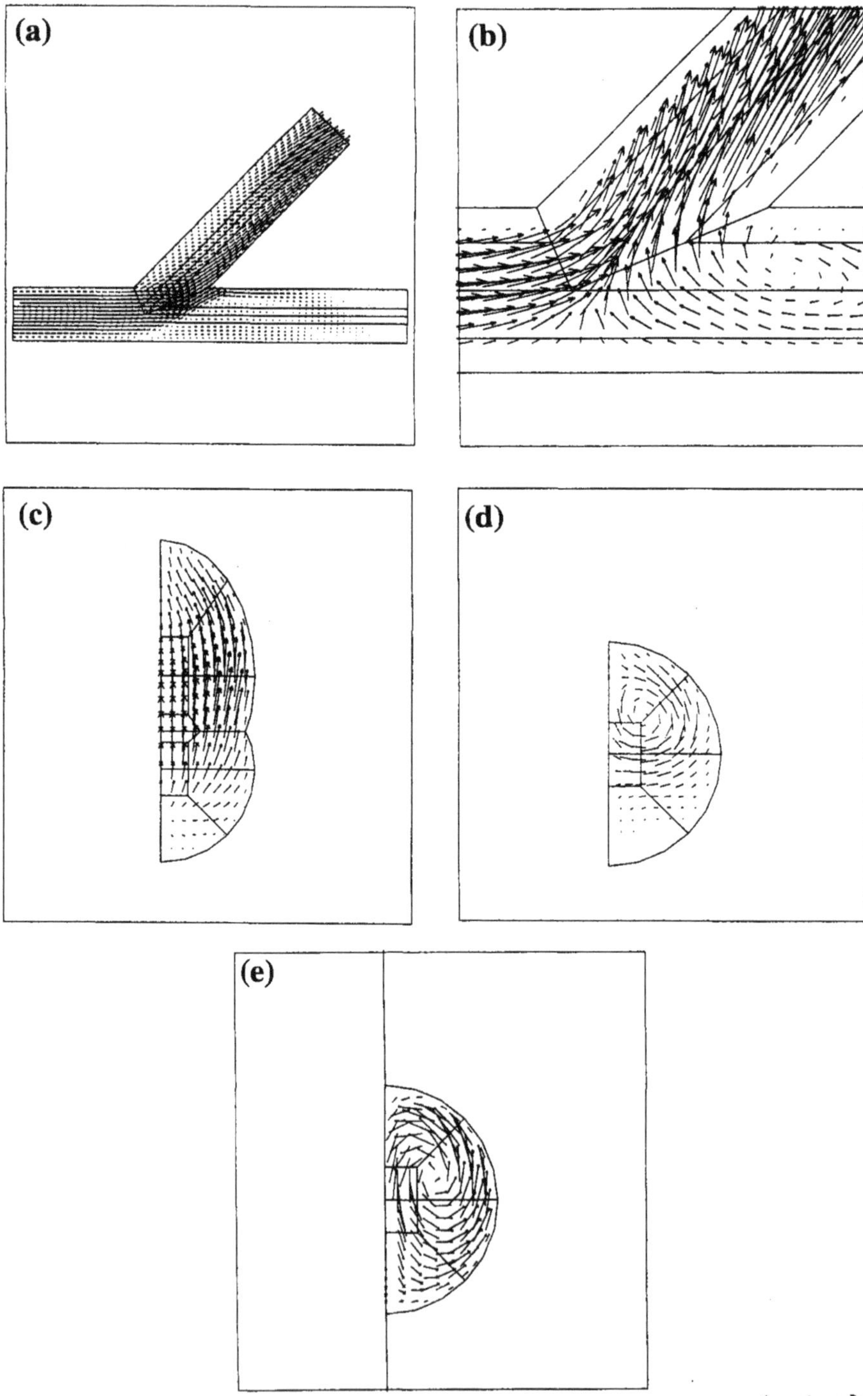

(continued)

Figure 14. Flow details of the model proximal anastomosis with 100% bypass flow; (a) full field in the plane of symmetry, (b) details of the flow in the anastomosis site in the plane of symmetry, (c) cross-plane secondary flow in the anastomosis site, (d) cross-plane secondary flow in the host artery distal of the anastomosis, and (e) cross-plane secondary flow in the graft.

---

inspection of the flow in the region of the host artery between the anastomosis and the occluded end revealed significant differences. In particular, a double vortex structure seen in the experimental flows was not predicted. Since the default advection model, the Hybrid model, is known to cause significant smearing when operating in its upwind mode, two other models were tried. These were the second order accurate Higher Upwind model (HUW) and a third order accurate model known as QUICK. However, all three models predicted roughly the same (erroneous) flow field in the distal region. It was found that a grid more than three times as dense as the original (5,460 control volumes) was required to achieve the desired flow pattern. It can be seen in Figure 13 that with this grid HUW advection model predicts that the center of the second vortex is somewhat below the tube's centerline, which corresponds to the experimental finding, while the formally more accurate QUICK model predicts that the vortex is above the tube's centerline. Further, it can be seen that the Hybrid model is unable to predict the second vortex, even at the higher grid density. Based on these findings, the remainder of the calculations were performed using a grid of 5,460 control volumes and HUW.

In Figure 14 are given details of predicted velocity vectors in various planes for the flow in the proximal anastomosis with 100% bypass flow. These compare well with experimental measurements and are presented to illustrate the three-dimensional nature of the flow in the anastomosis region and occluded end. An interesting feature of the flow in the graft tube is the occurrence of two counter-rotating vortices which are roughly aligned with the mean flow direction. These appear to feed energy into the flow near the wall and thus retard separation of the flow in that region. Similar streamwise vortex pairs have been seen in the daughter vessels of arterial bifurcations. It is interesting to note that aeronautical engineers have for some time induced streamwise vortices to occur over aerofoils to retard the onset of boundary-layer separation. It would appear that nature thought of this first.

D. Stenoses

The high incidence of atherosclerosis in the Western World has led to much research into the genesis and early detection of atheroma. As was mentioned in a previous section, researchers have tried to correlate certain hydrodynamic factors, chiefly levels of wall shear stress, to areas in the circulatory system where atheroma is found. Inspection of the literature reveals that the majority of numerical simulations concerned with atherogenesis have considered the flow in unaffected arterial bifurcations, while those concerned with the effect of the lesion once formed have restricted their attention to studies of the flow through partially blocked straight tubes. It is primarily the latter that will be discussed in this section.

There have been several studies of axisymmetric stenoses, both steady and pulsatile. For example, Luo and Kuang[71] recently considered the effects of blood's non-Newtonian nature on the steady flow through an axisymmetric stenosis. While few details are given of the numerical procedure used, it appears they used some form of FEM. As might be expected, comparison of the non-Newtonian and Newtonian models revealed differences in velocity profiles, and hence, the level of wall shear stress. Tu et al.[72] used a Galerkin FEM code to simulate pulsatile flow through an axisymmetric stenosis. They showed that their model was able to match the measurements of flow in a similar *in vitro* model.

Realizing that arterial stenoses are unlikely to be axisymmetric, a number of researchers have considered more physiologically realistic models. Tutty[73] considered the pulsatile flow through a plane channel with an indentation on one side as a model of the flow through an artery with an asymmetric stenosis. He considered both sinusoidal and more physiologically realistic volume flow waveforms. The predicted flows distal of the stenosis are complex for both types of volume flow waveform. The flows exhibit trains of counter-rotating vortices which are reminiscent of the vortex streets which form downstream of a cylinder in cross flow. The effect of the counter-rotating vortices is to produce shear stresses on each wall that change sign periodically in the axial direction. This would obviously be significant if it reflected actual conditions downstream of an asymmetric arterial stenosis. However, the experimental study of Ojha et al.,[74] and our own preliminary predictions discussed below, would suggest that while the flow downstream of an asymmetric stenosis in a circular tube is complex, it does not exhibit an axially periodic shear stress distribution.

As mentioned above, the authors have begun a study of flow through arteries with axisymmetric and asymmetric stenoses. Some preliminary finding have been given in Henry and Collins.[75] Initially, it is intended to simulate the well documented *in vitro* pulsatile flow studies of Ojha et al.[74] These will be used for code validation. Once this has been achieved, more realistic predictive models of stenoses will be considered. Sketches of the experimental model stenoses used by Ojha et al.[74] are given in Figure 15. Given in Figure 16 is part of a typical (sparse) multi-block grid used to model the asymmetric case. The assumed symmetry of these flows meant that it was necessary to model only one-half of the tube. Two preliminary calculations using FLOW3D, Release 3, have been performed, one for the 38% asymmetrical case, and one for the 65% axisymmetric case, and these will be discussed briefly in the text that follows. In these calculations the flow was assumed to be driven by a sinusoidally varying pressure at inlet. Specifically, the pressure was defined as

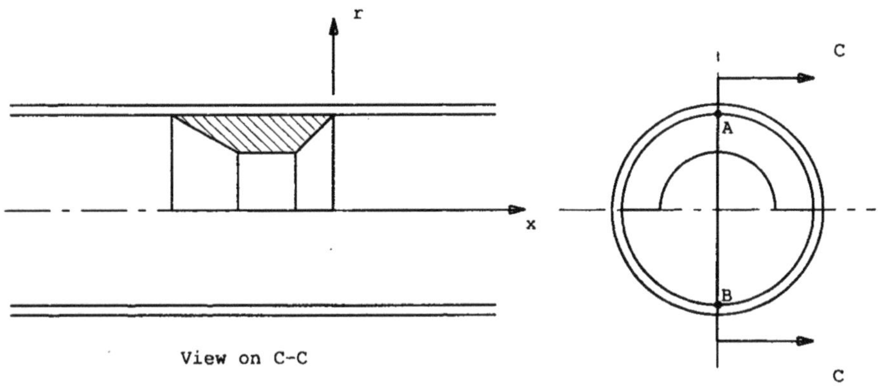

Figure 15. Model asymmetric stenosis.

$$p = \frac{p_{max}}{2}[Cos(nt) + 1] \qquad (8)$$

where p_{max} was set to 1.0 kPa, $n = 2\pi/P$, and P is the cyclic period, the latter being taken to be 0.75 seconds. Taking the diameter, D, of the tube to be 5.1 mm, and kinematic viscosity, ν, of blood to be 3.4×10^{-6} m^2 s^{-1}, the Womersley number $(D/2(2\pi/\nu P)^{1/2})$ of the flows was approximately equal to 4.0. The outlet was defined to be a plane of constant pressure. The flows produced using the above pressure distribution had peak Reynolds numbers in the range 700–900. In both cases to be discussed HUW differencing was used for the advection terms together with Backward Euler time stepping and a time step equal to one-twenty-fourth of the period.

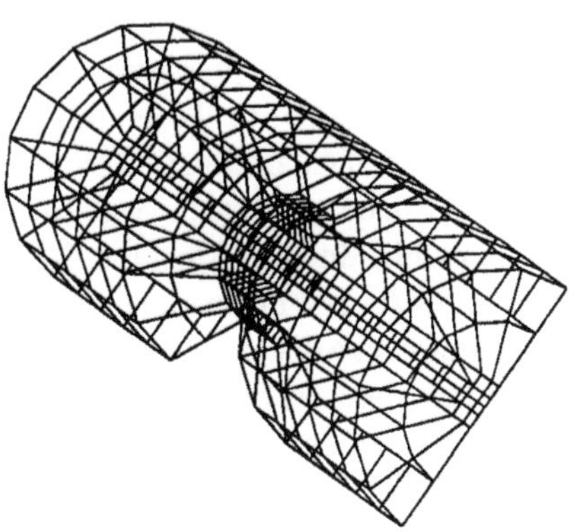

Figure 16. Typical grid for model asymmetric stenosis.

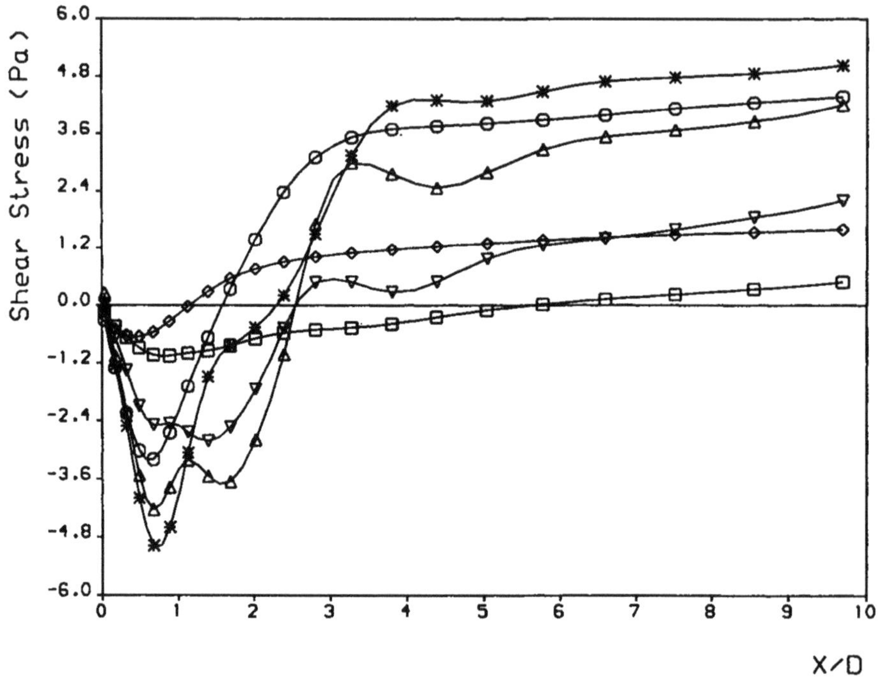

(continued)

Figure 17. Wall shear stress distal of an asymmetric stenosis (see Figure 18 for definitions of symbols); (a) along line A defined in Figure 15, and (b) along line B defined in Figure 15.

Given in Figure 17 are profiles of wall shear stress at various times in the cycle along a line bisecting the half-annular stenosis (denoted by A in Figure 16) and a line bisecting the unstenosed half of the tube's cross section (denoted by B in Figure 15). It can be seen that these lines are the intersection of the plane of symmetry with the tube wall. A schematic of the variation of pressure is given in Figure 18. Included in this figure are symbols corresponding to those used in the shear stress plots. These are used to indicate the time of each shear plot. Note that x is measured from the distal end of the stenosis. It can be seen that the wall shear stress on the stenosed half of the tube changes sign which indicates that the flow separates at the site of the stenosis and then reattaches further downstream. It can also be seen that the location of zero stress oscillates over a distance of several diameters throughout the flow cycle. However, in contrast, the shear on the unstenosed half is always positive, and hence, the flow remains attached on this half throughout the cycle. In fact there cannot be a sudden change of shear distribution from one-half of the tube to the other. What has been shown are the two extremes. The distribution of shear stress must change in some complex but relatively smooth manner from one extreme to

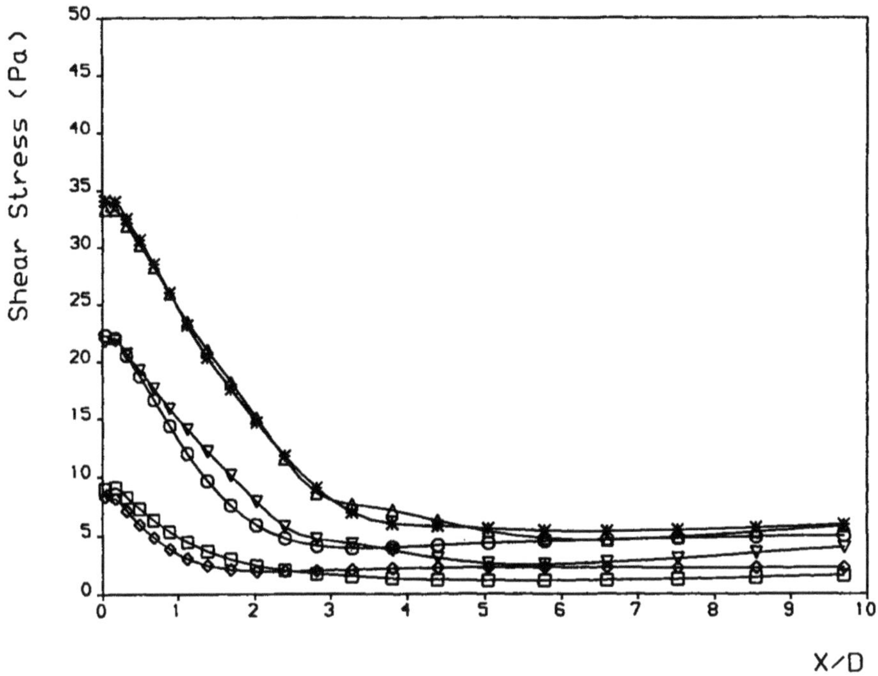

Figure 17. (Continued)

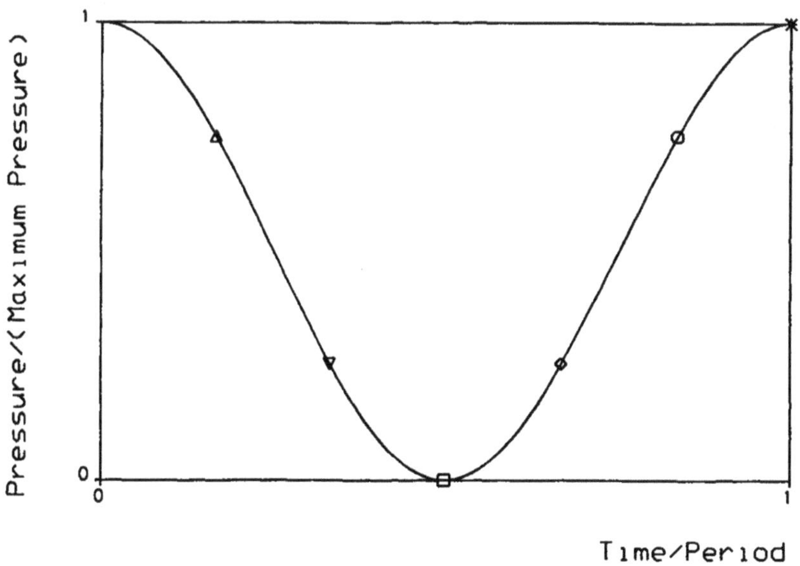

Figure 18. Schematic of applied inlet pressure. Symbols define points in cycle of shear plots in Figures 17 and 19.

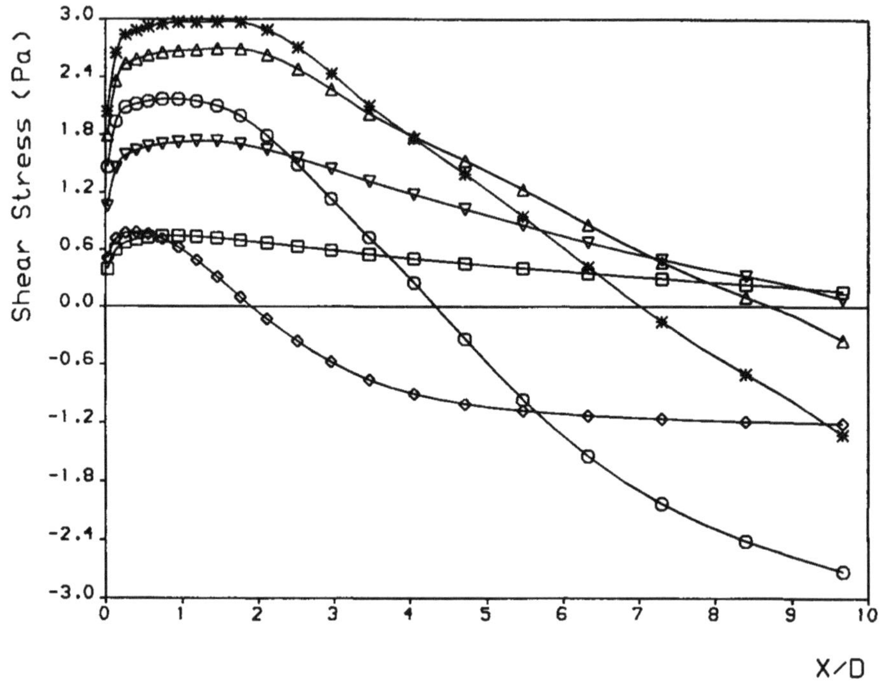

Figure 19. Wall shear stress distal of an axisymmetric stenosis (see Figure 18 for definitions of symbols).

the other. For comparison, similar stress profiles are given in Figure 19 for the axisymmetric stenosis.

The shear stress profiles given in Figure 19 for the axisymmetric case indicate that the separated region is longer for this case than for the asymmetric case. However, the two cases are not exactly comparable. They differ in the degree of stenosis, and this has the effect of producing different volume flow rates, and hence Reynolds numbers, for the same pressure drop. As might be expected, the shear stress distribution was found to be axisymmetric. However, it is recognized that assuming the flow to be symmetrical about a plane that bisects the half-annular stenosis, while useful for computational reason, may be over restrictive. Geometrical symmetry does not necessarily impose a similar condition on the flow. This point might be important if the flow becomes unstable in some part of the cycle. Forcing the flow in one-half of the tube to be a reflection of that in the other one-half, could conceivably suppress some modes of oscillation.

Velocity vector plots of the flow in the symmetric plane of the asymmetric and axisymmetric stenosed tubes are given in Figure 20. The cross plane plot of the asymmetric case included in this figure shows the secondary flow pattern at a station approximately two tube diameters distal of the stenosis. It can be seen that at this

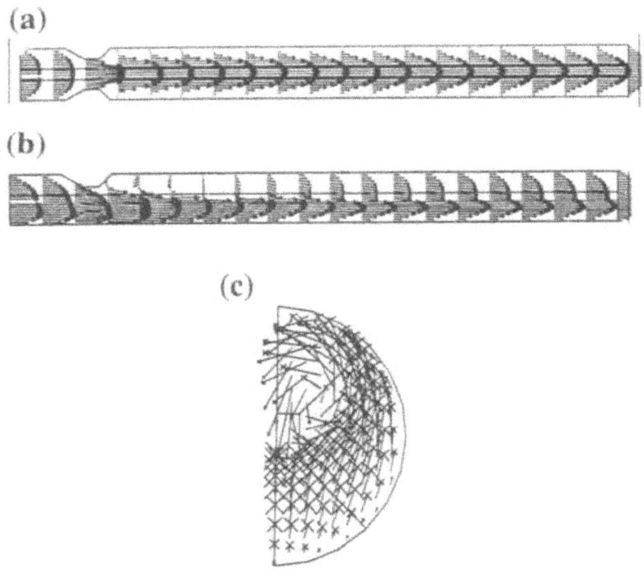

Figure 20. Predicted flow fields in model stenoses; (a) Axisymmetric case—flow in symmetry plane, (b) Asymmetric case—flow in symmetry plane, and (c) Asymmetric case—flow in cross plane two diameters distal of the stenosis.

location there is a net flow from the unstenosed part of the wall toward the stenosed part. It is noted that the wall shear stress plotted in Figure 17 is actually only the axial component of the stress. In axisymmetric flows the second component of the stress vector would always be zero. However, the cross plane plot given in Figure 20 shows that in the axisymmetric case there will in general also be a non-zero component of wall shear in the direction.

The preliminary predictions discussed above broadly agree with the experimental findings of Ojha et al.[74] However, the experimental flows appeared to become turbulent over some part of the cycle due to the separated region, distal of the stenosis, becoming unstable. The question of whether the flow in stenosed arteries truly becomes turbulent is both important and difficult to answer. It is important in that turbulent stresses may be of sufficient strength to promote hemolysis. However, as stated previously, defining turbulence in pulsatile flow is somewhat subjective. It may be that the flow remains laminar, but becomes highly distorted at certain times in the cycle. If these structures are repeated, albeit in a complex fashion, they should be predictable without recourse to turbulence modeling. Again as stated previously, flows that have laminar and turbulent regions cannot be computed using conventional turbulence models because they require the flow to be fully turbulent throughout the solution domain and for all time. However, if the flow is only going into the first stages of transition it may be possible to perform a DNS of the process.

IV. PROBLEMS OUTSTANDING

A. Coupled Behavior

Much of the interest in blood flow modeling is focused on conditions prevailing at the interface between the blood and the vessel wall. In reality the cardiovascular system is composed of a rather complex, fibrous, compliant material, and it would seem necessary to consider, in a coupled manner, the interaction between the flow and the wall if the hydrodynamic influences on atherosclerosis, for example, are going to be addressed comprehensively. An additional feature to be addressed in the case of the heart is that its walls are composed of muscle tissue, the contraction of which generates the wall motion. The authors have recently considered the basic requirements of a numerical code capable of addressing this coupled problem,[76] and have concluded that while such a code might still be a few years away from development, it is within reach, and the present authors are actively engaged in this development.[63]

A number of groups are actively working on the development of coupled fluids/solids numerical codes. Interest is not restricted to universities and other research institutions. A number of software vendors appear to expect sufficient interest in this type of code in the engineering community at large to make it a commercially viable venture.[77] It is evident, however, that whereas state-of-the-art CFD codes can adequately represent all the large scale behavior of blood flows, to simulate the various aspects of the compliance and muscle behavior of arterial vessels require the most comprehensive solid mechanics stress models available today. Hence, the simultaneous solution of the fluid and solid motion is not an easy task, and even when a coupled code is developed, it may still be that some parts of the cardiovascular system have geometries that are simply too complex to be accurately modeled even on the largest currently available computers. A realistic simulation of the moving leaflets of the mitral or tricuspid valves with the aim of analyzing regurgitant flow, say, are examples of such problems.

The authors are currently engaged in two projects aimed at developing codes capable of solving the coupled problem. The first involves solving for the wall motion in a special subroutine within FLOW3D. In this case, the wall is treated somewhat like a quasi-fluid in that the equations of motion of an incompressible solid are solved using the techniques employed by FLOW3D to solve the Navier-Stokes equations. The second project is being carried out in conjunction with researchers at Nuclear Electric's Engineering Analysis Centre, and will be based on their FEM based code called FEAT. FEAT, being FEM based, is more suited to adding a conventional FEM structural analysis solver.

Coupled calculations of flow in a compliant artery using infinitesimal strain theory have been successfully conducted using both codes. See Henry and Collins[63] for some preliminary coupled predictions using FLOW3D. Work is underway to extend the codes to more general geometries and to allow finite strain.

B. Organized Structures

Another area in which CFD could usefully be applied is in furthering the understanding of the large scale organized structures seen in experimental flows distal of stenoses, bifurcations, and grafts at certain points in the cardiac cycle. The visualization studies of Ojha et al.[74] of axisymmetric and asymmetric stenoses show clear evidence of what they describe as a Kelvin-Helmholtz type vortex being formed just before peak flow. The breakdown of this vortex was thought to lead to the production of some rather low energy turbulence. The existence of organized repeatable "coherent structure" in fully turbulent shear flow is now well documented and a well-known exposition has been made by Hussain.[78]

To identify the "coherent structures," Hussain gives two alternative decompositions of the turbulent flow field. "Doubled decomposition" is where the mean flow is regarded as coherent, and combined with the other coherent structures, the residual field being "incoherent," or random turbulence. In "triple decomposition," the three fields are "mean flow," "coherent structures," and "random turbulence." Lieber[79] has recently applied these concepts to disturbed pulsatile flow of direct relevance to hemodynamics. As with Hussain, the presence of "organized structures" is tested by phase decomposition and data sampling. "Triple decomposition" into mean, organized, and random components is used. We believe Lieber's term "organized structures" is a better description for the flows of interest here because "coherence" implies the presence of permanent random turbulence whereas normal hemodynamic flow is largely laminar.

It may be that the organized structures seen in the flow situations mentioned above can be studied using techniques similar to those developed for coherent structures in fully turbulent flow. Indeed, as the above structures are essentially laminar phenomena, a CFD code without conventional turbulence modeling should be able to resolve them. Such a simulation would clarify the role played by these structures in generating peak values of wall shear stress and whether they can be implicated in the destruction of red blood cells. We are also testing the same approach for purely laminar flows where an organized structure (a vortex) may only exist for part of the pulsatile cycle. Such a situation exists for example in the *in vitro* model SMVs of Shortland et al.[80] In fact, in these axisymmetric models, two, and in certain circumstances, three, simultaneous vortex rings are clearly apparent from experiment; the two vortex ring situation has been predicted by our CFD model. In general, the concept of organized structure in laminar flow may only be a procedure to aid visualization of complex three-dimensional transient flows: it may, however, provide an interpretation to wall shear stress fluctuations of significant in atherosclerosis.

C. Closure

Throughout this chapter we have been extolling the virtues of the computational approach and they are many. However, while we are both enthusiastic advocates of CFD, we would like to close with a note of caution. The sudden proliferation of commercially available CFD codes in recent years has lead to some concern of late[81] that CFD is in danger of getting a bad name. This is primarily because these codes are starting to be used by people with little or no understanding of fluid mechanics or numerical analysis. This is partly the fault of the software vendors' advertising copy which would lead the uninitiated to believe that CFD can tackle any problem regardless of size or complexity. While modern CFD codes are powerful tools that when applied judiciously can help both in the understanding of a wide variety of flow situations and provide quantitative data for engineering design, it is important to understand their limitations. It is easy to be beguiled by multicolored displays of velocity vectors overlaid with pressure contours, say, into believing that the results are physically correct. It must be remembered that the accuracy of a numerical solution of an unsteady flow depends on the grid density, and time step used, and on the accurate description of boundary conditions. A CFD code will solve for the conditions given without regard to whether or not they are physically reasonable. Simulating a highly three-dimensional flow with a two-dimensional model, or calculating steady turbulent flow through a device that in reality has a largely laminar, pulsatile flow are examples of questionable applications. Also, achieving some predetermined convergence condition is no guarantee that the solution is an accurate representation of reality. It cannot be emphasized enough that results of simulations for flows in complex geometries should be checked against experimental measurements whenever possible. This is because in these cases it is unlikely that there will be sufficient computing power to perform full grid and time-step dependency checks. It is for these reasons that we pay particular attention to code validation, thus permitting our results to be viewed as both qualitatively and quantitatively reliable.

ACKNOWLEDGMENT

This work was funded by a grant from the Wellcome Trust.

REFERENCES

1. Fry, D.L. Haemodynamic forces in atherogenesis. In: *Cerebro-vascular Disease*. Raven Press, New York, 1976, pp. 77–95.
2. Caro, C.G.; Fitz-Gerald, J.M.; Schroter, R.C. Atheroma and arterial wall shear: Observation, correlation and proposal of a shear: dependent mass transfer mechanism for atherogenesis. *Proc. R. Soc. Lond.* **1971**, *B171*, 109–159.

3. Giddens, D.P.; Bharadvaj, B.K.; Ku, D.N. Cerebral and peripheral hemodynamics. In: *Cardiovascular Ultrasonic Flowmetry* (Altobelli, S.A.; Voyles, W.F.; Greene, E.R., Eds.). Elsevier, New York, 1985, pp. 125–145.
4. Ku, D.N.; Giddens, D.P.; Zarnins, C.K.; Glagov, S. Pulsatile flow and atherosclerosis in human carotid bifurcation. Positive correlation between plaque location and low and oscillating shear stress. *Arteriosclerosis* **1985**, *5*, 293–302.
5. Jones, C.J.H.; Lever, M.J.; Ogasawara, Y.; Parker, K.H.; Hiramatsu, O.; Mito, K.; Tsujioka, K.; Kajiya, F. Blood velocity distributions within intact canine arterial bifurcations. *Am. J. Phys. (Heart and Circulatory Physiology)* **1992**, *32(5)*, H1592–1599.
6. Locket, J.F.; Hunter, J.C.; Collins, M.W. Problems in using holographic interferometry to resolve the four-dimensional character of turbulence. Part I: Theory and experiment; Part II: Image and data processing. *Int. J. Opt. Sensors* **1986**, *1*, 211–234.
7. Collins, M.W.; Coifalo, M. Computational fluid dynamics and its application to transport processes. *J. Chemical Technology and Biotechnology* **1991**, *52*, 5–47.
8. Moore Jr., J.; Delfino, A.; Meister, J.-J. Preliminary analysis of the effect of blood vessel movement on blood flow patterns. In: *Advances in Bioengineering, ASME* (Bidez, M.W., Ed.), 1992, pp. 537–539.
9. Baker, A.J. *Finite Element computational Fluid Mechanics*. Hemisphere, 1983.
10. Hutton, A.G.; Smith, R.M.; Hickmott, S. The computation of turbulent flows of industrial complexity by the finite element method-progress and prospect. *Int. J. Numerical Methods in Fluids* **1987**, *7*, 1277–1298.
11. Anderson, D.A.; Tannehill, J.C.; Pletcher, R.H. *Computational Fluid Mechanics and Heat Transfer.* Hemisphere, Washington, 1984.
12. Patankar. S.V. *Numerical Heat Transfer and Fluid Flow*. Hemisphere, Washington, 1980.
13. Hussaini, M.Y.; Zang, T.A. Spectral methods in fluid dynamics. *Ann. Rev. Fluid Mech.* **1987**, *19*, 339–367.
14. Patera, A.T. A spectral element method for fluid dynamics: Laminar flow in a channel expansion. *J. Comp. Phys.* **1984**, *54*, 468–488.
15. He, X.; Ku, D.N. Entrance flow development for pulsatile flow in a straight tube. In: *Advances in Bioengineering* (Bidez, M.W., Ed.), 1992, pp. 461–464, ASME, New York.
16. Wolfe, A. CFD software: Pushing analysis to the limit. *ASME Mechanical Engineering* **1991**, *113*, 48–54.
17. Thompson, J.F.; Warsi, Z.U.A.; Mastin, C.W. *Numerical Grid Generation: Foundations and Applications*. Elsevier, 1985, New York.
18. Roache, P.J. *Computational Fluid Dynamics*. Hermosa, Albuquerque, 1972.
19. Patankar, S.V.; Spalding, D.B. A calculation procedure for heat, mass and momentum transfer in three dimensional parabolic flows. *Int. J. Heat Mass Transfer* **1972**, *15*, 1787–1806.
20. Van Doormal, J.P.; Raithby, G.D. Enhancements of the SIMPLE method for predicting incompressible fluid flows. *Numer. Heat Transfer* **1984**, *7*, 147–163.
21. Issa, R.I. Solution of the implicitly discretized fluid flow equations by operator splitting. *J. Comp. Phys.* **1985**, *61*, 40–65.
22. Yih, C-S. *Fluid Mechanics*. McGraw-Hill, New York, 1969.
23. Voke, P.R.; Collins, M.W. Forms of the generalized Navier-Stokes equations. *J. Eng. Math.* **1984**, *18*, 219–233.
24. Burns, A.D.; Wilkes, N.S. A finite difference method for the computation of fluid flows in complex 3D geometries. *UKAEA Report*. AERA, Harwell, 1987.
25. Rhie, C.M.; Chow, W.L.A. A numerical study of the turbulent flow past an airfoil with trailing edge separation. *AIAA J.* **1983**, *21*, 1525–1532.
26. Demirdžić, I.; Perić, M. Finite volume method for the prediction of fluid flow in arbitrary shaped domains with moving boundaries. *Int. J. Num. Meth. Fluids* **1990**, *10*, 771–790.

27. Hawkins, I.R.; Wilkes, N.S. Moving grids in Harwell Flow3d. *UKAEA Report, AEA-InTech-0608*, 1991.
28. Zienkiewicz, O.C. *The Finite Element Method*. McGraw-Hill, New York, 1977.
29. Gresho, P.M.; Lee, R.L.; Sani, R.L. On the time dependent solution of the incompressible Navier-Stokes equations in two and three dimensions. In: *Recent Advances in Numerical Methods in Fluids*, Vol. 1 (Taylor, C.; Morgan, K., Eds.). Pineridge Press, Swansea, U.K., 1980.
30. Babuska, I.; Aziz, A.K. Survey Lectures on the Mathematical Foundation of the Finite Element Method. In: *Mathematical Foundation of the Finite Element Method* (Aziz, A.K., Ed.). Academic Press, New York, 1972.
31. Brezzi, F. On the existence, uniqueness, and approximation of saddle-point problems arising from Lagrange Multipliers. *R.A.I.R.O., Numer. Anal.* **1974**, *8*, 129–151.
32. Perktold, K. On numerical simulation of three-dimensional physiological flow problems. *Ber. Math.-Stat. Sektion. Forschungsges. Joanneum Graz, Nr. 280*, Graz, 1989.
33. Hilbert, D. An efficient Navier-Stokes solver and its application to fluid flow in elastic tubes. In: *Numerical Methods, Colloquia Societatis János Bolyai*. North-Holland, Amsterdam, 1897, pp. 423–431.
34. Perktold, K.; Resch, M.; Peter, R.O. Three-dimensional numerical analysis of pulsatile flow and wall shear stress in the carotid arterial bifurcation. *J. Biomech.* **1991**, *24*, 409–420.
35. Glowinski, R.; Pironneau, O. Finite element methods for Navier-Stokes equations. *Ann. Rev. Fluid Mech.* **1992**, *24*, 167–204.
36. Moin, P. The computation of turbulence. *Aerospace America*, **1992**, *January*, 42–46.
37. Voke, P.R.; Collins, M.W. Large-eddy simulations of turbulent flow in plain and distorted channels. *AERA Harwell Report HTFS RS 551*. Annual Research Symposium, University of Warwick, 1984.
38. Ciofalo, M.; Collins, M.W. Large-eddy simulations of turbulent flow and heat transfer in plain and rib-roughened channels. *Int. J. Num. Meth. Fluids*, **1992**, *15*, 453–489.
39. Ciofalo, M. Large-eddy simulations of turbulent flow and heat transfer in simple and complex geometries. Ph.D. Thesis, Department of Mechanical Engineering and Aeronautics, City University, London, 1992.
40. Ciofalo, M.; Collins, M.W.; Stasiek, J.A.; Chew, P. Computational fluid dynamics and alternative turbulence models applied to heat transfer and pressure drop performance in cross corrugated heat exchangers. Submitted to Engineering Applications of Computational Fluid Dynamics, European Conference, IMechE Headquarters, London, 7–8 September, 1993.
41. Henry, F.S.; Collins, M.W.; Ciofalo, M. Prediction of swirling flow in a corrugated channel. *Proceedings of the 2nd International Conference on Applications of Supercomputers in Engineering*, 1991, August 13–15.
42. Speziale, C.G. Analytical methods for the development of Reynoldsstress closures of turbulence. *Ann. Rev. Fluid Mech.* **1991**, *23*, 107–158.
43. Kleiser, L.; Zang, T.A. Numerical simulation of transition in wall-bounded shear flows. *Ann. Rev. Fluid Mech.* **1991**, *23*, 495–538.
44. Womersley, J.R. The mathematical analysis of arterial circulation in a state of oscillatory motion. *Technical Report W.A.D.C.-T.R. 56-614*, Wright Air Development Center, 1957.
45. Xu, X.Y.; Collins, M.W.; Jones, C.J.H. Flow studies in canine artery bifurcations using a numerical simulation method. *ASME J. Biomech. Eng.* **1992**, *114*, 504–511.
46. Peskin, C.S.; McQueen, D.M. A three-dimensional computational method for blood flow in the heart. 1. Immersed elastic fibers in a viscous incompressible fluid. *J. Comp. Phys.* **1989**, *81*, 372–405.
47. McQueen, D.M.; Peskin, C.S. A three-dimensional computational method for blood flow in the heart. 2. Contractile fibers. *J. Comp. Phys.* **1989**, *82*, 289–297.
48. Peskin, C.S. The fluid dynamics of heart valves: Experimental, theoretical, and computational methods. *Ann. Rev. Fluid Mech.* **1982**, *14*, 235–259.

49. Reif, T.H. A numerical-analysis of the backflow between the leaflets of a St-Jude medical cardiac-valve prosthesis. *J. Biomech.* **1991**, *24(3)*, 733–741.
50. Lei, M.; Vansteenhoven, A.A.; Vancampen, D.H. Experimental and numerical-analyses of the steady flow field around an aortic Bjork-Shiley standard valve prosthesis. *J. Biomech.* **1992**, *25(3)*, 213–222.
51. Kim, S.M.; Chandran, K.B.; Chen, C.J. Numerical simulation of steady flow in a two-dimensional total artificial heart model. *ASME J. Biomech. Eng.* **1992**, *114*, 497–503.
52. Chen, C.J.; Chen, H.C. Finite analytic numerical method for unsteady two-dimensional Navier-Stokes equations. *J. Comp. Phys.* **1984**, *53*, 209–226.
53. Salmons, S.; Jarvis, J.C. Cardiac assistance from skeletal muscle: A critical appraisal of the various approaches. *Br. Heart. J.* **1993**, *68*, 333–338.
54. Xu, X.Y.; Collins, M.W. A review of the numerical analysis of blood flow in arterial bifurcations. *Proc. IMechE Part H: J. Eng. Med.* **1990**, *204*, 205–216.
55. Forster, F.K.; Chikos, P.M.; Frazier, J.S. Geometric modeling of the carotid bifurcation in humans: Implications in ultrasonic Doppler and radiologic investigations. *J. Clin. Ultrasound* **1985**, *13*, 385–390.
56. Wille, S.O. Numerical simulation of steady flow inside a three-dimensional aortic bifurcation model. *J. Biomed. Eng.* **1984**, *6*, 49–55.
57. Yung, C.-N.; De Witt, K.J.; Keith, Jr., T.G. Three-dimensional steady flow through a bifurcation. *ASME J. Biomech. Eng.* **1990**, *112*, 189–197.
58. Rint, C.C.M.; van Steenhoven, A.A.; Janssen, J.D.; Reneman, R.S.; Segal, A. A numerical analysis of steady flow in a three-dimensional model of the carotid artery bifurcation. *J. Biomech.* **1990**, *23(5)*, 461–473.
59. Dinnar, U.; Enden, G.; Israeli, M.A. A numerical study of flow in a three dimensional bifurcation. Submitted to Cardiovascular Dynamics Society Meeting, Canada, 1988.
60. Perktold, K.; Peter, R.O.; Resch, M.; Langs, G. Pulsatile non-Newtonian blood flow in three-dimensional carotid bifurcation models: A numerical study of flow phenomena under different bifurcation angles. *J. Biomed. Eng.* **1991**, *13*, 507–515.
61. Lonsdale, R.D. An algorithm for solving thermalhydraulic equations in complex geometries: The ASTEC code. *UKAEA Report.* Dounreay, 1988.
62. Reuderink, P.J. Analysis of the flow in a 3D distensible model of the carotid artery bifurcation, Ph.D. Thesis, Eindhoven University of Technology, The Netherlands, 1991.
63. Henry, F.S.; Collins, M.W. A novel predictive model with compliance for arterial flows. *Advances in Bioengineering* (Tarbell, J.M., Ed.), 1993, pp. 131–135, ASME, New York.
64. Krijger, J.K.B.; Heethaar, R.M.; Hillen, B.; Hoogstraten, H.W.; Ravensbergen, J. Computation of steady 3-dimensional flow in a model of the basilar artery. *J. Biomech.* **1992**, *25(12)*, 1451–1465.
65. Richardson, P.D.; Christo, J.L. Flow separation opposite a side branch. In: *Biofluid Mechanics, Blood Flow in Large Vessels* (Liepsch, D., Ed.). Springer-Verlag, 1990, pp. 125–284, Berlin.
66. Xu, X.Y.; Collins, M.W. Prediction of flow separation in branched tubes. In: *Proceedings of the 8th International Conference on Laminar and Turbulent Flows*, July, Swansea, U.K., 1993.
67. Pietrabissa, R.; Inzoli, F.; Fumero, R. Simulation study of the fluid dynamics of aorto-coronary bypass. *J. Biomed. Eng.* **1990**, *12*, 419–424.
68. Kim, Y.H.; Chandran, K.B. A numerical study of steady flow across end-to-side vascular bypass graft anastomoses. In: *Advances in Bioengineering* (Bidez, M.W., Ed.), 1992, pp. 233–236, ASME, New York.
69. Ku, D.N.; Liepsch, D. The effect of non-Newtonian viscoelasticity and wall elasticity on flow at a 90° bifurcation. *Biorheology* **1986**, *23*, 359–370.
70. Henry, F.S.; Collins, M.W.; Hughes, P.E.; How, T.V. Numerical investigation of the flow in proximal and distal end-to-side anastomoses. In Press, *ASME J. Biomech. Eng.* **1996**.
71. Luo, X.Y.; Kuang, Z.B. Non-Newtonian flow patterns associated with an arterial stenosis. *ASME J. Biomech. Eng.* **1992**, *114*, 512–514.

72. Tu, C.; Deville, M.; Dheur, L.; Vanderschuren, L. Finite-element simulation of pulsatile flow through arterial-stenosis. *J. Biomech.* **1992**, *25(10)*, 1141–1152.
73. Tutty, O.R. Pulsatile flow in a constricted channel. *ASME J. Biomech. Eng.* **1992**, *114*, 50–54.
74. Ojha, M.; Cobbold, R.S.C.; Johnston, K.W.; Hummel, R.L. Detailed visualization of pulsatile flow fields produced by modelled arterial stenoses. *J. Biomed. Eng.* **1990**, *12*, 463–469.
75. Henry, F.S.; Collins, M.W. Computations of flow distal of axisymmetric and asymmetric stenoses. *Advances in Bioengineering* (Tarbell, J.M., Ed.), 1993, pp. 87–90, ASME, New York.
76. Henry, F.S.; Collins, M.W. The numerical analysis of wall-fluid interaction problems in arterial flows: A review. To be submitted for publication. n.d.
77. Puttré, M. FEA programs band together. *ASME Mech. Eng.* **1991**, *113*, 77–80.
78. Hussain, A.K.M.F. Coherent structures—Reality and myth. *Phys. Fluids* **1983**, *26*, 241–258.
79. Lieber, B.B. The decomposition of apparent stresses in disturbed pulsatile flow in the presence of large scale organized structures. *J. Biomech.* **1990**, *23(10)*, 1047–1060.
80. Shortland, A.P.; Jarvis, J.C.; Henry, F.S.; Black, R.A.; Collins, M.W.; Salmons, S. Physical and numerical simulations of blood flow within a skeletal muscle ventricle. World Symposium on Cardiomyoplasty and Biomechanical Assist, Paris, May 24–26, 1993.
81. Grescho, P.M. An attempt at a status report on a fast moving target: CFD. *Proceedings World User Association in Applied CFD*, Basel, Switzerland, May, 1992.

FLOW VISUALIZATION AND MEASUREMENT WITH THE PHOTOCHROMIC TRACER TECHNIQUE:
HEMODYNAMIC APPLICATIONS

G. G. Couch and M. Ojha

Advances in Hemodynamics and Hemorheology,
Volume 1, pages 113–143.

ISBN: 1-55938-634-7.

I. INTRODUCTION

To date, *in vitro* flow visualization and measurement has been the primary method of investigating hemodynamic flows due to practical difficulties encountered with *in vivo* approaches. However, conventional *in vitro* techniques have shortcomings that have limited their effectiveness in hemodynamic applications. Most techniques, including Pitot tubes and hot wire anemometry, are invasive and while they have been superceeded by more advanced and non-invasive methods like laser Doppler anemometry, they are essentially all single point measurement techniques. More recent methods, such as particle tracking and particle image velocimetry are essentially well-established methods of flow visualization that with the advent of inexpensive digital imaging systems have evolved into full-fledged quantitative techniques capable of full-field velocity measurement. However, even with modernization, these techniques have retained the undesirable characteristic of being invasive.

The inadequacies of these *in vitro* techniques become very apparent when applied to the complex and unsteady flows that are often seen in hemodynamic studies, particularly at bifurcations and downstream of stenoses. Many are unable to provide accurate velocity measurements near to the vessel wall. Furthermore, a single point measurement technique capable of accurate velocity measurement in the near wall region can provide only an estimate of the wall shear stress. To calculate the instantaneous wall shear rate and to characterize its dynamic behavior, simultaneous measurements of the instantaneous fluid velocity at multiple radial sites are needed.

The role of hemodynamics in arterial function, remodeling, and disease is an active area of research. It is well-recognized that arteries respond to local changes in blood flow, and that this response is facilitated by the ability of the endothelium to sense and respond to the viscous shearing forces or shear stress at the vessel wall. For example, the dynamic behavior of the wall shear stress may contribute to the development of several focal arterial disorders including intimal hyperplasia[1] and poststenotic dilatation.[2,3] Clearly, to investigate the etiology of these disorders requires a more detailed and accurate characterization of the hemodynamics in the near wall region than is possible with the conventional techniques.

II. THE PHOTOCHROMIC TRACER TECHNIQUE

The photochromic tracer technique is a unique tool that has led to significant contributions in the fields of fluid and biofluid mechanics. Originally developed at the University of Toronto,[4] the photochromic technique showed good promise, but due to poor trace quality, lacked the spatial resolution necessary for many applications. Subsequently, the technique has been steadily refined and improved.[5,6] In its present state, the photochromic tracer technique is an ideal tool for resolving the complex and unsteady flow structures associated with arterial blood flow, since it is non-invasive and exhibits high spatial and temporal resolution.

A. The Photochromic Method

Essentially, the basic operating principle of the photochromic system depends on the addition of a normally colorless indicator to the test fluid. This indicator undergoes a reversible photochemical reaction, becoming opaque, when irradiated with ultraviolet light. The basic experimental system consists of a pulsed nitrogen laser, a mechanical pump flow synthesizer, a 35 mm camera with electronic flash, a microcomputer controller, optics, and a test section (Figure 1). A set of focused ultraviolet beams obtained from the laser are used to produce fine traces in the flow field of the test section. Initially, the traces are undeformed or straight. After a suitable time interval, the traces, deformed by the flow field, are photographed and the flow field including the near-wall velocity profile can be deduced.

The microcomputer controller provides for the synchronization of the laser, pump, camera, and flash. It enables the deformed traces to be recorded at a given

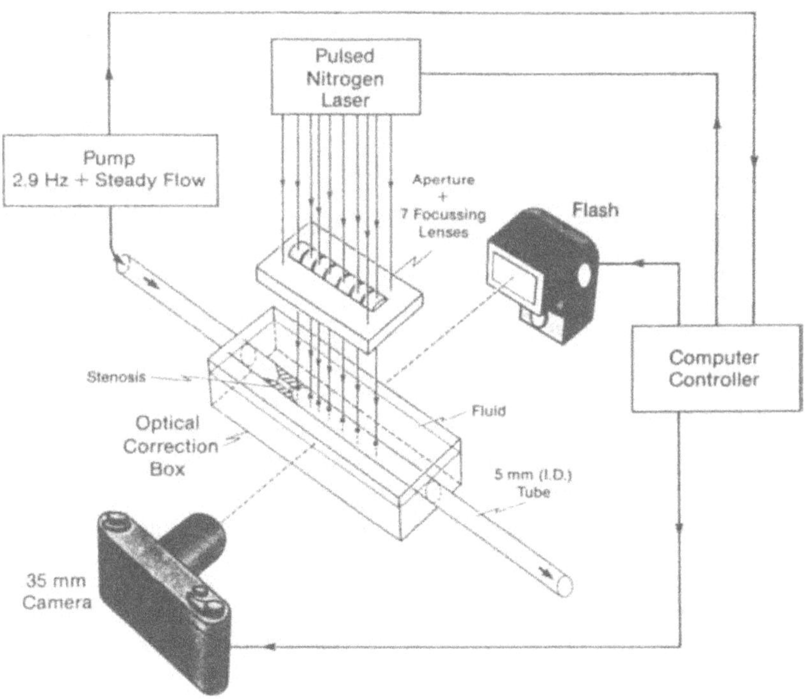

Figure 1. Schematic representation of the major components of the flow measurement system.[2]

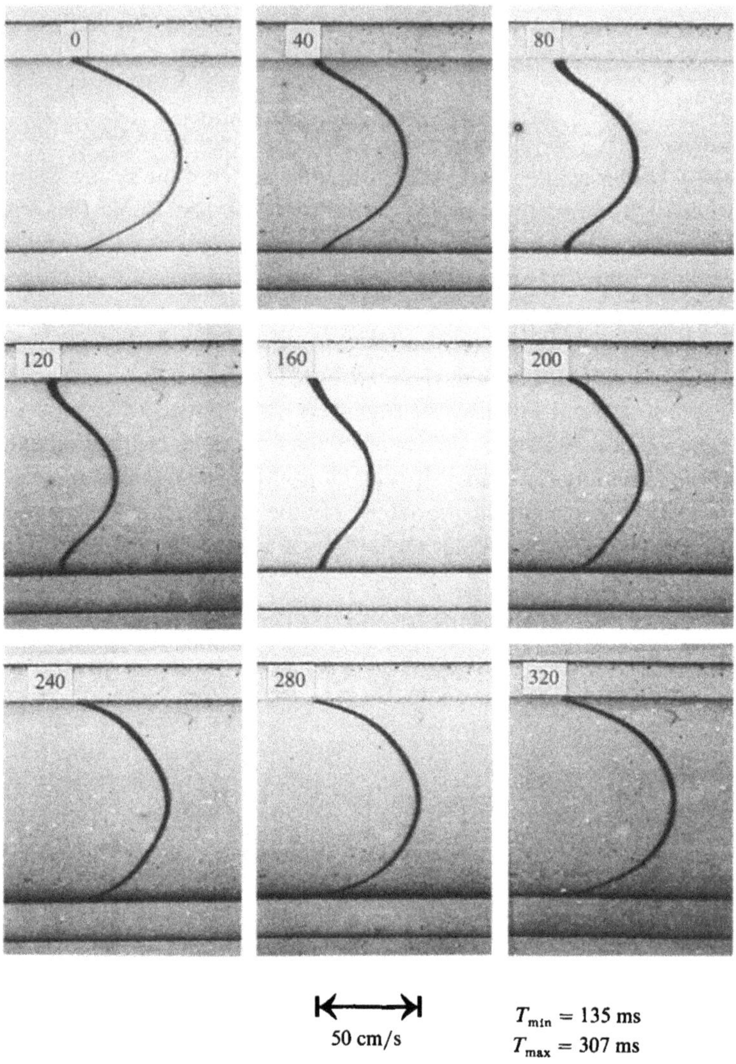

Figure 2. Sequence of photographs illustrating the displacement of photochromic traces 5.1 ms after formation in a straight tube of 5.1 mm internal diameter. Flow is a 2.9 Hz sinusoidal flow superimposed on a steady flow. Each profile represents the axial velocity.[13]

time after formation within the flow cycle. Figure 2 illustrates this process with a sequence of photochromic traces recorded in a straight tube of 5.1 mm internal diameter. The flow cycle consisted of a steady flow with a superimposed 2.9 Hz sinusoidal flow. Each trace was recorded 5.1 ms after formation. The displacement of the trace represents the average axial velocity over the interval between formation and recording. Precise alignment of the laser beam perpendicular to the axis of the conduit is not required; it is the net displacement that is used to determine the velocity.

B. Extending the Photochromic Method with Grids

To date, most applications of the photochromic tracer technique have been performed with a set of non-intersecting traces. Although this is sufficient for flows which are essentially one-dimensional, and will permit the determination of the instantaneous axial wall shear stress under both laminar and turbulent flow conditions, multidimensional flows present additional complications. It is impossible to track individual elements of a trace within a multidimensional flow field; only the essentially one-dimensional velocity field in the near wall region can be determined accurately. These shortcomings are eliminated by using two sets of parallel traces that intersect to form a grid in the flow field.

The superposition of grids within a flow field was originally proposed by de Gennes[7] for the purpose of measuring velocity and velocity gradients. Falco and Chu[8] produced photochromic grids, and demonstrated the potential of this technique. However, photochromic grids have seen little practical application, primarily due to the difficulty of producing traces of sufficient quality for measurement.

An advanced nitrogen laser was developed that produces ultraviolet beams with a divergence of 0.1×0.3 milliradians and a pulse energy of about 1 millijoule, and is capable of generating grids that are well-defined.[9] We have used this new laser to produce a 6×6 photochromic grid which can be used to determine the velocity and vorticity fields (see Figures 3 and 4) in the near and far wall regions,[10] which was not possible with the use of non-intersecting traces.

C. Analysis of Photochromic Traces and Grids

To obtain accurate measurements of the fluid velocity and other quantities from the photochromic traces, each photographic frame is digitized with a film scanner

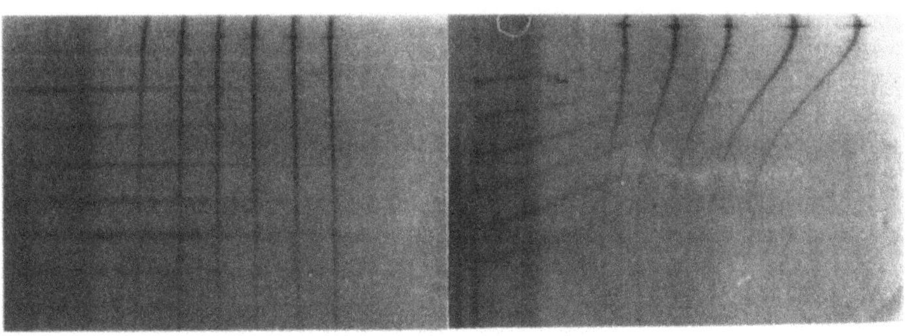

Figure 3. (a) Photochromic grid immediately after formation, and (b) after deformation by flow field.

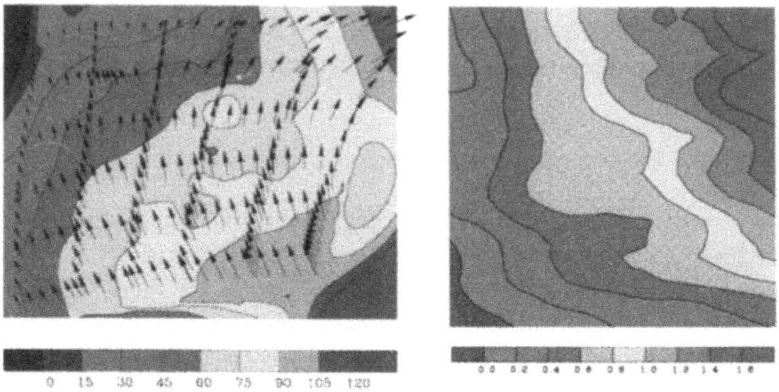

Figure 4. (a) Velocity field, and (b) vorticity field computed from the photochromic grids depicted in Figure 3.

(Nikon LS3500) with a maximum resolution of 5120 × 4096 pixels at 256 gray levels (Figure 5a). A custom image processing and analysis program developed for a SPARCstation IPX allows the coordinates of each of the trace profiles to be extracted (Figure 5b). The displacement profiles are then used to calculate the wall velocity gradient and the wall shear stress (Figure 6).

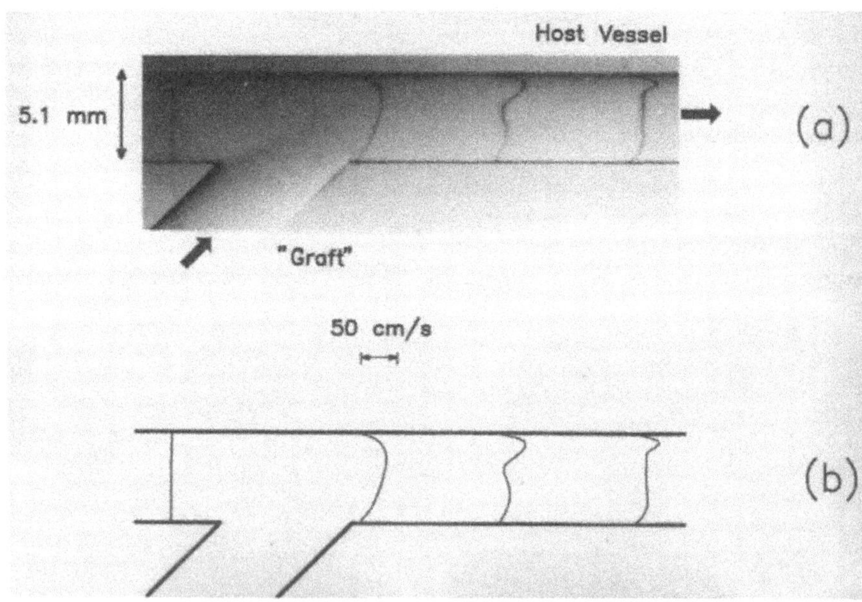

Figure 5. (a) Photochromic traces in a model of an end-to-side anastomosis, and (b) profiles of photochromic traces obtained by image processing and analysis.

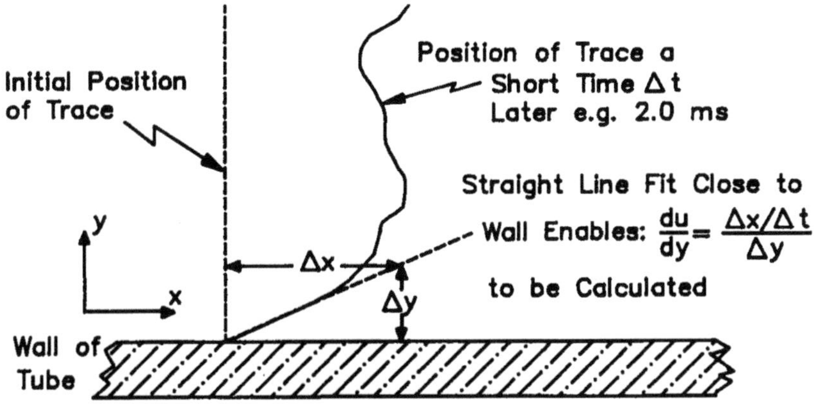

Figure 6. Illustration of the method used for determining wall shear stress.[2]

Photochromic grids are analyzed in a similar fashion, except that considerably more information can be derived from the flow field. A procedure for determining the vorticity and velocity fields from photochromic grids is described by Couch et al.[10]

III. APPLICATION OF THE PHOTOCHROMIC TRACER TECHNIQUE TO HEMODYNAMICS

Hemodynamic factors are known to affect arterial function and remodeling and are suspected of contributing to the onset of disease. In particular, shear stress acting on the arterial wall appears to play a significant role in physiological and pathological remodeling. Several focal disorders are suspected of being initiated by unusual or non-physiological levels of shearing. Atherosclerosis, anastomotic intimal hyperplasia leading to bypass graft occlusion, and poststenotic dilatation leading to aneurysm formation, are three notable examples of disorders that may be induced by shear stress.

To develop a clear understanding of the mechanisms of shear-induced arterial remodeling and disease, it is necessary to determine the shear stress acting on the endothelium, to identify the mechanisms by which endothelial cells sense and transduce changes in the shear stress, and to determine the response of the underlying tissue.

The following are descriptions of ongoing *in vitro* studies that focus on accurately estimating the wall shear stress in regions where focal disorders tend to occur and identifying flow structures or characteristics that may contribute to the development of specific vascular disorders. The complex flow fields found at these sites, make flow visualization and measurement difficult and render conventional *in vitro* techniques inaccurate. In each of these studies, the photochromic tracer technique was the primary investigative tool and proved to be well-suited despite the complexity of the flow fields at these sites.

A. Flow through Stenosed Arteries

Background and Motivation

The detection and quantitative assessment of constricted or stenosed arteries is of great importance, since it allows for the possibility of appropriate intervention early in the progression of disease. It is well-known that for a reduction of 80% in the arterial cross-section, the mean volumetric flow is reduced significantly, and turbulence may be triggered. Stenoses of this severity can be detected non-invasively by Doppler ultrasound spectral analysis.[11,12] Less severe stenoses are not as easily detected non-invasively, especially in the case of minor stenoses which are characterized by an area reduction of 50% or less. This inability can be partially attributed to the lack of detailed information on the hemodynamics of pulsatile flow through such a stenosed vessel. An improved understanding would allow for better interpretation of Doppler ultrasound spectra, and could provide for earlier and more reliable detection.

To this end, the photochromic tracer method was used to examine pulsatile flow in vessels with mild to moderate degrees of stenosis.[13] Both axisymmetric and asymmetric stenoses were examined.

Stenosis and Flow Waveform Modeling

Figure 7 depicts the geometry of the stenoses used in these experiments. They were made from transparent acrylic rods (Plexiglas), and glued inside of a Pyrex glass tube. Axisymmetric stenoses of 45%, 65%, and 75%, along with an asymmetric stenosis of 38% of the original cross-sectional area were used. The fact that mild stenoses tend to be asymmetric *in vivo* provided the motivation for investigating the asymmetric stenosis. The flow section included a 1.2 m long glass tube with an internal diameter of approximately 5.0 mm. The test section containing the stenosis was located 160 internal vessel diameters downstream of the entrance. A flow waveform was generated by combining the output of a 2.9 Hz piston pump with the output of a steady flow gear pump. The test fluid consisted of deoderized kerosene (Shell-Sol 715) with a density of 0.755 g/cm^3 and a kinematic viscosity of 1.43 cP at 20 °C, to which a photochromic indicator, 1′,3′,3′ trimethylindoline-6-nitrobenzospiropyran (TNSB), was added.

Flow Separation in a Mild Stenosis

A mild stenosis is defined to be one that triggers localized disturbances or unsteadiness in isolated regions of the vessel and not disturbances that affect the

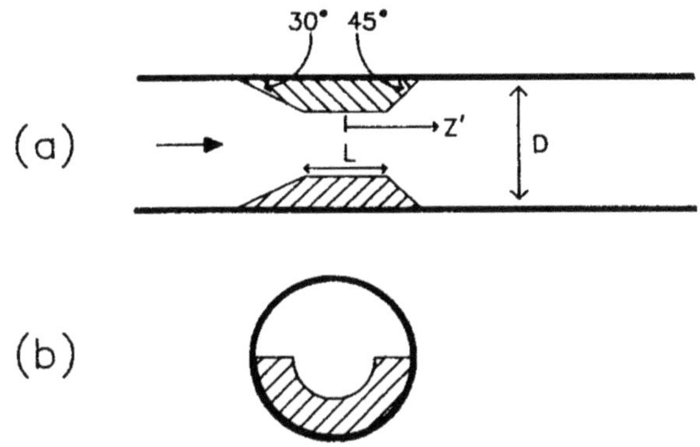

Figure 7. (a) Geometry of axisymmetric, and (b) asymmetric stenoses used in experiments. For both, $L = 1.5$ mm and $D = 5.0$ mm. The normalized distance from the center of the stenosis is given by $Z = Z'/D$.[13]

flow in the entire vessel cross-section. Both the 38% asymmetric and the 45% axisymmetric stenoses fall into this category.

Photographic recordings of the traces were made at 20 ms intervals throughout the flow cycle. These recordings were used to measure the centerline velocity waveforms at three sites. In the case of the 45% axisymmetric stenoses (Figure 8b), the sinusoidal nature of the flow remained apparent downstream. With the 38% asymmetric stenosis, the sinusoidal nature of the flow was only apparent near to the stenosis. Further downstream, the asymmetry of the stenosis affected the flow significantly, as expected. In the vicinity of the reattachment point and during the deceleration phase of the flow cycle, substantial fluctuations of the axial velocity were observed over the vessel cross-section. The variation of the axial velocity with downstream position was greater than seen with the axisymmetric stenosis, indicating significant non-axial components of the velocity.

Flow separation downstream of the stenosis was easily visualized by examining the distortion of the traces over the flow cycle and using the position of the reattachment point to characterize the length of the separation region. The presence of multiple traces helps to identify the position of the reattachment point by "bracketing" its position for a given interval. In this manner, the separation point was seen to oscillate over a large distance during the flow cycle. In the case of the 45% axisymmetric stenosis, as the flow began to accelerate, the reattachment point moved rapidly upstream from a position approximately 4 vessel diameters from the stenosis, until the separation region vanished. Late in the acceleration phase and throughout deceleration, the separation region reappeared, and the reattachment point reversed direction and slowly returned downstream to its initial position just prior to the start of a new acceleration phase. During deceleration, the thickness of the separation region gradually increased.

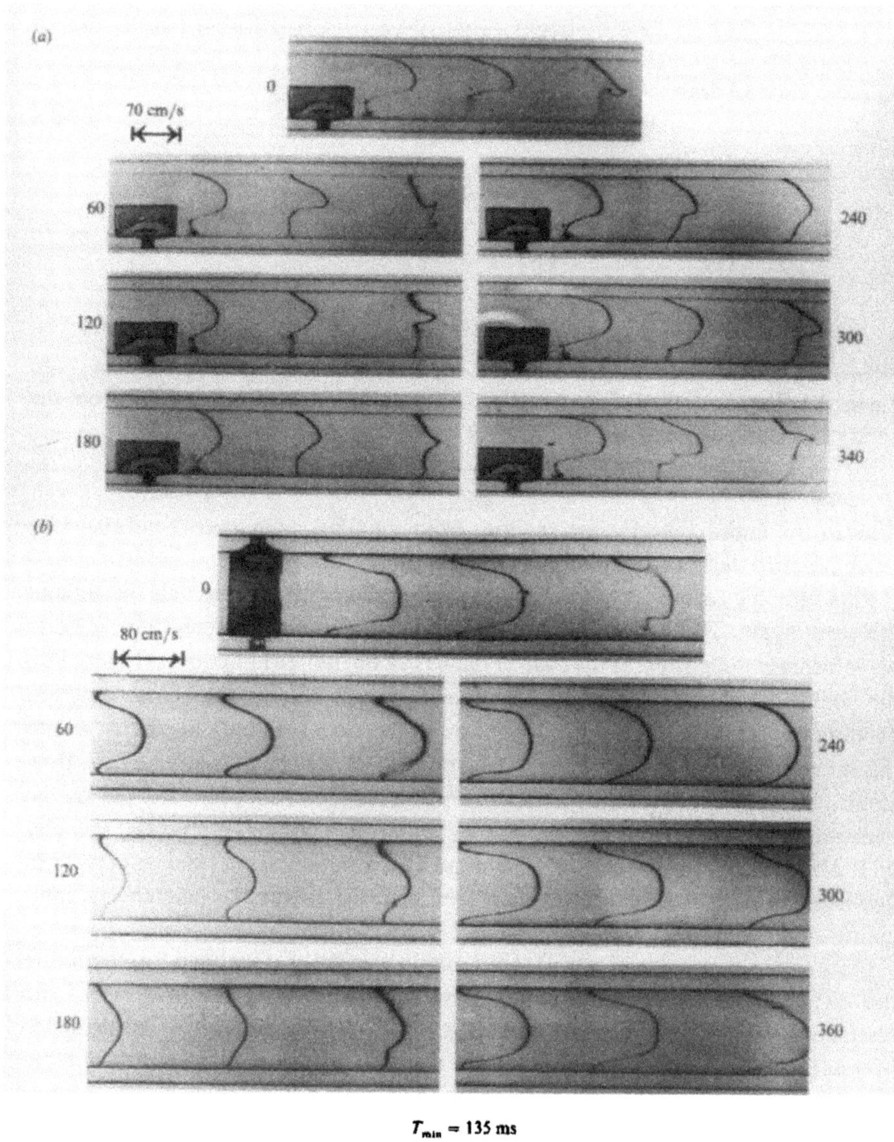

Figure 8. Photographs selected from the flow cycle at the indicated times for (a) the 38% asymmetric stenosis, and (b) the 45% axisymmetric stenosis. The times given are in milliseconds.[13]

As expected, a different flow separation pattern was observed for the asymmetric 38% stenosis (Figure 8a). Unlike the 45% axisymmetric stenosis, separation was observed throughout the flow cycle along the stenosed side of the vessel. The length of the separation region increased from 0.7 to 4.0 vessel diameters during the latter part of the acceleration phase. At the end of deceleration, the reattachment point returned rapidly to the furthest upstream position. Further downstream, approximately 2 vessel diameters from the stenosis, strong non-cyclical fluctuations in the separation boundary were seen during deceleration and into early acceleration. These non-periodic effects were probably caused by waves, vortices, and other flow structures generated in the region of the reattachment point.

Flow Separation in a Moderate Stenosis

A moderate stenosis is defined as one for which turbulence is generated across the entire vessel, but does not result in a physiologically significant reduction in flow. The 65 and 75% stenoses fall into this category. In Figures 9a and 9b the photochromic traces clearly illustrate the transition to turbulence. In both stenoses the jet region extended approximately 3 vessel diameters downstream. Within this region, the flow was essentially two dimensional and stable, as evidenced by the nearly sinusoidal variation of the centerline velocities. The separation zone lying within this stable region behaved in a similar manner to that observed for the mild axisymmetric stenosis. However, the 75% stenosis was characterized by the presence of a permanent separation zone, unlike the transient zone seen with the 45 and 65% stenoses.

The motion of the separation boundary for the 75% stenosis was visualized by generating dye streaks within the separation region (Figure 10). These dye streaks were created by ultraviolet light from a 100W mercury lamp (Phillips CS 100 W) focused onto the tube wall. A 16 mm movie camera (Locam) was used to record the motion of the streaks at a rate of 400 frames per second. Figure 11 presents a schematic representation of the observed motion of the separation boundary.

The first frame illustrates the instant in the flow cycle just prior to minimum flow when turbulence due to the disintegration of waves and streamwise vortices in the region of the reattachment point vanishes. Past the reattachment point the flow was still highly unstable throughout much of the flow cycle. As the flow accelerated the separation zone boundary developed a wave that started at the stenosis and traveled downstream, reducing the overall thickness of the zone. As the cycle approached maximum acceleration, the reattachment point moved upstream restricting the separation zone to approximately 2.8 vessel diameters from the stenosis. Simultaneously with this event was the generation of waves and streamwise vortices. As these structures were transported downstream they triggered a transition to turbulence in the vicinity of the reattachment point. This turbulence persisted well into the deacceleration phase of the flow cycle. In fact, there was insufficient time for

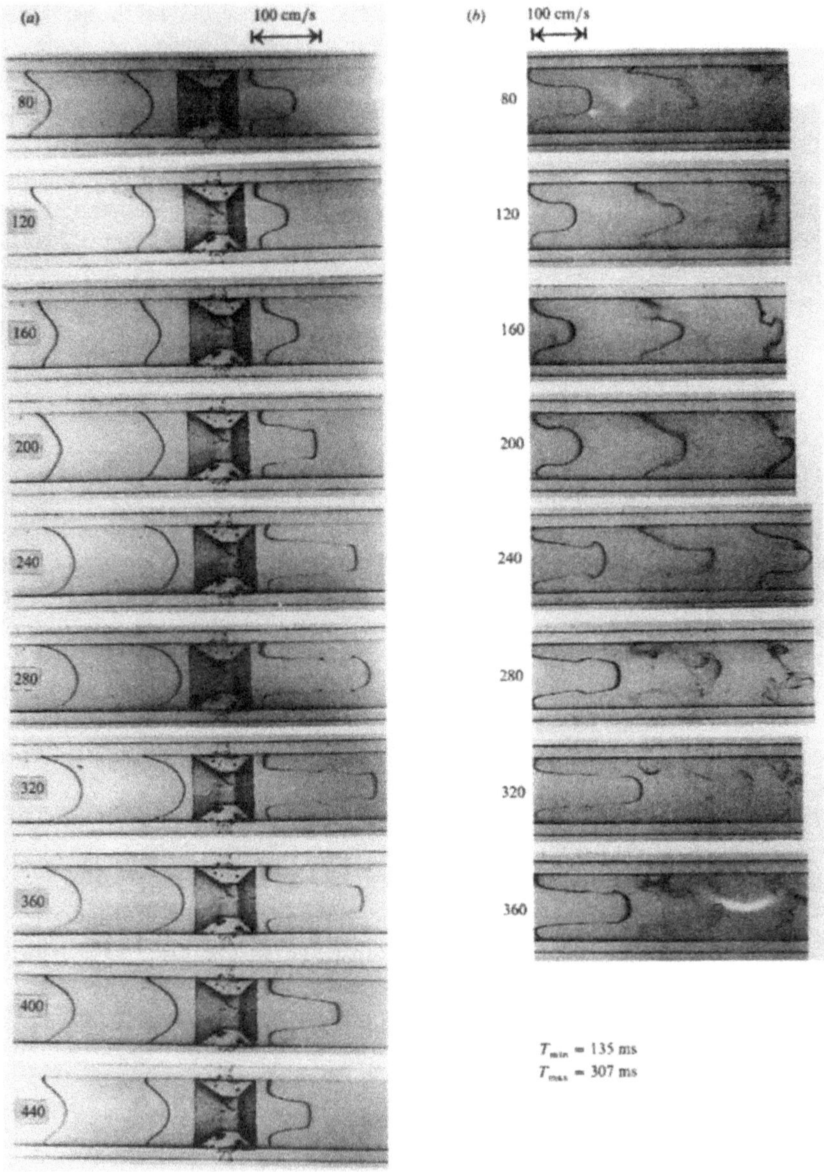

Figure 9. Photographs selected from the flow cycle at 40 ms intervals for the 75% axisymmetric stenosis (a) upstream and in the immediate poststenotic region, and (b) downstream of the poststenotic region.[13]

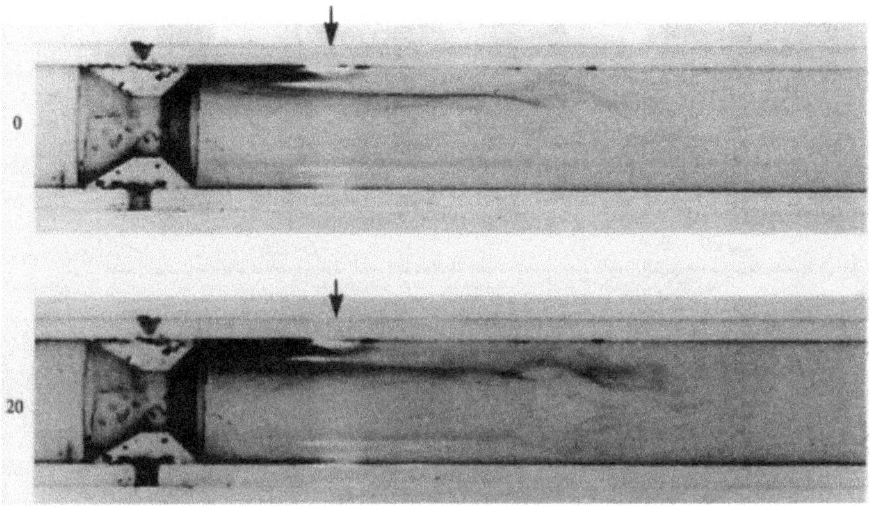

Figure 10. Photographs of the 75% axisymmetric stenosis illustrating the behavior of the flow separation region. The first frame was recorded 45 ms after peak flow, with the second frame taken 20 ms later. The arrows indicate the position of the illumination from the mecury lamp.[13]

complete flow relaminarization within the acceleration phase of the flow cycle, unlike the 65% stenosis.

The poststenotic flow field can be divided into four zones for any given phase of the flow cycle. Although the velocity field in each zone is time-dependent, to simplify the description, the flow cycle is divided into two phases, generation of turbulence and relaminarization. The first zone is that of the stable jet. In this zone, the flow was three dimensional with circumferential motion being significant during the generation of turbulence. During this phase, three toroidal vortices appearing as typical Kelvin-Helmholtz vortex roll-ups were seen and sometimes accompanied by wave motions.

In the second zone, the flow remained in a transitional state and lacked symmetry. This lack of symmetry was more pronounced during the turbulent phase due to the strong interaction among the jet flow, the vortices, and the waves. This led to the formation of the third zone, wherein the coherent structures break down as a result of this stronger interaction. The turbulent region involved the entire vessel cross-section and extended downstream for approximately three diameters. In the final zone, dissipation of the turbulence was seen during the entire flow cycle.

Prior studies of pulsatile flow through a mild stenosis have suggested that the vortical structures found downstream were shed at the stenosis early in the acceleration phase of the flow cycle. These suggestions were based on flow visualization with hydrogen bubbles and from fluid velocity measurement with LDA.[14,15]

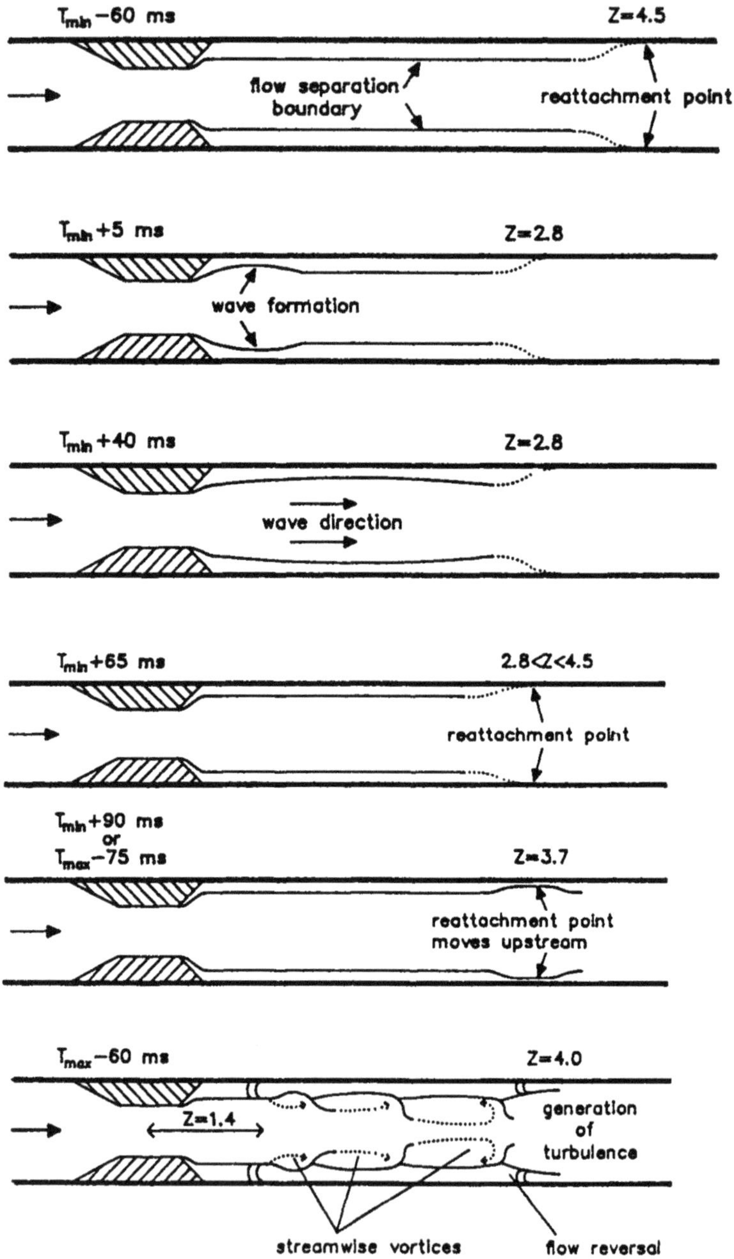

Figure 11. Schematic illustration of the behavior of the flow separation region during the flow cycle in the 75% axisymmetric stenosis.[13]

However, flow visualization and measurement with the photochromic tracer technique has provided evidence to the contrary. These structures were formed during early deceleration in the region of either high shearing or inflectional velocity profiles. It would appear that the flow structures observed in prior studies are remnants of the original structures formed in the deceleration phase of the previous flow cycle.

The shear stress upstream of the stenoses were seen to range from –1.7 to 10.0 dynes/cm^2 with a mean of 4.2 dynes/cm^2, and show excellent agreement with the theoretical predictions based on the Womersley rigid-tube model. Downstream of the stenoses, the wall shear stress measured depended on stenosis severity and axial position, and in the case of the asymmetric stenosis, angular position as well. It is of interest to note that a large positive wall shear stress was observed on the stenosis free wall at the exit of the asymmetric 38% stenosis. This can be attributed to the flow being skewed in the direction opposite to the stenosis.

These results from the photochromic studies provide evidence of the difficulty of using LDA for measuring wall shear stress. Given the large fluctuations in the thickness of the viscous sublayer observed with the photochromic technique it is difficult, if not impossible, to ensure that the LDA sample volume is positioned within this sublayer. In addition, significant error can arise from attempting to measure a non-periodic velocity gradient with a single-point velocity measurement technique. Thus, the photochromic tracer technique provides a more detailed and accurate description of the wall shear stress field than was previously possible, that is, obtained by the use of LDA.[14,15]

B. Poststenotic Dilatation

Background and Motivation

Poststenotic dilatation (PSD) is an example of an arterial disorder considered to arise as a result of non-physiological or unusual hemodynamics. It manifests itself downstream of a stenosis as an expansion of the arterial wall. This condition is seen downstream of various conditions such as subclavian artery compression caused by thoracic outlet syndrome, aortic and arterial coarctations, and aortic valve stenosis.

PSD has been investigated for nearly one and one-half centuries, yet the mechanisms responsible for its development are still poorly understood. Determining the mechanisms responsible for poststenotic dilatation is particularly important for two reasons. First, it may provide an understanding of the pathogenesis of aneurysms, a common consequence of PSD. Second, it may provide substantial insights into the basic mechanisms responsible for arterial remodeling. Many explanations of its etiology have been offered, but few have been supported by experimental evidence. PSD was originally thought to arise from an abnormally high transmural pressure. A similar explanation suggested that the kinetic energy of the jet flow from the stenosis was converted into a lateral wall pressure. These explanations for the observed dilatation are not valid as it is clear that the transmural pressure distal to the stenosis is always less than the proximal transmural pressure. Robicsek et al.[16] and Hugh and Fox[17] both suggested cavitation as the cause. However, the pressure gradients required for cavitation are extremely unrealistic in a physiological setting.

Roach[18,19] and Boughner and Roach[20] suggested that vibration arising from the poststenotic turbulence may induce dilatation through structural fatigue of the elastin layers of the arterial wall, however, Gow et al.[21] have recently shown that vibration alone does not produce arterial dilatation.

Originally, arterial wall shear stress was thought to be unimportant in the etiology of PSD, a conclusion largely attributable to inaccurate assessments of the shear stress in the poststenotic region. Rodbard et al.[22] argued that if endothelial cells are sensitive to shear stress, they could initiate arterial dilatation. They suggested that a localized region of abnormally high shear stress is produced downstream of the stenosis as a result of the turbulent flow, and that it is the response of the endothelial cells to this region that initiates PSD. However, they made no measurements of the wall shear stress. Instead, they based their arguments on the fact that under identical conditions of steady flow through a straight tube, the wall shear stress would be much higher for fully developed turbulent flow than for laminar steady flow.

Wall Shear Stress Measurement in the Poststenotic Region

By determining the instantaneous wall shear stress in the poststenotic region with the photochromic method as described in this section, Ojha et al.[2] showed that the transition to turbulence not only caused the mean wall shear stress to increase but produced a strong fluctating component as well. The authors suggest that the development of PSD may be related to these large non-periodic fluctuations of the wall shear stress around the reattachment point.

The primary objective of this *in vitro* study[2] was to examine the temporal and spatial variations of wall shear stress in a model of an artery with an axisymmetric 65% cross-sectional area reduction stenosis in a vessel of 5 mm internal diameter. The flow waveform consisted of a 2.9 Hz sinusoidal component offset by a steady flow. Based on the vessel diameter and the average axial velocity, the mean and the peak-to-peak Reynolds numbers were 500 and 940, respectively. The Womersley parameter, α was 7.9. These parameters were chosen to approximate the conditions seen in medium-sized human arteries during the systolic phase of the cardiac cycle.

A lense array was used to focus the UV beam from the nitrogen laser, and produce seven dye traces simultaneously in the vessel. From these traces (Figure 12), the flow could be visualized and the velocity and wall shear stress calculated in the manner previously described. Transition to turbulence was seen to occur in the interval shortly before and after peak flow. This transition was initiated by the development of a high shear layer between the central jet flow and the separation zone. As the shear rate exceeded a critical value, the layer "rolled-up" and formed

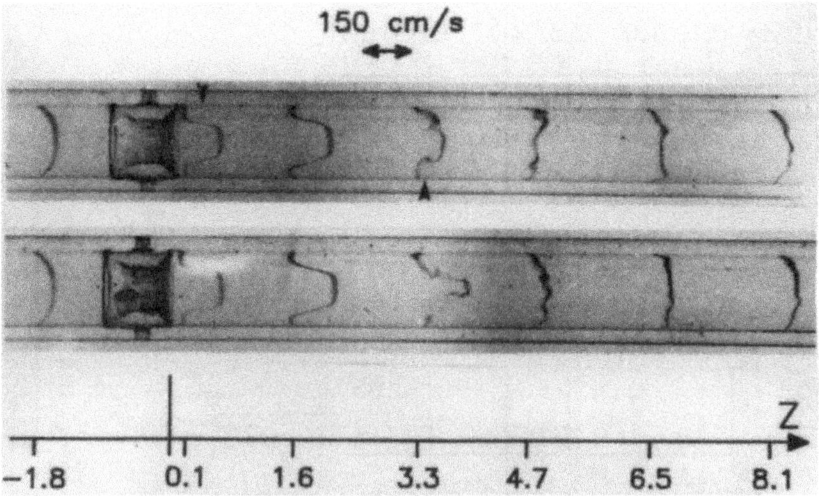

Figure 12. Photograph of the dye traces 2.1 ms after formation in the 65% axisymmetric stenosis. The frames were recorded at 3 and 13 ms after peak flow. The arrow nearest to the stenosis indicates the position of the high shear layer while the second arrow shows the position of the vortex that interacts with the jet flow and leads to turbulence.[2]

vortices. As described previously, these vortices were shed approximately three vessel diameters downstream of the stenosis and resulted in turbulence.

Results of the wall shear stress measurement are presented in Figure 13 for the turbulent phase of the flow cycle. Shortly after peak flow, the mean wall shear stress in the turbulent region downstream was significantly greater than upstream of the stenosis (10 vs. 23 dynes/cm^2). When specific locations were examined, significant fluctuations of the wall shear stress were observed. Peak values were greater than eight times the maximum shear stress observed upstream of the stenosis.

Angiographic Study of Poststenotic Dilatation

In addition, an attempt was made to establish the clinical significance of these hemodynamic measurements. Seven angiograms of patients with PSD arising from thoracic outlet syndrome were selected to avoid any difficulty in interpretation that is inherent with atherosclerotic arteries. The maximum diameter, its location with respect to the downstream edge of the constriction, and the overall length of the PSD zone were measured from each of the angiograms. For comparison, each parameter was normalized by the mean proximal arterial diameter.

From the seven cases studied, the location of maximum dilatation was 2.0 ± 0.3 vessel diameters from the downstream edge of the stenosis. The PSD zone extended

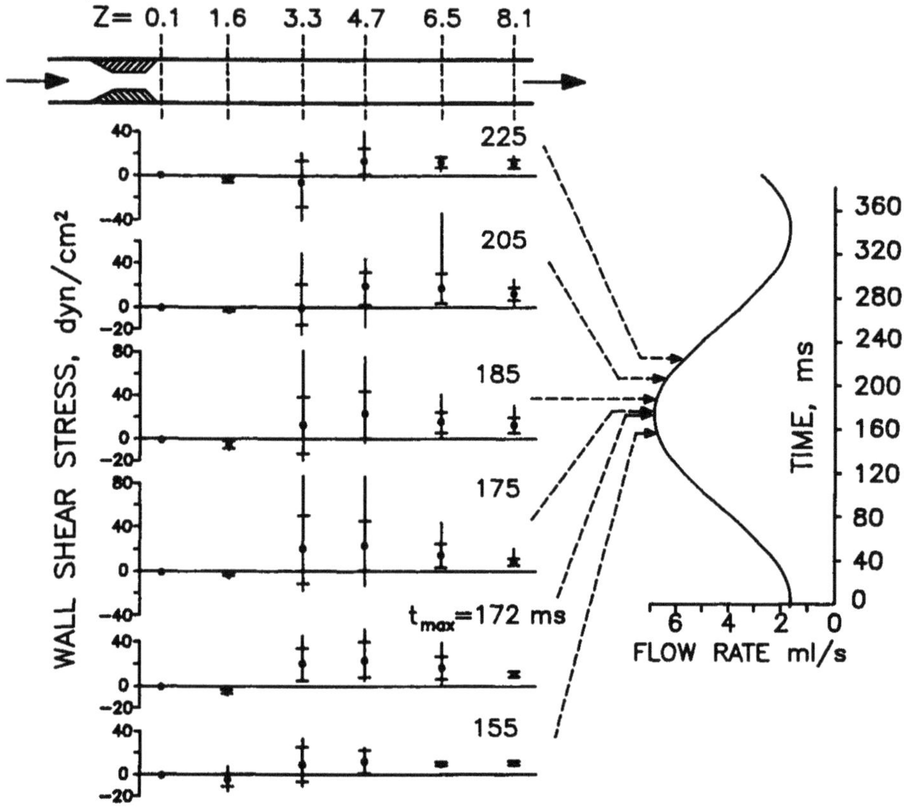

Figure 13. The variation of the wall shear stress downstream of the 65% axisymmetric stenosis during the portion of the flow cycle that turbulence is present. Each graph shows the mean, standard deviation, and the range of the shear stress recorded at each site and the time at which the measurements were made.[2]

from just beyond the stenosis edge to 4.4 ± 1.2 vessel diameters, while the maximum diameter of the dilated region was 1.6 ± 0.2 times that of the proximal diameter.[2]

Is Wall Shear Stress a Factor in Poststenotic Dilatation?

In vitro measurements with the photochromic tracer technique in this and previous studies clearly confirm the presence of significant non-periodic fluctuations (over positive and negative values with a zero or negligible mean value) in the wall shear stress between 1.6 and 3.3 vessel diameters downstream of stenoses. Further downstream, these fluctuations continue albeit with a non-zero mean value.

The wall shear stress value measured with the photochromic technique represents only the axial component, as the total wall shear stress is given by the vector sum of the axial and circumferential components. In the presence of a significant non-axial flow component, as is seen distal to a stenosis, the measured wall shear stress will be an underestimate of the total value, especially in the region of the reattachment point.

The angiographic studies indicated that the site of maximum dilatation occurred approximately 2 vessel diameters downstream of the stenosis. Although the severity of the constrictions could not be determined with any accuracy, and given that the PSD region will tend to grow as the artery degenerates, the *in vitro* studies demonstrate that the site of maximum shear stress fluctuation lies within the observed PSD zone, suggesting that the hemodynamic factor that best correlates with PSD is the wall shear stress fluctuation.

A large volume of work has demonstrated that hemodynamic factors, particularly wall shear stress, can elicit major cellular and structural responses,[23] thus lending support to this hypothesis.

Is Turbulence a Factor in Poststenotic Dilatation?

In a subsequent study,[24] an attempt was made to relate the local hemodynamic conditions to the development of PSD in the common carotid arteries of rabbits. As in the previous studies, the photochromic tracer technique was used for *in vitro* flow visualization and measurement. Flow conditions were selected to simulate the common carotid artery of a rabbit, the mean and peak-to-peak Reynolds numbers were 140 and 240, respectively, and the Womersley parameter α was 3.1. Three different models of stenosis geometry were investigated. The first was a stenosis with a 55% diameter reduction and a throat length of 0.2 vessel diameters, while the remaining were 70% diameter reductions with throat lengths of 0.3 and 4.0 vessel diameters.

For all of the stenosis models, the flow disturbances generated in the previous cycle were observed to dissipate completely during the new acceleration phase. Flow separation was seen throughout most of the flow cycle, and, in the vicinity of the reattachment point, a relatively smooth transition in the velocity field could be observed during the acceleration phase. During deceleration, some unsteadiness in the velocity field could be seen near to the reattachment point. However, except for the 70% stenosis, no turbulence was observed.

With the 70% stenosis, transition to turbulence was triggered near to peak flow approximately 10 vessel diameters downstream of the stenosis. This turbulent region or plug extended upstream to 6 vessel diameters distal to the stenosis as the flow decelerated. The shorter throat length stenosis generated more intense turbulence that extended further upstream compared to the longer stenosis.

In a complementary *in vivo* study, stenoses were surgically produced in male New Zealand White rabbits. Three different geometries were used: a short 50% diameter reduction stenosis and a long and a short 60% stenosis. After three weeks, the rabbits were sacrificed and casts of the carotid arteries were made. Luminal diameter measurements were made from these arterial casts. The average upstream lumen diameter was determined to be 2.09 ± 0.03 mm. Significant PSD appeared downstream of both long and short 60% stenoses. For the short stenosis, maximum dilatation of 49.0 ± 3.9% was seen 5.91 ± 0.67 mm downstream. In the case of the long stenosis, the maximum dilatation of 43.6 ± 3.6% occurred 5.92 ± 0.22 mm downstream. In contrast, the PSD seen distal to the short 50% stenosis was substantially less severe. The site of maximum dilatation of 22.1 ± 3.7% was seen 4.94 ± 0.39 mm downstream.

Clearly, large scale dilatation characterized by a 40–50% diameter increase could be produced consistently in 2 mm diameter rabbit common carotid arteries with 60% stenoses. Variation of stenosis throat length stenosis (1 mm versus 6 mm) yielded little difference in the severity of dilatation. Stenosis severity (50 vs. 60% diameter reduction) was the most significant determinant of the ultimate degree of PSD.

The *in vitro* flow studies, which involved reproducing *in vivo* geometry and pulsatile flow conditions, confirmed that turbulence was either absent or limited to a narrow region much further downstream than the observed site of PSD. Thus, turbulence is unlikely to be responsible for the initiation of PSD.

A more likely scenario for the initiation of PSD involves an endothelial response to abnormal shearing. Endothelial cells, within the stenosis throat or in the downstream region in the vicinity of the reattachment point, may release vasodilators in response to extreme shear stress. These agents are trapped in the separation zone and transferred to the vessel wall in the PSD region via the recirculating flow. In a preliminary investigation of this hypothesis, endothelium-derived relaxing factor (EDRF) appeared not to be an agent: inhibition of EDRF synthesis with the arginine analogue N^G-nitro-L-arginine-methyl ester (L-NAME) did not affect the formation of PSD.

C. Distal Anastomotic Intimal Hyperplasia

Vascular Reconstruction with Bypass Grafting

Each year, more than half a million patients in Canada and the United States undergo treatment for vascular disease with most requiring some form of surgical intervention. These interventions often include vascular reconstruction with natural or synthetic graft material and are divided approximately equally between the coronary circulation system and the peripheral vascular system. Consequently, arterial bypass grafting has become one of the most commonly performed surgical procedures in the developed world, particularly in North America.

Typically, the procedure involves bypassing the affected or obstructed portion of an artery in an attempt to improve the circulation of blood. The procedure consists of suturing or anastomosing the inlet of the bypass vessel upstream of the affected region of the host vessel, and the outlet of the bypass vessel downstream of the affected region. A commonly used bypass graft configuration consists of a side-to-end anastomosis at the inlet of the graft and an end-to-side anastomosis at the outlet.

The immediate results of vascular reconstruction are excellent as evidenced by acceptable function, as well as low morbidity and mortality rates. However, in the long term, these procedures have been subject to complications that often result in failure of the reconstruction.

The Role of Intimal Hyperplasia in Bypass Graft Failure

The primary causes of vascular graft failure are thrombosis, infection, degradation, and intimal hyperplasia. Postoperative failure within hours or days is attributed to thrombus formation resulting from insufficient levels of blood flow, infection, or imperfections arising from surgical injury or poor graft incorporation. Until recently, the mechanisms responsible for late failure occurring several months or years postoperatively were not clearly understood. It is now recognized that the development of intimal hyperplasia at the distal anastomosis of a bypass graft is the major obstacle to long-term patency. This mode of failure is referred to as distal anastomotic intimal hyperplasia.

Echave et al.[25] estimated that approximately 30–50% of late graft failures are attributable to the development of intimal hyperplasia. Although thickening of the intima is a natural response of the vessel wall designed to promote healing and to provide a mechanism by which the vessel adapts to changes in the flow of blood, it can be responsible for graft failure due to complete occlusion. In a graft this response may continued unchecked into the lumen of the vessel. Subsequently, thrombus formation may obstruct the lumen and interfere with the function of the graft.

Factors Contributing to the Development of Intimal Hyperplasia

The mechanisms responsible for the development of intimal hyperplasia are only partially understood despite extensive study. It is hypothesized that the intimal hyperplasic response is triggered by injury to the arterial wall, particularly to the endothelial layer. This initiates the replication of smooth muscle cells within the media followed by migration across the internal elastic lamina into the intima. Once the smooth muscle cells reach the intima, they continue to proliferate and form an extracellular matrix composed mostly of collagen. Several factors may contribute to the development of intimal hyperplasia: Surgical injury, surface thrombogenesis, compliance mismatch, and hemodynamic forces have all been implicated.[1,26,27]

Histological studies have shown that the development of intimal hyperplasia at the anastomotic suture lines of a bypass graft is a vascular response designed to heal the surgical injury. Surgical injury can also result in thrombus formation at the anastomotic suture lines, and this may contribute to acute graft thrombosis early in the postoperative period. However, there is no evidence that non-occlusive thrombus promotes intimal hyperplasia. In addition, it is unlikely that intimal hyperplasia due solely to surgical injury is responsible for late graft occlusion.

The mismatch of elastic properties between the graft vessel and host artery is considered by many to be an important factor in intimal hyperplasia. Madras et al.[28] suggested that any difference in compliance of the materials at the suture line would cause energy dissipation leading to the development of cyclical stresses in the arterial wall. Previously, Leung et al.[29] demonstrated that cyclical stresses are capable of inducing hyperplasia of smooth muscle cells. Subsequently, compliance mismatch has become an important area of research in the failure of vascular prostheses.

Many studies have examined the role of hemodynamics in arterial function and remodeling, and vascular disease. It is recognized that the arterial wall can adapt to local changes in the flow of blood, and that this adaptation is facilitated by the ability of endothelial cells to sense and respond to the viscous shearing force or shear stress experienced by the vessel wall. Even atherosclerosis itself is suspected of being caused by a shear-induced response of the endothelium.[30]

Some of the earliest evidence that hemodynamics might contribute significantly to the development of intimal hyperplasia was provided by Imparato et al.[26] who found that intimal hyperplasia tended to develop in regions exposed to flow with unusually low or high velocities. LoGerfo et al.[31] reported that, although the geometry of the inlet side-to-end anastomosis and outlet end-to-side anastomosis of a bypass graft are similar, the degree of intimal hyperplasia is significantly greater at the outlet. Sottiurai et al.[32] and Bassiouny et al.[33] showed that, at the outlet of the bypass, intimal hyperplasia develops primarily at the heel and toe of the anastomosis, and on the bed opposite to the anastomosis (refer to Figure 14 for the definition of these locations).

In a subsequent examination, Bassiouny et al.[33] commented on the morphology and histology of intimal hyperplasia on the bed of an anastomosis. Although the cellular and extracellular composition was quite similar to that seen at the suture line, the extent of hyperplasia did not depend on the choice of graft material. This observation contradicts the suggestion that compliance mismatch plays a significant role in the development of intimal hyperplasia. Furthermore, the proposal of LoGerfo et al.[34] that platelet activation by the material of the vascular prosthesis is responsible for intimal hyperplasia is also refuted.

Bassiouny et al.[33] performed complementary *in vivo* and *in vitro* studies which indicated that the distribution of intimal hyperplasia on the bed appeared to correlate with the position of the stagnation point of the flow entering the host artery, particularly around peak flow. They hypothesized that low and perhaps oscillating wall shear stress was the factor triggering hyperplasia. However, this and similar hypotheses made by others were all based on questionable wall shear stress characterization.

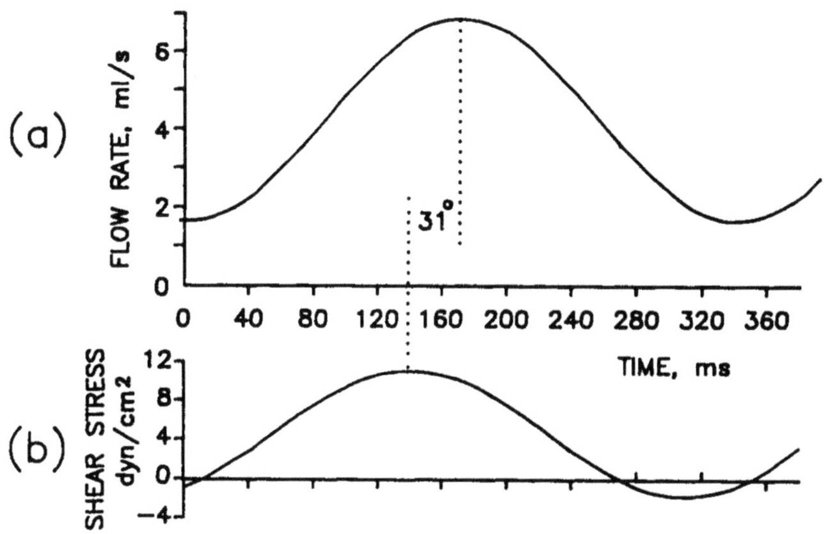

Figure 14. (a) 2.9 Hz sinuosoidal flow waveform offset by a steady flow, and (b) the predicted wall shear stress for a straight tube based on the theory due to Womersley.[35]

Does Wall Shear Stress Contribute to the Development of Intimal Hyperplasia?

Nearly all studies on the effects of hemodynamics in a bypass graft have focused on the outflow end since failure is more likely to occur at this point. Using the photochromic technique, Ojha et al.[35,3,1] has examined the hemodynamics in both the inflow side-to-end and outflow end-to-side anastomoses. Based on clinical evidence that the outlet of the graft is more prone to failure than the inlet, Ojha et al.[3] proposed that the inlet be used as a control in a comparative study. By identifying the differences in the flow fields present at the inlet and outlet, this study attempted to determine the hemodynamic factors contributing to distal anastomotic intimal hyperplasia.

A 2.9 Hz sinusoid waveform offset by a steady flow was used to produce a mean and a peak-to-peak Reynolds number of 575 and 720, respectively (Figure 14). As noted earlier, these flow parameters were chosen to approximate the conditions found in a medium-sized human artery. An anastomosis was fabricated from ultraviolet transparent Plexiglas.® Two cylindrical channels of internal diameter 5.1 mm were bored such that the end of one channel intersected near the middle of the other channel at an angle of 45°. The same model could be used to simulate both the inflow and outflow anastomosis simply by reversing the direction of flow. When simulating the inflow anastomosis, the distal end of the host vessel was fully occluded. In the case of the outflow anastomosis, the proximal end of the host vessel

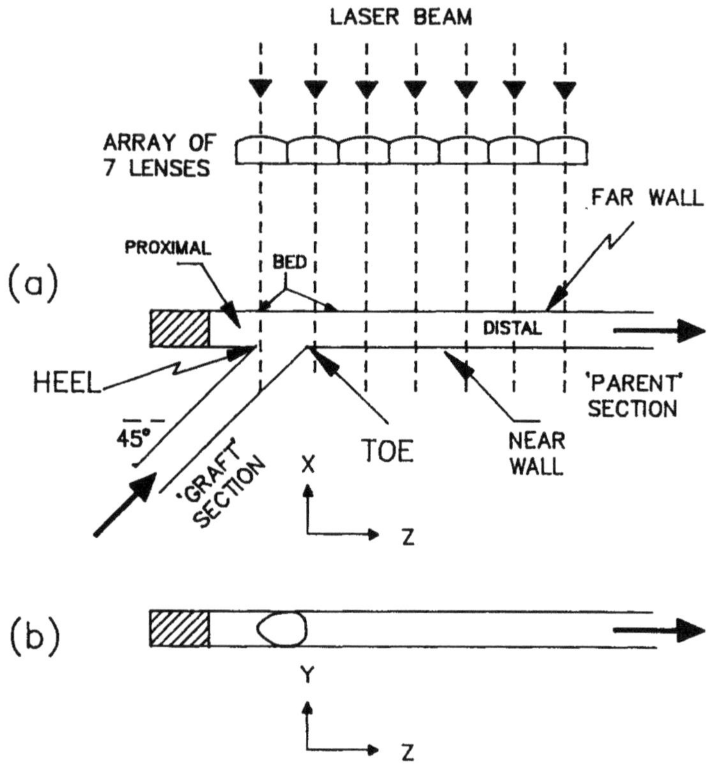

Figure 15. Arrangement of lense array and end-to-side anastomosis model for flow visualization in (a) the x-z plane, and (b) the y-z plane.[35]

was sealed. Figure 15 illustrates the design of the anastomosis model and the experimental setup for recording the photochromic traces in two perpendicular planes. The resulting photographs are shown in Figure 16.

At the proximal junction, as the flow approached the inlet of the graft, it was divided into two components at the heel with the major portion of the flow entering the graft. The remaining fluid recirculated in the occluded distal end of the host vessel before entering the graft. In the distal region a significant helical flow developed near to the heel, and in turn, created a weaker counterrotating vortex closer to the occlusion. Adjacent to the bed of the host vessel across from the junction a region of separation was induced by the change in direction of the flow and the increased cross-sectional area of the vessel. The position and overall length of this separation region fluctuated with the flow cycle. During the early portion of the acceleration phase of the flow cycle, the separation point moved downstream reaching a position approximately one-half of a vessel diameter beyond the toe at midacceleration. During the remainder of the acceleration phase and early in the

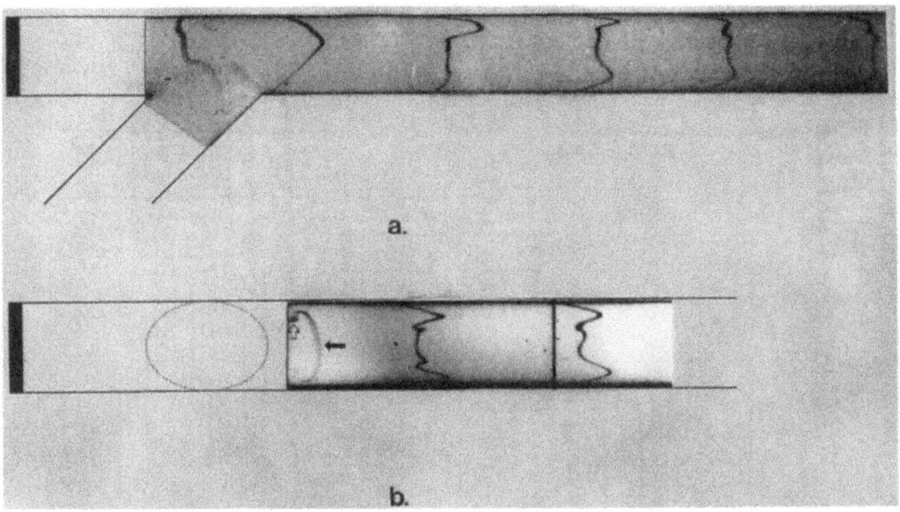

Figure 16. Photographs of the the dye traces in (a) the x-z plane, and (b) the y-z plane.

deceleration phase, the separation point returned upstream, eventually reaching a position on the bed just ahead of the toe.[3]

Figure 17a depicts the variation of the wall shear over the flow cycle at four different sites for the proximal junction: at the heel, at the toe, on the bed opposite to the heel, and on the bed opposite to the toe. In each case, the variation appeared to be sinuosoidal. On the bed across from the heel, the amplitude of the wall shear stress was the least of the measured sites, with a negative magnitude throughout the flow cycle. This is attributable to the permanent presence of separation at this point. On the bed across from the toe the wall shear stress fluctuated over positive and negative values as the separation point repeatedly cycled through this site. At the toe, the wall shear stress remained positive throughout most of the flow cycle with a peak value nearly four times as large as would be measured in a straight tube under similar conditions. The largest variations of wall shear stress were observed at the heel. Here the shear stress was observed to oscillate over a wider range of mostly positive values; the peak value recorded was approximately seven times that for a straight tube.

At the distal or end-to-side junction, as the flow entered the host vessel from the outlet of the graft, it was again divided into two components, a forward component and a relatively minor reverse component, due in this instance to the occlusion at the proximal end. Proximal to the heel, the minor reverse flow led to low velocities in the near wall region, even on the bed opposite to the heel. The forward flow remains coherent until it impacts the bed of the host vessel, at which point the high inertia of the fluid forces the flow to spread circumferentially along the vessel wall.

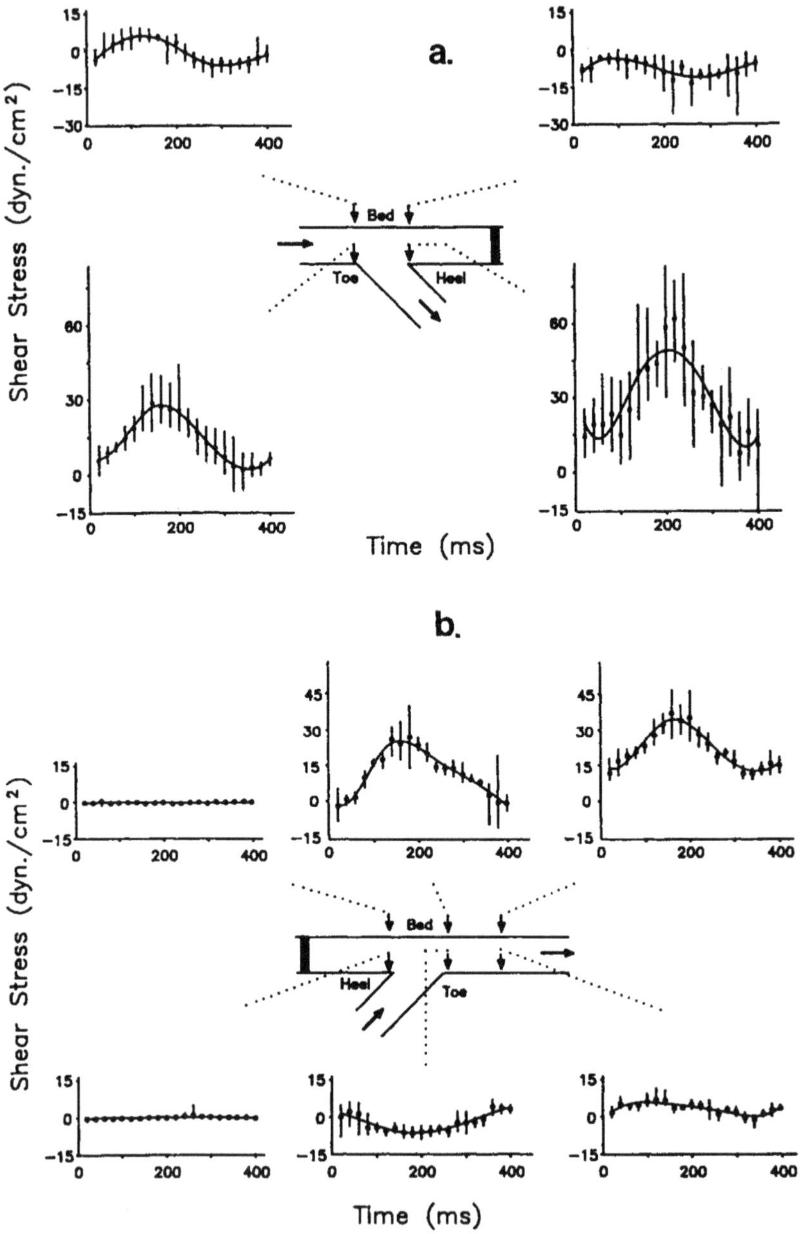

Figure 17. The variation of the wall shear stress over the flow cycle for (a) the inflow side-to-end anastomosis, and (b) the outflow end-to-side anastomosis. Each graph shows the mean and range of the wall shear stress together with a spline profile fitted through the mean values over the flow cycle.

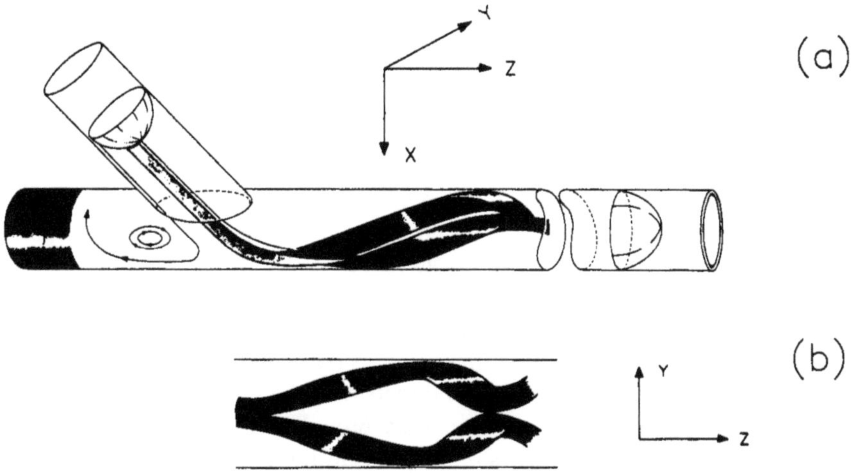

Figure 18. Conceptual sketches of the spiraling path of the high inertia fluid emerging from the graft and impacting on the bed of the end-to-side anastomosis; (a) side view illustrating the development of a double helix flow structure; and (b) plan view.[35]

Centrifugal forces induced by this abrupt change of direction causes the fluid to continue to spiral along the inner wall of the host artery as it travels downstream (Figure 18). This secondary motion was similar to that seen in curved vessels and remained evident for 4–5 vessel diameters downstream.

The three-dimensional nature of this spiraling motion gave rise to significant spatial variations in the wall shear stress within the anastomosis itself and in adjacent regions. Over the entire flow cycle, at and across from the heel on the bed, the near wall velocities were approximately zero, due to the relatively minor reverse flow component. Flow separation was seen just distal to the toe and was induced by the inertia of the fluid entering the host vessel from the graft. Consequently, low fluid velocities were measured near the vessel wall in this region. Across from the separation zone, the inertia of the fluid emerging from the graft and impacting on the bed of the host artery had the opposite effect. As a result, large velocity magnitudes and gradients were measured in the near wall region.

The variations in wall shear stress are reported in Figure 17b for six different sites: at the heel, at the toe, distal to the toe, and the three opposing positions on the bed. At the heel and across from the heel on the bed the wall shear stress remained nearly zero for most of the flow cycle. This is consistent with the relatively minor reverse flow seen in this region. For the remainder of the locations, the wall shear stress varied in an approximately sinusoidal manner. In the separation region near to the toe, the wall shear stress remained negative for much of the flow cycle. Across from the toe on the bed, the wall shear stress revealed the effect of the incoming flow impacting on the bed. At this site, the shear stress reached a peak value of almost three times the value seen in a straight tube under similar flow conditions. Further downstream, the added effect of the spiraling motion only increased the level of shear stress to four times the normal value. Throughout the flow cycle, the wall shear stress remained positive. Across from this site on the bed, the opposite effect was observed. Although the wall shear stress remained positive in value, the peak value was less than one-half of the normal value.

If the wall shear stress measurements are compared to the regions of preferential development of intimal hyperplasia as reported by Sottiurai et al.[32] and Bassiouny et al.,[33] several observations can be made. First, these results lend further support to the hypothesis that low shear stress promotes the development of intimal hyperplasia. Regions of low shear stress are clearly evident at the heel and toe of the outflow anastomosis and correspond well with the reported development of intimal hyperplasia at the suture line. Also providing further credence to this hypothesis is the absence of low shear regions at the corresponding sites in the inflow anastomosis.

Although low shear stress appeared to correlate with the development of intimal hyperplasia initiated by surgical injury in the region of the suture line at the heel and toe of the outflow anastomosis, there is no such correlation evident with the reported distribution of intimal hyperplasia on the bed across from the graft exit. Sottiurai et al.[27,32] proposed that high levels of wall shear stress and even turbulent flow may be responsible for the development of intimal hyperplasia on the bed. The results of these *in vitro* flow studies clearly indicate that neither low nor high values of wall shear stress are responsible.

Is the Dynamic Behavior of Wall Shear Stress Significant?

On the bed across from the outlet of the graft, the location where the flow impacts and divides into forward and reverse components is referred to as the stagnation point. At this point, the wall velocity gradient and shear stress are zero. In a subsequent study performed by Ojha,[1] the stagnation point was found to oscillate over the region on the bed which correlates with the distribution of hyperplasia as reported by Sottiurai et al.[32] and Bassiouny et al.[33]

To examine the oscillation of the stagnation point, the photochromic tracer technique was again used. Ojha[1] reported large spatial and temporal variation of the wall shear stress over the region of oscillation. Although large spatial gradients of the wall shear stress are not physiologically unusual and have been encountered at the heel of the inflow anastomosis and other bifurcations, motion of the stagnation point does appear to be abnormal. These observations lead to the hypothesis that the large temporal gradients of the shear stress created by the motion of the stagnation point may induce intimal hyperplasia by causing an unusual deformation or injury particularly to the endothelial layer.

It was further suggested that the development of hyperplasia is a response of the arterial wall designed to reduce the sharp temporal variations in the shear stress to a physiologically acceptable level. This is achieved by the hyperplasia acting as a blunt flow-divider and restricting the motion of the stagnation point on the bed. The large temporal gradients of the shear stress may induce tissue proliferation by causing injury or unusual deformation. Moreover, it would seem that this remodeling is a response of the tissue in an attempt to normalize the sharp temporal variations in the shear stress over large positive and negative values. Specifically, with the development of intimal hyperplasia, the stagnation point would be confined to the tip of the tissue overgrowth which would lead to major reductions in the temporal changes of the shear stress.

IV. CONCLUSIONS

The photochromic tracer method has proved to be an effective tool for *in vitro* flow visualization and measurement, and as a noninvasive technique capable of full-field velocity and vorticity measurement, it is superior to many conventional techniques. The ability to make detailed velocity measurements has facilitated the investigation of the hemodynamics of flow through a constricted vessel to provide for improved interpretation of the information provided by Doppler ultrasound, an important clinical technique.

When combined with *in vivo* studies, the photochromic tracer method was used to disprove a long-standing theory regarding the development of post-stenotic dilatation. Visualization of the post-stenotic flow field revealed that poststenotic dilatation could occur in the absence of turbulence, a contradiction of the generally accepted notion that turbulence-induced vibration was the cause of the dilatation.

Perhaps the most important advantage offered by the photochromic tracer method is the accuracy with which wall shear stress can be measured. This has permitted the novel observation of the sharp temporal gradients of the shear stress due to the unsteadiness of the stagnation point, a significant factor in the development of intimal hyperplasia in an end-to-side anastomosis. Stagnation point flow under steady or unsteady flow conditions has not been well characterized for this geometry, largely due to the complexity of the flow field. To resolve the small changes in the velocity profile needed to accurately calculate the sharp temporal variation of the shear stress would have required a prohibitively small sample volume for most experimental and even numerical methods.

These applications confirm that the photochromic tracer technique is an ideal tool for investigating the complex flow fields that arise in hemodynamics. The importance of this experimental tool is underscored by the fact that the intricate complexity of these flows have resisted resolution by computational means.

ACKNOWLEDGMENTS

We gratefully acknowledge the financial support of the Heart and Stroke Foundation of Ontario, the Medical Research Council of Canada, and the Whitaker Foundation.

REFERENCES

1. Ojha, M. Spatial and temporal variations of the wall shear stress within an end-to-side arterial anastomosis model. *J. Biomech.* **1993**, *26*, 1377–1388.
2. Ojha, M.; Johnston, K.W.; Cobbold, R.S.C. Evidence of a possible link between poststenotic dilatation and wall shear stress. *J. Vascular Surgery* **1990**, *11*, 127–135.
3. Ojha, M.; Cobbold, R.S.C.; Johnston, K.W. Hemodynamics of a side-to-end proximal arterial anastomosis model. *J. Vascular Surgery* **1993**, *17*, 646–655.
4. Popovich, A.T.; Hummel, R.L. A new method for non-disturbing turbulent flow measurements close to a wall. *Chem. Eng. Sci.* **1967**, *22*, 21–25.
5. Ojha, M. An experimental investigation of pulsatile flow through modelled arterial stenoses. Ph. D. Thesis, University of Toronto, 1987.
6. Ojha, M.; Hummel, R.L.; Cobbold, R.S.C.; Johnston, K.W. Development and evaluation of a high resolution photochromic dye method for pulsatile flow studies. *J. Phys. E: Sci. Instrum.* **1988**, *21*, 998–1004.
7. de Gennes, P.G. Principe de nouvelles mesures sur les ecoulements par echauffement optique localise. *J. Phys.: Lett.* **1977**, *38*, L1–L3.
8. Falco, R.E.; Chu, C.C. Measurement of two-dimensional fluid dynamic quantities using a photochromic grid tracing technique. *Proceedings of SPIE: International Conference on Photomechanics and Speckle Metrology* **1987**, 706–710.
9. Park, H.; Ojha, M.; Hummel, R.L. *J. Phys. E: Sci. Instrum.* **1993**.
10. Couch, G.G.; Park, H.; Ojha, M.; Hummel, R.L. Flow visualization using photochromic grids. *Proceedings of SPIE: 20th International Congress on High Speed Photography and Photonics* **1992**, 678–685.
11. Barnes, R.W.; Bone, G.E.; Reinerson, J.E.; Skymaker, F.E.; Hokenson, D.E. Noninvasive ultrasound carotid arteriography: Prospective validation by contrast arteriography. *Surgery* **1976**, *80*, 328–335.
12. Johnston, K.W.; Baker, W.H.; Burnham, S.J.; Hayes, A.C.; Kupper, C.A.; Poole, M.A. Quantitative analysis of continous-wave Doppler spectral broadening for the diagnosis of carotid disease: Results of a multicenter study. *J. Vascular Surgery* **1986**, *4*, 493–504.
13. Ojha, M.; Cobbold, R.S.C.; Johnston, K.W.; Hummel, R.L. Pulsatile flow through constricted tubes: An experimental investigation using photochromic tracer methods. *J. Fluid Mech.* **1989**, *203*, 173–197.
14. Ahmed, S.; Giddens, D.P. Pulsatile poststenotic flow studies with laser Doppler anemometry. *J. Biomech. Eng.* **1984**, *17*, 695–705.
15. Lieber, B. Order and random structures in pulsatile flow through constricted tubes. Ph.D. Thesis, Georgia Institute of Technology, Atlanta, 1985.
16. Robicsek, F.; Sanger, P.W.; Taylor, F.H.; Magistro, R.; Foti, E. Pathogenesis and significance of post-stenotic dilatation in great vessels. *Annals of Surgery* **1958**, *147*, 835–844.
17. Hugh, A.E.; Fox, J.A. Circulation cavitation: "Bubbles in the blood." *Lancet* **1963**, *2*, 717–718.
18. Roach, M.R. Poststenotic dilatation in arteries. In: *Cardiovascular Fluid Dynamics* (Bergel, D.H., Ed.). Academic Press, San Diego, 1972, pp. 111–139.
19. Roach, M.R. Hemodynamic factors in arterial stenosis and poststenotic dilatation. In: *Hemodynamics in Pathology* (Stebhens, W.F., Ed.). C.C. Thomas Publishers, Springfield, 1979, pp. 439–464.
20. Boughner, D.R.; Roach, M.R. Effect of low frequency vibration on the arterial wall. *Circulation Research* **1971**, *29*, 136–144.
21. Gow, B.S.; Legg, M.J.; Yu, W. Lee, L.L. Does vibration cause poststenotic dilatation of arteries *in vivo* and influence atherogenesis in cholesterol-fed rabbits? *J. Biomech. Eng.* **1992**, *114*, 20–45.
22. Rodbard, S.; Ikeda, K.; Montes, M. An analysis of the mechanisms of poststenotic dilatation. *Angiology* **1967**, *18*, 349–369.

23. Langille, B.L.; Gotlieb, A.I. The structure and function of vascular smooth muscle. In: *Blood Vessel Wall and Thrombosis, Vol. 1* (Machovich, R., Ed.). CRC Press Inc., Boca Raton, 1988, pp. 141–158.
24. Ojha, M.; Langille, B.L. Evidence that turbulence is not the cause of poststenotic dilatation in rabbit carotid arteries. *Arteriosclerosis and Thrombosis* **1993**, *13*, 977–984.
25. Echave, V.; Koornick, A.R.; Haimov, M. Intimal hyperplasia as a complication of the use of polytetrafluoroethylene graft for femoral-popliteal bypass. *Surgery* **1979**, *86*, 791.
26. Imparato, A.M.; Bracco, A.; Kim, G.E.; Zeff, R. Intimal and neointimal fibrous proliferation causing failure of arterial reconstruction. *Surgery* **1972**, *72*, 1007–1017.
27. Sottiurai, V.S.; Lin, S.S.; Feinberg, E.L.; Bringaze, W.L.; Tran, A.T.; Batson, R.C. Distal anastomotic intimal hyperplasia: Biogenesis and etiology. *Eur. J. Vasc. Surg.* **1988**, *2*, 245–256.
28. Madras, P.N.; Ward, C.A.; Johnson, W.R.; Singh, P.I. Anastomotic hyperplasia. *Surgery* **1981**, *90*, 922.
29. Leung, D.M.; Glasgov, S.; Mathews, M.B. Cyclic stretching stimulates synthesis of matrix components by arterial smooth muscle cells *in vitro*. *Science* **1976**, *191*, 575–577.
30. Stary, H.C.; Blankenhorn, D.H.; Chandler, A.B.; Glagov, S.; Insull, W.; Richardson, M.; Rosenfield, M.E.; Schaffer, S.A.; Schwartz, C.A.; Wagner, W.D.; Wissler, R.W. A definition of the intima of human arteries and its atherosclerotic-prone regions: A report from the committee on vascular lesions of the council on atherosclerosis, American Heart Association. *Arteriosclerosis & Thrombosis* **1992**, *12*, 120–134.
31. LoGerfo, F.W.; Quist, W.C.; Nowak, M.D.; Crawshaw, H.M.; Haudenschild, C.C. Downstream anastomotic hyperplasia: A mechanism for failure in Dacron arterial grafts. *Annals of Surgery* **1983**, *197*, 479–483.
32. Sottiurai, V.S.; Yao, J.S.T.; Batson, R.C.; Sue, S.L.; Jones, R.; Nakamura, Y.A. Distal anastomotic intimal hyperplasia: Histopathological character and biogenesis. *Annals of Vascular Surgery* **1989**, *3*, 26–33. Sottiurai, V.S. Biogenesis and etiology of distal anastomotic intimal hyperplasia. *International Angiology* **1990**, *9*, 56–59.
33. Bassiouny, H.S.; White, S.; Glagov, S.; Choi, E.; Giddens, D.P.; Zarins, C.K. Anastomotic intimal hyperplasia: Mechanical injury or flow induced. *J. Vascular Surgery* **1992**, *15*, 708–717.
34. LoGerfo, F.W.; Soncrant, T.; Teel, T.; Dewey, C.F. Boundary layer separation in models of side-to-end arterial anastomoses. *Arch. Surgery* **1979**, *114*, 1369.
35. Ojha, M.; Ethier, C.R.; Johnston, K.W.; Cobbold, R.S.C. Steady and pulsatile flow fields in an end-to-side arterial anastomosis model. *J. Vascular Surgery* **1990**, *12*, 747–753.

THE APPLICATION OF MAGNETIC RESONANCE TO BLOOD FLOW STUDIES

David N. Firmin and Raad H. Mohiaddin

Advances in Hemodynamics and Hemorheology,
Volume 1, pages 145–200.

ISBN: 1-55938-634-7.

I. OVERVIEW OF FLOW IN NUCLEAR MAGNETIC RESONANCE

The relationship between flow and motion and the Nuclear Magnetic Resonance (NMR) signal has been studied since 1948[1] only two years after the first experimental studies of NMR itself.[2,3] This first studies concentrated more on molecular motion than bulk flow; Bloembergen et al.,[1] for example, investigated the effect of molecular motion on the signal line width, while Hahn[4] was the first to note the effect of diffusion. A literature survey on NMR flow reveals that the early research merely set out to analyze the effects of flow and did not detail a particular application. The first to study the effect of coherent flow was Suryan,[5] and applications were soon developed utilizing his finding that flow effectively reduced the relaxation time T_1. There were two main areas of interest: to increase the signal to noise ratio in high resolution NMR spectroscopy, and to help to accurately measure the value of T_1 for various liquids.

The first attempts to measure flow were not until 1959 and interestingly, despite the numerous potential applications, many of these early studies were developed with the measurement of blood flow in mind. In the years that followed a wide variety of flow measurement techniques were developed all of which could be categorized into one of two types:

1. Time-of-flight.
2. Phase shift.

On the development of NMR imaging techniques, methods began to be described for obtaining and presenting flow data in the form of an image. Again the methods could be categorized into time-of-flight and phase shift types, many of which have now been demonstrated and validated both *in vitro* and *in vivo*.

In the past there have been numerous review articles on NMR flow effects and techniques. The first comprehensive work was by Zhernovoi and Latyshev[6] who described the theory of the subject and also included many experimental. Later, an article by Jones and Child[7] also reviewed the subject extremely thoroughly; their work was split into two main sections covering the characteristics of the NMR signal from a flowing liquid and also a number of possible applications. The main concern of the article was, however, with non-imaging studies although many of the techniques described form the basis of more recent flow imaging techniques. The effects of diffusion and flow on the NMR signal in the presence of a magnetic field gradient was dealt with in more detail by Singer[8] in an overview in which the idea of blood flow imaging was also introduced. There have also been a number of recent articles[9,10,11] that have dealt more specifically with flow imaging methods and the effect of flow on the signal intensities of the image.

A. Early Investigations into Effect of Flow on the NMR Signal

Many of the early analytical investigations of NMR flow effects involved work on continuous wave (CW) systems. In such systems the absorption of the radio-frequency (r.f.) radiation is measured, either by sweeping the r.f. through the larmor frequency of the sample, or by sweeping the magnetic field with the frequency fixed. For these CW systems the signal is proportional to the population difference between the allowed energy states. For a stationary sample, this population difference can be altered from its equilibrium value by saturation due to previous r.f. applications at the larmor frequency without a long enough recovery time. For a flowing sample, however, the population can be altered in two further ways: first by the washout of saturated spins and washin of polarized equilibrium spins to the detector, and second, at higher flows, by making use of the washin of unpolarized spins from outside the magnetic field. The former will result in an increase in the population difference and will increase the signal amplitude, while the latter will result in a decrease in the population difference and signal amplitude. The early studies took into account one or other of these effects and sometimes both. Later studies involved pulsed NMR systems, however, on such systems the above factors affect the signal in very much the same way.

B. Non-imaging NMR Flow Measurement

Time-of-flight Methods

There are two basic approaches used in the time-of-flight category of flow measurement techniques: the first is to correlate the variation in signal that results from the washin and washout of different spin populations with the flow rate, and the second is to tag spins by exciting or saturating them and then to follow their motion.

In 1959 two attempts were described to measure flow using the first approach and both of these were motivated by the goal of blood flow measurement. Initially, Bowman and Kudravcev[12] described a number of experimental arrangements by which they observed that the NMR signal was affected by flow. They demonstrated that slight adjustments to the experimental arrangement could completely change the results and suggested that this was due to the signal's dependence on a number of phenomena: relaxation effects, saturation effects, r.f. absorption coefficients, and, in particular, on the geometrical arrangement of the coil. In two of the experiments described the resultant graphs of signal versus flow were totally opposed: in one the signal increased with flow while in the other it decreased with flow. As the authors explained, however, the most important factor was that the results were reproducible with the same experimental arrangement. In conclusion they suggested that a method of NMR blood flow measurement may be conceivable in the future but they did not suggest any experimental arrangement by which this would be achieved. Later in the same year Singer described the first attempt to measure blood flow by NMR.[13] He studied the absorption curve obtained from a mouse's tail placed between the poles of a magnet in an NMR spectrometer. The author stopped and started the flow by using a tourniquet and showed that the absorption curve increased in amplitude with flow. This effect he explained as resulting from unsaturated blood entering the volume of the spectrometer coils. Singer went on to analyze this method[14] and derived equations relating the velocity v and the volume flow Q to the signal absorption with flow, A_f, and without flow, A:

$$v = L(A_f - A)/A\,T_1 \qquad (1)$$

or:

$$Q = V(A_f - A)/A\,T_1 \qquad (2)$$

where L and V are the length and volume of the r.f. coil respectively.

In their review, "Nuclear Magnetic Resonance in a Flowing Liquid," Zhernovoi and Latyshev[6] described a blood flowmeter which was connected up to the femoral artery of a dog and the periodic variation in signal amplitude versus time recorded. This particularly invasive approach involved diverting the blood out of the body and through the flowmeter before it was returned to the circulation, and was liable to dramatically affect the very blood flow being measured.

Later in the early 1970s Battocletti et al. developed a number of NMR blood flowmeters for pulsatile flow measurement primarily in the arm.[15,16] The flowmeters consisted of a magnetizer and a detector magnet that contained a transmitter and a receiver coil arrangement into which an arm could be placed. Arterial blood was partially magnetized as it passed through the magnetizer coil. The degree of magnetization depended on the period that blood remained within the magnetizer and therefore depended on the flow rate. This partially magnetized arterial blood then moved down the arm to the detection system where the NMR signal, which was directly related to the magnetization, was detected. The system was analyzed *in vitro* and *in vivo*, using on each occasion electromagnetic flowmeters for comparison. The *in vitro* studies demonstrated a linear relationship between the NMR signal and the peak-to-peak flow rate. The *in vivo* studies showed periodic NMR signals with a constant phase relationship to those measurements from the electromagnetic flowmeter.

Similar "whole body" studies on blood flow measurement were made by Halbach et al.[17,18] using larger superconducting magnets. Their method was to repeatedly excite the blood protons in the sensitive region such that saturation resulted and the signal was reduced. However, during periods of fast systolic flow "fresh" unsaturated protons flowed into the detector region and thus increased the signal. The main problem with these methods was that of spatially localizing the signals. Methods were described including one that involved defining a plane by introducing a modulated field gradient such that signals were only obtained where there was a null in the field at a particular plane of interest. Traces obtained from arterial pulsatile flow in limbs, the brain, and the region of the heart, were presented. The calibration procedures for pulsatile flow involved a comparison with an electromagnetic flowmeter *in vitro*.[19]

For the second time-of-flight approach where spins are tagged and their motion followed the velocity equals the distance moved divided by the measured time between the tag and the detection. Singer, in his paper of 1959[13] where he attempted to measure the blood flow in a mouse's tail by an absorption method, also described the possibility of using a receiver coil separated from, and downstream of, the transmitter coil, so that the delay could be measured between the time of the excitation and the received signal; this delay, being the time for blood to flow between the two coils, could then be used to measure the flow velocities. The method was proposed as a way of measuring flow in arms or fingers, for example. Various attempts were made to implement the techniques,[20] however, it was not until 1970 that Morse and Singer[21] eventually overcame problems of poor signal to noise and described the implementation of this type of approach for the measurement of venous blood flow in the human arm. Two small localized coils were placed on the skin just above a surface vein. For this study, the arm of a subject was placed between the poles of a 3.6 Tesla electromagnet; a 6 mm diameter transmitter was then used to perform an adiabatic fast passage (reversal of polarization) on the protons under the coil. The second coil was used to measure NMR absorption which was reduced as the reverse polarized bolus of blood passed underneath.

Another approach for continuous flow monitoring was proposed by Zhernovoi and Latyshev.[6] Their technique, called the pulse frequency method, involved the use of an alternating demagnetizing current fed through a demagnetizing coil upstream of the detector. The current was only applied when a signal was detected and in this way the demagnetizing field was pulsed on and off at a rate which was proportional to the flow through the tube. The pulse rate was detected and converted to a voltage which was output to a recording device. Another method, the time pulse method, involved pulsed demagnetization at a constant rate. The flow rate was measured from the phase shift between the demagnetizing pulse and the detected signal.

Potential industrial applications formed an important source of motivation for the development of such an automatic flow metering devices. Genthe et al.[22] described such a system consisting of a polarizing magnet followed by a tagging and detector magnet arrangement. A slightly off-resonance applied r.f. field resulted in an adiabatic change in direction of the magnetization of the flowing solution. Tagging was accomplished by momentarily altering the applied magnetic field at the entrance to the detector so that the r.f. field was on-resonance. Successive application of the tagging field created a near sinusoidal modulation pattern of the magnetization in the fluid and hence the detected signal. The phase of the modulation was related to the velocity. This system was incorporated into a number of industrial installations, including a paint manufacturing plant where it was used to measure the flow of the solvents xylene, butyl alcohol, and naphtha, as well as into an agricultural chemical process where it was used to measure the flow of concentrated hydrochloric acid. The independence of the NMR system of fluid properties such as conductivity, temperature, viscosity, density, gas or solid entrainment and opacity and, in addition, having no moving parts and no mechanical or electrical contact with the fluid, meant that it was ideal for this type of application.

Phase Methods of NMR Flow Measurement

Phase methods of NMR flow measurement use the fact that the signal obtained from a sample moving in the direction of a magnetic field gradient will exhibit a motion-related phase shift. The early work using these methods was concerned with flows that are slow in comparison with those found in arteries and veins. Apart from a fine layer in contact with the vessel wall, however, these studies have helped in the more recent development of faster flow and flow imaging techniques.

The first to suggest an implementation of this type of method was Hahn in 1960 when he suggested a method of measuring slow sea water motion (approx 10^{-3} cm/sec) by detecting the phase shift of the signal, resulting from flow in the direction of a field gradient.[23] His suggestion was that a phase shift as low as 0.3 radians could be measured by beating the acquired signal against a reference signal. The implication was that the measurements should be made in the earth's very homogeneous magnetic field by using an initial polarizing field that would be turned off to initiate the precession. A gradient G would be applied by passing a current through an appropriate coil of wire, the current was reversed at time $t = \tau$ and the echo formed at $t = 2\tau$. Hahn formulated an equation for the phase at this time:

$$\phi = \gamma G v \tau^2 \tag{3}$$

where γ is the gyromagnetic ratio and v is the velocity of motion. This equation now forms the basis of the phase-based methods of flow imaging. The theory was later validated by Lucas et al. when convection currents down to 10^{-5} m/sec were measured.[24]

Singer,[8,25] in his reviews of previous work on flow measurement, described a method similar to that used earlier by Packer[26] for plotting the phase of spins with a range of velocities. The technique was termed "phase graphing" and he used it to demonstrate the effect of coherent flow on the Carr Purcell sequence when pulsed gradients were applied between each r.f. application. The potential to jointly display the flow and the anatomical information by the use of color on the same image was also suggested. However, no suggestions were made for a method of obtaining the flow information.

The phase of the signal obtained from different forms of multiple echo sequences was used by Grover and Singer[27] and also Garroway[28] to study flow velocity distribution as opposed to mean flow. In both studies a gradient was applied in the direction of flow and a Fourier transformation was used to transform the echo envelope into a plot of velocity distribution. Garroway tested his method on two rectangular tubes with laminar flow of different velocities and the results fitted in reasonably well with the theoretically expected rectangular step-like curve. Grover and Singer went on to investigate flow in the human finger and obtained a complex velocity distribution heavily weighted at the zero velocity end. This result was explained by the fact that much of the flow would be capillary with very low flow velocity components in the direction of the phase shift producing gradient. The zero weighting was presumably also complemented by all the stationary tissue within the finger.

II. SPATIALLY RESOLVED NMR FLOW IMAGING

The methods of spatially resolving and imaging flow information have broadly speaking used the same flow effects as described earlier in this chapter in combination with an imaging sequence. The interest in flow in NMR imaging has not been solely directed toward the goal of quantitative flow measurement. A large amount of effort has also been devoted to understanding the appearance of a flowing fluid on an image, as this can often be indicative of the type of flow present and therefore give important information in the diagnosis of a particular disorder. Also, the development of MR angiography techniques has also required a full understanding of these effects.

In 1984, soon after the development of the first clinical MR scanners, there was a great increase of interest in the quest for an MR method of imaging flow. Review articles were published and a number of techniques described. Work on blood flow imaging, effects of flow on conventional images, measurement of vessel lumen area, and the characterization of atherosclerotic lesions were well covered by Crooks and Kaufman[29] and a thorough review that analyzed the blood flow effects including a number of experimental results was made by Axel.[30]

A. The Effect of Flow on the Appearance of the NMR Image

Over the years since NMR imaging was first developed there has been much interest in the effect of flow on the appearance of the image. Grant and Back[31] were the first to investigate experimentally the effects of flow on the intensity of two-dimensional NMR images. They termed these flow studies "NMR rheotomography" ("rheo" being the Greek word for flow). Two flow effects were recognized as having the potential to alter the signal intensity on an image. It was first noted that, when using a spin-echo sequence, only those spins that experienced both r.f. pulses could contribute to the spin-echo such that, in the case of high flow, the signal was reduced because fewer spins experienced both pulses. The other fact that was noted was that flow in the direction of an imaging gradient could result in signal dephasing. The resultant signal loss due to this second effect was demonstrated in the outflow section of a U-tube where secondary flows existed across the diameter.

Flow phantoms were also used by George et al.[32] who studied the effect of continuous and pulsatile flow of aqueous solutions of $MnCl_2$ and whole bovine blood on the NMR imaging signal. Flow-related enhancement was initially observed with increasing continuous flow rates, however, at higher flow rates this was replaced by an overall signal loss, particularly when the Reynolds Number exceeded 2,000 where turbulence would be expected.

Flow-related signal enhancement was studied in greater detail by Waluch and Bradley[33] who found there to be two contributary factors. The two factors identified were the simple washin effects described previously by Singer and Crooks[34] and an additional effect seen only on images from even echoes. The second enhancement was termed even echo rephasing and, although it had been identified relatively early in the development of NMR spectroscopy,[35] this was the first occasion where the phenomenon was identified on an image.

Bradley and colleagues[36,37] undertook in-depth studies of the appearance of different types of flowing blood on standard NMR images. Experiments were set up using flow phantoms with various flow rates and tube diameters so that laminar or turbulent flow could be investigated. Again washin enhancement and even echo rephasing were studied as well as signal loss due to flow during the period between the 90° and 180° pulses of a spin echo sequence.

One artifact that can result because of in-plane flow is that of signal misregistration. This was addressed by Von Schulthess and Higgins[38] who studied the effect that results due to blood flow between the times of slice selection, phase encoding, and frequency encoding.

Non-structured flow artifacts can also result in the presence of pulsatile blood flow as a consequence of the signal phase shift resulting from the motion of blood during the application of imaging gradients. This phase shift is added to the spatial phase encoding; if the motion of blood is the same on successive applications of the gradients then the additional phase shift will remain constant and will not affect the image. However, if the motion of blood changes, the resultant phase shift will introduce a varying error to the spatial phase encoding. The Fourier transform will then get "confused" and the blood signal will end up being spread out along the phase encoding axis. Perman et al.[39] investigated this type of flow artifact and demonstrated that even echo rephasing corrected the error and removed the artifact as long as the motion was reasonably simple (i.e., it did not contain significant acceleration or other high order derivatives of position).

B. Time-of-flight Flow Imaging Methods

As with the non-imaging time-of-flight methods, there are two categories of techniques. The first category known as washin/washout flow imaging methods normally relies on the saturation or partial saturation of material in a selected slice or volume being replaced by fully magnetized spins before the next excitation and thus increasing the signal; the second involves some form of tagging and then imaging to follow the motion of the tagged material.

The first description of experiments to spatially resolve flow information from the NMR signal was by Garroway[28]. He used a 90°-τ-90° pulse sequence with a gradient applied perpendicular to the direction of flow through a tube during the sampling. The Fourier transform of the signal from the first 90° pulse gave a measure of the spin density across the vessel and that of the second 90° pulse gave a velocity profile of spins across the vessel. The first 90° pulse saturated stationary spins so that they would not generate a signal from the second 90° pulse. However, spins that were flowing would produce a signal related to their velocity. Thulborn et al.[40] and Singer and Crooks[34] also adopted very similar approaches and investigated varying the time τ between the two 90° pulses to change the potential range of the flow measurement. The latter attempted flow measurements in the internal jugular veins with the technique. A major problem with these methods was that only low flow velocities could realistically be measured with practical slice thicknesses.

The first to describe a true tagged time-of-flight approach to imaging blood flow were Feinberg et al.[41] Their method involved a variation on a double echo spin-echo sequence, the first 180° selected slice was displaced by 3mm from the initial excitation slice and the second was displaced by 9mm. The first 180° selection overlapped sufficiently with the 90° selection to produce a good anatomical image. The second 180° pulse selection did not overlap with that of the 90° or the first 180° pulse selection, and therefore produced no anatomical image but gave high signal from the blood that had experienced all the preceding r.f. pulses (i.e., that had passed between the different selected planes). Flow in the carotid and vertebral arteries of a volunteer's neck was identified with the technique although flow velocities could not be defined accurately.

Methods have also been described where the time-of-flight flow movement can be visualized directly on an image.[42,43] The methods involved slice selection and frequency encoding being applied in the same axis. In this way material that had

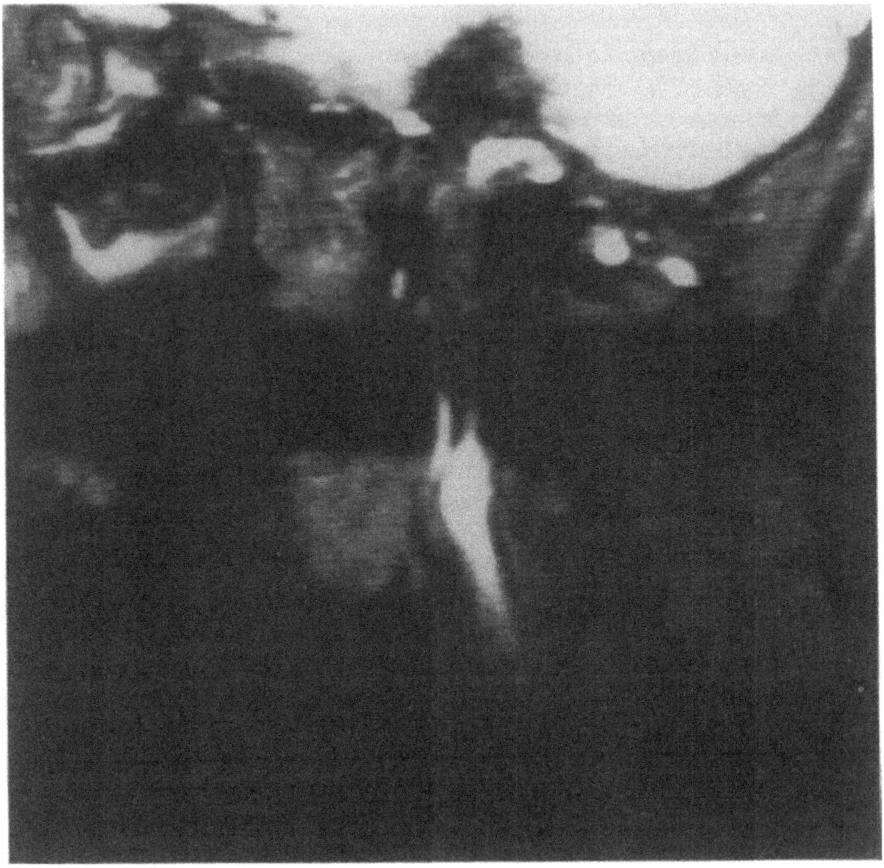

Figure 1. In-flow of arterial blood (up) and venous blood (down) into a presaturation band in a 2D cardiac gated image. The saturation has taken place just above the carotid bifurcation during systole. The time between the saturation pulse and the echo time is 18 ms. The distance the blood has moved is slightly more than 1.2 cm, resulting in a velocity of about 60 cm/s.

Source: Reproduced with permission of Raven Press from Haacke, E.M.; Smith, A.S.; Lin, W.; Lewin, J.S.; Finelli, D.A.; Duerk, J.L. *Top. Magn. Reson. Imaging* **1991**, *3*, 34.

moved in this axis between selection and reading would be displaced relative to the stationary material. This technique is therefore making use of signal misregistration, an effect that is often seen as a problem in other methods of flow imaging. With the flow information encoded in one axis as described, the other one or two axes may be spatially encoded by use of stepping phase encoding gradients. When tested *in vivo* the technique could be repeated rapidly throughout the cardiac cycle and was well able to measure very high flow velocities.[44]

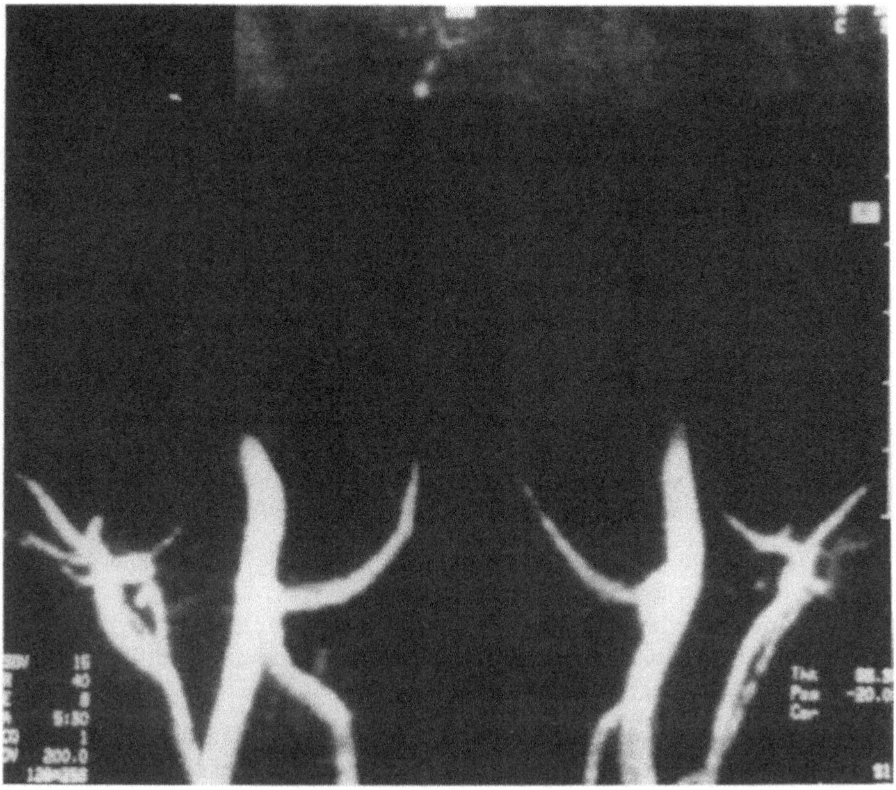

Figure 2. Inflow during a 3D TOF experiment carried out in the upper neck/head region reveals many arterial vessels. The peak flow represented in the internal carotid is more than 1m/s.

Source: Reproduced with permission of Mosby-Year Book, Inc. from Firmin, D.N.; Dumoulin, C.L.; Mohiaddin, R.H. In: *Magnetic Resonance Angiography, Concepts and Applications* (Potchen, E.J.; Haacke, E.M.; Siebert, J.E.; Gottschalk, A. Eds.). 1993.

Two approaches using saturation pulses have been used. The first uses a saturation pulse to saturate any stationary tissues within a region so that magnetized blood flowing into the region can be imaged and flow quantified by measuring the distance of inflow. In general, although results have been demonstrated using two-dimensional cardiac gated techniques (Figure 1) the best results have been obtained with three-dimensional ungated methods (Figure 2) although the flow measurement is not well defined and is in the form of a weighted mean. The other approach is to saturate a band of tissue, for example, in a transverse plane, then to follow the progress of this dark band in the coronal or sagittal planes.[45]

The major limitation of these saturation methods is that they are limited by the T_1 of the various tissues being saturated. The appearance of the saturated blood or tissue will change with time eventually making it difficult to measure accurately distances traveled. Also, motion during the sampling gradients results in signal and thus image distortion.[46] For arterial flow measurements where cardiac gating is required, only two-dimensional images can realistically be acquired. This means that details of the flow profile can only be studied in one dimension at the most.

C. Phase Flow Imaging Methods

Although considerable knowledge had been gained on the measurement of flow from the phase of the NMR signal, from the non-imaging studies described earlier, it was not until 1984 that experimental attempts were described for using such methods to imaging flow.[47,48] Two years earlier, Moran[49] had in fact suggested the theory of introducing bipolar velocity phase encoding pulses to the three axes of a standard spin-echo imaging sequence to produce what he described as six dimensions of information, these being three spatial and three velocity. The imaging methods that were developed fell broadly into two categories:

1. Phase Velocity Mapping methods that mapped the phase of the signal directly in order to measure the flow.
2. Fourier Flow Imaging methods that phase encoded flow velocity so that a number of flow phase encoding steps would separate the different flow velocities present at a particular location.

The same theory forms the basis for both of these categories of methods: The NMR signal from a sample of spins moving in the direction of a bipolar magnetic gradient pulse pair (Figure 3) will accumulate a phase shift given by:

$$\phi = \gamma v \Delta A_g \tag{4}$$

where A_g is the area of one gradient lobe, Δ is the time between the centers of the two gradient lobes, v is the velocity, and γ is the gyromagnetic ratio. A quantitative

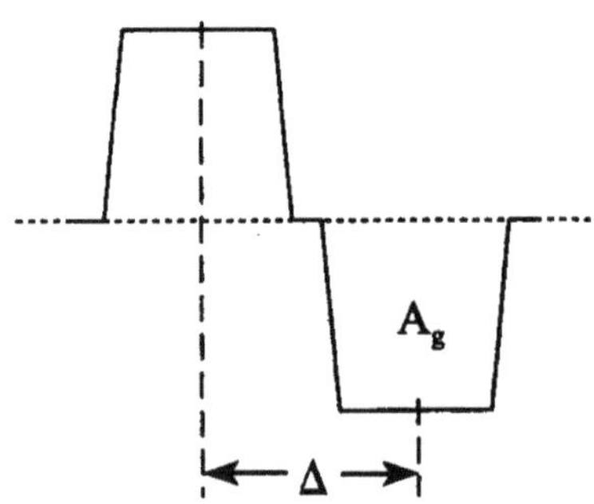

Figure 3. A magnetic gradient waveform with a bipolar time profile used to produce velocity dependent phase shifts.

measure of the velocity of the sample can therefore be determined if the phase shift can be measured.

Two different approaches to utilizing the phase shift to produce a quantitative flow image are discussed below along with a number of clinical applications and a variety of factors can result in errors in the flow measurement.[50]

Phase Velocity Mapping

The initial demonstrations of phase velocity mapping were described by van Dijk[47] and Bryant et al.[48] in 1984; they had both measured velocity directly from the phase of the signal originating from each imaging voxel. These early techniques suffered due to problems in repeating the sequence rapidly and signal loss due to shear and other more complex flows. These problems were reduced and the methods were made clinically more useful, partly by the use of a field-echo sequence[51,52]

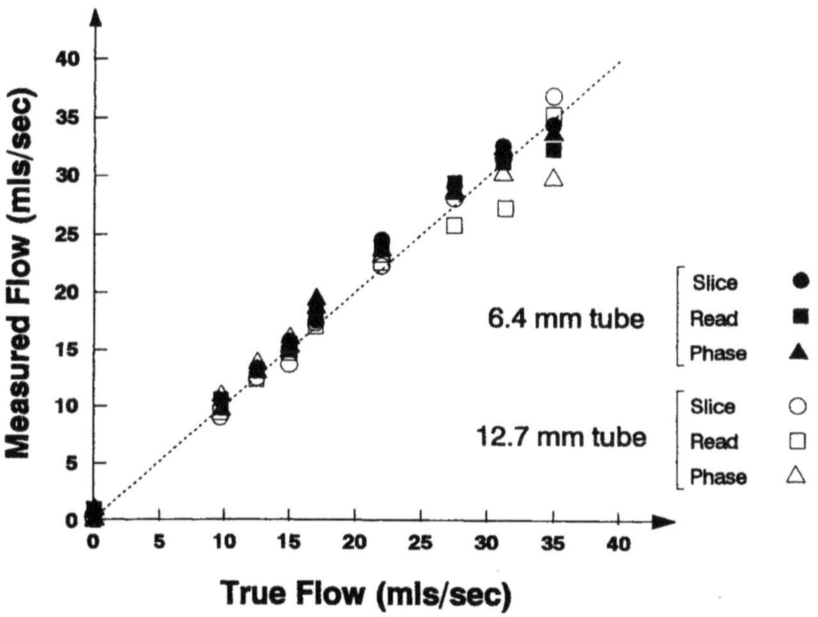

Figure 4. Plot comparing the measurement of volume flow using the FEER velocity mapping technique and true flow. The measurements were made in the slice select, read, and phase encode directions using both 6.4 mm and 12.7 mm diameter tubes in each direction. Many of the measured flow rates exceeded the critical Reynolds Number (Re_{max} = 8,000) indicating that turbulence, while reducing the signal magnitude, does not initially affect the phase velocity measurement when using direct phase mapping techniques.

Source: Reproduced with permission of Academic Press from Firmin, D.N.; Nayler, G.L.; Kilner, P.J.; Longmore, D.B. *Magn. Reson. Med.* **1990**, *14*, 230.

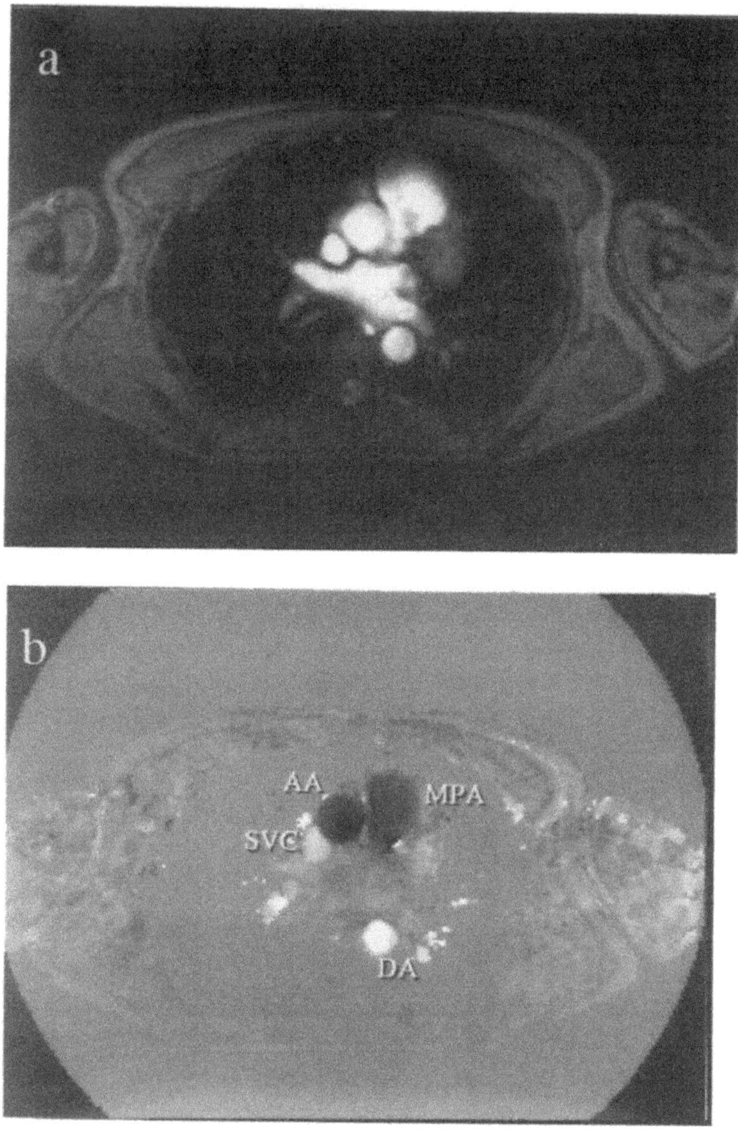

Figure 5. (a) A field even echo rephasing (FEER) magnitude image of a transverse slice through the major vessels above the heart of a normal volunteer showing high signal from the blood in the slice. (b) shows the corresponding velocity phase map acquired at a timing of 270 msec following the R-wave. Flow toward the head in the ascending aorta (AA) and main pulmonary artery (MPA) is displayed in darker shades of grey tending toward black and flow toward the feet in the descending aorta (DA) and superior vena cava (SVC) can be seen in lighter shades of grey tending toward white.

and, more importantly, by the introduction of velocity compensated gradient waveforms.[53,54] The technique has been validated both *in vitro* and *in vivo*[55] and is now providing useful information in clinical and physiological flow studies. *In vitro* results are presented in Figure 4 which shows phase velocity mapping volume flow measurements versus true flow measurements obtained with a bucket and stop-watch technique. The results of flow through two different tube diameters in three different directions (slice select, phase encoding, and frequency encoding), demonstrate the inherent accuracy of the phase mapping method. One point of interest is that many of the measurements were made on flows exceeding the critical Reynolds Number (Re_{max} = 8,000) thus confirming that the accuracy of the measurements is maintained in this situation.

For validation *in vivo*, Figure 5 shows the magnitude and phase velocity images of one of a series of time frames of transverse images just above the heart. At this systolic timing flow can be seen in the ascending and descending aortae, the pulmonary artery, and the SVC. The measurements of these flows made throughout the cardiac cycle are presented in Figure 6 and the stroke volume can be measured by integrating under the aortic flow curve. The result of the *in vivo* validation[55] was a very good correlation between this measurement of the stroke volume of the heart and a previously validated multi-slice volume method (Figure 7).[56]

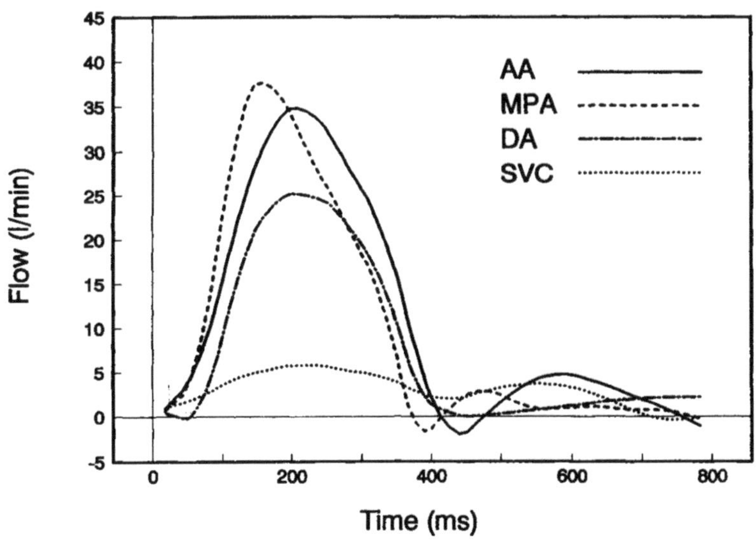

Figure 6. Instantaneous volume flow curves of the ascending aorta (AA), main pulmonary artery (MPA), descending aorta (DA), and superior vena cava (SVC) calculated from the complete cine velocity map acquired in the same plane (see Fig. 5).

Source: Reproduced with permission of the American Heart Association from Mohiaddin, R.H.; Longmore, D.B. *Circulation* **1993**, *88*, 264.

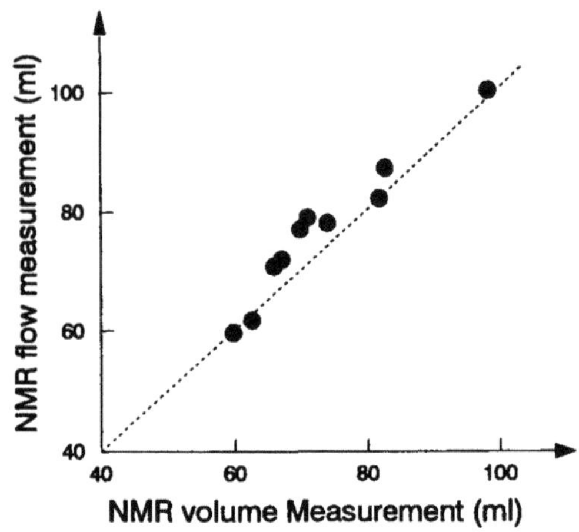

Figure 7. The correlation of left ventricular stroke volume measurements. The standard error of the estimate is 3.3 ml.

Source: Reproduced with permission of Mosby-Year Book, Inc. from Firmin, D.N.; Dumoulin, C.L.; Mohiaddin, R.H. In: *Magnetic Resonance Angiography, Concepts and Applications* (Potchen, E.J.; Haacke, E.M.; Siebert, J.E.; Gottschalk, A., Eds.). 1993.

Fourier Flow Imaging

Experimental results using the Fourier flow imaging method were first described by Redpath et al.[57] in 1984; eight flow phase encoding steps were used to image a circle of fluid filled tubing rotating in the image plane. Different segments of the circle, each corresponding to different velocity ranges, were seen on the eight resultant images. A year later Feinberg et al.[58] applied the method both *in vitro* and *in vivo* but increased the velocity resolution and simplified the reconstruction by increasing the number of flow phase encoding steps and removing the spatial phase encoding. The accuracy of the method was demonstrated using a phantom while the *in vivo* study, which showed the flow in the descending aorta, highlighted the problem of very high signal from stationary tissue imaged in the spatial dimension. In 1988 Hennig et al.[59] described a development of this method where the signal from stationary tissue was saturated and the sequence repeated much more rapidly. The one major problem with the Fourier methods of blood flow imaging, is the time required to obtain a reasonably high resolution in the velocity axis, and this precludes the use of more than one spatial dimension.

Rapid Phase Flow Imaging Methods

The previously described methods acquire flow data over a period of time that is long in comparison to the cardiac cycle, rapid variations in flow cannot be followed and the measured flow is a weighted average over the scan acquisition time. Very rapid flow imaging techniques have been developed, either by combin-

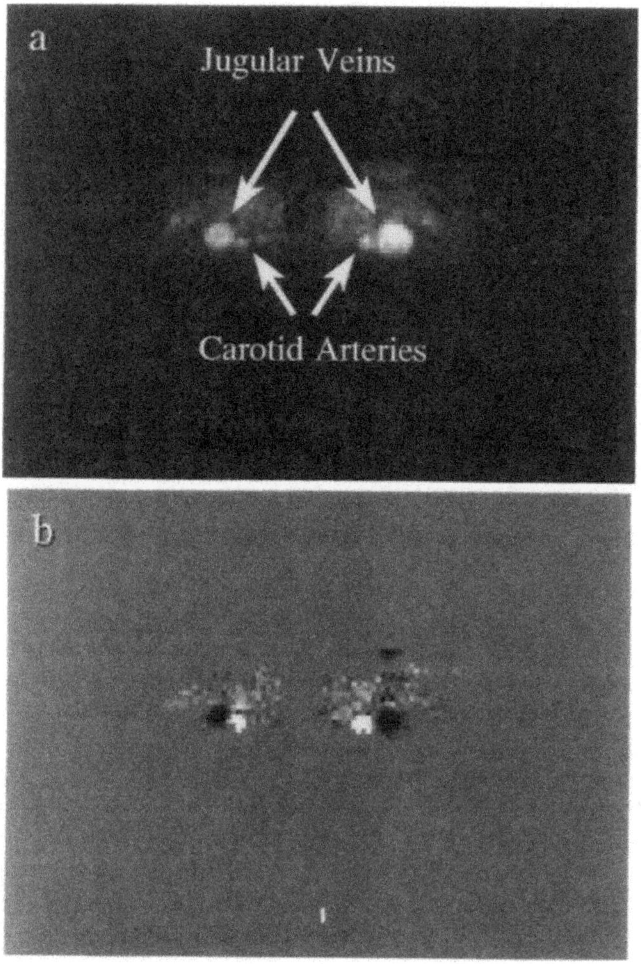

Figure 8. (a) A magnitude reconstruction of echo planar data acquired with one excitation. The selected region includes the carotid arteries and jugular veins which can be seen exhibiting high blood signal. (b) An example of a velocity phase map reconstructed from data acquired with one excitation of each of a modified and unmodified EPI flow sequence. Flow in the carotid arteries can be seen tending toward white while that in the jugular veins tends toward black with stationary material mid grey.

Source: Reproduced with permission of Academic Press from Firmin, D.N.; Nayler, G.L.; Kilner, P.J.; Longmore, D.B. *Magn. Reson. Med.* **1990**, *12*, 316.

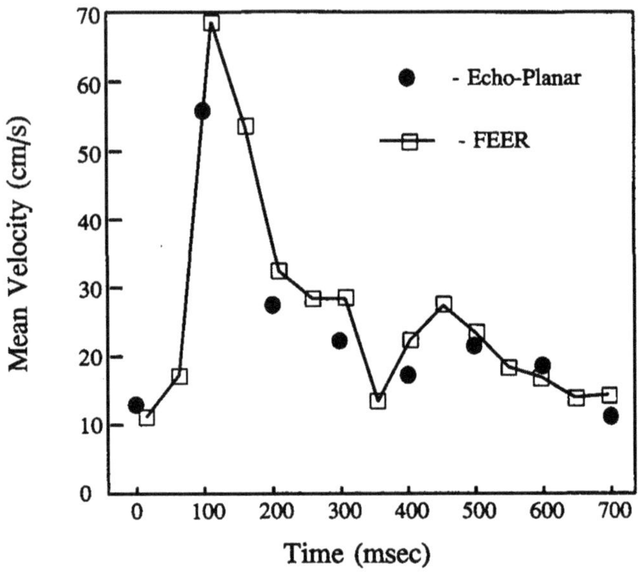

Figure 9. A comparison of the flow versus time plots for the right carotid artery of a normal volunteer obtained from echo-planar velocity images and FEER velocity images. The close comparison between the two plots suggests that echo-planar velocity measurements are as accurate as the FEER measurements. It should be noted, however, that the two types of flow information are somewhat different.

Source: Reproduced with permission of Academic Press from Firmin, D.N.; Nayler, G.L.; Kilner, P.J.; Longmore, D.B.*Magn. Reson. Med.* **1990,** *12,* 316.

ing a phase mapping type approach with a very fast imaging method such as echo-planar[60] or by imaging only one spatial dimension.[61] The echo-planar method approach is technically difficult and not easily implemented on current standard machines, however, techniques to minimize the problems have been developed. Figure 8 shows echo-planar velocity images of a normal neck acquired in 40 msec of two consecutive cardiac cycles. Figure 9 shows a comparison of echo-planar velocity mapping with FEER velocity mapping. The two techniques agree well although it should be remembered that one is an average over time and the other is not. An additional problem is that the echo-planar sequence is inherently phase sensitive to flow even if additional flow compensation is applied. This has been used to advantage for more qualitative flow imaging showing flow disturbances for example.[60] More recently improved echo-planar and spiral echo-planar sequences have been developed with a relatively low flow sensitivity.[62] Figure 10 shows an example of a series of four spiral echo-planar phase velocity images taken from a cine set acquired at 50 ms intervals over two cardiac cycles (one cardiac cycle

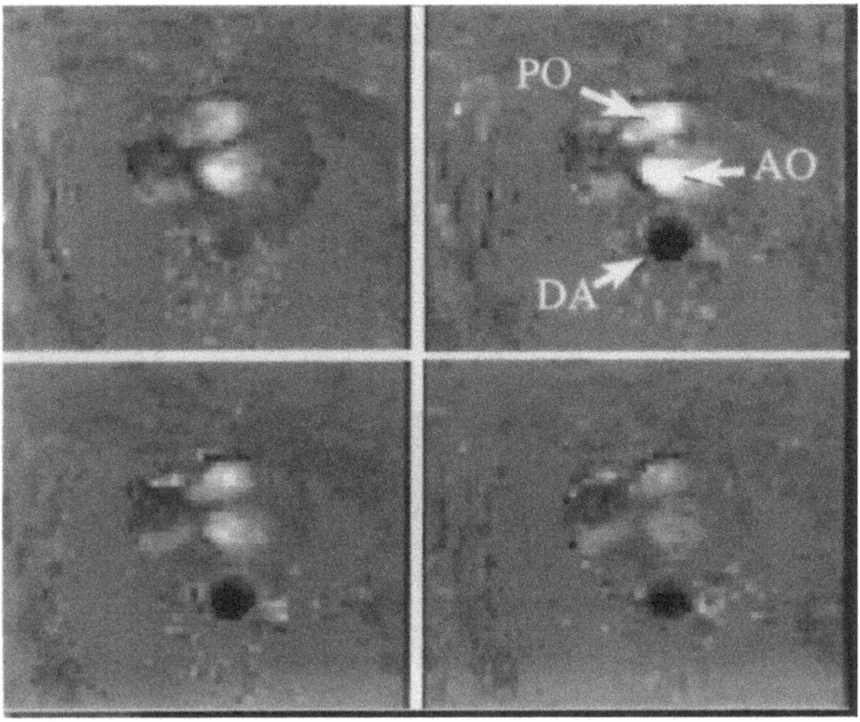

Figure 10. Four frames from a series of 16 flow images acquired over two cardiac cycles with a single shot spiral echo-planar sequence. PO—Pulmonary Outflow; AO—Aortic Outflow; and DA—Descending Aorta.

Source: Reproduced with permission of Williams and Wilkins from Gatehouse, P.D.; Firmin, D.N.; Collins, S.; Longmore, D.B. *Magn. Reson. Med. 31*, 504.

reference and one velocity encoded), and Figure 11 demonstrates real time acquisition of flow in the descending aorta throughout a valsalva maneuver, and before and after exercise.

The one-dimensional rapid acquisition mode, RACE (Real time ACquisition and velocity Evaluation),[61] can be used to measure flow perpendicular to the slice. The technique can be repeated rapidly throughout the cardiac cycle in order to give near real time flow information. Figure 12 shows an example of flow images acquired using the RACE technique. One problem with this type of approach is that data is acquired from a projection through the patient; this means that any signal overlapping with the flow signal will combine and introduce errors to the flow measurement. Several strategies have been suggested for localizing the signal in order to avoid this: they include spatial presaturation, projection dephasing (applying a gradient to suppress stationary tissue), collecting a cylinder of data, and multiple oblique measurements.

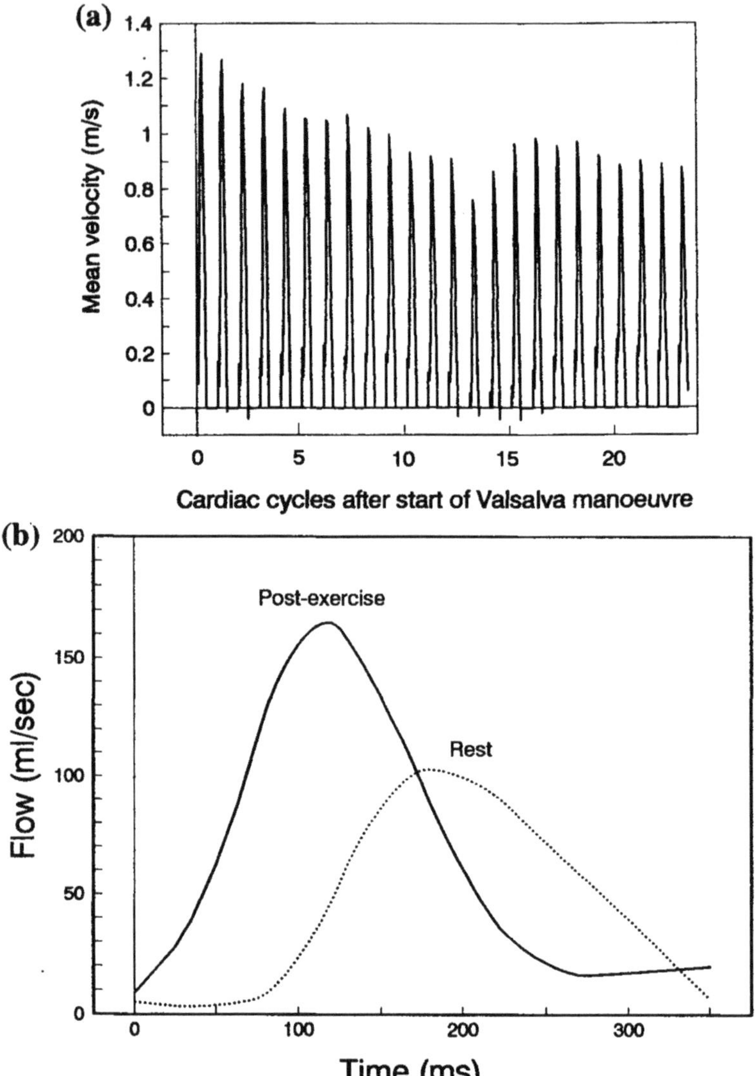

Figure 11. (a) Real time measurements of descending aortic mean blood flow velocity over 25 cardiac cycles during a valsalva maneuver and (b) time-related flow curves in the descending thoracic aorta acquired in a single heart beat at rest and immediately following exercise. Data acquired using a single shot spiral echo-planar sequence.

Source: (a) Reproduced with permission of Williams and Wilkins from Gatehouse, P.D.; Firmin, D.N.; Collins, S.; Longmore, D.B. *Magn. Reson. Med., 31,* 504.
(b) Reproduced with permission of the Society of Magnetic Resonance from Mohiaddin, R.H.; Gatehouse, P.D.; Firmin, D.N. *Magn. Reson. Imag.* **1995**, *5*:159–163.

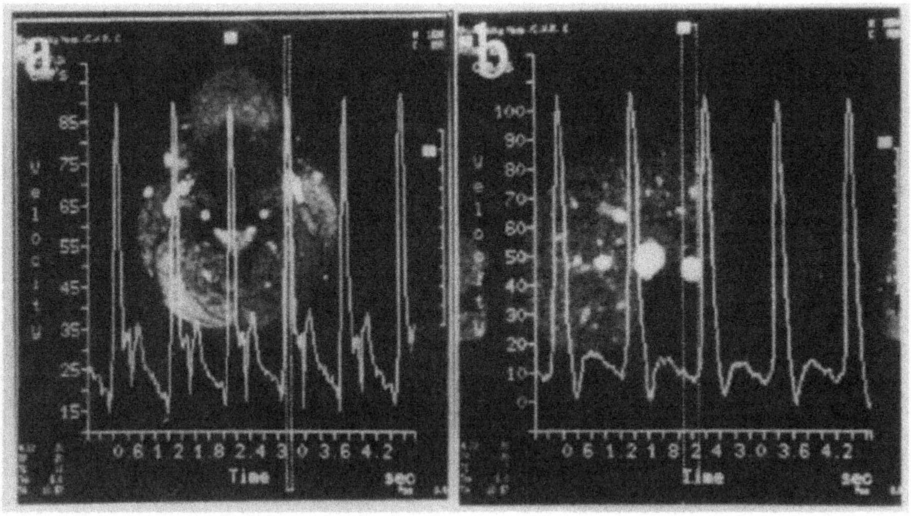

Figure 12. RACE profile through five cardiac cycles in the neck (a) and four cardiac cycles in the abdomen (b). The carotid velocity is similar to that seen with Doppler ultrasound, never quite dropping to zero. The aortic flow is rapid and drops to zero and then returns to a fairly constant low flow rate.

Source: Reproduced with permission of Raven Press from Haacke, E.M.; Smith, A.S.; Lin, W.; Lewin, J.S.; Finelli, D.A.; Duerk, J.L. *Top. Magn. Reson. Imaging,* **1991**, *3,* 34.

III. CLINICAL APPLICATIONS OF MAGNETIC RESONANCE FLOW IMAGING

In the previous sections a number of techniques have been described for imaging and measuring flow by magnetic resonance. The development of these methods has greatly enhanced the potential of MRI as a physiological tool in cardiology. As yet much of the clinical use has been with the method of phase-shift velocity mapping and for this reason alone many of the following observations and applications will concentrate on this approach. The following sections concentrate on the use of magnetic resonance to measure blood flow in various regions of the body.

A. Thoracic Aorta

Quantitative analysis of aortic flow using magnetic resonance imaging has been a subject of considerable interest in health and disease. This is partly because the aorta is large and it is relatively immobile making it suitable for all types of magnetic resonance flow measurement (Figures 5 and 6). Blood flow in the ascending and descending thoracic aorta is phasic. Although the average resting flow measured by MR velocity mapping in the ascending and descending thoracic aorta of normal subjects is 6.0 l/min and 3.9 l/min, respectively, instantaneous peak systolic flow in these vessels can reach more than 40 l/min and 30 l/min, respectively.[63] Normal systolic flow in the ascending aorta is plug flow with a skewed velocity profile which has higher velocities around the inside of the arch. Throughout diastole the blood continues to move with simultaneous forward and reverse channels.[64,65,66] In normal subjects, the reverse flow channel is closely associated with the left coronary sinus and it is tempting to speculate that it augments flow in the left coronary artery by imparting momentum to the blood which is destined to enter it. In patients with coronary artery disease, the reverse flow channel is smaller and may enter any of the coronary sinuses.[67] In aortic valve regurgitation, the magnitude of the reverse flow is understandably increased (Figure 13) and aortic or pulmonary regurgitation may be quantified from the backflow of blood in the proximal great vessels assessed by velocity mapping.[68] Abnormal flow patterns with secondary vortices and larger than normal reverse flow have been demonstrated in patients with aortic aneurysm using vector mapping of multidirectional magnetic resonance velocity mapping.[69]

B. Aortic Flow Wave Velocity

Aortic flow wave velocity can be calculated by magnetic resonance velocity mapping from the delay between the leading edge or onset of the flow wave in the ascending and descending limbs of the thoracic aorta.[70,71] This parameter is closely related to aortic compliance[70] which may prove to be useful for the detection and monitoring of arterial disease.[72,67,73,74]

C. Aortic Dissection

Aortic dissection is readily detected by spin echo imaging and its extent can be displayed including involvement of other vessels. MRI using spin echo imaging compare favorably with echocardiography and x-ray computed tomography in the evaluation of aortic dissection.[75,76,77] However, a thin intimal flap may not be shown in these images unless static blood in the false lumen provides natural contrast with the true lumen. If there is any doubt, then the flap will be more easily seen using a gradient echo sequence, and velocity mapping will confirm the diagnosis by demonstrating the differential flow velocities in each lumen (Figure 14).[78,79]

Figure 13. Gradient echo images in a coronal plane acquired during ventricular diastole (a) in a patient with Marfan's syndrome and aortic valve regurgitation. (b) Flow volume curves in the ascending aorta (AA) and descending thoracic aorta (DA) measured from the complete cine acquisition acquired in a transverse plane perpendicular to the AA and DA. Note the large retrograde net flow during diastole. (Next page.)

Source: Reproduced with permission of the American Heart Association from Mohiaddin, R.H.; Longmore, D.B. *Circulation* **1993**, *88*, 264.

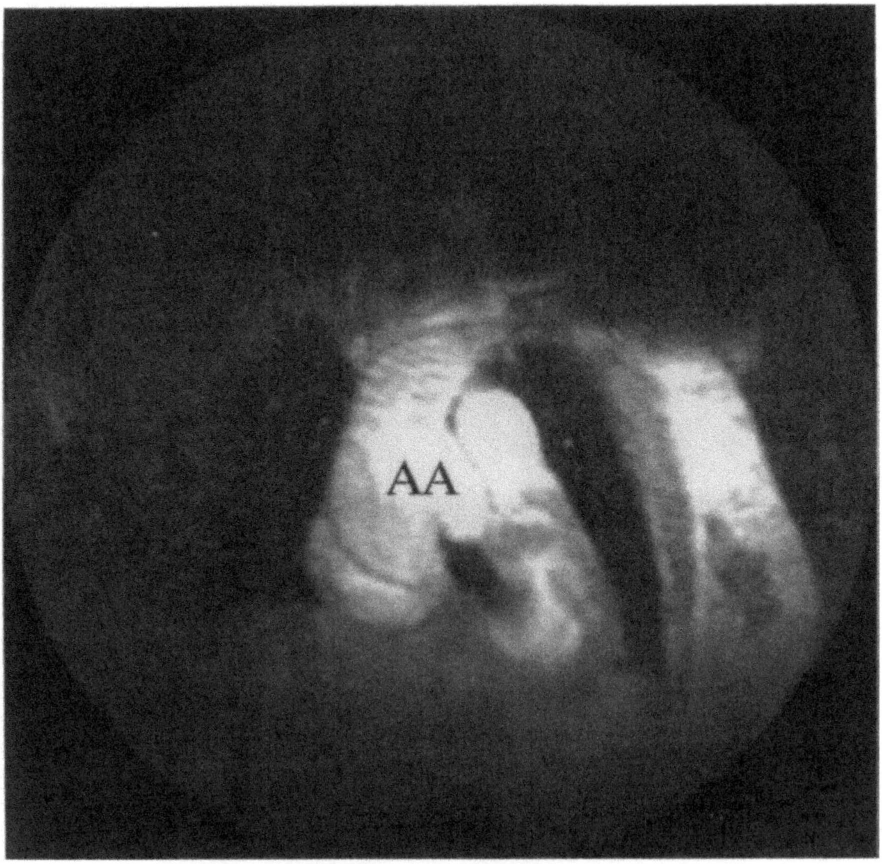
AA

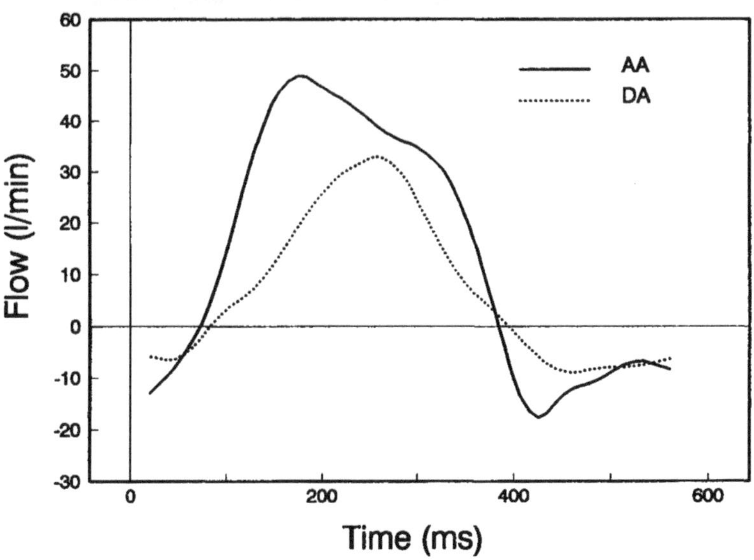
AA
DA
Flow (l/min)
60
50
40
30
20
10
0
-10
-20
-30
0
200
400
600
Time (ms)

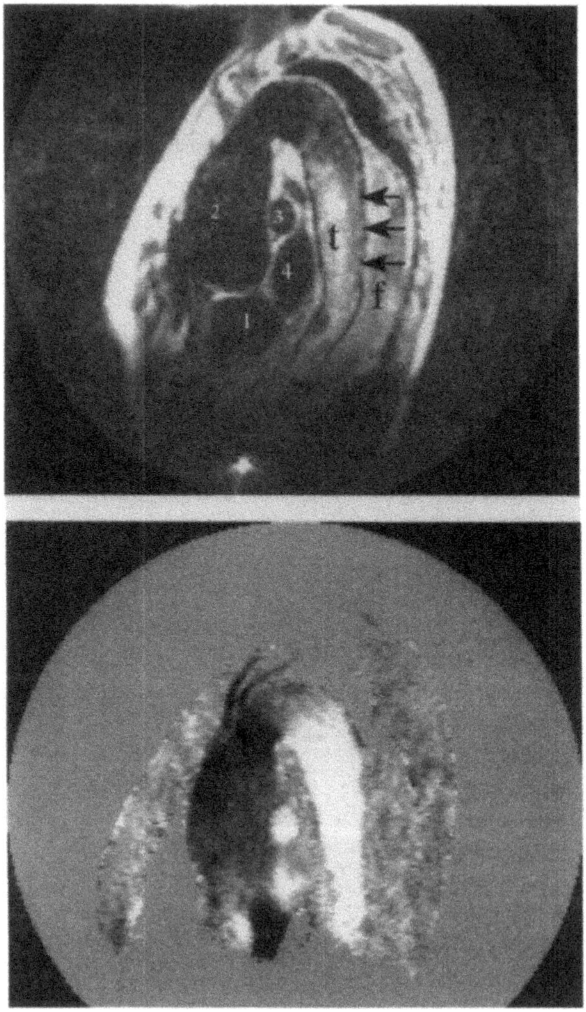

Figure 14. (a) A spin echo image in an oblique plane through the ascending aorta, aortic arch, and descending thoracic aorta showing a dilated atherosclerotic thoracic aorta with an intimal flap (arrows) separating the true lumen (*t*) from the thrombosed false lumen (*f*). (b) The systolic velocity image shows high velocity in the true lumen and zero velocity in the false lumen. 1: left ventricle, 2: ascending aorta, 3: right pulmonary artery, 4: left atrium.

Source: Reproduced with permission of the American Heart Association from Mohiaddin, R.H.; Longmore, D.B. *Circulation* **1993**, *88*, 264.

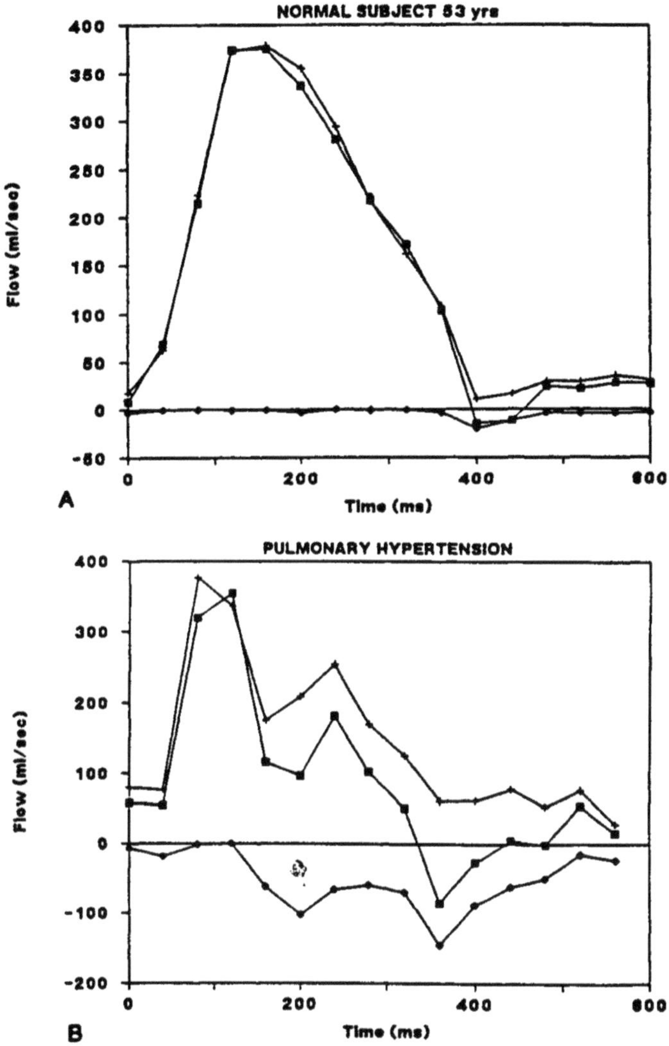

Figure 15. Main pulmonary artery flow volume curves measured by cine magnetic resonance velocity mapping in a normal subject (a), and in a patient with pulmonary arterial hypertension (b). ■ = net flow, + = forward flow, and ○ = reverse flow. In the patient study, the net flow, the forward, and the reverse flow were irregular and the reverse flow was relatively large.

Source: Reproduced with permission of Mosby-Year Book Inc. from Bogren, H.G.; Klipstein, R.H.; Mohiaddin, R.H.; Firmin, D.N.; Underwood, S.R.; Rees, R.S.O.; Longmore, D.B. *Am. Heart J.* **1989**, *118*, 990.

D. Central Pulmonary Arteries

The retrosternal position of the central pulmonary arteries makes it difficult to assess pulmonary blood flow by Doppler echocardiography especially in the presence of skeletal or lung abnormalities. MR velocity imaging is not technically constrained and is capable of accurate and detailed velocity maps in the central pulmonary arteries. It is seen that plug flow occurs during most of systole and that it is skewed toward the posterior two-thirds of the artery. A small backflow channel develops posteriorly toward the end of systole and backflow continues through early diastole. Pulmonary flow profiles have been studied less well in patients but MR velocity mapping has confirmed an abnormally early forward systolic peak and increased reverse diastolic flow in patients with pulmonary hypertension.[80,81,82] This abnormal pattern (Figure 15) may be caused by reflected waves from the distal vasculature which has a high impedance.

Patients with a single lung transplant are unique in that the cardiac output is ejected into two pulmonary vascular beds with different characteristics, and in these patients the differential blood flow depends on the relative resistance in each lung. Velocity mapping can assess the total and differential pulmonary blood flow which may be useful for monitoring these patients.[83,84] The ratio of blood flow in the transplanted and the native lungs is about 3:1 and the flow profile in the artery of the transplanted lung shows forward flow during systole and most of diastole, while that of the native lung shows a narrow early systolic peak and reverse flow during most of diastole[83,85] (Figure 16). Dysfunction of the pulmonary artery and vein anastomoses following lung transplantation are possible complications which can be monitored using magnetic resonance imaging with velocity mapping. These complications usually necessitate angiography and cardiac catheterization.

E. Caval Veins

The caval veins are also relatively large, and reliable velocity maps and flow measurements can readily be obtained.[86] The volumes of flow in the superior and

---

Figure 16. (a) A spin echo image of the pulmonary artery bifurcation of a patient with left lung transplantation. (b) Flow curves of the main (MPA), right (RPA), and left (LPA) pulmonary arteries of the same patient calculated from the complete cine velocity mapping. Blood flow in the transplanted left pulmonary artery is qualitatively and quantitatively different from the that in the native right pulmonary artery. (c) Flow-volume curves of the MPA, RPA, and LPA pulmonary arteries calculated from the complete cine velocity mapping acquired in a healthy volunteer. Flow in the RPA and LPA is qualitatively and quantitatively similar. 1: main pulmonary artery, 2: right pulmonary artery, 3: left pulmonary artery, 4: ascending aorta. (Next page.)

Source: Reproduced with permission of the American Heart Association from Mohiaddin, R.H.; Longmore, D.B. *Circulation* **1993**, *88*, 264.

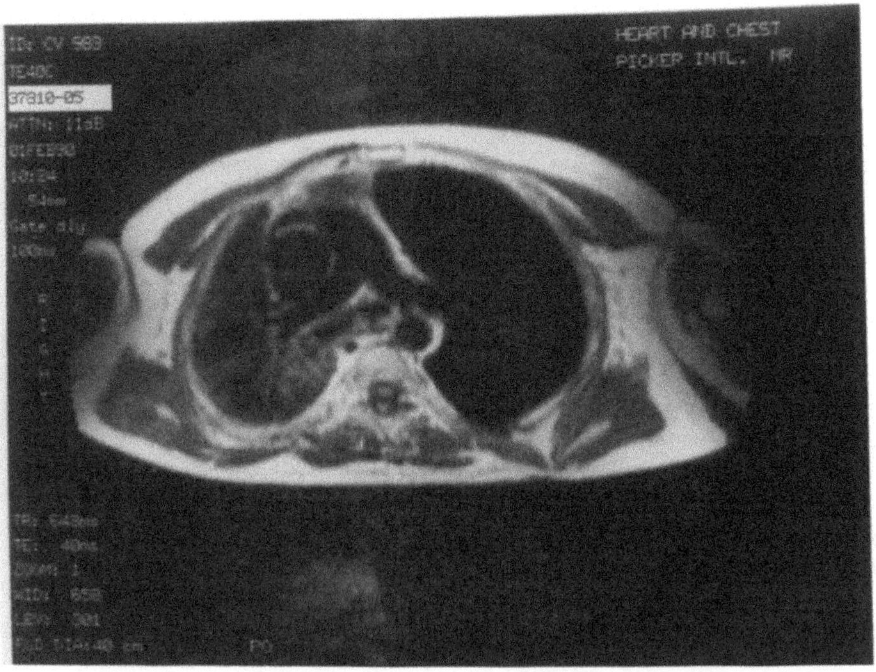
HEART AND CHEST
PICKER INTL. MR
37810-05

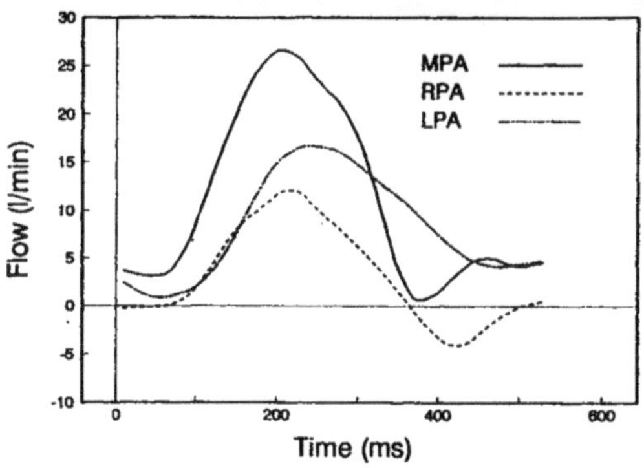
MPA
RPA
LPA
Flow (l/min)
Time (ms)
30
25
20
15
10
5
0
-5
-10
0
200
400
600

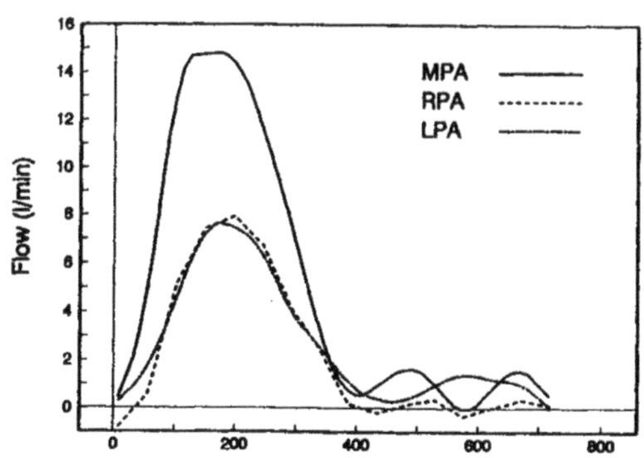
MPA
RPA
LPA
Flow (l/min)
Time (ms)
16
14
12
10
8
6
4
2
0
0
200
400
600
800

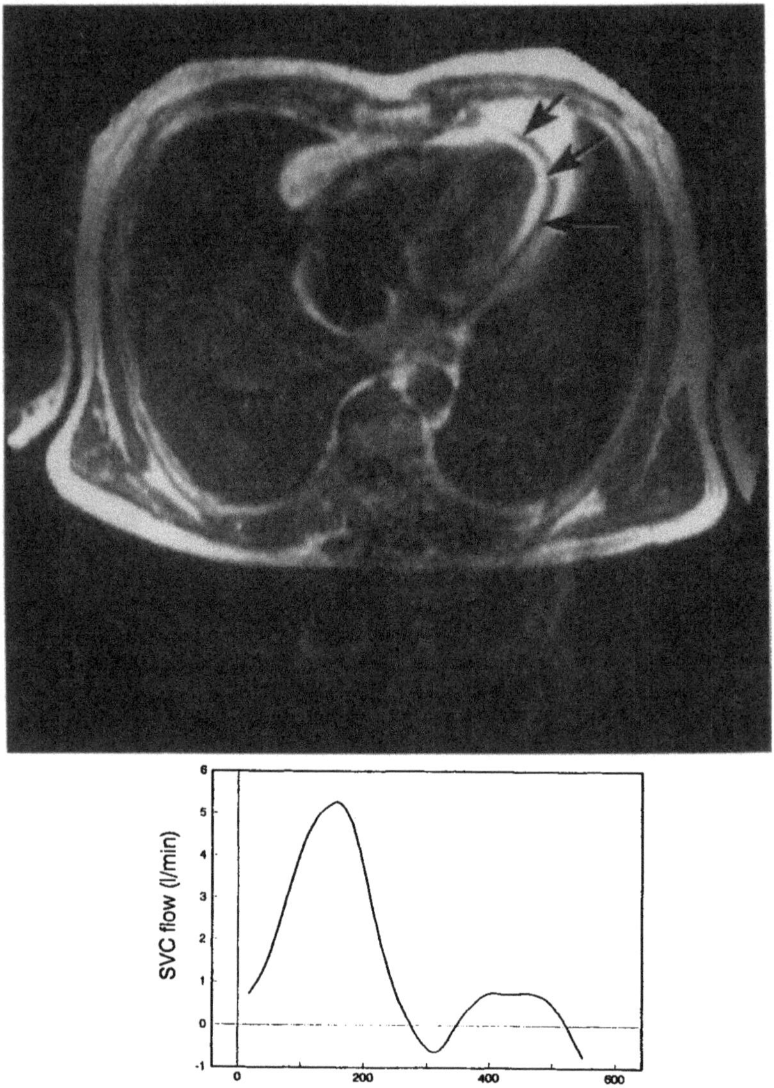

Figure 17. (a) A spin echo image in a transverse plane at mid-ventricular level in a patient with constrictive pericarditis showing pericardial thickening (arrows). (b) Superior vena caval flow curve of the previous patient measured from the complete cine velocity map acquisition throughout the cardiac cycle. The diastolic peak is attenuated which implies impaired right ventricular filling. 1: left ventricle, 2: right ventricle, and 3: right atrium.

Source: Reproduced with permission of RSNA Publications from Mohiaddin, R.H.; Wann, S.L.; Underwood, S.R.; Firmin, D.N.; Rees, R.S.O.; Longmore, D.B. *Radiology* **1990**, *177*, 537.

inferior venae cavae are 35 and 65% of cardiac output, respectively.[86] The normal pattern of caval flow has two forward peaks in ventricular systole and diastole, but this pattern is disturbed by disease. Any condition that causes impaired filling of the right ventricle reduces the diastolic peak, a pattern seen in constrictive and restrictive cardiac disease (Figure 17).[86] Tricuspid regurgitation attenuates the systolic peak of caval flow, sometimes to the extent that reverse flow occurs (Figure 18).[86] This is perhaps less helpful, because the severity of regurgitation can be assessed by cine imaging or from a comparison of right and left ventricular stroke volumes. Nevertheless, a normal systolic flow peak suggests that tricuspid regurgitation, if it is seen, is not significant. MR is commonly requested for the assessment of pericardial disease and therefore the ability to measure caval flow is an important adjunct, providing an estimate of the functional significance of the disease. In patients with obstruction of the superior vena cava, absence of flow can be confirmed and reverse flow in the azygous vein can be measured (Figure 19).[86]

F. Pulmonary Veins

Normal pulmonary venous flow measured by MR velocity mapping shows two peaks of forward flow, one during ventricular systole and the other in diastole.[87,88]

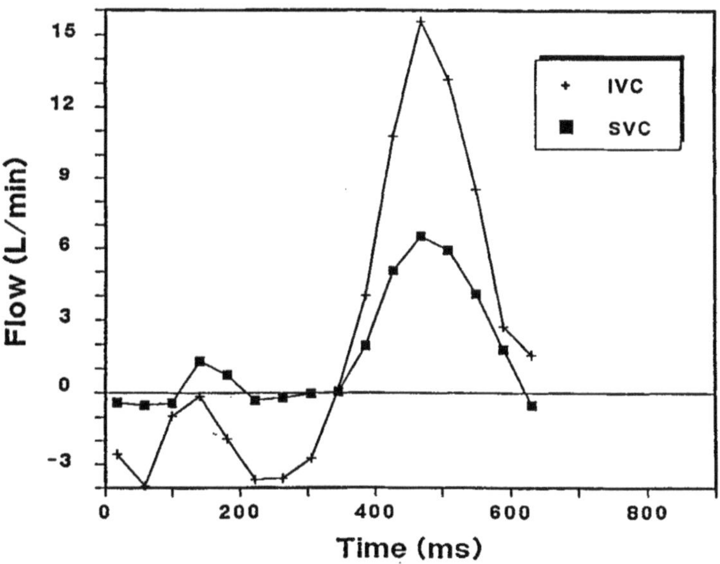

Figure 18. Superior and inferior vena caval flow curves in a patient with tricuspid valve regurgitation. The systolic peak is attenuated and there is retrograde flow in the inferior vena cava in systole.

Source: Reproduced with permission of RSNA Publications from Mohiaddin, R.H.; Wann, S.L.; Underwood, S.R.; Firmin, D.N.; Rees, R.S.O; Longmore, D.B. *Radiology* **1990**, *177*, 537.

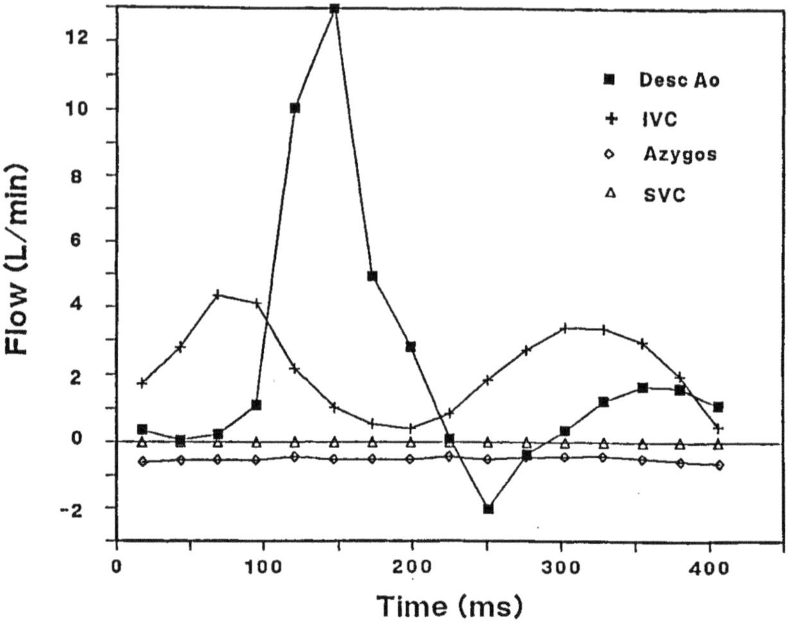

Figure 19. Flow curves in the superior and inferior venae cavae in a patient with a rhabdomyosarcoma and obstruction of the superior vena cava showing high flow with a normal pattern in the inferior vena cava, absent flow in the superior vena cava, and retrograde flow in the azygos vein.

Source: Reproduced with permission of RSNA Publications from Mohiaddin, R.H.; Wann, S.L.; Underwood, S.R.; Firmin, D.N.; Rees, R.S.O.; Longmore, D.B. *Radiology* **1990**, *177*, 537.

A small back flow during atrial systole occurs. A similar reverse flow has been demonstrated in the pulmonary veins by transoesophageal Doppler echocardiography during atrial systole and the trans-mitral "A" flow peak.[89] A non-compliant left ventricle produces high left atrial pressure during atrial systole causing the retrograde flow in the pulmonary veins to become larger than the flow through the mitral valve. An attenuated systolic forward flow peak has been demonstrated in patients with mitral valve regurgitation and the degree of this attenuation correlates well with the severity of regurgitation.[90]

G. Ventricular Filling

Blood flow through the mitral and tricuspid valves is an important parameter in the assessment of left and right ventricular diastolic function.[91,92] Normal flow through the mitral and tricuspid valves takes place in two phases which can clearly be recognized by magnetic resonance velocity mapping (Figure 20). The initial passive flow in early diastole (E wave) is produced because the ventricle relaxes

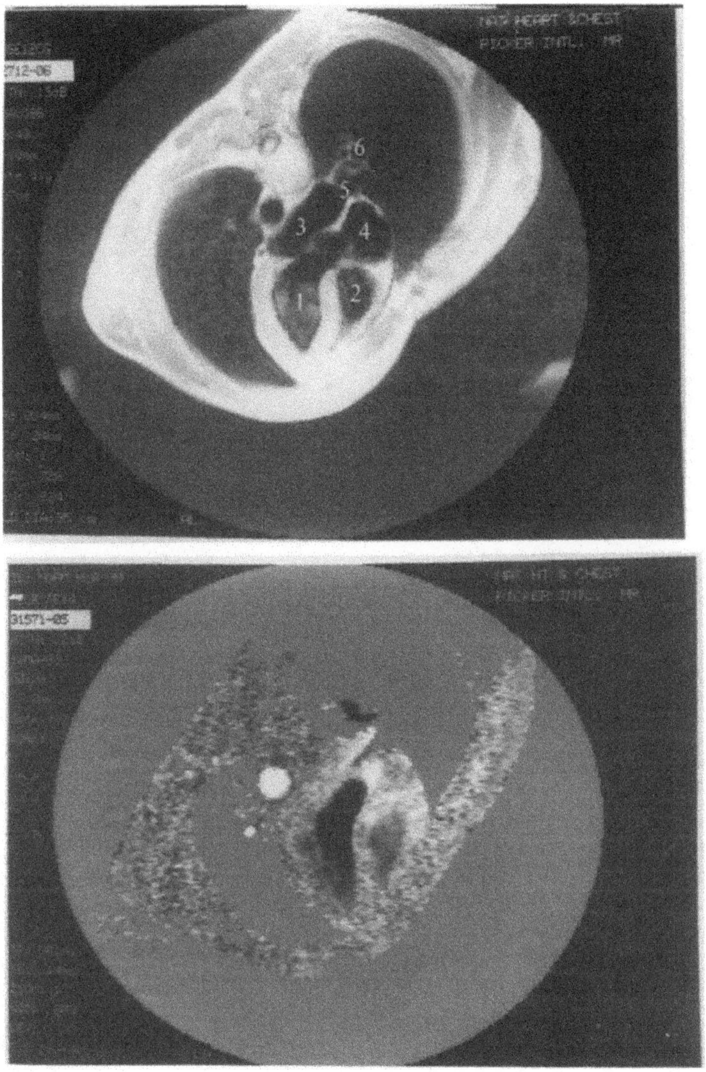

(continued)

Figure 20. Spin echo images in the horizontal long axis of the left ventricle of a healthy volunteer acquired at end ventricular systole (a), with velocity mapping acquired during mid ventricular systole (b, left), and early ventricular diastole (b, right). Velocity is encoded vertically on the images and ventricular emptying is seen in black. In diastole, filling toward the apex is seen in white. Signal intensity is proportional to velocity of flow and stationary tissue is depicted in mid-grey. (c) Mitral valve and pulmonary vein blood velocity throughout the cardiac cycle measured from the cine velocity map acquisition. 1: left ventricle, 2: right ventricle, 3: left atrium, 4: right atrium, 5: pulmonary vein, and 6: pulmonary artery branches.

Source: Reproduced with permission of Raven Press from Mohiaddin, R.H.; Amanuma, M.; Kilner, P.J.; Pennell, D.J.; Manzara, C.C.; Longmore, D.B. *J. Comput. Assist. Tomogr.* **1991**, *15*, 237.

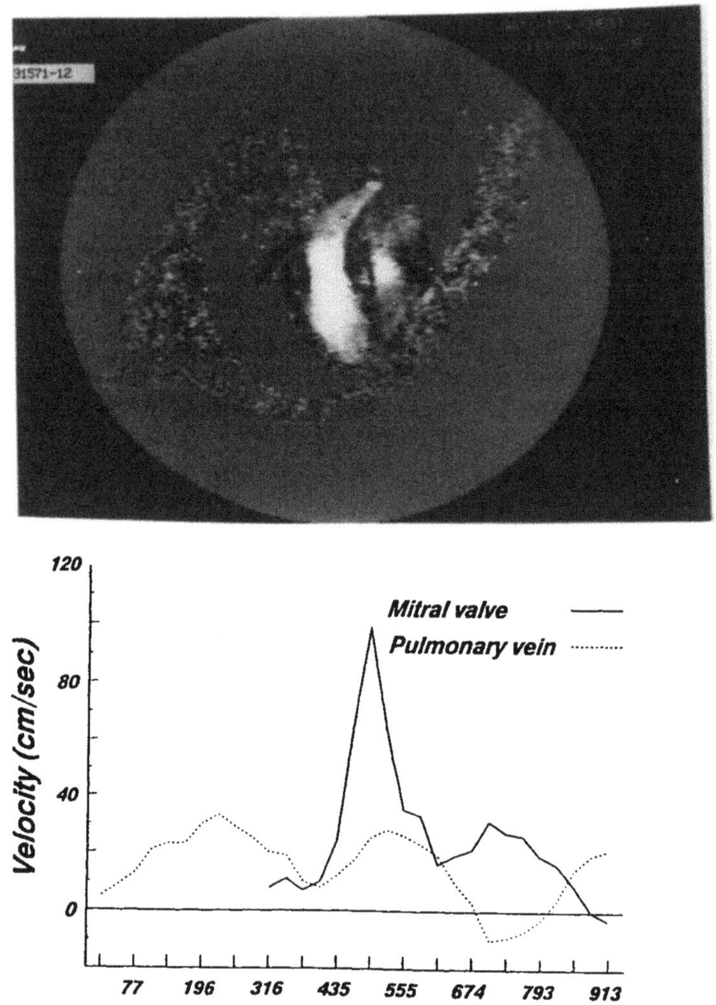

Figure 20. (Continued)

and allows blood to flow from the slightly higher pressure in the left atrium. This flow normally takes place very quickly and there is a mid-diastolic reduction or cessation of flow before a second phase of flow caused by atrial contraction (A wave).[87] Magnetic resonance velocity mapping of the E wave agrees well with those obtained by Doppler echocardiography, but magnetic resonance underestimates A wave velocity. This underestimation is due to beat length variability which occurs in the T-P time interval.[93] In mitral stenosis, initial diastolic flow persists at a high velocity because the narrow orifice of the valve cannot relieve the pressure

(a)

(b)

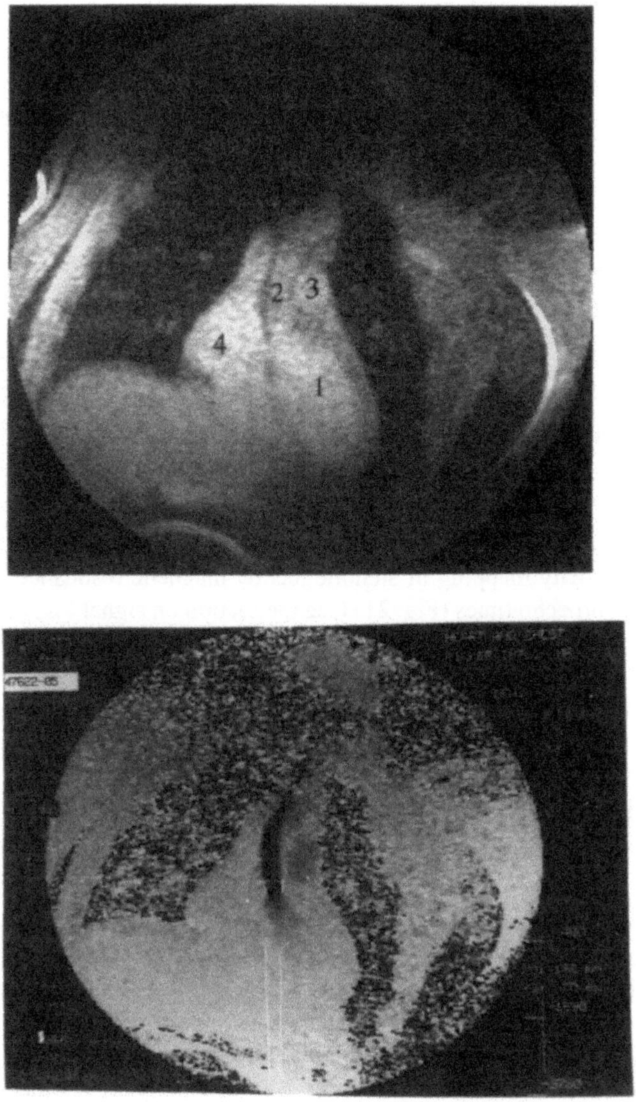

Figure 21. A gradient echo image (a) and a corresponding velocity map (b) in a patient with aortic stenosis. The unusual oblique plane was necessary to orientate the abnormal jet direction vertically for velocity encoding. Velocity profile displayed in the center of the jet recorded a peak velocity of 3.3m/s (44mmHg). 1: the left ventricle, 2: ascending aorta, 3: pulmonary artery, and 4: right atrium.

Source: Reproduced with permission of the American Heart Association from Mohiaddin, R.H.; Longmore, D.B. *Circulation* **1993**, *88*, 264.

difference between the atrium and the ventricle. In addition the normal smooth (laminar) flow through the widely open valve becomes a narrow turbulent jet through the stenotic orifice. In patients with reduced ventricular compliance, for example, in ischaemic heart disease, the A wave becomes more prominent and the E/A ratio is reduced.[93]

H. Valvular Stenosis

A stenotic valve may be assessed by measuring the flow velocity in the jet of blood passing through the stenosis. In the case of a given flow an increasingly narrow stenosis leads to an increase in the velocity of flow through the orifice. The relationship between the velocity of the jet and the difference between the pressures on either side of the stenosis can be approximated by the modified Bernoulli equation which in its simplest form is:

$$\Delta P = 4V^2 \tag{5}$$

where ΔP is the pressure drop across the stenosis (mmHg) and V is the velocity (m/s).

Accurate velocity mapping of stenotic jets by magnetic resonance requires the use of very short echo times (Fig. 21) (see the section on signal loss below).[87,94,95] The velocity map may be *through-plane* with the jet passing perpendicularly through the chosen imaging plane, or *in-plane* when the imaging plane is chosen to encompass the length of the jet (Fig. 22). In-plane imaging yields a greater number of pixels for analysis of velocity but if the jet is small, in-plane imaging is less reliable because of partial volume effects and movement of the jet out of the imaging plane. It is preferable to acquire data in both planes.

I. Flow in the Coronary Arteries

Non-invasive examination of the epicardial coronary arteries and measurement of blood flow in these arteries is of extreme importance to clinical cardiology. The development of fast imaging techniques has improved the ability of magnetic resonance to image directly the proximal portions of both the left and right coronary arteries but these are not comparable in quality with X-ray angiograms.[96,97,98,99] It is doubtful, however, that the place for MR is simply as another method of demonstrating coronary anatomy subject to interpretation but revealing no information about flow in the vessels. The feasibility of magnetic resonance imaging for measurement of flow in the epicardial coronary arteries has been demonstrated (Fig. 23).[100,101] This is an important development and opens a new opportunity for cardiovascular MRI. Coronary artery bypass grafts can be assessed using magnetic resonance velocity mapping.[102] Flow measurement within the grafts would be clinically valuable but is not always possible because of signal loss due to sternal suture and clips left after surgery. Metallic objects cause larger artefact in field echo sequences than in spin echo imaging sequences.

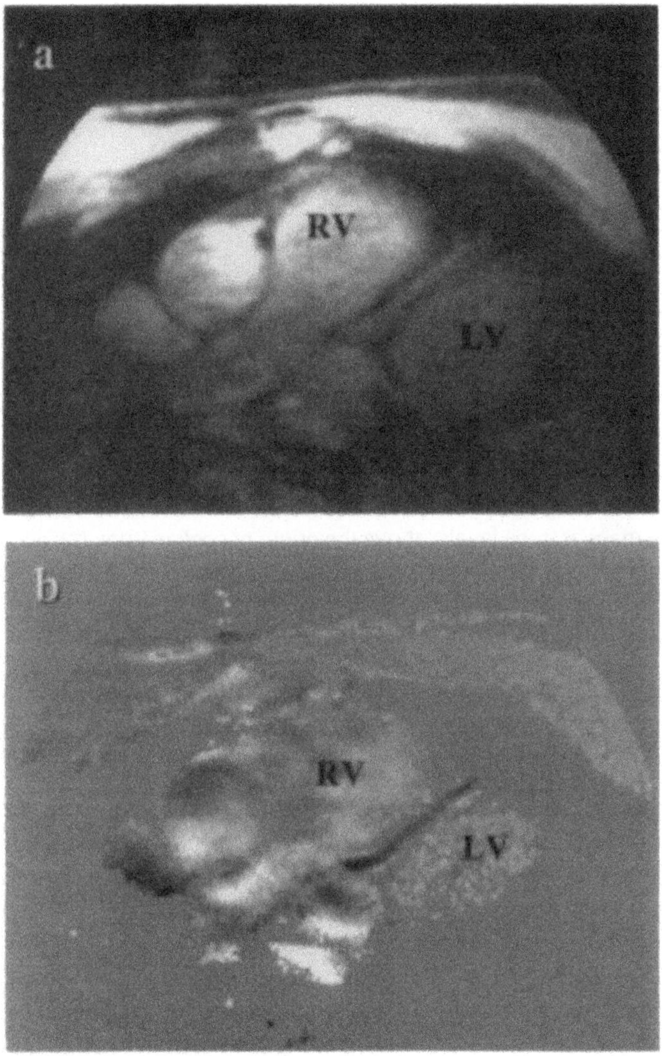

Figure 22. In-plane early diastolic magnitude images (a) and velocity maps (b) with superimposed velocity profile of a normal left anterior descending artery (LAD). The peak flow was measured at 12 cm/s (LV, left ventricle; RV, right ventricle).

Source: Reproduced with permission of Williams & Wilkins from Keegan, J.; Firmin, D.N.; Gatehouse, P.D.; Longmore, D.B. *Magn. Reson. Med.* **1994**, *31*, 526.

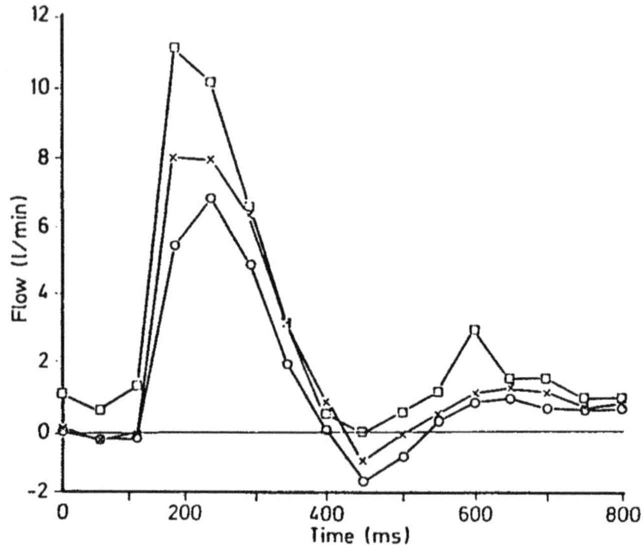

Figure 23. Flow-volume curves for the abdominal aorta obtained from three sets of cine velocity maps at three levels: above the origin of coeliac trunk (□), above (x) and below (○) the origins of renal arteries.

Source: Reproduced with permission Springer Verlag from Amanuma, M.; Mohiaddin, R.H.; Hasegawa, M.; Heshiki, A.; Longmore, D.B. *Eur. J. Radiol.* **1992**, *2*, 559.

J. Abdominal Aorta

Aortic cross-sectional area reduces gradually caudally but peak blood flow velocity throughout the abdominal aorta (110 ± 25 cm/sec) remains relatively constant.[103] Flow in the abdominal aorta and mesenteric arteries has the typical "plug" flow pattern characteristic of flow in large arteries.[103,104] The maximum aortic flow is during mid-ventricular systole and decreases toward the end of systole. Below the origin of the renal arteries, however, net aortic flow is reversed toward the renal arteries during most of diastole.[103] This reversed flow component is predominantly distributed along the posterior wall of the abdominal aorta. Flow in the superior mesenteric artery consists of a predominant systolic peak of forward flow. In diastole, however, net flow in the superior mesenteric artery nearly always remains forward (Figure 24) and this could reflect the low resistance in the vascular bed of this artery. Patients with chronic mesenteric ischemia show abnormal flow.[105] Total and differential renal blood flow has also been measured by magnetic resonance velocity mapping and the total renal blood flow has been shown to be accurate when compared with conventional methods.[106]

(a)

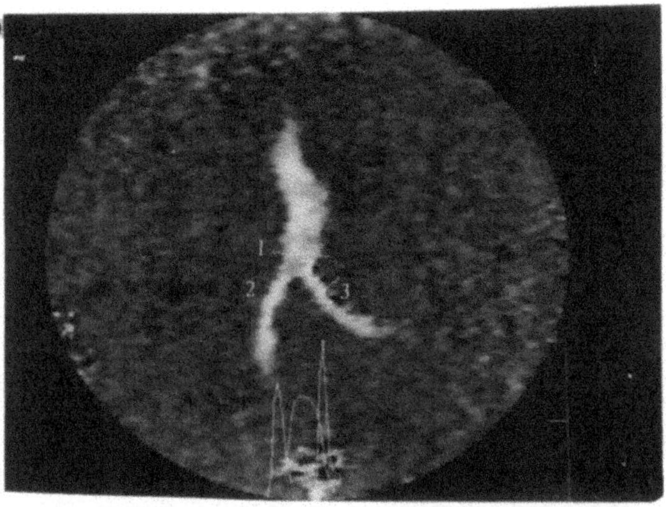

(b)

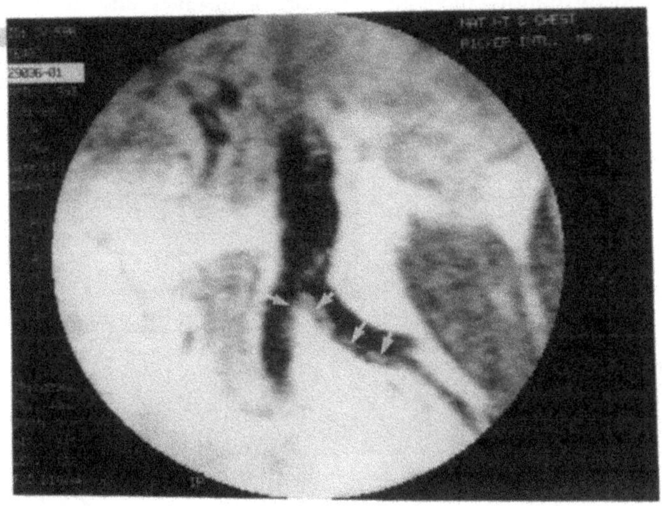

Figure 24. Multiple atheromatous plaques causing stenoses of both common iliac arteries and the origin of the left internal and external iliac arteries (arrows). (a) Spin echo image. (b) Magnetic resonance velocity map in the same plane as (c) showing velocity profiles across (1) abdominal aorta, (2) right, and (3) left iliac arteries. There is increased peak velocity in both iliac arteries compared with the aorta. The peak velocity is greater in the left iliac artery which demonstrates that the stenosis on the left is greater than that on the right.

Source: Reproduced with permission from Mohiaddin, R.H.; Sampson, C.; Firmin, D.N.; Longmore, D.B. *Eur. J. Vasc. Surg.* **1991**, *5*, 383.

K. Peripheral Arteries

Peripheral atherosclerosis produces clinical problems either by reducing blood flow or by the release of emboli from ulcerated plaques. Arterial stenosis can be detected and its severity can be assessed from measurement of changes in the velocity profile across the stenosis using magnetic resonance velocity mapping (Fig. 25). By the conservation of mass, total flow at every cross section of the vessel must be constant. When the area of the vessel is decreasing (stenosis), the average velocity must increase which requires an acceleration and a pressure difference in the direction of flow. Using magnetic resonance velocity mapping it is also possible to measure the flow rate and the flow volume curve in a vessel calculated from the mean velocity and the cross sectional area of the vessel. We have demonstrated that the flow ratio in paired vessels, like the iliac arteries, is always > 0.85 in healthy volunteers (Figure 26) and <0.85 in patients with tight stenosis of one iliac artery (Figure 27a).[107] In patients studied pre- and post-angioplasty, the pre-angioplasty ratio was 0.31 and post-angioplasty improved to 0.79 (Figure 27b). This finding was supported by improvement in the walking distance and in the peripheral Doppler pressure measurement (the posterior tibial/brachial artery index was right = 1, left = 0.88 on the pre-angioplasty recordings, and right = 1, left = 0.96 on the post-angioplasty recordings).[107] When large changes in flow occur as in the case studied pre- and post-angioplasty, quantitative flow may still give an indication of

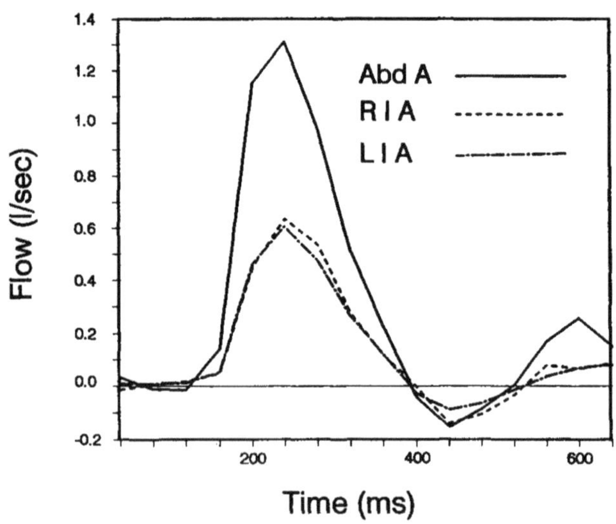

Figure 25. Flow curve in the abdominal aorta and right and left iliac arteries of a volunteer. Flow is qualitatively similar in the three arteries and quantitatively similar in the R and L iliac arteries.

Source: Reproduced with permission from Mohiaddin, R.H.; Sampson, C.; Firmin, D.N.; Longmore, D.B. *Eur. J. Vasc. Surg.* **1991**, *5*, 383.

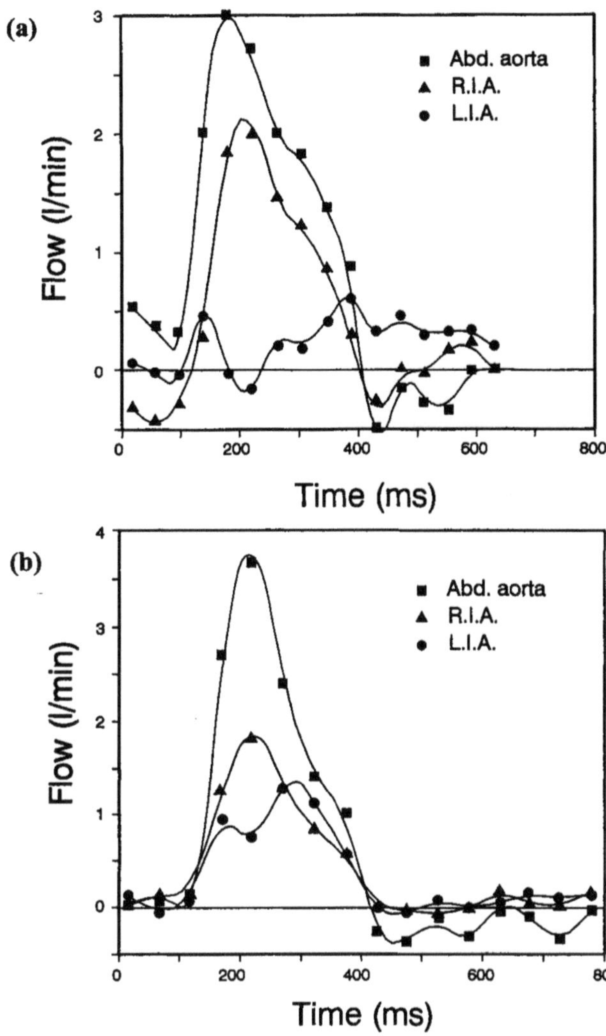

Figure 26. The flow in the abdominal aorta and R and L iliac arteries of a patient, (a) pre-angioplasty and (b) post-angioplasty. The flow in the left artery has improved and it is interesting to note that flow in the right artery has decreased.

Source: Reproduced with permission from Mohiaddin, R.H.; Sampson, C.; Firmin, D.N.; Longmore, D.B. *Eur. J. Vasc. Surg.* **1991**, *5*, 383.

the effectiveness of the procedure but of greater interest is the flow curve which is altered in diseased vessels when compared with the normal. The shape of the flow curve is another indication of the presence of a diseased vessel especially when there is disease of both iliac arteries as ischemia may be balanced and quantitative flow may be equal in the two vessels.

(a)

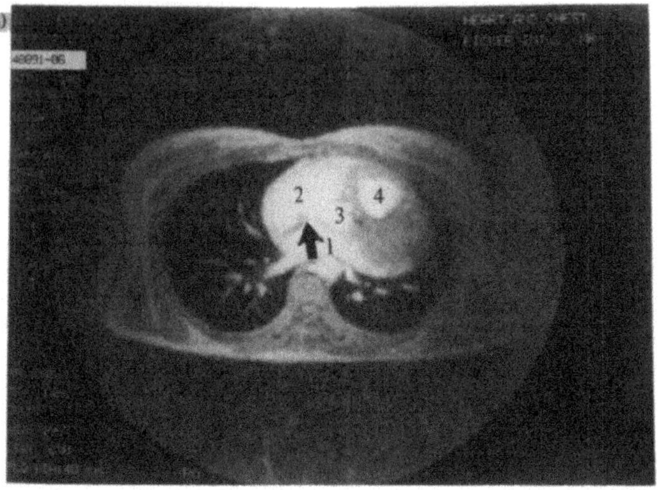

(b)

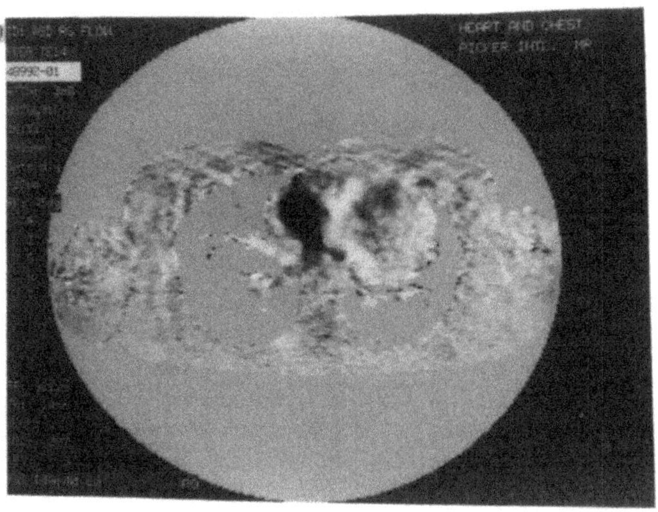

(continued)

Figure 27. (a) Gradient echo image showing a large atrial septal defect (arrow). (b) Velocity map of the same slice encoded from bottom to top of the image (posterior to anterior). The left-to-right shunt through the defect is seen black. (c) Flow volume curve of the main pulmonary artery (Qp) and the ascending aorta (Qs) in the same patient (Qp/Qs ratio = 1.9). 1: left atrium, 2: right atrium, 3: aorta, and 4: main pulmonary artery.

Source: Reproduced with permission of the American Heart Association from Mohiaddin, R.H.; Longmore, D.B. *Circulation* **1993**, *88*, 264.

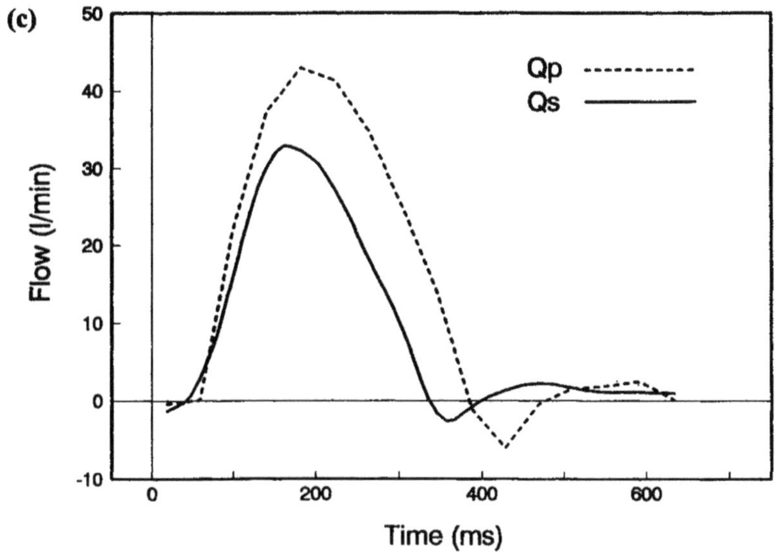

Figure 27. (Continued)

L. Congenital Heart Disease

Magnetic resonance velocity mapping has been very successful in grown up patients with congenital heart disease.[63,94,108,109] Intra- and extra-cardiac shunting can be measured in a number of ways by MR velocity mapping. Flow directly through atrial and ventricular defects can be visualized (Figures 28a and 28b), but the best method has been to measure pulmonary (Qp) to systemic (Qs) ratio directly from aortic and pulmonary flow (Figure 28c).[108,110,111] This is especially helpful in complex lesions where the possibility of surgery in such patients depends partly upon pulmonary flow which is difficult to measure by other techniques. From measurements of flow in the aorta, pulmonary artery, and right and left pulmonary arteries it is possible to calculate flow through each of the defects separately. Other structures of interest for the measurement of flow are surgically created shunts and conduits and baffle obstruction following the Mustard procedure.[94,109] The technique is also useful for following up patients after Fontan's operation and its modifications.[112] The hemodynamic significance of aortic coarctation or recoarctation can be assessed non-invasively by magnetic resonance velocity mapping.[63] The modified Bernoulli equation can be used to calculate the pressure difference across the diseased segment from peak jet velocity in the coarctation. Abnormalities in aortic volume flow and aortic flow waveforms distal to the coarctation can also be measured and could represent an additional index for monitoring the hemodynamic significance of coarctation or recoarctation in patients (Figure 29).[63]

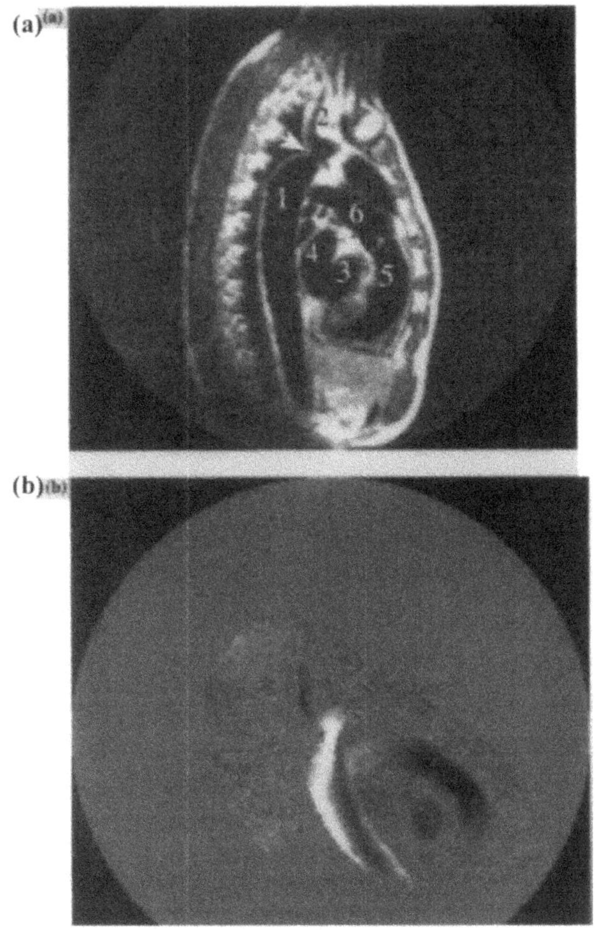

Figure 28. (a) A spin echo image (TE 40ms) acquired in a sagittal plane in a patient with aortic coarctation (large arrow), and a corresponding velocity map (b) acquired during systole with velocity encoding displayed vertically on the image. The sagittal plane was rotated during velocity map acquisition to align the velocity-encoding direction with the jet. The velocity maps indicate zero velocity as mid gray, caudal velocities in lighter shades of gray, and cranial velocities in darker shades of gray. Coarctation jet velocity is seen in white with a peak velocity of 4 m/s in the center of the jet. A reverse flow is seen in black anterior to the jet. 1: descending aorta, 2: left subclavian artery, 3: left ventricle, 4: left atrium, 5: right ventricle, and 6: Pulmonary trunk.

Source: Reproduced with permission of Elsevier Science from Mohiaddin, R.H.; Kilner, P.J.; Rees, R.S.O.; Longmore, D.B. *J. Am. Coll. Cardiol.* **1993**, *22,* 1515.

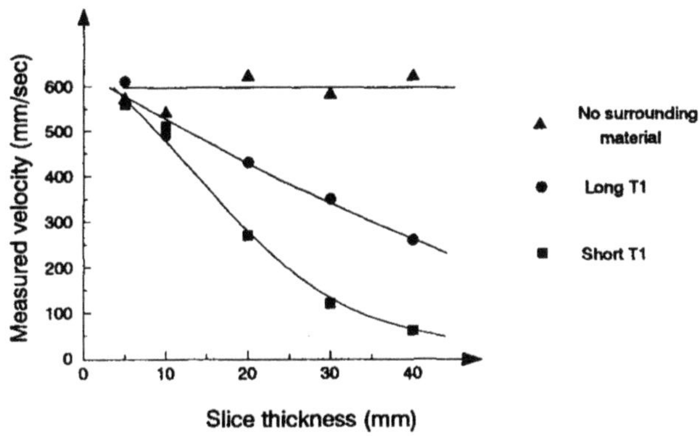

Figure 29. Plot showing the error in flow measurement due to the partial volume effect. As the slice thickness is increased more signal from surrounding fluid is included in each voxel and the potential error is thus increased. The plot includes results with three different surrounding media: air, a fluid with a short T_1 (100 msec) and a fluid with a long T_1 (1,000 msec). With air as the surrounding medium, reasonably constant flow measurements were obtained with increasing slice thickness. However, with a fluid surrounding the tube the partial volume effect resulted in a reduced flow measurement. The error depended on the relative signals between the flowing and stationary fluids; saturation of the surrounding media with a long T_1 resulting in a smaller error than that which occurred when the surrounding media was of a short T_1.

Source: Reproduced with permission of Academic Press from Firmin, D.N.; Nayler, G.L.; Kilner, P.J.; Longmore, D.B.*Magn. Reson. Med.* **1990**, *14*, 230.

IV. THE ACCURACY OF FLOW IMAGING TECHNIQUES

A. Velocity Aliasing

Because the measurement of phase is limited to a range of 2π radians, there is also a limit to the range of phase-related velocity measurement that can be made without ambiguity. This is not normally a major problem, however, as long as some knowledge of the types of velocity that can be expected; the difference in flow sensitivity between the two sequences used can be simply tailored to suit the particular application under study. The range of flow velocities that can be measured by the phase velocity mapping technique covers the complete physiological range from approximately 0.1 mm/sec[113] to several meters per second.[114,94]

Any method of MR velocity imaging that uses phase shifts is prone to similar problems. The Fourier flow method, for example, suffers from aliasing in the velocity axis of the image similar, in principle, to the aliasing of material outside the field of view that occurs when spatial phase encoding is employed.

B. Misalignment of the Direction of Flow Encoding

Misalignment between the direction of flow encoding and the direction of the flow will affect all phase methods of flow measurement. The relationship between the true flow and the measured flow is:

$$F_{\text{true}} = F_{\text{meas}} / \cos\theta \tag{6}$$

where F_{true} and F_{meas} are the true and measured flows, respectively, and θ is the angle of misalignment. This cosine relationship results in small errors; an angle of misalignment of as much as 20° would produce only a 6% error while a more realistic angle of 5° would cause an error of less than 1%. For measurements of volume flow, where mean velocity is multiplied by the area of the vessel, the reduction in measured velocity is largely compensated for by the increase in vessel area on the image.

C. Partial Volume Effect

Where in-plane flow is being studied in vessels which are relatively small there can be a problem of averaging flow phase data through the slice. This can result in a flow measurement error the extent of which depends on the amplitude of the signal contribution from the surrounding tissue which, in turn, depends on a number of factors such as the T_1 and T_2 of the tissue and the T_R, T_E, and excitation pulse angle of the imaging sequence. The result of this error is demonstrated in Figure 30 which shows a plot of the measured mean flow velocity versus slice thickness for three different surrounding media: air, a solution of long T_1 (approx 1 sec), and a solution of short T_1 (approx 100 msec) respectively. The extent of this problem very much depends on the application, however, in-plane flow measurements, particularly from small and tortuous vessels, should be treated with a degree of caution.

The Fourier flow imaging techniques[57,58,59] are not affected by these problems, by their very nature the velocity phase encoding separates out all the different velocity components whatever proportion of stationary material is present.

D. Signal Misregistration

Flow or motion between the times of slice selection, phase encoding, and frequency encoding can result in signal misregistration, where the signal from flowing material may be mispositioned with respect to any stationary material on the image. Errors of two forms can result: the first is that the blood signal cannot be compared directly with features on an anatomy image because this signal might well be out of position and distorted, the second occurs if the misregistration results in the flow signal overlaying any stationary tissue, a type of partial volume error can result where the phases of the stationary and flowing material are averaged.

Phantom experiments demonstrating these errors are presented in Figure 31 where the misregistration is between the time of phase encoding and frequency

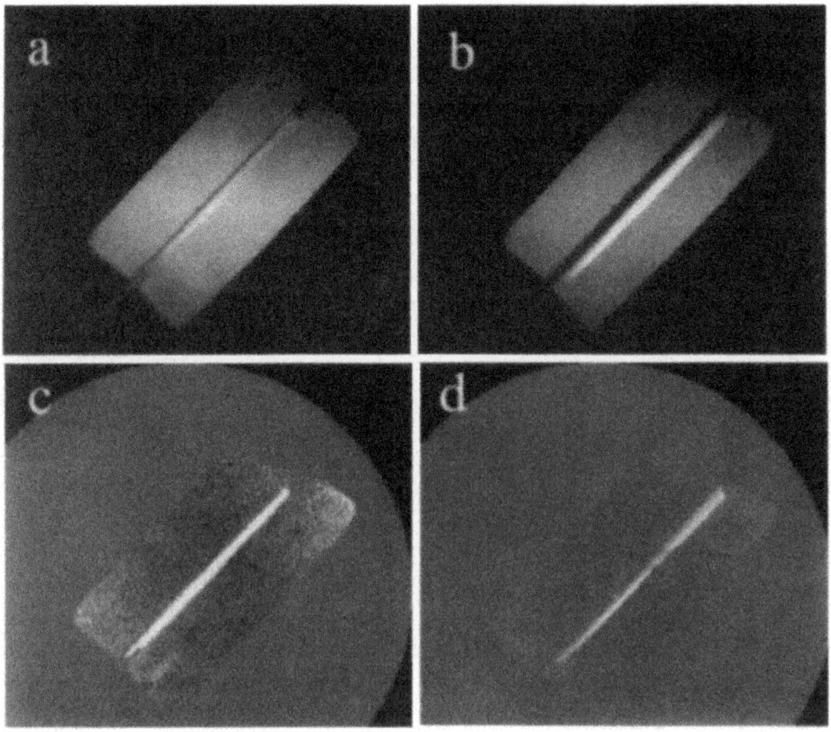

Figure 30. Demonstration of the effect of in-plane misregistration. (a) and (b) are magnitude images acquired with short (3 msec) and long (33 msec) intervals between phase encoding and reading, respectively. In (b) the signal from the fluid, which is flowing at an angle of 45° to the phase encode and read directions, can be seen to be misregistered over the signal from the stationary fluid surrounding the tube. The corresponding flow images are shown in (c) and (d), the signal averaging present due to the large misregistration in (d) resulting in a mean flow measurement of only 53% of that in (c).

Source: Reproduced with permission of Academic Press from Firmin, D.N.; Nayler, G.L.; Kilner, P.J.; Longmore, D.B. *Magn. Reson. Med.* **1990**, *14*, 230.

encoding and Figure 32 where the effect is between times of slice selection and frequency encoding. The errors introduced into quantitative flow measurement when the flow signal overlays the stationary signal can be treated as if it were a partial volume error, as described above, and therefore only applies to the phase mapping method. Misregistration effects occur in all flow imaging methods in which there are delay intervals between the different axes of spatial encoding. The effects of misregistration can be minimized in two ways: first, by designing a sequence that is very short and therefore has very short periods between the different

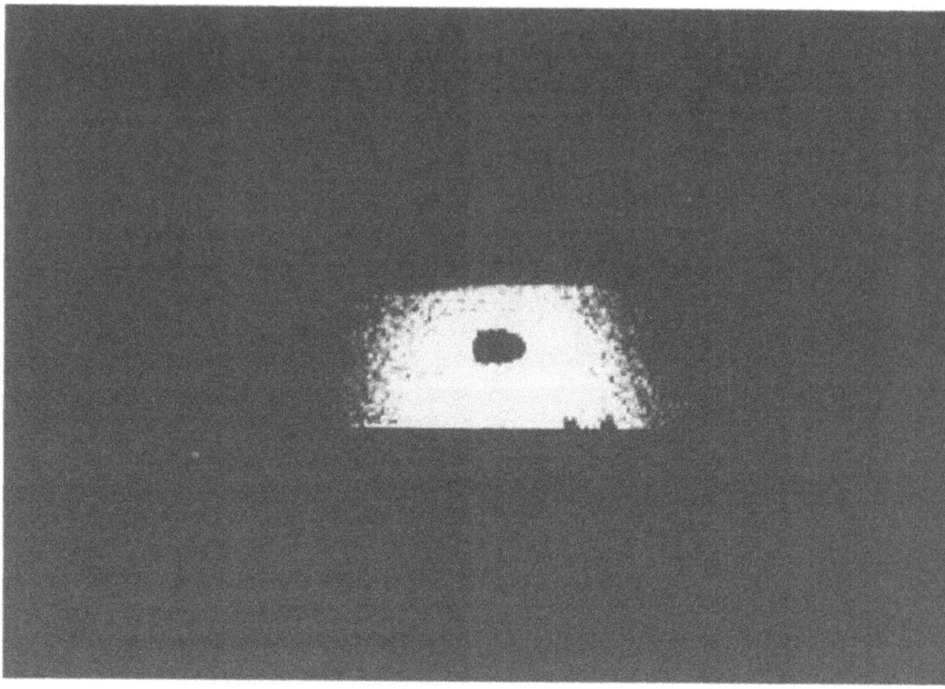

Figure 31. Magnitude image demonstrating the misregistration due to flow between the slice selection and data collection, when the direction of flow was at an angle of 45° to the slice select and read axes. The deformed, crescent shaped, flow signal resulting from the velocity profile is seen to overlay the stationary fluid signal.

Source: Reproduced with permission of Academic Press from Firmin, D.N.; Nayler, G.L.; Kilner, P.J.; Longmore, D.B. *Magn. Reson. Med.* **1990**, *14*, 230.

spatial encodings and, second, by arranging the image plane such that the flow is perpendicular to the slice and zero motion occurs in the other spatial encoding axes. Additionally, the phase encoding gradient can be designed to correct for any misregistration due to flow in its direction.

E. Signal Loss

One of the most significant factors that can affect the accuracy of flow measurement methods is that of flow-related signal loss. This is normally a result of loss of phase coherence within a voxel and, eventually, will result in an inability to detect the encoded phase of the flow signal above the random phase of the background noise. Partial signal loss alone, however, should not greatly affect the accuracy of the phase mapping flow measurement and, in the case where signal loss is the result of a spread of phase within a voxel, the mean phase will be detected. Both phase

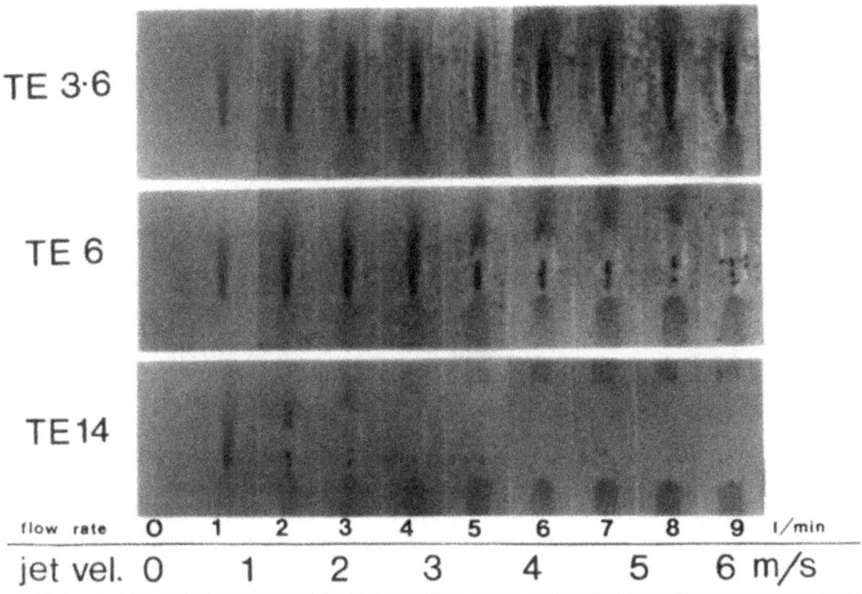

Figure 32. *In vitro* jet velocity mapping. Velocity maps of flow (increased from left to right) through stenosed tube. Only the 3.6ms TE sequence allows mapping of high velocity jets, up to a maximum tested velocity of 6.0 m/sec.

Source: Reproduced with permission of RSNA Publications from Kilner, P.J.; Firmin, D.N.; Rees, R.S.O.; Martinez, J.; Pennell, D.J.; Mohiaddin, R.H.; Underwood, S.R.; Longmore D.B. *Radiology* **1991**, *178*, 229.

velocity mapping and Fourier velocity imaging are susceptible to signal loss of one form or another, although this can normally be minimized by appropriate gradient profile design.

Two approaches have been suggested for gradient designs to minimize signal loss and the success of each is dependent on the flow type that is present. The first approach is to design complex gradient waveforms that nullify phase shifts due to the first few successive derivatives of position.[115] A problem with this method is that for every additional successive derivative an additional gradient section is required and the length of the sequence is thus extended. Probably a better method, particularly where turbulent flow is present with many higher orders of position, is to use a symmetrical gradient waveform that nullifies phase shifts due to all the odd order derivatives of position[53] and then to shorten the sequence as much as possible to reduce the effects of the even order derivatives.

Spatial variation of higher order motion components is thought to be the greatest contributing factor to signal loss and the following theory emphasizes the importance of shortening the sequence in order to reduce these effects.

The general equation describing the phase of the signal from a spin isochromat is:

$$\phi = \gamma \int G(t)x(t)dt \tag{7}$$

where $x(t)$ and $G(t)$ are functions describing the variation in the position of the spin isochromat and the gradient amplitude with time respectively. The function $x(t)$, which in fact describes the spin motion can generally be split into a series of different orders of position given by:

$$x(t) = x(0) + x(0)'t + x(0)''t^2/2! + \dots\dots\dots\dots + x(0)^{[n]}t^n/n! \tag{8}$$

From equations 9 and 10 the phase shift due to the nth order of motion, $x^{[n]}$, can be described by:

$$\phi = S_n \cdot x^{[n]} \tag{9}$$

where S_n is the sensitivity of the signal phase to the nth term and is given by:

$$S_n = G_f \cdot c_n \cdot \tau^{(n+1)} \tag{10}$$

where G_f is a factor relating to the gradient waveforms shape and amplitude, c_n is a constant of proportionality relevant to the nth motion term, and τ is the duration of the gradient waveform.

In imaging there is normally a constraint such that:

$$G_f \cdot \tau = C \tag{11}$$

where C is a constant (i.e., a higher gradient strength is required to produce the same effect in a shorter time). It therefore follows that:

$$S_n = C \cdot c_n \cdot \tau^n \tag{12}$$

demonstrating that progressively higher motion terms become progressively more significant with longer durations of the gradient profile. The signal loss is therefore reduced as this duration is shortened and this enables the quantitative study of a wide range of flow types.[116,114] This is well demonstrated in Figure 21 by comparing the maximum measurable flow jet velocity, when flow is accelerated through a narrowing in a tube.

Signal loss of the type described is much less of a problem for the Fourier flow imaging method. In this case the Fourier transform is used to separate out constituent velocities.

F. Measurement of Lumen Area

For volume flow measurement where a measure of the lumen cross-sectional area is required an additional error may be introduced. For a vessel whose diameter is measured to be 20 pixels, the minimum uncertainty in the diameter is 5% and in the area is 10%. Consequently, the accuracy of any volume flow measurement will at best be +/– 10%.

G. Method of Cardiac Synchronization

The method of cardiac synchronization has been shown to have implications on the accuracy of pulsatile blood flow measurement using phase velocity mapping.[117,118] There have been two approaches: the first, known as prospective cardiac gating or cardiac triggering, was initially developed and used for the early cardiac MR scanning and can be used for any type of MR imaging sequence. The second, known as retrospective gating, has been developed for synchronizing fast repetition cine acquisitions to the cardiac cycle. Both techniques have their own advantages and disadvantages as will be discussed below.

Both prospective and retrospective forms of gating involve monitoring the ECG or other cardiac related signal with the regular detection of a prominent part, such as the QRS complex of the ECG or the arterial flow pulse of a pulsimeter waveform. With the prospective system the detection is used to initiate one or a number of sequence repetitions following a predefined delay and separated by a repetition time TR.[119] The sequence repetitions can be of the same phase encoding acquisition for a number of different slices, the same phase encoding acquisition for a number of frames from a cine scan or a number of different phase encoding acquisitions contributing toward one image. If the sequence can be repeated very quickly then a complete image can be acquired within a fraction of a cardiac cycle; in this case a delay is used following the cardiac detection to the start of the sequence repetitions to define the period of the cardiac cycle over which the image is acquired. For extremely fast techniques such as echo-planar, several images can be acquired within a cardiac cycle and rather as for the phase encoding steps of the slower methods, either multiple slices or multiple cine frames can be acquired at defined times following the cardiac detection.

One problem with this prospective method of triggering is that of cardiac arrythmia which invalidates the assumption that the cardiac cycles used to construct the image are identical. Because of this, for cine flow imaging, there is an enforced requirement to leave a period of delay between the last cine frame and the next trigger, and this results in a relatively high signal (sometimes known as the lightning artefact) on the first cine image frame because of the increase longitudinal relaxation that occurs during the delay. The additional problem is that part of the heart cycle cannot be imaged, this being the end-diastolic part when using the R-wave of the ECG to trigger.

The retrospective method of cardiac gating was developed to enable cine field echo and flow images to be acquired throughout the entire cardiac cycle, to remove the lightning artefact and to reduce the other artifacts that result because of cardiac arrhythmia.[120,121,122] The method involves the continuous repetition of the sequence at a constant *TR* while also monitoring and recording a cardiac waveform. There are two ways of implementing the technique: first by advancing the phase encoding at a regular interval that is longer than the longest cardiac cycle, or second by changing the phase encoding step immediately after each cardiac detection. The first method is the easier to implement; the cardiac cycle is monitored for a period of approximately 30 seconds and the mean cardiac cycle duration calculated. Then from the *TR* of the predefined cine sequence, the number of time frames that will fit into the mean cardiac cycle plus 20% is calculated; the phase encoding is then stepped regularly after this number of frames. The phase encoding steps now occur at arbitrary times in the cardiac cycle and data reordering is necessary before image reconstruction. The disadvantage of this approach is that the scan time is extended by approximately 20% and the data for particular phase encoding steps are likely to be acquired over two different cardiac cycles. The second method of retrospective gating overcomes these problems; the phase encoding gradient is incremented immediately after each cardiac detection and depending on the cardiac duration a variable number of sequence repetitions and data acquisitions may be employed for each phase encoding gradient. However, the technique requires very flexible hardware enabling the phase encoding gradient to be incremented on the fly, an option not available on the majority of systems.

For phase velocity mapping, when prospective gating is used data are acquired for the different velocity sensitivities on successive cardiac cycles. This enables high temporal resolution flow data to be acquired throughout the majority of the cardiac cycle. With retrospective gating, in order to minimize the scan time, a method of alternately acquiring the different velocity sensitivity data on successive sequence repetitions has been used. This has the effect of doubling the *TR* of the sequences and has been shown to result in a high frequency filtering of the pulsatile blood flow waveform. Methods of reducing or removing this error include using a shorter *TR*, minimizing the interpolation window and running two separate scans with the order of velocity encoding reversed before resorting the data appropriately.[117] Alternatively, the method of acquisition could be changed so that the velocity sensitivity of the sequence is changed less frequently; on successive cardiac cycles or on successive periods 20% longer than the mean cardiac cycle depending on the method of retrospective gating method that is being used.

ACKNOWLEDGMENTS

The authors would like to thank the staff of the Magnetic Resonance Unit, Royal Brompton Hospital, and the National Heart and Lung Institute for all their helpful advice and discussions. We are also particularly grateful to the Coronary Artery Disease Research Association (CORDA, The Heart Charity) for its financial support.

REFERENCES

1. Bloembergen, N.; Purcell, E.M.; Pound, R.V. Relaxation effects in nuclear magnetic resonance absorption. *Phys. Rev.* **1948**, *73*, 679–712.
2. Bloch, F.; Hansen, W.W.; Packard, M.E. Nuclear induction. *Phys. Rev.* **1946**, *69*, 127.
3. Purcell, E.M.; Torrey, H.C.; Pound, R.V. Resonance absorption by nuclear magnetic moments in a solid. *Phys. Rev.* **1946**, *69*, 37–38.
4. Hahn, E.L. Spin echoes. *Phys. Rev.* **1950**, *80*, 580–594.
5. Suryan, G. Nuclear resonance in flowing liquids. *Proc. Indian Academy of Science (A)* **1951**, *33*, 107–111.
6. Zhernovoi, A.I.; Latyshev, G.D. *Nuclear Magnetic Resonance in a Flowing Liquid* (translated from Russian by Turton, C.N.; Turton, T.E.). Consultants Bureau, New York, 1965.
7. Jones, D.W.; Child, T.F. NMR in flowing systems. *Adv. Magn. Reson.* **1976**, *8*, 123–148.
8. Singer J.R. NMR diffusion and flow measurement and an introduction to spin phase graphing. *J. Phys. E: Sci. Instrum.* **1987**, *11*, 281–291.
9. Axel L. Blood flow effects in magnetic resonance imaging. *AJR* **1984**, *143*, 1157–1166.
10. Bradley, W.G. Flow phenomenon in MR imaging. *AJR* **1988**, *150*, 983–994.
11. Alfidi, R.J.; Masaryk, T.J.; Haacke, E.M.; Lenz, G.W.; et al. MR angiography of peripheral, carotid, and coronary arteries. *AJR* **1987**, *149*, 1097–1109.
12. Bowman, R.L.; Kudravcev, V. Blood flowmeter utilizing nuclear magnetic resonance. *IRE Trans. Med. Elec.* **1959**, *6*, 267–269.
13. Singer, J.R. Blood flow rates by nuclear magnetic resonance measurements. *Science* **1959**, *130*, 1652–1653.
14. Singer, J.R. Flow rates using nuclear or electron paramagnetic resonance techniques with applications to biological and chemical processes. *J. Appl. Phys.* **1960**, *31*, 125–127.
15. Battocletti, J.H.; Linehan, J.H.; Larson, S.J.; Sances, A.; et al. Analysis of a nuclear magnetic resonance blood flowmeter for pulsatile flow. *IEEE Trans. Biomed. Eng.* **1972**, *BME-19*, 403–407.
16. Battocletti, J.H.; Sances, A.; Larson, S.J.; Evans, S.M.; et al. Clinical applications and theoretical analysis of NMR blood flowmeter. *Biomed. Eng.* **1975**, *10*, 12–15.
17. Halbach, R.E.; Battocletti, J.H.; Sances, A.; Larson, S.J.; et al. Blood flow detection using the flat crossed-coil nuclear magnetic resonance flowmeter. *IEEE Trans. Biomed. Eng.* **1981**, *BME-28*, 40–42.
18. Halbach, R.E.; Battocletti, J.H.; Salles-Cunha, S.X.; Sances, A. The NMR blood flowmeter—design. *Med. Phys.* *8* 444–451.
19. Salles-Cunha, S.X.; Halbach, R.E.; Battocletti, J.H.; Sances, A. The NMR blood flowmeter—applications. *Med. Phys.* **1981**, *8*, 452–458.
20. Singer, J.R. Measuring fluid velocity by nuclear resonance. *Electronics* **1960** *33*, 77–78.
21. Morse, O.C.; Singer, J.R. Blood velocity measurements in intact subjects. *Science* **1970**, *170*, 440–441.
22. Genthe, W.K.; Vander Heyden, W.R.; Battocletti, J.H.; McCormick, W.S.; et al. NMR applied to flow measurement. An obstructionless, digital flowmeter. *Instr. Tech.* **1968**, *15*, 53–58.
23. Hahn, E.L. Detection of sea-water motion by nuclear precession. *Geophys. Res.* **1960**, *65*, 776–777.
24. Lucas, P.G.J.; Penman, D.A.; Tyler, A.; Vavasour, E. Measurement of fluid flow profiles using pulsed NMR. *J. Phys. E: Sci. Instr.* **1977**, *10*, 1150–1152.
25. Singer, J.R. Blood flow measurements by NMR of the intact body. *IEEE Trans. Nucl. Sci.* **1980**, *NS-27*, 1245–1248.
26. Packer, K.J. The study of slow coherent molecular motion by pulsed nuclear magnetic resonance. *Mol. Phys.* **1969**, *17*, 355–368.
27. Grover, T.; Singer, J.R. NMR spin-echo flow measurements. *J. Appl. Phys.* **1971**, *42*, 938–940.

28. Garroway, A.N. Velocity measurements in flowing fluids by NMR. *J. Phys. D: Appl. Phys.* **1974**, *7*, L159–L163.
29. Crooks, L.E.; Kaufman, L. NMR imaging of blood flow. *Brit. Med. Bul.* **1984**, *40*, 167–169.
30. Axel, L. Blood flow effects in magnetic resonance imaging. *AJR* **1984**, *143*, 1157–1166.
31. Grant, J.P.; Back, C. NMR rheotomography: Feasibility and clinical potential. *Med. Phys.* **1982**, *9*, 188–193.
32. George, C.R.; Jacobs, G.; MacIntyre, W.J.; Lorig, R.J.; et al. Magnetic resonance signal intensity patterns obtained from continuous and pulsatile flow models. *Radiology* **1984**, *151*, 421–428.
33. Waluch, V.; Bradley, W.G. NMR even echo rephasing in slow laminar flow. *J. Comput. Assist. Tomogr.* **1984**, *8*, 594–598.
34. Singer, J.R.; Crooks, L.E. Nuclear magnetic resonance blood flow measurements in the human brain. *Science* **1983**, *221*, 654–656.
35. Carr, H.Y.; Purcell, E.M. Effects of diffusion on free precession in nuclear magnetic resonance experiments. *Phys. Rev.* **1954**, *94*, 630–638.
36. Bradley, W.G.; Waluch, V.; Lai, K.S.; Fernandez, E.J.; et al. The appearance of rapidly flowing blood on magnetic resonance images. *AJR* **1984**, *143*, 1167–1174.
37. Bradley, W.G.; Waluch, V. Blood flow: Magnetic resonance imaging. *Radiology* **1985**, *154*, 443–450.
38. Von Schulthess, G.K.; Higgins, C.B. Blood flow imaging with MR: Spin phase phenomena. *Radiology* **1985**, *157*, 687–695.
39. Perman, W.H.; Moran, P.R.; Moran, R.A.; Bernstein, M.A. Artifacts from pulsatile flow in MR imaging. *J. Comput. Assist. Tomogr.* **1986**, *10*, 473–483.
40. Thulborn, K.R.; Waterton, J.C.; Radda, G.K. Proton imaging for in vivo blood flow and oxygen consumption measurements. *J. Magn. Reson.* **1981**, *45*, 188–191.
41. Feinberg, D.A.; Crooks, L.E.; Hoenninger, J.; Arakawa, M.; et al. Pulsatile blood velocity in human arteries displayed by magnetic resonance imaging. *Radiology* **1984**, *153*, 177–180.
42. Shimizu, K.; Matsuda, T.; Sakurai, T.; Fujita, A.; et al. Visualization of moving fluid: Quantitative analysis of blood flow velocity using MR imaging. *Radiology* **1986**, *159*, 195–199.
43. Axel, L.; Shimakawa, A.; MacFall, J. A time-of-flight method of measuring flow velocity by magnetic resonance imaging. *Mag. Res. Imaging* **1986**, *4*, 199–205.
44. Matsuda, T.; Shimizu, K.; Sakurai, T.; Fujita, A.; et al. Measurement of aortic blood flow with MR imaging: Comparative study with Doppler US. *Radiology* **1987**, *162*, 857–861.
45. Edelman, R.R.; Mattle, H.P.; Kleefield, J.; Silver, M.S. Quantification of blood flow with dynamic MR imaging and presaturation bolus tracking. *Radiology* **1989**, *171*, 551–556.
46. Izen, S.H.; Haacke, E.M. Measuring non-constant flow in magnetic resonance imaging. *IEEE Trans. Med. Imaging* **1990**, *9*, 450–460.
47. van Dijk, P. Direct cardiac NMR imaging of heart wall and blood flow velocity. *J. Comput. Assist. Tomogr.* **1984**, *8*, 429–436.
48. Bryant, D.J.; Payne, J.A.; Firmin, D.N.; Longmore, D.B. Measurement of flow with NMR imaging using a gradient pulse and phase difference technique. *J. Comput. Assist. Tomogr.* **1984**, *8*, 588–593.
49. Moran, P.R. A flow zeugmatographic interlace for NMR imaging in humans. *Magn. Reson. Imaging* **1982**, *1*, 197–203.
50. Firmin, D.N.; Nayler, G.L.; Kilner, P.J.; Longmore, D.B. The application of phase shifts in NMR for flow measurement. *Magn. Reson. Med.* **1990**, *14*, 230–241.
51. Young, I.R.; Bydder, G.M.; Payne, J.A. Flow measurement by the development of phase differences during slice formation in MR imaging. *Magn. Reson. Med.* **1986**, *3*, 175–179.
52. Ridgway, J.P.; Smith, M.A. A technique for velocity imaging using magnetic resonance imaging. *Brit. J. Radiol.* **1986**, *59*, 603–607.
53. Nayler, G.L.; Firmin, D.N.; Longmore, D.B. Blood flow imaging by cine magnetic resonance. *J. Comput. Assist. Tomogr.* **1986**, *10*, 715–722.

54. Haacke, E.M.; Lenz, G.W. Improving MR image quality in the presence of motion by using rephasing gradients. *AJR* **1987**, *148*, 1251–1258.
55. Firmin, D.N.; Nayler, G.L.; Klipstein, R.H.; Underwood, S.R.; et al. *In vivo* validation of MR velocity imaging. *J. Comput. Assist. Tomogr.* **1987**, *11*, 751–756.
56. Longmore, D.B.; Klipstein, R.H.; Underwood, S.R.; Firmin, D.N.; et al. Dimensional accuracy of magnetic resonance in studies of the heart. *Lancet* **1985**, *1*, 1360–1362.
57. Redpath, T.W.; Norris, D.G.; Jones, R.A.; Hutchinson, M.S. A new method of NMR flow imaging. *Phys. Med. Biol.* **1984**, *29*, 891–898.
58. Feinberg, D.A.; Crooks, L.E.; Sheldon, P.; Hoenninger III, J.; et al. Magnetic resonance imaging and velocity vector components of fluid flow. *Magn. Reson. Med.* **1985**, *2*, 555–566.
59. Hennig, J.; Mueri, M.; Brunner, P.; Friedburg, H. Quantitative flow measurement with the fast Fourier flow technique. *Radiology* **1988**, *166*, 237–240.
60. Firmin, D.N.; Klipstein, R.H.; Hounsfield, G.L.; Paley, M.P.; et al. Echo-planar high-resolution flow velocity mapping. *Mag. Reson. Med.* **1989**, *12*, 316–327.
61. Mueller, E.; Laub, G.; Grauman, R.; Loeffler, W. RACE - Real time ACquisition and Evaluation of pulsatile blood flow on a whole body MRI unit. In: *Proceedings, 7th Ann. Meeting Soc. Magn. Reson. Med.* **1988**, *2*, 729.
62. Gatehouse, P.D.; Firmin, D.N.; Collins, S.; Longmore, D.B. Real time blood flow imaging by spiral scann phase velocity mapping. *Magn. Reson. Med.* **1994**, *31*, 504.
63. Mohiaddin, R.H.; Kilner, P.J.; Rees, R.S.O.; Longmore, D.B. Magnetic resonance volume flow and jet velocity mapping in aortic coarctation. *J. Am. Coll. Cardiol.* **1993**, *22*, 1515–1521.
64. Klipstein, R.H.; Firmin, D.N.; Underwood, S.R.; Rees, R.S.O.; et al. Blood flow patterns in the human aorta studied by magnetic resonance. *Br. Heart J.* **1987**, *58*, 316–323.
65. Bogren, H.G.; Underwood, S.R.; Firmin, D.N.; Mohiaddin, R.H.; et al. Magnetic resonance velocity mapping in aortic dissection. *Br. J. Radiol.* **1988**, *61*, 456–462.
66. Kilner, P.J.; Yang, G.Z.; Mohiaddin, R.H.; Firmin, D.N.; et al. Helical and retrograde secondary flow patterns in the aortic arch studied by three-directional magnetic resonance velcoity mapping. *Circulation* **1993**, *88*, 2235–2247.
67. Bogren, H.G.; Mohiaddin, R.H.; Klipstein, R.H.; Firmin, D.N.; et al. The function of the aorta in ischemic heart disease: A magnetic resonance and angiographic study of aortic compliance and blood flow patterns. *Am. Heart J.* **1989**, *118*, 234–247.
68. Dulce, M.C.; Mostbeck, G.H.; O'Sullivan, R.N.; Cheitlin, M.; et al. Severity of aortic regurgitation: Interstudy reproducibility of measurements with velocity-encoded cine MR imaging. *Radiology* **1992**, *185*, 235–240.
69. Mohiaddin, R.H.; Yang, G.Z.; Kilner, P.J. Visualization flow by vector analysis of multidirectional cine magnetic resonance velocity mapping. *J. Comput. Assist. Tomogr.* **1994**, *18*, 383–392.
70. Mohiaddin, R.H.; Firmin, D.N.; Longmore, D.B. Age-related changes of human aortic flow wave velocity measured non-invasively by magnetic resonance imaging. *J. Applied Physiol.* **1993**, *74*, 492–497.
71. Dumoulin, C.L.; Doorly, D.J.; Caro, C.G. Quantitative measurement of velocity at multiple positions using comb excitation and Fourier velocity encoding. *J. Mag. Reson. Med.* **1993**, *29*, 44–52.
72. Mohiaddin, R.H.; Underwood, S.R.; Bogren, H.G.; Firmin, D.N.; et al. Regional aortic compliance studied by magnetic resonance imaging: The effects of age, training, and coronary artery disease. *Br. Heart J.* **1989**, *62*, 90–96.
73. Stefanadis, C.; Wooley, C.F.; Bush, A.C.; Kolibash, A.J.; et al. Aortic distensibility abnormalities in coronary artery disease. *Am. J. Cardiol.* **1987**, *59*, 1300–1304.
74. Dart, A.M.; Lacombe, F.; Yeoh, J.K.; Cameron, J.D.; et al. Aortic distensibility in patients with hypercholesterolaemia, coronary artery disease, or cardiac transplantation. *Lancet* **1991**, *338*, 270–273.

75. Goldman, A.P.; Kotler, M.N.; Scanlon, M.H.; Ostrum, B.J.; et al. Magnetic resonance imaging and two dimensional echocardiography. Alternative approach to aortography in diagnosis of aortic dissecting aneurysm. *Am. J. Med.* **1986**, *80*, 1225–1229.
76. Goldman, A.P.; Kotler, M.N.; Scanlon, M.H.; Ostrum, B.; et al. The complementary role of magnetic resonance imaging, Doppler echocardiography, and computed tomography in the diagnosis of dissecting thoracic aneurysms. *Am. Heart J.* **1986**, *111*, 970–981.
77. Nienaber, C.A.; Spielmann, R.P.; von Kodolitsch, Y.; Siglow, V.; et al. Diagnosis of thoracic aortic dissection: Magnetic resonance imaging versus transoesophageal echocardiography. *Circulation* **1992**, *85*, 434–447.
78. Bogren, H.G.; Underwood, S.R.; Firmin, D.N.; Mohiaddin, R.H.; et al. Magnetic resonance velocity mapping in aortic dissection. *Br. J. Radiol.* **1988**, *61*, 456–462.
79. Chang, J.M.; Friese, K.; Caputo, G.R.; Kondo, C.; et al. MR measurement of blood flow in the true and false channel in chronic aortic dissection. *J. Comput. Assist. Tomogr.* **1991**, *15*, 418–423.
80. Bogren, H.G.; Klipstein, R.H.; Mohiaddin, R.H.; Firmin, D.N.; et al. Pulmonary artery distensibility and blood flow patterns: A magnetic resonance study of normal subjects and of patients with pulmonary arterial hypertension. *Am. Heart J.* **1989**, *118*, 990–999.
81. Kondo, C.; Caputo, G.R.; Masui, T.; Foster, E.; et al. Pulmonary hypertension: Pulmonary flow quantification and flow profile analysis with velocity-enhanced cine MR imaging. *Radiology* **1992**, *183*, 751–758.
82. Frank, H.; Globits, S.; Neuhold, A.; Kneussl, M.; et al. Detection and quantification of pulmonary artery hypertension with MR imaging: Results in 23 patients. *AJR* **1993**, *161*, 27–31.
83. Mohiaddin, R.H.; Paz, R.; Theodoropolus, S.; Firmin, D.N.; et al. Magnetic resonance characterization of pulmonary arterial blood flow following single lung transplantation. *J. Thoracic Cardiovasc. Surg.* **1991**, *101*, 1016–1023.
84. Silverman, J.M.; Julien, P.J.; Herfkens, R.J.; Pelc, N.J. Quantitative differential pulmonary perfusion: MR imaging versus radionuclide lung scanning. *Radiology* **1993**, *189*, 699–701.
85. Mohiaddin, R.H.; Paz, R.; Theodoropolous, R.; Longmore, D.B.; et al. Flow pulsatility and wall distensibility measured by magnetic resonance imaging in the central pulmonary arteries of patients with single lung transplantation. American College of Cardiology, Dallas. *J. Am. Coll. Cardiol.* **1992**, *19*, 297A (Abstract).
86. Mohiaddin, R.H.; Wann, S.L.; Underwood, S.R.; Firmin, D.N.; et al. Vena caval flow: Assessment with cine MR velocity mapping. *Radiology* **1990**, *177*, 537–541.
87. Mohiaddin, R.H.; Amanuma, M.; Kilner, P.J.; Pennell, D.J.; et al. Magnetic resonance phase-shift velocity mapping of mitral and pulmonary venous flow. *J. Comput. Assist. Tomogr.* **1991**, *15*, 237–243.
88. Mohiaddin, R.H.; Amanuma, M.; Longmore, D.B. Magnetic resonance measurement of pulmonary venous flow and distensibility. *Am. J. Noninvasive Cardiol.* **1992**, *6*, 13–18.
89. Nishimura, R.A.; Abel, M.D.; Hatle, L.K.; Tajik, J.A. Relation of pulmonary vein to mitral flow velocities by transoesophageal Doppler echocardiography: Effect of different loading conditions. *Circulation* **1990**, *81*, 1488–1497.
90. Van Rossum, A.; Sprenger, K.H.; Peels, F.C.; et al. *In vivo* validation of quantitative flow imaging in arteries and veins using magnetic resonance phase shift techniques. *Eur. Heart J.* **1991**, *12*, 117–126.
91. Appleton, C.P.; Hatle, L.K.; Popp, R.L. Relation of mitral flow velocity patterns to left ventricular diastolic function: New insight from a combined haemodynamic and Doppler echocardiographic study. *J.Am. Coll. Cardiol.* **1988**, *12*, 426–440.
92. Myreng, Y.; Smiseth, O.A. Assessment of left ventricular relaxation by Doppler echocardiography. Comparison of isovolumic relaxation time and transmitral flow velocities with time constant of isovolumic relaxation. *Circulation* **1990**, *81*, 260–266.
93. Karwatowski, S.P.; Brecker, S.; Yang, G.Z.; St John Sutton, M.; et al. Assessment of mitral diastolic flow in ischaemic heart disease: A comparison of Doppler echocardiography and magnetic

resonance imaging. *Berkeley: Society of Magnetic Resonance in Medicine* **1992**, *11*, 2521 (Abstract).

94. Kilner, P.J.; Firmin, D.N., Rees, R.S.O., Martinez, J.E., et al. Valve and great vessel stenosis: Assessment with MR jet velocity mapping. *Radiology* **1991**, *178*, 229–235.
95. Kilner, P.J.; Manzara, C.C.; Mohiaddin, R.H.; Pennell, P.J.; et al. Magnetic resonance jet velocity mapping in mitral and aortic valve stenosis. *Circulation* **1993**, *87*, 1239–1248.
96. Edelman, R.R.; Li, W. Fast flow quantification in human coronary arteries using a segmented K-space phase contrast method. *Society of Magnetic Resonance in Medicine* **1992**, *11*, 602 (Abstract).
97. Poncelet, B.; Kantor, H.; Weisskoff, R.M.; Holmvang, F.; et al. Quantitation of the coronary flow with echo-planer. *Berkeley: Society of Magnetic Resonance in Medicine* **1992**, *11*, 604 (Abstract).
98. Edelman, R.R.; Manning, W.; Burstein, D.; Paulin, S. Coronary arteries: Breath-hold MR angiography. *Radiology* **1991**, *181*, 641–643.
99. Wang, S.J.; Nishimura, D.G.; Macovski, A. Fast angiography using fast selective inversion recovery. *Mag. Res. Med.* **1992**, *23*, 109–121.
100. Edelman, R.R.; Manning, W.J.; Gervino, E.; Li, W. Flow velocity quantification in human coronary arteries with breath hold MR angiography. *J. Magn. Reson. Imag.* **1993**, *3*, 699–703.
101. Keegan, J.; Firmin, D.N.; Gatehouse, P.D.; Longmore, D.B. The application of breath hold phase velcoity mapping technqiue to the measurement of coronary artery blood flow: Phantom data and initial *in vivo* results. *Magn. Reson. Med.* **1994**, *31*, 526–536.
102. Debatin, J.F.; Strong, J.A.; Sostman, H.D.; Negro-Vilar, R.; et al. MR characterization of blood flow in native and grafted internal arteries. *J. Magn. Reson. Imag.* **1994**, *3*, 443–450.
103. Amanuma, M.; Mohiaddin, R.H.; Hasegawa, M.; Heshiki, A.; et al. Abdominal aorta: Characterization of blood flow and measurement of its regional distribution by cine magnetic resonance phase-shift velocity mapping. *Eur. J. Radiol.* **1992**, *2*, 559–564.
104. Maier, S.E.; Meier, D.; Boesiger, P.; Moser, U.T.; et al. Human abdominal aorta: Comparative measurement of blood flow with MR imaging and multigated Doppler US. *Radiology* **1989**, *171*, 487–492.
105. Li, K.C.P.; Whitney, W.S.; McDonnell, C.H.; Fredrickson, J.O.; et al. Chronic mesenteric ischemia: Evaluation with phase-contrast cine MR imaging. *Radiology* **1994**, *190*, 175–179.
106. Wolf, R.L.; King, B.F.; Torres, V.E.; Wilson, D.M.; et al. Measurement of normal renal artery blood flow—Cine phase-contrast MR imaging vs clearance of P-aminohuppurate. *AJR* **1993**, 995–1002.
107. Mohiaddin, R.H.; Sampson, C.; Firmin, D.N.; Longmore, D.B. Magnetic resonance morphological, chemical shift and flow imaging in peripheral vascular disease. *Eur. J. Vasc. Surg.* **1991**, *5*, 383–396.
108. Rees, R.S.O.; Firmin, D.N.; Mohiaddin, R.H.; Underwood, S.R.; et al. Application of flow measurements by magnetic resonance velocity mapping to congenital heart disease. *Am. J. Cardiol.* **1989**, *64*, 953–956.
109. Martinez, J.E.; Mohiaddin, R.H.; Kilner, P.J.; Khaw, K.; et al. Obstruction in extracardiac ventriculopulmonary conduits: Value of magnetic resonance imaging with velocity mapping and Doppler echocardiography. *J. Am. Coll. Cardiol.* **1992**, *20*, 338–344.
110. Sieverding, L.; Jung, W.I.; Klose, U.; Apitz, J. Noninvasive blood flow measurement and quantification of shunt volume by cine magnetic resonance in congenital heart disease: Preliminary results. *Paediatric Radiology* **1992**, *22*, 48–54.
111. Mohiaddin, R.H.; Underwood, R.; Romeira, L.; Anagnostopoulos, C.; Karwatowski, S. P.; Laney, R.; Somerville, J. Comparison between cine magnetic resonance velocity mapping and first-pass radionuclide angiocardiography for quantitating intracardiac shunts. *Am. J. Cardiol.* **1995**,*75*, 529–532.
112. Rebergen, S.A.; Ottenkamp, J.; Doornbos, J.; Van der Wall, E.E.; et al. Postoperative pulmonary flow dynamic after Fontan surgery: Assessment with nuclear magnetic resonance velocity mapping. *J. Am. Coll. Cardiol.* **1993**, *21*, 123–131.

113. Young, I.R.; Hall, A.S.; Bryant, D.J.; Thomas, D.G.T.; et al. Assessment of brain perfusion with MR imaging. *J. Comput. Assist. Tomogr.* **1988**, *12*, 721–727.
114. Firmin, D.N.; Kilner, P.J.; Pennell, D.J.; Underwood, S.R.; et al. Short echo time (3.5 to 6 msec) field even echo rephasing sequences for improved accuracy and reliability of blood flow velocity and acceleration measurements. *Proceedings, 8th Annual Meeting of the Society of Magnetic Resonance Medicine* **1989**, *2*, 888.
115. Pattany, P.M.; Phillips, J.J.; Chiu, L.C.; Lipcamon, J.D.; et al. Motion artifact suppression technique (MAST) for MR imaging. *J. Comput. Assist. Tomogr.* **1987**, *11*, 369–377.
116. Firmin, D.N.; Kilner, P.J.; Pennell, D.J.; Underwood, S.R.; Mohiaddin, R.H.; Rees, R.S.O.; Longmore, D.B. Short echo time (3.5 to 6 msec) field even echo rephasing sequences for improved accuracy and reliability of blood flow velocity and acceleration measurements. *Proceedings, 8th Annual Meeting Society Magnetic Resononance Medicine* **1989**, *2*, 888.
117. Buonocore, M.H.; Bogren, H. Factors influencing the accuracy and precision of velocity-encoded phase imaging. *Magn. Reson. Med.* **1992**, *26*, 141–154.
118. Søndergaard, L.; Ståhlberg, F.; Thomsen, C.; Spraggins, T.A.; et al. Comparison between retrospective gating and ECG triggering in magnetic resonance velocity mapping. *Magn. Reson. Imaging* **1993**, *11*, 533–537.
119. Lanzer, P.; Botvinick, E.; Schiller, N.B.; Crooks, L.E.; et al. Cardiac imaging using gated magnetic resonance. *Radiology* **1984**, *150*, 121–127.
120. Glover, G.H.; Pelc, N.J. A rapid-gated cine MRI technique. In: *Magnetic Resonance Annual* (Kressel, H.Y., Ed.). Raven Press, New York, 1988, pp. 299–333.
121. Lenz, G.W.; Haacke, E.M.; White, R.D. Retrospective cardiac gating: A review of the technical aspects and future directions. *Magn. Reson. Imaging* **1989**, *7*, 445–455.
122. Bohning, D.E.; Carter, B.; Liu, S.; Pohost, G.M. PC-based system for retrospective cardiac and respiratory gating of NMR data. *Magn. Reson. Med.* **1990**, *16*, 303–316.

TURBULENCE IN THE CARDIOVASCULAR SYSTEM:
ITS PHYSICAL NATURE, PATHOPHYSIOLOGICAL ROLE, MEASUREMENT, AND ANALYSIS

Takami Yamaguchi

Advances in Hemodynamics and Hemorheology,
Volume 1, pages 201–227.

ISBN: 1-55938-634-7.

I conjecture that the origin of life and the origin of problems coincide.
—Karl Popper, *Unended Quest—An Intellectual Autobiography*, p. 178[1]

I. INTRODUCTION

It is gradually being recognized that the Reynolds number in the large arteries, particularly in the aortas of large animals, such as humans, can exceed 2,000, based on the average velocity. If we take the peak value as a representative velocity, the Reynolds number undoubtedly exceeds 10,000, thus, the blood flow becomes turbulent at least within a certain duration of time in the systolic phase. Moreover, the arteries and veins have such complicated geometrical configurations, as branching, curves, tapering, and so forth, that the flow inside can be disturbed even though it is not fully turbulent. Stenosis, kinking, and other wall irregularities occurring on the atherosclerotic arterial wall may inevitably increase these disturbances of the blood flow. In a word, the blood flow in the largest arteries of some large animals should be regarded as turbulent under normal physiological conditions.

Once flow becomes turbulent, mixing of the fluid increases to a great extent as compared to the laminar flow even though the general conditions remain the same. This increase of the internal mixing yields an apparent shear stress in the fluid (called Reynolds stress[2]) and also augments the mechanical stress exerted on the vessel wall by the flow. We have to pay close attention to this augmentation of the shear stress because there are two aspects to be distinguished, namely, increase of the average shear stress, and the increase of fluctuating components. Though cellular components of the blood are, in general, considered to be affected mainly by the increase of the inner stress, that is the Reynolds stress, the contribution of the surface nature of the vessel wall can not be neglected. Affected cellular components of the blood and the endothelium of the vessel wall in turn produce various biologically active agents, such as endothelium-derived relaxing factor (EDRF), prostaglandin, and so forth. These are currently recognized as the principal causes of the physiological and pathological changes of the configuration of the vessels and make both the normal and the abnormal reaction of the whole system very complicated.

Recent rapid advancement of medical technology has introduced a number of artificial organs which interacts with the blood and the blood flow directly or indirectly. Of particular interest from the viewpoint of turbulence are the artificial heart valve, the artificial heart, and vascular grafts. The biocompatibility of these artificial organs is closely related to various flow phenomena. The mechanical valves, due to their non-natural geometrical shapes, are known to cause severe turbulence, which is thought to be responsible for both acute complications of the surgical operation and the late insufficiencies of the replaced valves.

As introduced above, turbulence in the living body has been customarily thought of as having rather adverse influences on the normal body. However, some studies indicate that it may be a part of normal physiological state of the blood flow. It is therefore highly conceivable that turbulence, if it always exists in the large arteries, must have a favorable role in maintaining the integrity of physiological functions, since it is not just the recent time in archeological ages when our ancestors dropped off the tail and were given such a large body.

Turbulence in the living body, therefore, has to be studied with regards to any physiological as well as pathological significance. As for its physiological nature, the pulsatility of the blood flow itself in the living body imposes many interesting characteristics. Indeed the unsteady turbulence itself has been of pure fluid mechanical concern and has been studied theoretically as well as experimentally for a long time. Especially the mechanism of generation, development, decay, and disappearance of turbulence with respect to the pulsatility of the blood flow has drawn the initial attention and has been studied using a wide variety of analyzing tools. These were the studies that gave us an insight into the time-varying alterations of internal structures of the fluid motion in unsteady turbulence. Since the studies of turbulence in the field of proper fluid mechanics is said to be probably the only one remaining unsolved field in classic physics, the studies of turbulence in the living body may occupy a definite part of this ever advancing field.

In the present review, we would like to discuss the current state and the future direction of the research of turbulence in the living system. In the following, unless otherwise specially mentioned, turbulence in the living system in our mind is one that occurs in non-diseased physiologically normal structures in the living system, particularly in the aorta.

II. TRANSITION, DEVELOPMENT, AND THE PHYSIOLOGICAL NATURE OF TURBULENCE IN THE AORTA

A. Transition to the Turbulent State in the Unsteady Flows

It is widely known that the steady flow in a pipe becomes turbulent at (to be exact, can not be continuously turbulent below) a Reynolds number of 2,000–2,300. Here, $Re = UL/\nu$, U is a representative velocity, L is a representative length, and ν is the kinematic viscosity. This Reynolds number is usually called the critical Reynolds number (Re_{cr}). In unsteady flows, the critical Reynolds number is thought of as a function of some other non-dimensional parameters. The relationship was first studied using pulsatile pipe flows with zero mean velocity by number of investigators.[3–5] Their results showed that the critical Reynolds number defined by using the Stokes-layer thickness as the representative length becomes constant. This statement can be paraphrased in that the critical Reynolds number, using the characteristics length of the flow field such as the diameter of the pipe, is a function of some non-dimensional representation of the frequency of the oscillatory flow. This could typically be the Wormerseley's alpha (frequency) parameter ($\alpha = r\sqrt{\omega/\nu}$, where r is the radius, ω is the angular frequency).

Applying these results obtained from pure oscillating flows to the aortic flow in the canine aorta, Nerem and Seed[6] proposed the following relation:

$$Re_{cr} = K\,\alpha. \tag{1}$$

They found the constant K to be 150 for the ascending aorta and 250 for the descending aorta. In purely experimental study of Sergeev,[3] the K value was estimated to be 710.

Most flows in the living body have a definite average component, thus cannot be regarded as purely oscillating flows. Sarpkaya[7] studied the instability of pulsatile non-zero mean flow in a straight pipe. His results showed that the stability of such flows depend on two non-dimensional parameters, the frequency parameter α and the amplitude parameter β. The amplitude parameter β is defined as the ratio of the peak velocity to the average velocity. Similar results were reported by Dantan et al.[8] on the aortic velocity both theoretically and experimentally. Although a clear agreement has not been obtained in this class of pulsating flows, it is generally believed that pulsating flows become more stable when the α parameter becomes large. Nerem and Seed[6] discussed the instability of the flow in the aorta using a boundary layer stability theory. They attributed the enhancement of stability under the condition of large α parameters to the shortage of time for the seed of turbulent fluctuations to develop. Stettler and Hussain[9] found that the stability enhancement was limited to a certain range of α values. More recently, Bluestein et al.[10] proposed an approach to define the α parameter using a preferred frequency in the target flow field. From this they were able to construct the stability diagram.

In spite of these studies on the subject, however, we have to conclude that clear agreement has not yet been reached on the critical conditions of laminar-turbulent transitions in arterial systems. Experimental results so far obtained, regardless from *in vivo* or *in vitro* (model) experiments, could cover a limited range of the combinations of flow conditions, such as α or β. Indeed, the aortic blood flow has very large β values, particularly under normal physiological conditions. Some experiments, which are better designed to cover a wide physiological range, will reveal the clear relationship governing the laminar-turbulent transition.

B. Development of Turbulence in the Intact Arteries

Turbulence in the arterial system inherits the large-scale unsteadiness from the overall blood flow. Since there is no other available energy source other than the heart beat and the resulting blood flow in the arteries, the time course of turbulence is directly connected to the basic temporary characteristics of the overall blood flow. Figure 1 shows a set of velocity records (top left panel) obtained in the center of the canine ascending aorta using a hot-film anemometer. The ensemble average velocity waveform is shown in the bottom left. The turbulence components (top right) are the difference between each measured velocity record and the ensemble average, and the turbulence intensity (bottom right) is defined as the square root of the ensemble average of the squared turbulence components.[11,12]

The ensemble averaging is an averaging method dealing with the unsteady nature of the arterial turbulence. In an unsteady stochastic process, such as turbulence in the aorta, the only statistically correct way of averaging is the ensemble average. Time average should not be used to define the turbulence components because an unsteady stochastic process is never an ergodic process. This point will be discussed in detail in the later section on the measurement and analysis of turbulence.

As clearly shown in Figure 1, the turbulence intensity is almost nil at the onset of each heart beat and remains almost insignificant during the accelerating phase of the blood flow. It is noteworthy that the turbulence intensity is also a statistical parameter which suffers from the variability of the data other than turbulence. The turbulence intensity then starts to increase in the decelerating phase of the blood

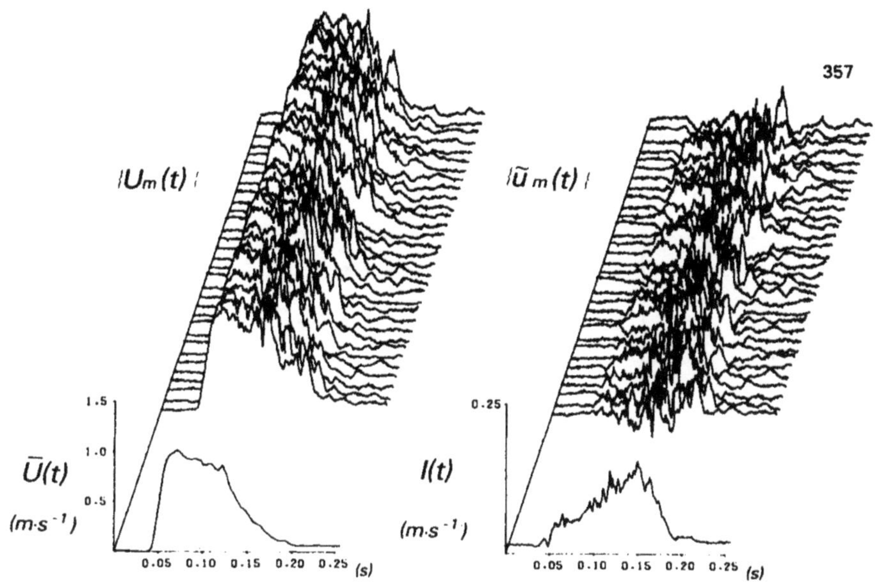

Figure 1. Illustration of the data and its analysis procedure for turbulent velocity fluctuations in the canine ascending aorta measured using a hot-film anemometer. The top left panel shows a set of velocity records obtained in the center of the aorta. They are segmented by conditional sampling. The ensemble average velocity waveform is shown in the bottom left. The turbulence components are shown in the top right panel, each of which is the difference between each measured velocity record and the ensemble average velocity. The turbulence intensity shown in the bottom right panel is defined as the square root of the ensemble average of the squared turbulence components.

Source: With permission of Pergamon Press.[12]

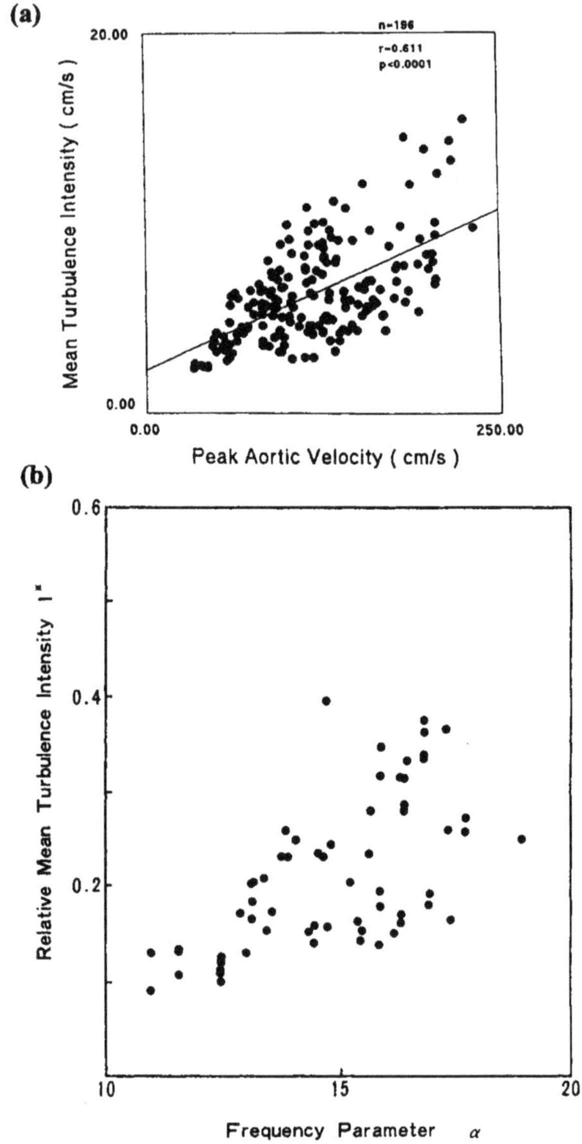

Figure 2. (a) The mean turbulence intensity plotted against the peak velocity of the aortic flow. The mean turbulence intensity is the square root of the total energy of turbulent fluctuations summed for the decelerating phase.

Source: With permission of Pergamon Press.[24]

(b) The mean turbulence intensity (the same as in part (a)) plotted against the Wormerseley's α parameter. The amplitude ratio (peak velocity / mean velocity) is larger than 4.13.

Source: With permission of the American Society of Mechanical Engineers.[11]

flow velocity, shows a peak in the middle of the decelerating phase, decreases after the peak, and finally decays in the diastolic phase. Figure 2a shows the mean turbulence intensity (which is the square root of the total energy of turbulent fluctuations during the decelerating phase) plotted against the peak velocity of the aortic flow. There is clearly a linear relationship between those two values. In other words, turbulence energy in the aorta occupies a fixed percentage of the total ejection energy of the heart contraction. In Figure 2b, the turbulence intensity is plotted against the Wormerseley's α parameter, under a condition that the amplitude ratio is larger than a certain value. The mean intensity is almost positively correlated with the α parameter, that is, the turbulence energy becomes larger when the heart rate increases.[11] This is apparently inconsistent with the condition of the laminar-turbulent transition discussed in the previous section. It should be noted, however, that the intensity is not necessarily directly related to the mechanism of the transition.

III. THE PATHOPHYSIOLOGICAL ROLE OF TURBULENCE IN THE ARTERIAL SYSTEM

A. Turbulence and Vascular Murmurs

The heart and vascular murmurs audible on the body surface have been important measures of the clinical diagnosis of various disorders. Most murmurs are thought to be related to disturbances of blood flow and therefore are useful in evaluating the nature and the severity of the underlying pathological states. Many investigations have been reported on the generation of murmurs with respect to turbulence, either directly or indirectly.

As is theoretically clarified, the aerodynamic noise related to turbulence is not the direct result of turbulent velocity fluctuations. The sounds are produced by unsteady movement of turbulent eddies.[13,14] That is, if there is no motion or there is only a steady movement of the turbulent eddies, no sound pressure wave is generated. In this context, the conclusion of Sabbah and Stein[15] is insufficient in the point that they directly correlated the turbulent velocity fluctuations to the sound generation, although this is not totally incorrect. Abdallah and Hwang[16] discussed that the sound pressure fluctuations are not directly due to the turbulent velocity fluctuations, but due to the eddies produced in the turbulent flow field being convected down. This interacts with the walls to produce the pressure fluctuations detected as the sound. These results agree with the generally accepted knowledge of aerodynamic noise acoustics and therefore seems to be reasonable.

According to these results, the frequency of the murmur detected on the body surface is not merely a function of the turbulent velocity fluctuations. It should be a function of the scale of the eddies and the convective component of the mean velocity by which the eddies are transported. The representative length scale of the eddies should be related to a representative scale of the flow field, such as the size of stenosis, for example. The relationship between the convective velocity and the observed turbulent fluctuations can be corrected by the method reported by Yamaguchi et al.[12] This method may be used to reconstruct the geometry of the flow field from the murmur which is audible on the body surface.

B. Cardiovascular Disease and Turbulence

Atherosclerosis

Many hypotheses have been proposed to explain the apparent preferential localization of atherosclerosis by hemodynamic or fluid mechanical effects of the blood flow on the arterial wall. Some placed an emphasis on the existence of turbulence in the artery. Stehbens[17] conducted an extensive series of studies on atherogenesis and proposed that the fatigue of the wall element and the resultant disruption of the structural component, particularly elastin and collagen, are the real cause of atherosclerosis. It is difficult, however, to explain overall atherogenic procedures by this hypothesis, although it seems to be based on sounder observation than the widely believed "response to injury" hypothesis.

Provided that the increase of the wall shear stress caused by turbulence in the arteries is responsible for the injury of the vascular wall, the injury may occur particularly in the largest artery, that is, the aorta. This is where the overall blood flow velocity is the fastest in the human body and it usually becomes turbulent under physiological conditions. However, the human atherosclerotic lesion, at least the raised plaque type lesion, preferentially occur in the abdominal aorta,[18] where the Reynolds number is estimated to be about an order of magnitude lower than that in the ascending aorta. From pathological findings, it has been established that human atherosclerotic disease occurs at the locations where the wall shear stress is low by the fluid mechanical considerations.[19,20] Browse[21] also questioned the widely presumed assumption of adverse influence of turbulence in atherogenesis.

Though it is beyond the scope of this chapter to extensively review the subject of atherogenesis with regard to fluid mechanics, we should distinguish two classes of turbulence in the arterial system in the discussion. First, we now know that the normal circulation sees turbulence or disturbed flow in the larger arteries. As is mentioned in the introduction, we must not jump to the conclusion that this class of turbulence is harmful to the living system. Second, turbulence related to artificial structures such as prosthetic valves has much higher intensity than that in physiological turbulence. This kind of turbulence is known to induce many disorders and should therefore be discussed separately.

Hypertension

Hypertension itself is an important cause of stroke and is also a strong risk factor in atherosclerosis. Judging from the strong relationship between the blood flow and atherosclerosis, hypertension could affect the blood flow and hence turbulence.

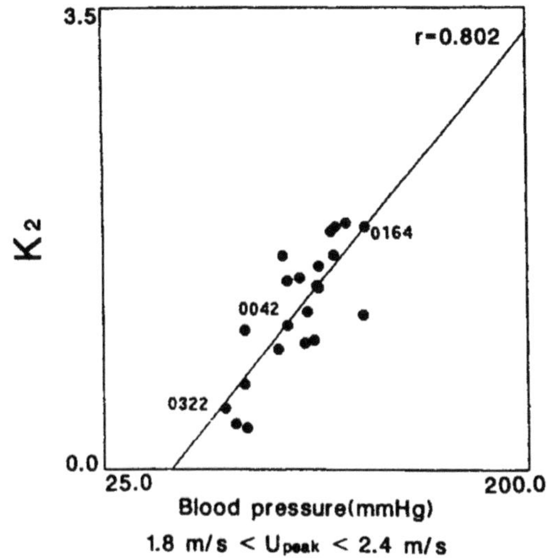

Figure 3. Relationship between the spectral parameter of turbulence in the canine ascending aorta and the blood pressure (transmural pressure). The second order characteristic wavenumber K_2 in the ordinate is a value characterizing the spectral bandwidth. K_2 showed a very good correlation with the blood pressure when turbulence is well developed.

Source: With permission of Pergamon Press.[24]

Spence et al.[22,23] reported a difference of some anti-hypertensive agents in clinical cases as well as in animal experiments. Their conclusion is that agents such as Hydralazine increase the heart rate and the average blood velocity, and β-blocking agents decreases the heart rate and lower the blood velocity in the long term. They found that atherosclerosis developed in the group that received the former therapy (Hydralazine), and suggested that turbulence occurring in the artery played an important role. These are not based on the actual measurement of turbulence in the arteries.

Hanai et al.[24] reported a positive relationship between several parameters of turbulence in the canine ascending aorta and the blood pressure (transmural pressure). They found that the average intensity and some parameters representing the spectral broadening of turbulent velocity fluctuations were linearly related to the blood pressure under relatively high flow conditions as is shown in Figure 3. They discussed this phenomenon with respect to the non-linear stress-strain relationship of the arterial walls and the visco-elastic dumping of the fluctuational motion of the blood near the wall.

These two results apparently coincide if turbulence in the aorta under high-flow condition has some pathological role in atherogenesis. Nevertheless, as is pointed out in the former section, atherosclerosis predominantly occurs in the abdominal aorta in human cases and turbulence has been measured mainly in the ascending or thoracic aorta. Moreover, applying equation (1), the critical Reynolds number which governs laminar-turbulent transition in the arteries is high when the heart rate decreases. Some more detailed studies demarcating laminar-turbulence transition under many physiological conditions, such as cardiac output, the pulse rate, and so forth, are necessary to resolve this interesting question.

Valvular and Vascular Stenosis

Turbulence accompanying stenosis of the heart valves and the arteries has been studied extensively. Since most of the studies were already reviewed in the literature[25,26] and sometimes were of little interest from the viewpoint of turbulence, we shall concentrate on rather recent works here. Yearwood et al.[27] found that the peak velocity and the axial Reynolds stress (which is equal to the square of the turbulence intensity) of the turbulent jet formed downstream of a stenosis are very susceptible to the configuration of the valvular stenosis. Therefore these parameters can be used to evaluate the shape factors of the valvular stenosis in the heart. There are many studies on this subject after the introduction of the pulsed Doppler ultrasound measurement, particularly with color Doppler. Sugawara et al.[28] proposed a method to estimate the valvular stenosis and regurgitation using a general knowledge of turbulent free jets inside the heart.

Vascular stenosis has also interested a number of researchers because of its non-invasive diagnostic significance. As the discovery of the coherent structure in turbulence was one of most striking topics in the field of fluid mechanics, attempts to identify such structures in the stenotic blood flow were performed. It would be of clinical value if there are any coherent structures sensitive to the stenosis geometry. Lieber and Giddens[29] and Lieber et al.[30] found a prominent peak in the turbulence spectrum which was calculated by a combination of ensemble averaging and digital filtering techniques. The apparent stress induced by turbulent velocity fluctuations could be, according to their studies, underestimated by the ensemble averaging technique alone. This is because the large-scale coherent structure may be strongly correlated with the basic frequency of the blood flow and can therefore be classified with the average component by using ensemble averaging technique.

Poststenotic Dilatation

In her extensive review, Roach[31] summarized that "turbulence (indicated by a murmur) is the *sine qua non*" for the development of poststenotic dilatation (PSD), and suggested that a particular frequency range of the vibration induced by turbulence could make the arterial wall distensible and whence cause the PSD. Her conclusion was based on her group's *in vitro* and *in vivo* findings and has also been supported by other investigators.

However, this idea might be presently reconsidered because most of those works were carried out before, or at least paying little or no attention to, the discovery of the endothelium derived relaxing factor (EDRF) and the success of the *in vitro* culture of the endothelium. Recently, Gow et al.[32] reported that the external vibration imposed on the rabbit artery *in vivo* failed to produce either PSD or lipid accumulation around the stimulated area.

It is now well established that the increase of wall shear stress induces the endothelial production of EDRF or nitrogen oxide (extensively reviewed in Vanhoute[33]). The arterial endothelium was also shown to respond to turbulent fluctuating shear stresses under *in vitro* cultured conditions, and in a different manner from the response to the laminar increase of the shear stress.[34] This consideration strongly indicates an active role of the endothelium in the PSD process, and the contribution of shear stress is important and not the vibration. Indeed, Masuda[35] reported that the rabbit arteries denuded of their endothelium did not dilate when the wall shear stress level was elevated several times above the normal level by means of arterio-venous shunt. In his earlier experiments, the intact artery easily dilated when the average wall shear stress was increased by the same method.

Thrombosis

Turbulence and the thrombosis has been studied from many viewpoints. A factor of time seems to be important when the thrombosis is discussed with respect to turbulent blood flow. It is well known that a certain reaction time is necessary to complete the coagulation process since it consists of a long chain of cascaded biochemical reactions. Since the blood never stops under physiological conditions, this time factor is directly related to the local velocity of the blood flow and the geometrical scale of the flow tract.

It is of course important that the coagulation process be initiated by some chemical and/or physical stimuli to start the process of thrombosis. Turbulence, particularly abnormally high intensity turbulence such as that found in the vicinity of either mechanical or biological artificial heat valves, and other prosthesis could cause mechanical stimuli or damage blood constituents and could trigger the thrombosis. There are enormous numbers of studies on the subject of the blood flow near the prosthetic valves, which are not directly within the scope of the current chapter. However, it is noteworthy that there is a coexistence of the turbulent or high shear stress region and slow blood flow region. This has been suggested to be necessary to promote thrombotic process in the artificial organs.[36]

Therefore, we need to know the whole flow field not only close to the turbulent region but also close to the stagnant region relatively far downstream from these rapidly agitating region of turbulence. In this context it is necessary to broaden the area of interest in the evaluation of the contribution of turbulence to the thrombotic process, and therefore the computational approach should provide a powerful tool.[37,38]

Hemolysis and Hemolytic Anemia

Hemolysis is also a serious complication of prostheses in the cardiovascular system. Among the prostheses, artificial heart valve replacements, particularly mechanical valves, are known to provoke severe hemolysis and, consequently, hemolytic anemia, hemoglobinuria, and so on. Sallam and Hwang[38] reported that there is a threshold value (400 Pa) for the hemolysis induced by an increase of the Reynolds shear stress. Below this value no hemolysis occurs. Once the Reynolds stress exceeds this threshold the amount of hemolysis is almost linearly related to the Reynolds stress. This threshold value can be easily exceeded in the blood flow around various artificial organs particularly the mechanical artificial valves.

Nevertheless, the Reynolds shear stress inside the blood is not the only cause of the hemolysis. Artificial surface can cause hemolysis as frequently found in vascular grafts. Nakahara and Yoshida[40] found that the Reynolds stress and not the artificial surface was the determining factor of the hemolysis. On the other hand, Khanna et al.[41] reported that turbulence in the patent ductus arteriosus did not cause hemolysis. Indeed, the shunt flow which is frequently accompanied with turbulent jet formation, for instance, ventricular septal defect and patent ductus arteriosus, is scarcely accompanied by hemolysis or hemolytic anemia by clinical observations. Therefore, a combination of foreign surfaces of the artificial organs and the increase of the Reynolds shear stress may be necessary to promote clinically symptomatic hemolysis. To answer this, it would be necessary to estimate the true Reynolds shear stress (the sum of the products of three dimensional velocity components) by modern methods, which has not been performed for the normal circulation.

C. Turbulence and Oxygen Transport

Turbulence increases mass transport by enhancing mixing in the fluid. The effect of turbulent mixing differs according to the length scale of the transport of the target mass species and that of minimum length scale of turbulent motion of the fluid. The latter is estimated by the Kolmogoroff's microscale (of the length) η defined by:

$$\eta = (\nu^3/\varepsilon)^{1/4}, \tag{2}$$

where ν is the kinematic viscosity, and ε is the dissipation rate per unit mass.[42] The energy dissipation rate ε is an order of magnitude of the energy supply rate as follows:

$$\varepsilon \approx u^3/l, \tag{3}$$

where l is the Taylor's integral scale, which represents the largest size eddy of turbulence. Therefore the microscale η becomes small when the velocity u becomes large provided the kinematic viscosity does not change. This means that the mixing length scale becomes small when turbulence develops. Huxley and Kutchai[43] and Gad-El-Hak et al.[44] reported that the increase of the turbulent mixing can destroy the diffusion boundary layer near an erythrocyte and can therefore augment the oxygen transport between the plasma and the erythrocyte. However, the estimated Kolmogoroff's length scale is still one or two orders of magnitude larger than that of the erythrocyte in turbulence of the Reynolds number of several thousands, which is the largest Reynolds number observed in the larger arteries. Consequently, the effect of turbulent mixing, at least to the oxygen transport, would not be significant under physiological conditions. It is noteworthy that this discussion can be applied to many problems in which the role of the oxygen transport is of causative significance, whether enhancement or suppression is concerned.

D. Turbulence and the Endothelium

As is discussed, atherogenesis is one of most interesting subjects in terms of the possible biological sequelae of hemodynamics. Recently the endothelium of the arterial wall was found to react to various mechanical stimuli, particularly the wall shear stress, and this response was found to play an active role in such a pathological process. Since turbulence increases the wall shear stress as compared to the laminar flow with the same mean flow rate, the reaction of the endothelium to the turbulent or fluctuational shear stress could be of significance. Davies et al.[34] found that the turbulent shear stress exerted onto the confluently cultured endothelial cells induced an increase in cell turnover. They attributed it to a microscale random distribution of the shear stress which ends up as a differential force between the cells. These points were discussed by Fung and Liu[45] from the mechanical viewpoint and also reviewed by Davies and Tripathi[46] from the cell biological viewpoint. Fung and Liu emphasized the importance of intracellular stresses which coincide with the direction of the cell biological pursuit for the intracellular mechanism of the shear stress transduction. Rosati and Garay[47] found that the cultured smooth muscle cell (not the endothelial cell) respond to turbulent flow and showed an increase in cholesterol uptake. Under normal conditions, smooth muscle cells do not have direct contact with the blood flow, and therefore the physiological significance of their findings is obscure. However, some other reports suggest a possible relationship between disturbed flow and the proliferation of vascular tissues.[48]

IV. MEASUREMENT AND ANALYSIS OF TURBULENCE IN THE VASCULAR SYSTEM

Turbulent flow is usually characterized by high-frequency fluctuations of velocity and pressure. Much more important is that turbulence is a strongly non-linear stochastic process. These characteristics should be kept in mind when we try to measure and to analyze turbulent fluctuations. We are going to discuss some of representative measurement devices and methods of analysis.

A. Velocity Measurements

Historically, the turbulent velocity fluctuations in the animal and human arterial system were measured using a thermal (hot-film) anemometer. In model studies, the laser Doppler anemometer (LDA) later introduced proved to provide much better position resolution, but the overall frequency response was approximately the same. Unfortunately, the LDA, except a fiberoptic catheter type could not be applied to *in vivo* arterial blood flow. The catheter type at the moment does not have a broad enough frequency response to quantify turbulent velocity fluctuations. Two clinical measuring techniques, ultrasonic Doppler flowmetry and magnetic resonance imaging (MRI) are both non-invasive and can yield a graphical presentation of a map of the spatial distribution of the velocity fluctuations. However, as described above, it is necessary to pay attention to the original definition of turbulence when using substantially one-time-only measurement devices, which will be dealt with in detail in the later section of data analysis.

Thermal (Hot-Film, Hot-Wire) Anemometers

The hot-wire and the hot-film anemometers are the most widely established devices used to evaluate turbulent velocity fluctuations. The velocity is measured through the heat loss from a small body in the flow. In the air flow, a hot-wire is usually used; a hot-film is specially designed to be used in liquid flow, particularly in an electrically conducting liquid. The hot-film probe we used (DISA) was made of a sputtered and etched nickel thin film on a quartz substrate which was covered by a quartz thin membrane (Figure 4a).

There are two types of electrical circuits to heat the sensor, specifically, the constant temperature anemometer (CTA) type and the constant current type. Of these two, CTA is usually used for blood flow measurement. The CTA circuit is basically composed of a bridge containing a sensor and a feedback servo amplifier (Fig. 4b).

When Ling et al.[49] and Schultz et al.[50] first applied the hot-film anemometer to blood flow, the device was shown to be able to measure high frequency velocity fluctuations. Many studies since then have been performed with respect to the blood flow turbulence.[6,12,51–54] Even after the introduction of LDA and other non-invasive methods, the hot-film anemometer still is an important technique in spectral analysis and has until recently been used with careful checking of its basic characteristics.[55]

Laser Doppler Anemometer

As mentioned above, the laser Doppler anemometer (LDA) is non-invasive but usually needs transparent walls and fluid, and is therefore mainly used in model studies. Flow fields downstream of artificial valves have been extensively studied by several groups using LDA. Some model studies on vascular stenosis were also

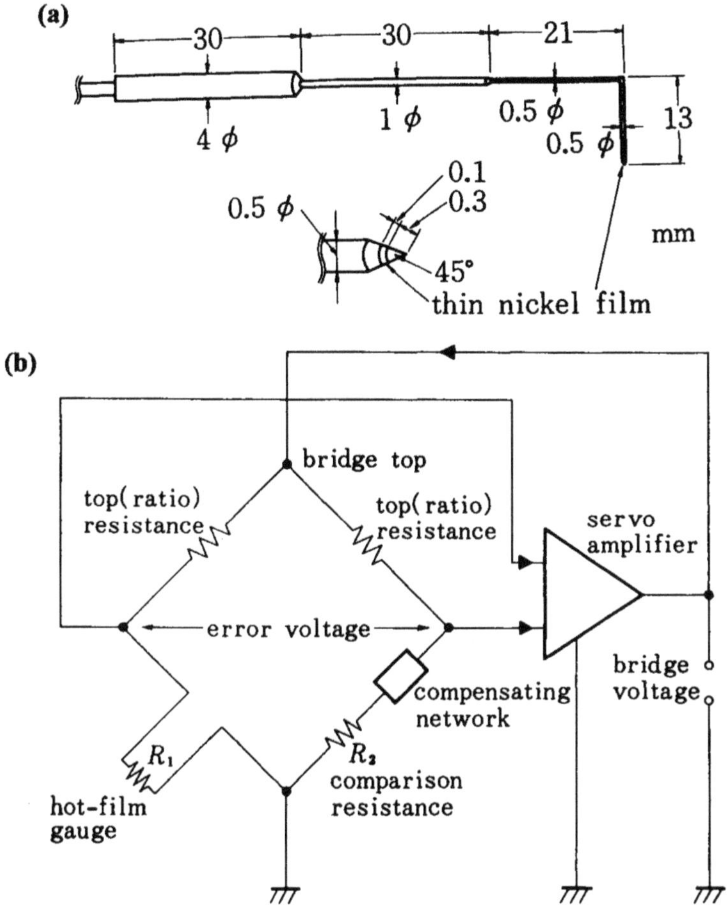

Figure 4. (a) A schematic drawing of the structure of a hot-film probe (DISA), made from a nickel thin film sputtered and etched on a quartz substrate and covered by the quartz thin membrane. (b) A schematic representation of the electrical circuits of the constant temperature anemometer (CTA) type thermal anemometer. It is composed of a bridge containing the sensor and a feedback servo amplifier.

Source: With permission of Springer-Verlag.[70]

conducted using LDA.[56] There are some limitations such as seeding density, and so on, for obtaining the precise measurement. Also, the quality of the higher frequency signal is sensitive to the device and measurement conditions. The frequency response bandwidth seems to be sometimes lower than that of the hot-film anemometer. To measure the intravascular blood velocity, a meticulous fiber-optics based LDA has been developed.[57] Unfortunately the reported fiberoptic LDA does not have a broad enough frequency response in order to measure turbulent velocity fluctuations.

Ultrasonic Measurements

The Doppler ultrasound measurement is one of the most rapidly advancing areas in the medical and biological instrumentation. It is inherently non-invasive and applicable to non-transparent wall of the real arteries, therefore enormous numbers of clinical application studies have been compiled. However, from the viewpoint of turbulence measurement, it suffers from a limited frequency response. It is correct to say that most of the kinetic energy is contained in the larger turbulence eddies, so that the global nature of turbulence can be detected by the ultrasound measurement. Nevertheless, phenomena coupled with the smaller scale eddies cannot be precisely measured and resolved using any of ultrasonic Doppler type flow velocity measurement techniques.

Color Doppler mapping is another popular method utilized in clinical diagnosis of blood flow disorders. As in the case of pulsed ultrasound measurement, it seems that the current conventional understanding of turbulence based on these measurement devices is quite different from what is defined and investigated in the usual turbulence studies in engineering or physics. Conventionally, we know that different kinds of averaging operation, such as time averaging, spatial averaging, and so-called ensemble averaging, will give different measure of turbulence, especially in unsteady and anisotropic turbulence. As is well known and is frequently neglected in clinical studies, the time average and the ensemble average give a completely different picture on what the average component of an unsteady stochastic process is, for example, in turbulence in the cardiovascular system. The same can also be said of the spatial and the proper stochastic average. The spatial distribution of the velocity of refracting particles is illustrated by the color Doppler method, and the frequency scattering is measured due to this spatial distribution in the pulsed Doppler method. They need to be checked against theoretical turbulence properties, which is now being conducted by some careful authors.[58,59] Based on these foundations, the ultrasonic measurement can undoubtedly yield useful knowledge of turbulence in the cardiovascular system, particularly in the clinical situations.

MRI

Another recently advancing area in cardiovascular fluid mechanics is the use of the magnetic resonance imaging (MRI). There are many active attempts to develop methods to obtain information related to various parameters of the blood flow. So far, large-scale features of blood flow, such as the mean flow rate, the average velocity profile, and so forth, in the stenotic flows are successfully measured by the MRI method.[60,61] It appears that the MRI blood flow measurement which has been investigated does not possess a broad enough frequency response for the measurement of turbulence. However, the random nature of turbulence can yield some information imposed on the MRI signals, and it can be utilized to characterize turbulent flows in the cardiovascular system.[62,63]

B. Analysis of Turbulent Signals

Ensemble Averaging Technique

Turbulence is a strongly non-linear stochastic or random process which needs to be analyzed on the basis of statistics. "A random process . . . is an ensemble of real-valued . . . functions which can be characterized through its probability structure. . . . In practice, a sample function (or some time history record of finite length from a sample function) may be thought of as the observed results of a single experiment."[64]

In the case of turbulence in the arterial system occurring repetitively at every heart beat, a segmented (time history) record of the measured velocity corresponding a heart beat can be regarded as a sample function. This is obtained from a single experiment under a combination of physiological as well as physical conditions. "A particular sample function . . . , in general, would not be suitable for representing the entire random process ... to which it belongs."[64] Only "for the class of ergodic random process, it is possible to derive desired statistical information about the entire random process from appropriate analysis of a single arbitrary sample function."[64] Unfortunately, an unsteady process such as turbulence in the cardiovascular system can never be an ergodic process. It is therefore a misunderstanding that the ensemble averaging technique and time averaging procedure can be compared as equal alternatives.[65]

Since the intervals between successive heart beats vary in a random manner due to a number of physiological determinants, we cannot assume there is an exactly regular periodicity in it. Therefore, it is also misleading to define the average velocity component by using so-called phase averaging. Thus, the only possible and plausible definition of the average of turbulent velocity fluctuations in the cardiovascular system is the ensemble averaging based on the idea of the conditional sampling.[66]

Let $u(t)$ be the measured time series of the velocity at a point under a combination of fixed physiological conditions, and $u_n(t)$ ($n = 1, 2, \ldots, N$) be sample records each of which is corresponding to a heart beat, sliced off from the $u(t)$ using an appropriate conditional sampling criterion. The ensemble average is defined as:

$$\bar{u}(t) = \langle u_n(t) \rangle = \frac{1}{N} \sum_{n=1}^{N} u_n(t), \tag{4}$$

hereafter $\langle \bullet \rangle$ denotes the ensemble averaging operation.

In our case the criterion to slice the velocity data was a simple amplitude threshold of the velocity. The data segments which compose an ensemble are shown in Figure

1. By defining the average component of the velocity using equation (4), we can define the random components of the measured velocity, that is the turbulent components as follows:

$$u'_n(t) = u_n(t) - \bar{u}(t), \tag{5}$$

which again composes an ensemble, and the intensity of turbulence (RMS) is defined using the ensemble averaging as:

$$I(t) = \sqrt{\langle u'^2_n(t) \rangle} = \sqrt{\frac{1}{N}\sum_{n=1}^{N} u'^2_n(t)}. \tag{6}$$

It should be clearly understood that the intensity (which is the square root of turbulence energy) is a function of time. This is what we meant by noting that turbulence in the artery is not an ergodic process. The intensity of turbulence defined by equation (6) can give a time course of the turbulence onset, development, decay, and disappearance.

Spectrum Analysis

Conventional spectrum analysis. Measured turbulent velocity fluctuations consist of various frequency components which can be decomposed by the spectrum analysis technique. Spectrum analysis is a class of statistical analysis and therefore the same discussion should be applied as in the case of what kind of averaging technique to apply. The definition of the conventional power spectra based on ensemble averaging is:

$$P(f) = \langle \frac{1}{T} \left| \int_T u'_n(t) \cdot e^{-j \cdot 2\pi \cdot f \cdot t} dt \right|^2 \rangle, \tag{7}$$

where f is the frequency, T is the width (in time) of a window function (usually equals to the length of the sample data segment), j is the imaginary unit, and the ensemble averaging operation is performed at every frequency component over an ensemble. This type of spectra can give overall characteristics of turbulence[67] with little respect to its unsteady nature such as evolution of the spectral structure.

Short segment unsteady analysis. The unsteady nature stressed in the previous sections should also be carefully incorporated into the spectral analysis of turbulence. The analysis of unsteady stochastic process is a hot topic in the signal analysis studies and therefore there are a number of methods available, but none of them are exclusively established. Giddens and Kitney[68] first reported an application of the autoregressive spectral analysis for turbulence of poststenotic flow fields. We developed a method of short-segmented spectrum analysis of unsteady turbulence using an autoregressive or maximum entropy method of spectrum decomposition based on an ensemble averaging technique.[69,70] Although the autoregressive method is known to give smooth and stable spectral estimations when used for relatively short data sets, it could not be directly applied to our data ensemble because of the algorithm used. The original and reported methods were based on time averaging in its definition of the most fundamental statistical measure, that is, PARCOR coefficients. We modified the definition of the PARCOR coefficients to be based on the ensemble average. This allowed us to analyze a class of unsteady stochastic process whose spectral characteristics rapidly change in time. An example is shown in Figure 5.[70] In this case, a heart beat cycle was divided into 12 overlapping short-segments and each segment underwent spectral analysis using our method (Figure 5a). It can be clearly seen that the spectral bandwidth, which is an indicator of the development of turbulence, increases following the total turbulence energy increase in the development phase (Figure 5b). However, in the decaying phase, the bandwidth decreases faster than the total energy decrease reflecting the faster energy dissipation in rather the eddies of smaller scales than those of the larger scales (Figure 5c).

Frequency versus wavenumber spectra. It is usual practice to non-dimensionalize the turbulence spectra by using a representative velocity U and a representative length d. The resulting non-dimensional spectra becomes a function of the non-dimensional wave number or the Strouhall number defined as:

$$k' = St = (2\pi) f d/U. \tag{8}$$

The physical meaning of equation (8) is, in the context of turbulence spectrum analysis, that k' is the number of eddies included in a representative length $(2\pi)\,d$, which is convected by the velocity U. The famous Taylor's hypothesis of "frozen turbulence" is behind this idea of non-dimensionalization. It is assumed in the hypothesis that the turbulent eddies of various spatial sizes are to be convected to the point of observation without any deformation. This hypothesis has been successfully proven to be true in most cases. The frequency spectra directly measured can be transformed into the non-dimensional wave number spectra by using equation (8). However, this transformation from the frequency spectra to the wave number spectra is valid only when the representative velocity or the convective velocity is steady in time. Obviously this is not the case in turbulence in the arterial system.

Equation (8) assumes that the length scale, dx, of a structure convected in front of the eyes of an observer within a time interval, dt, at the convection velocity $U_c(t)$ is expressed by:

$$dx = U_c(t)\,dt. \tag{9}$$

If the $U_c(t)$ is the constant U, then the transform from the time scale to the spatial scale is linear, the order of operation does not matter and equation (8) can be used.

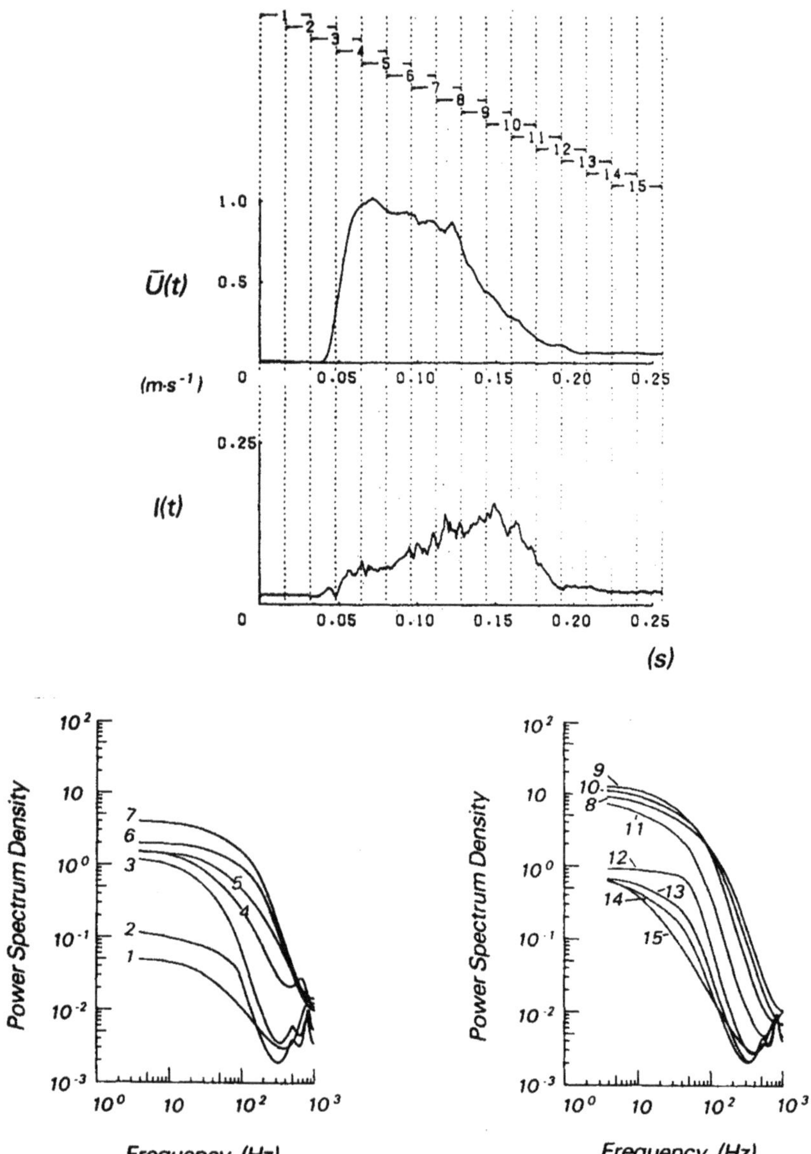

Figure 5. (a) The heart beat cycle was divided into 12 overlapping short segments and each segment was analyzed by the newly devised auto-regressive/maximum entropy method of spectral analysis. (b) Spectra calculated for the development phase of turbulence (defined by the turbulence intensity curve.) The spectral bandwidth increases following the total turbulence energy increase in the development phase. (c) Spectra calculated for the decaying phase of turbulence. In the decaying phase, the bandwidth decreases faster than the total energy decreases.

Source: With permission of Springer-Verlag.[24]

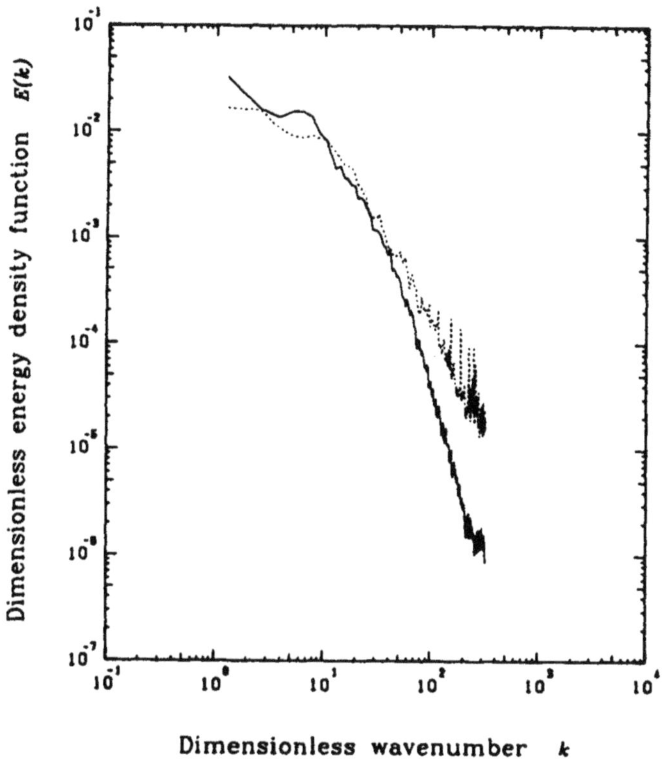

Figure 6. Comparison between the two types of wavenumber spectra. The dotted line represents the spectrum computed by conventional definition, assuming a steady representative velocity. The solid line is calculated using the transformed velocity fluctuations which is a function of distance and not of time. A coherent structure (solid line) is shown which could not be detected by the conventional spectral analysis.
Source: With permission of Pergamon Press.[12]

Nevertheless, if the convection velocity is not steady, we have to convert the time interval to the spatial length using the unsteady convection velocity first, and the wavenumber spectra should be calculated using the transformed velocity fluctuations which are a function of distance and not of time. By introducing this transformation, we could obtain a coherent structure which could not seen in the conventional spectral analysis as shown in Figure 6.[12]

C. Statistical Analysis

Kikkawa et al.[71] reported a statistical analysis of turbulence in the aorta using momentum functions. Expanding the idea of average (first order) and intensity (second order) of turbulence, statistical moment functions were defined and some interesting parameters (which are also functions of time as the first and second moment functions are), such as the skewness and the kurtosis, were calculated and examined. Some statistical tests of the distribution of the velocity fluctuations were also performed and showed that the probability distribution of the velocity fluctuations become Gaussian around when the turbulence intensity showed a maximum, that is, when turbulence most developed. In other words, turbulence in the aorta is fully developed under normal physiological conditions.

V. CONCLUDING REMARKS

We have so far discussed various aspects of turbulence mainly occurring in intact arteries under normal physiologic conditions. Increasing numbers of studies are showing that turbulence found in the blood flow is of physiological and pathophysiological significance. In particular, extensive studies have been carried out with respect to artificial organs such as the valvular prosthesis, the artificial heart, and so forth, and these have not been acknowledged in the present chapter in detail. They are, in general, dealing with abnormally high intensity turbulence which cannot be found in intact arteries.

Interestingly enough, in the blood flow in the larger arteries, particularly the aorta of relatively large-scale animals, resting physiological conditions are such that the blood flow is just critical in terms of the laminar-turbulent transition. Once the blood flow exceeds the critical condition, which is some complex combination of the Reynolds number and other non-dimensional numbers related with the time factors, turbulent fluctuations develop utilizing the total energy of the average blood flow. As is discussed in the text, these critical levels can be easily reached with just slight exercise or any other pathophysiological state, such as anemia. As Parker[54] pointed out in his insightful article, it seems that blood flow, at least in the aorta of larger mammals, is maintained at the critical level. The laminar-turbulent transition produces a discontinuous jump of physical or mechanical influence of the blood flow on the arterial wall. Consequently, the blood flow may be kept at the most effective condition to augment the sensitivity of the vascular wall to sense the increase or the decrease of the blood flow. Recent studies using cultured endothelial cells suggest the active contribution of these cells to such a mechanism.

Although we could not discuss this in detail in the present chapter, the computational fluid mechanics (CFM) method in the study of turbulence of the blood flow can be a very important tool. To the best of our knowledge, turbulence in the intact arteries has not been thoroughly studied by the CFM method except for flow around the prosthetic valves. One of the reasons for this is the theoretical difficulty in prescribing the boundary conditions. For some presently available CFM methods of turbulence computations, such as the κ-ε two equation model, we do not know what is the inlet condition in terms of turbulence quantities in the unsteady average flow cases. When we try to analyze the aortic turbulence, for example, we need to prescribe the turbulence intensity level on the outlet of the left ventricle in order to compute the development of turbulence in the aorta. A hybrid methodology, for example, an incorporation of experimental observation into the computation would be necessary to help to deepen our understanding of the mechanism of the onset, the development, and the decay of turbulence. As a preliminary step for the purpose of systemic flow studies, we have developed a method to model the real configuration of the cardiovascular system using physiological casts.[72] By this method, we can combine the models of the left ventricle, the aorta, and some of its branches as modules, and build a CFM model.[73] It is hoped that the nature of turbulent velocity fluctuations observed and measured can be interpreted in a more precise fluid mechanical description.

Newly developed statistical analysis methods are of potential help to elucidate the fundamental mechanics of turbulence. Of particular interest is the wavelet analysis[74] in which both theoretical and applied studies are extensively reported. Since turbulence in the arterial system requires an analysis in the time-wavenumber space, the wavelet analysis can help our precise understanding of the mechanics of unsteady turbulence as well as its physiological and pathophysiological significance. In particular the micro scaled structure of the velocity fluctuations are interesting both from the fluid mechanical and the biological viewpoints. The time-frequency (or time-wavenumber) structure of turbulence which can be revealed applying this method can consequently provide a set of conditions for the CFM studies. From the viewpoint of the blood flow versus the endothelial cell interactions, such an analysis can give a bridge between the macroscopic fluid mechanics and the microscopic studies of cellular mechanics.

Turbulence studies in the cardiovascular system has been conducted depending on novel measurement devices and sophisticated analysis methods. The recent introduction of some non-invasive means to measure the *in vivo* nature of the blood flow, a rapid accumulation of the cell physiological knowledge of the vascular tissues, and a powerful computational mechanical approach should be integrated. These can give us further insight and understanding, not only of the fluid mechanical, but also of the biological mechanism driving the cardiovascular system in terms of the interactions between the blood flow and the living system.

ACKNOWLEDGMENTS

The author acknowledges invaluable help in performing the studies of turbulence in the aorta given by my former colleagues, Professor S. Kikkawa, Professor K. Tanishita, Dr. T. Yoshikawa, and Professor M. Sugawara. The assistance of Dr. K.H. Parker was most helpful in completing the major part of works reviewed in the present chapter. The kind guidance and the friendship particularly during the stay in the Imperial College by Professor C.G. Caro, Dr. R.C. Schroter, and Dr. M.J. Lever is also gratefully acknowledged. The author also thanks Dr. T.W. Taylor for his help in the preparation of this manuscript. Finally, I sincerely thank my wife Uiko for her patient assistance during these days.

A part of the studies cited was supported by the Grant-in-Aid for Scientific Research "Biomechanics of Structure and Function of Living Cells, Tissues, and Organs" (#04237101) and Grant #04454537 from the Ministry of Education, Science, and Culture of Japan.

REFERENCES

1. Popper, J. *Unended Quest—An Intellectual Autobiography.* Collins, Glasgow, 1982, p. 178.
2. Hinze, J.O. *Turbulence,* 2nd ed. McGraw-Hill, New York, 1975, p. 22.
3. Sergeev, S.I. *Fluid Oscillations in Pipes at Moderate Reynolds Numbers, Fluid Dynamics* (translated). 1966, Vol. 1, pp. 21–22.
4. Merkli, P.; Thomann, H. Transition to turbulence in oscillating pipe flow. *J. Fluid Mech.* **1975,** *68,* 567–575.
5. Hino, M.; Sawamoto, M.; Takasu, S. Experiments on transition to turbulence in an oscillatory pipe flow. *J. Fluid Mech.* **1976,** *75,* 193–207.
6. Nerem, R.M.; Seed, W.A. An *in-vivo* study of the nature of aortic flow disturbances. *Cardiovasc. Res.* **1972,** *6,* 1–14.
7. Sarpkaya, T. Experimental determination of the critical reynolds number for pulsating poiseuille flow. *Trans. Am. Soc. Mech. Eng. Ser. D J. Basic Eng.* **1966,** *88,* 589–598.
8. Dantan, P.; De Jouvenel, F.; Oddou, C. Transition laminaire-turbulent end ecoulement pulsatile application a la circulation sanguine. *J. Phys.* **1976,** *37,* 157–160.
9. Settler, J.C.; Hussain, K.M. On transition of the pulsatile pipe flow. *J. Fluid Mech.* **1986,** *170,* 169–197.
10. Bluestein, D.; Einav, S.; Hwang, N.H.C. Stability of flow through heart valves: A comparison between the traditional approach and a novel approach. *1992 Advances in Bioengineering American Society of Mechanical Engineers* **1992,** *22,* 427–431.
11. Yamaguchi, T.; Kikkawa, S.; Yoshikawa, T.; Tanishita, K.; Sugawara, M. Measurement of turbulence intensity in the center of the canine ascending aorta with a hot-film anemometer. *J. Biomech. Eng.* **1983,** *105,* 177–187.
12. Yamaguchi, T.; Kikkawa, S.; Parker, K.H. Application of Taylor's hypothesis to an unsteady convective field for the spectral analysis of turbulence in the aorta. *J. Biomech.* **1984,** *17,* 889–895.
13. Lighthill, M.J. On sound generated aerodynamically I. General theory. *Proc. R. Soc. London, Series A* **1952,** *211,* 564–587.
14. Powel, A. Theory of vortex sound. *J. Acous. Soc. Am.* **1964,** *33,* 177–195.
15. Sabbah, H.N.; Stein, P.D. Turbulent blood flow in humans its primary role in the production of ejection murmur. *Circ. Res.* **1976,** *38,* 513–525.
16. Abdallah, S.A.; Hwang, N.H.C. Arterial stenosis murmurs: An analysis of flow and pressure fields. *J. Acous. Soc. Am.* **1988,** *83,* 318–334.
17. Stehbens, W.E. In: *Hemodynamics and the Blood Vessel Wall.* Charles C. Thomas Publisher, Springfield, Illinois, 1979, p. 316.
18. Mitchell, J.R.A.; Schwartz, C.J. In: *Arterial Disease.* Blackwell Scientific Publications, Oxford, 1965, pp. 50–67.
19. Yoshida, Y.; Yamaguchi, T.; Caro, C.G.; Glagov, S.; Nerem, R.M., Eds. *Role of Blood Flow in Atherogenesis.* Springer-Verlag, Tokyo, 1987.
20. Yoshida, Y.; Sue, W.; Yamane, T.; Okano, M.; Oyama, T.; Mitsumata, M.; Suda, K.; Yamaguchi, T.; Ooneda, G. Structural differences of arterial walls which are either vulnerable or resistant to atherosclerosis. *Acta Medica et Biologica* **1990,** *38(Suppl.),* 1–19.
21. Browse, N.L.; Thomas, M.L.; Young, A.E. Factors which influence the location of atherosclerosis. Is turbulence good or evil? *Br. J. Surg.* **1975,** *62,* 654.

22. Spence, J.D. Hemodynamic effects of antihypertensive drugs. Possible implications for the prevention of atherosclerosis. *Hypertension* **1984**, *6*, 163–168.
23. Spence, J.D.; Perkins, D.G.; Kline, R.L.; Adams, M.A.; Haust, M.D. Hemodynamic modification of aortic atherosclerosis. Effects of propranolol vs hydralazine in hypertensive hyperlipidemic rabbits. *Atherosclerosis* **1984**, *50*, 325–333.
24. Hanai, S.; Yamaguchi, T.; Kikkawa, S. Turbulence in the canine ascending aorta and the blood pressure. *Biorheology* **1991**, *28*, 107–116.
25. Young, D.F. Fluid mechanics of arterial stenosis. *J. Biomech. Eng.* **1979**, *101*, 157–175.
26. Stein, P.D.; Sabbah, H.N. Hemorheology of turbulence. *Biorheology* **1980**, *17*, 301–319.
27. Yearwood, T.L.; Misbach, G.A.; Chandran, K.B. Experimental fluid dynamics of aortic stenosis in a model of the human aorta. *Clin. Phys. Physiol. Meas.* **1989**, *10*, 11–24.
28. Sugawara, M.; Seo, Y.; Hongo, H. Fluid dynamics of free jets: Quantification of regurgitations, shunts, and stenosed flows. In: *Blood Flow in the Heart and Large Vessels* (Sugawara, M.; Kajiya, F.; Kitabatake, A.; Matsuo, H., Eds.). Springer-Verlag, Tokyo, 1989, pp. 173–177.
29. Lieber, B.B.; Giddens, D.P. Apparent stresses in disturbed pulsatile flows. *J. Biomech.* **1988**, *21*, 287–298.
30. Lieber, B.B.; Giddens, D.P.; Kitney, R.I.; Talhami, H. On the discrimination between band-limited coherent and random apparent stresses in transitional pulsatile flow. *J. Biomech. Eng.* **1989**, *111*, 42–46.
31. Roach, M.R. Poststenotic dilatation in arteries. In: *Cardiovascular Fluid Dynamics*, Vol. 2 (Bergel, D.H., Ed.). Academic Press, London, 1972, pp. 111–139.
32. Gow, B.S.; Legg, M.J.; Yu, W.; Kukongviriyapan, U.; Lee, L.L. Does vibration cause poststenotic dilatation *in vivo* and influence atherogenesis in cholesterol-fed rabbits? *J. Biomech. Eng.* **1992**, *114*, 20–25.
33. Vanhoutte, P.M. *Relaxing and Contracting Factors*. Humana, Clifton, NJ, 1988.
34. Davies, P.F.; Gordon, E.J.; Gimbrone, Jr., M.A.; Remuzzi, A.; Dewey, Jr., C.F. Turbulent fluid shear stress induces vascular endothelial cell turnover *in vitro*. *Proc. Natl. Acad. Sci. USA* **1986**, *83*, 2114–2117.
35. Masuda, H.; Kawamura, K.; Sugiyama, T.; Kamiya, A. Effects of endothelial denudation in flow-induced arterial dilatation. *Front. Med. Bio. Eng.* **1993**, *5*, 57–62.
36. Olsen, D.B.; Unger, F.; Oster, H.; Lawson, J.; Kessler, T.; Kolff, J.; Kolff, W.J. Thrombus generation within the artificial heart. *J. Thor. Cardiovas. Surg.* **1975**, *70*, 248–255.
37. Stevenson, D.M.; Yoganathan, A.P. Numerical simulation of steady turbulent flow through trileaflet aortic heart valves.—I. Computational scheme and methodology. *J. Biomech.* **1985**, *18*, 899–907.
38. Stevenson, D.M.; Yoganathan, A.P.; Williams, F.P. Numerical simulation of steady turbulent flow through trileaflet aortic heart valves.—II. Results on five models. *J. Biomech.* **1985**, *18*, 909–926.
39. Sallam, A.M.; Hwang, N.H.C. Human red blood cell hemolysis in a turbulent shear flow: Contribution of Reynolds shear stress. *Biorheology* **1984**, *21*, 783–797.
40. Nakahara, T.; Yoshida, F. Mechanical effects on rates of hemolysis. *J. Biomed. Mat. Res.* **1986**, *20*, 349–353.
41. Khanna, A.; Tandon, R.; Basu, A.K.; Shrivastaba, S. Red blood cell survival in patients with ventricular septal defect and patent ductus arteriosus. *Clin. Ped.* **1981**, *20*, 349–353.
42. Tennekes, H.; Lumley, J.L. *A First Course in Turbulence*. The MIT Press, Cambridge, MA, 1972, pp. 20–24.
43. Huxley, V.H.; Kutchai, H. Effect of diffusion boundary layers on the initial uptake of O_2 by red cells theory versus experiment. *Microvas. Res.* **1983**, *26*, 89–107.
44. Gad-El-Hak, M.; Morton, J.B.; Kutchai, H. Turbulent flow of red cells in dilute suspensions effect on kinetics of O_2 uptake. *Biophys. J.* **1977**, *18*, 289–300.
45. Fung, Y.C.; Liu, S.Q. Elementary mechanics of the endothelium of blood vessels. *J. Biomech. Eng.* **1993**, *11*, 1–12.

46. Davies, P.F.; Tripathi, S.C. Mechanical stress mechanism and the cell an endothelial paradigm. *Circ. Res.* **1993**, *72*, 239–245.
47. Rosati, C.; Garay, R. Flow-dependent stimulation of sodium and cholesterol uptake and cell growth in cultured vascular smooth muscle. *J. Hyper.* **1991**, *9*, 1029–1033.
48. Borow, K.M.; Glagov, S. Discrete subvalvular aortic stenosis: Is the presence of upstream complex blood flow disturbances an important pathogenic factor? *J. Am. Coll. Cardiol.* **1992**, *19*, 825–827.
49. Ling, S.C.; Atabek, H.B.; Fry, D.L.; Patel, D.J.; Janicki, J.S. Application of heated-film velocity and shear probe to haemodynamic studies. *Circ. Res.* **1968**, *23*, 789–801.
50. Schultz, D.L.; Tunstall-Pedoe, D.S.; G de J Lee; Gunning, A.J.; Bellhouse, B.J. Velocity distribution and transition in the arterial system. In: *Circulatory and Respiratory Mass Transport* (Wolstenholme, G.E.W.; Knight, J., Eds.). J&A Churchill, London, pp. 172–199.
51. Clark, C.; Schultz, D.L. Velocity distribution in aortic flow. *Cardiovas. Res.* **1973**, *31*, 601–613.
52. Stein, P.D.; Sabbah, H.N. Measured turbulence and its effect on thrombus formation. *Circ. Res.* **1974**, *35*, 608–614.
53. Giddens, D.P.; Mabon, R.F.; Cassanova, R.A. Measurements of disordered flows distal to subtotal vascular stenosis in the thoracic aortas of dogs. *Circ. Res.* **1976** *39*, 112–119.
54. Parker, K.H. Instability in arterial blood flow. In: *Cardiovascular Flow Dynamics and Measurements* (Hwang, N.H.C.; Norman, N.A., Eds.). University Park Press, Baltimore, 1977, pp. 633–663.
55. Paulsen, P.K.; Hsenkam, J.M.; Nygaard, H.; Gormsen, J. Analysis of the dynamic properties of a hot-film anemometer system for blood velocity measurements in humans. *Med. Biol. Eng. Comp.* **1987**, *25*, 195–200.
56. Deshpande, M.D.; Giddens, D.P. Turbulence measurements in a constricted tube. *J. Fluid Mech.* **1980**, *97*, 65–89.
57. Kajiya, F.; Tomonaga, G.; Tujioka, K.; Ogasawara, Y.; Nishihara, H. Evaluation of local blood flow velocity in proximal and distal coronary arteries by laser doppler method. *J. Biomech. Eng.* **1985**, *107*, 10–15.
58. Wending, F.; Jones, S.A.; Giddens, D.P. Simulation of doppler ultrasound signals for a laminar, pulsatile, nonuniform flow. *Ultrasound Med. Biol.* **1992**, *18*, 179–193.
59. Vattyam, H.M.; Shu, M.C.S.; Rittgers, S.E. Quantification of doppler color flow images from a stenosed carotid artery model. *Ultrasound Med. Biol.* **1992**, *18*, 195–203.
60. Rittgers, S.E.; Fei, D.Y.; Kraft, K.A.; Fatouros, P.P.; Kishore, P.R. Velocity profiles in stenosed tube models using magnetic resonance imaging. *J. Biomech. Eng.* **1988**, *110*, 180–184.
61. Ku, D.N.; Biancheri, C.L.; Pettigrew, R.I.; Perifer, J.; Markou, C.P.; Engels, H. Evaluation of magnetic resonance velocimetry for steady flow. *J. Biomech. Eng.* **1990**, *112*, 464–472.
62. Spielmann, R.P.; Schneider, O.; Thiele, F.; Heller, M.; Bucheler, E. Appearance of poststenotic jets in MRI: Dependence on flow velocity and on imaging parameters. *Mag. Reson. Imaging* **1991**, *9*, 67–72.
63. Kwiat, D.; Einav, S.; Elad, D. Possible detection of turbulent blood flow using multiparametric encoding gradients in MRI. *Med. Phys.* **1991**, *18*, 316–323.
64. Bendat, J.S.; Piersol, A.G. *Random Data: Analysis and Measurement Procedures*. Wiley-Interscience, New York, 1971, pp. 67–68.
65. Walburn, F.J.; Sabbah, H.N. An experimental evaluation of the use of an ensemble average for the calculation of turbulence in pulsatile flow. *Annals Biomed. Eng.* **1983**, *11*, 385–399.
66. Bradshaw, P. *Turbulence Second Corrected and Updated Edition* (Bradshaw, P., Ed.). Springer-Verlag, Berlin, Heidelberg, New York, 1978, p. 43.
67. Yamaguchi, T.; Kikkawa, S.; Tanishita, K.; Sugawara, M. Spectrum analysis of turbulence in the canine ascending aorta measured with a hot-film anemometer. *J. Biomech.* **1988**, *21*, 489–495.
68. Giddens, D.P.; Kitney, R.I. Autoregressive spectral estimation of poststenotic blood flow disturbances. *J. Biomech. Eng.* **1983**, *105*, 401–404.

69. Yamaguchi, T.; Kikkawa, S.; Parker, K.H. Simulation of nonstationary spectral analysis of turbulence in the aorta using a modified autoregressive or maximum entropy (AR/ME) method. *Med. Biol. Eng. Comp.* **1987**, *25*, 533–542.
70. Yamaguchi, T. Turbulence in the aorta. In: *Blood Flow in the Heart and Large Vessels* (Sugawara, M.; Kajiya, F.; Kitabatake, A.; Matsuo, H., Eds.). Springer-Verlag, Tokyo, 1989, pp. 37–49.
71. Kikkawa, S.; Yamaguchi, T.; Ishida, M.; Sugawara, M. Statistical properties of unsteady aortic turbulence. In: *Forum on Unsteady Flows in Biological Systems* (Friedman, M.H.; Wiggert, D.C., Eds.). The American Society of Mechanical Engineers, New York, 1985, pp. 21–24.
72. Taylor, T.W.; Yamaguchi, T. Three-dimensional graphics and computational model construction of vascular chambers using physiological cast measurements. *Adv. Bioeng., BED* **1992**, *22*, 469–472.
73. Yamaguchi, T.; Taylor, T.W. A parametrically defined computational fluid mechanical model for the study of the flow in arterial bifurcations. *Adv. Bioeng., BED* **1992**, *22*, 237–240.
74. Farge, M. Wavelet transforms and their applications to turbulence. *Ann. Rev. Fluid Mech.* **1992**, *24*, 395–457.

PULMONARY ARTERY BLOOD FLOW AND HEMODYNAMICS

Belinda Ha, William Henry, Carol Lucas, Hsing-Wen Sung, and Ajit Yoganathan

Advances in Hemodynamics and Hemorheology,
Volume 1, pages 229–324.

ISBN: 1-55938-634-7.

I. OVERVIEW OF PULMONARY CIRCULATION

A. Introduction

Ever since the pulmonary circulation was "discovered" by William Harvey (1628),[1] much work has been done on its physiology and pathophysiology. The advent of technology such as pulsed Doppler ultrasound, and more recently, magnetic resonance imaging has made available important tools to further increase our understanding of the pulmonary circulation in health and disease. The explosion of data generated by the new techniques has emphasized the need for an interdisciplinary approach to understanding the circulatory dynamics of the pulmonary circulation, an approach once favored by Thomas Young (1773–1829).[2] Thomas Young was part of the great eighteenth and nineteenth century science tradition of paying little regard to the distinction between biological and physical science. Significant contributions to our understanding of the circulation of blood were made by Euler, Daniel Bernoulli, Poiseuille, and Young himself by applying the theory of hydraulics to the circulatory system.

From a clinical point of view, the subject of pulmonary circulation mechanics is important due to the frequency and significance of pulmonary vascular changes in congenital heart diseases, as well as pulmonary hypertension in chronic rheumatic and thromboembolic diseases.[1] In contrast, the systemic circulation has long been an area of study in terms of hypertension, atherosclerosis, and other degenerative diseases.

From a physical point of view, the main pulmonary artery, as a major large vessel that is both curved and torqued, poses a number of interesting technical considerations. Although the systematic theoretical and experimental exploration of flow in curved conduits is of fairly recent origin, it has long been known that the flow is considerably more complex than that in straight conduits.[2] Experimentalists have studied the effects of curvature on flow in curved pipes[3,4,5] and the effect of curvature changes on blood flow patterns.

In order to improve our understanding of the altered fluid dynamics induced by congenital heart defects, an interdisciplinary approach is needed. The details of the rapidly changing pulmonary vascular geometry in the young need to be understood in terms of its effects on pulmonary fluid dynamics before the effects of abnormal pulmonary anatomy present in congenital heart defects can be understood. While a major portion of this work will involve experimentation, the development of mathematical models from the results of experiments can further aid in our understanding of physiological or medical observations.

In summary, an interdisciplinary approach will be used to present current knowledge on the pulmonary circulation. Clinical observations pave the way for *in vivo* animal and/or *in vitro* model experiments that will result in a fluid mechanical understanding of the pulmonary circulation. In addition, the use of applied mathematics to explain or predict experimental results can further contribute to the interpretation of the physiological or medical phenomenon. These areas pose an exciting challenge to fluid mechanics that we have only begun to meet.

B. Anatomical and Structural Organization

Normal Anatomy

The pulmonary vascular bed, which begins at the right ventricle and ends at the left atrium, may be classified sequentially according to their position in the lung as presegmental arteries, segmental arteries, intra-acinar arteries, microcirculation (consisting of precapillaries, capillaries, and postcapillaries) and veins.[6] Figure 1 shows the relationship between the respiratory tree and the vascular tree. The preacinar or first generation segmental arteries accompany the bronchi which supply the approximately 20 bronchopulmonary segments, while the first generation intra-acinar arteries accompany the terminal bronchioles which supply the acini. The microcirculation spans the alveoli and the veins correspond to the related arteries.

Reid and coworkers have also developed a unified, quantitative approach based on the composition of the vessel walls to determine the morphology of the pulmonary vascular bed in infants as well as adults who have normal and abnormal pulmonary hemodynamics.[7] Their technique involves determining the distribution and wall thickness of elastic and muscular arteries throughout the normal human lungs. The data obtained is then related to the age of the patient, as well as the diameter and position of the vessels in the pulmonary vascular tree. Moving distally along any path beyond the hilum and pleura, vessels have been sequentially classified as elastic, transitional, muscular, partially muscular, and finally nonmuscular or endothelial. The pulmonary arteries form a continuum from elastic to muscular to non-muscularized vessels, with a gradual transition between morphologic types and with a gradual, rather than abrupt, diameter change.[8] The elastic vessels are the capacitance vessels in the lungs and the muscular vessels are the resistance vessels.[9] It is primarily the elastic vessels which increase in length with increasing somatic growth, while the resistance vessels remain relatively constant in their length.[8]

West has defined another classification system based on how pulmonary vessels are affected by respiration.[10] Alveolar vessels may be defined as those that are affected by alveolar pressure changes. Extra-alveolar vessels are vessels that are not exposed to alveolar pressure but are exposed to interstitial and intrapleural pressure alterations. During normal inflation of the lung, interalveolar vessels are

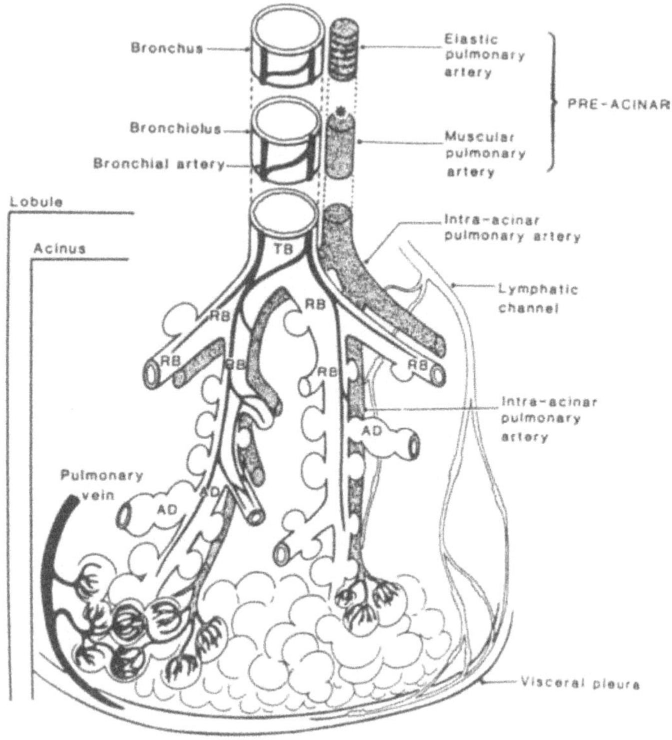

Figure 1. Diagram of the airway and arterial pathway showing an elastic artery which accompanies cartilaginous lobar, segmental, and subsequent bronchi. The pulmonary artery becomes muscularized at the seventh–ninth division from the segmental hilum, as indicated by * in this figure. TB (terminal bronchioles); RB (respiratory bronchioles); AD (alveolar duct). Reproduced from Haworth.[24]

compressed and extra-alveolar vessels are dilated, affecting their respective blood volumes.

Differences Between Pre- and Postnatal Structure

A number of developmental and adaptive changes occur in the pulmonary vascular bed in the transition from intrauterine to extrauterine life. The adaptive changes lead to an abrupt and marked decrease in pulmonary vascular resistance at birth to be followed by remodeling of the vasculature.[11,12]

Since pulmonary hemodynamics in the prenatal period differ substantially from those in the postnatal period, the differences in the wall structure of the pulmonary vessels may be attributed to the functional differences of the pulmonary circulation. The pulmonary circulation before birth is a system with high resistance and low blood flow; after birth, the system becomes low resistance with high blood flow.

Elastic arteries. During fetal and immediate postnatal life, the pulmonary trunk has a structure identical to that of the aorta.[11] This applies both to the wall thickness which is approximately equal in the two vessels, and to their elastic configuration. The vessels comprise a dense and regular arrangement of parallel elastic membranes separated by a few smooth muscle cells and collagenous fibers.

After birth, the elastic tissue gradually diminishes and is replaced by collagen.[13] From the fourth month upwards, and in particular after the first year of life, the elastic laminae become more fragmented. This change is accompanied by a relative decrease in the thickness of the media as compared to the aorta.

The elastic intrapulmonary arteries are scarce and gradually increase in number.[14] They have a regular elastic configuration and very often, the largest vessels exhibit a thick layer of intimal cellular proliferation over considerable distances. This layer tends to disappear in the course of the first postnatal year.

Muscular pulmonary arteries and partially muscular arteries. The differences between these vessels and their postnatal counterparts are particularly striking. In fetuses of 20 week gestation and in newborn infants, the muscular pulmonary arteries and arterioles have thick walls with narrow lumen.[11] The thickness of the media is essentially 15–25% of the external arterial diameter.[15] In the normal human lung, the relative pulmonary arterial medial thickness decreases rapidly during the first days of life, particularly in arteries < 250 μm in diameter.[16,17] It reaches a mature adult level in all arteries during the first three months of life.[15,16] Within the first 2–3 weeks, the muscular pulmonary arteries become considerably larger in diameter, in particular the smallest arteries and arterioles which were almost closed during fetal life.[12] The gradual increase in size of the lumen was accompanied by a decrease in medial thickness so that at an age variably given as six months to 1.5 years, the media/diameter ratio has attained the same values existing in adult life (Figure 2).

Geometry

The normal geometry of the pulmonary artery bifurcation varies with age. The normal pulmonary vascular tree consists of vessels that bifurcate and curve, with varying diameters. Moreover, recent studies using computer tomography[18] and magnetic resonance imaging[19–21] to image cardiac structures have shown that changes in the geometry/dimensions of the pulmonary arteries and right ventricle occur in the presence of disease such as pulmonary hypertension. The combined effects of age and disease on the complex geometry of the pulmonary artery bifurcation add to the complexities of flow in the pulmonary artery and in most cases contribute additional energy losses.

A systematic study on the effects of geometry on pulmonary blood flow and on the effects of maturation on pulmonary geometry and pulmonary hemodynamics is being performed jointly by our two research groups. Quantitative studies on

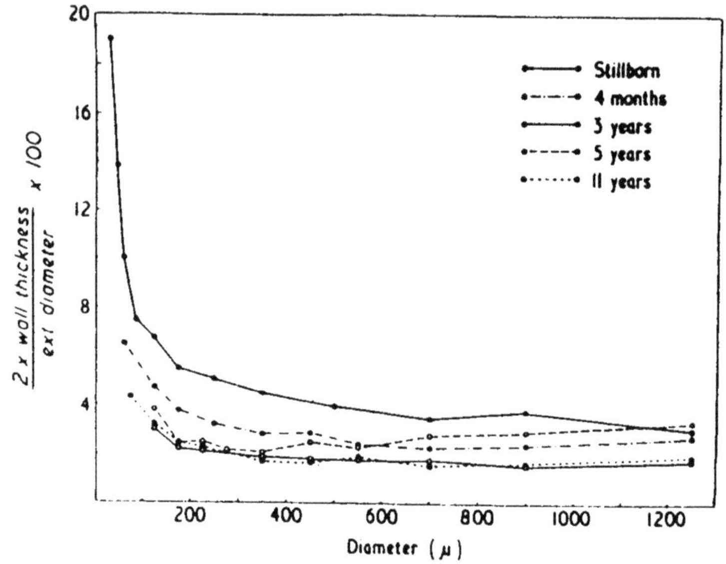

Figure 2. Thickness of the medial coat of the pulmonary artery at birth and throughout childhood. Adult values are reached by four months; at 3–5 years, muscle does not penetrate such small arteries as at other ages shown. Reproduced from G. Davies and L. Reid.[210]

pulmonary geometry are being aided by computer reconstructions and analyses of three-dimensional right heart and pulmonary vasculature CT image data. Data were obtained by imaging methyl methacrylate casts of right heart and pulmonary vasculature of lambs ranging from two days to six months in age.

Marked changes in taper, curvature, and degree of angulation of the main pulmonary artery are observed following birth.[22] In the fetus, the pulmonary trunk continues into the ductus arteriosus, and the left and right pulmonary arteries appear as smaller branches off the trunk. After birth, when the ductus arteriosus closes, the pulmonary trunk is still very large and is dome-shaped with the branch arteries rising posteriorly (Figure 3). For several weeks after birth, the marked discrepancy between the large main pulmonary trunk and branches gradually diminishes as the branches increase in size. With further growth, the branches enlarge and assume adult configuration while the main pulmonary artery exhibits tapering to the point of its bifurcation.[23]

Branching Pattern

The arteries and airways develop together.[24] About three-fourths of the branches present at birth are formed between the tenth and fourteenth weeks of gestation.

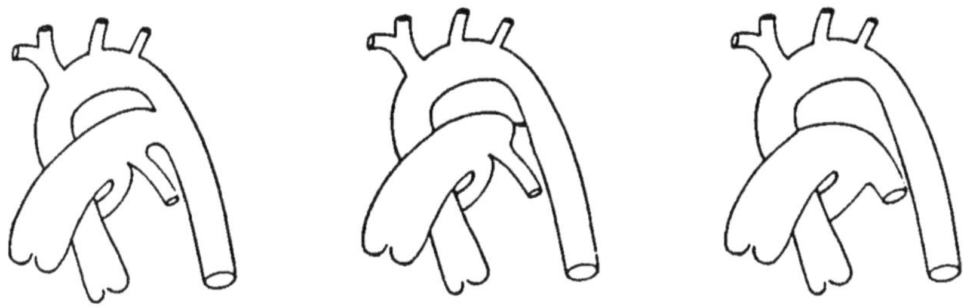

Figure 3. The changes in the configuration of the pulmonary trunk and its primary branching from fetal to adult life. Reproduced from Rudolph.[22]

The longer segments of the lung continue developing until the sixteenth week. Pre-acinar airway and arterial branching is also complete at this age. In contrast, the intra-acinar arteries develop relatively late in fetal life and continue to increase markedly after birth as the alveolar ducts and alveoli develop.[16] The intrapulmonary veins development parallels that of the arteries. Studies by Reid and coworkers have shown that the adult pattern of pulmonary artery branching is completed by the nineteenth week of fetal life.[12] The branches of the pulmonary arterial tree, including the elastic as well as the muscular arteries, tend to follow a dichotomous system. This pattern is not maintained because of the small, muscular, supernumerary vessels that arise perpendicularly from the elastic and larger muscular arteries.

C. Physiology

Fetal Circulation

Information on the fetal circulation has been derived principally from animal experiments, by a variety of methods, that is, cineangiography,[25] blood gas analysis,[26,27] measurement of blood flow by electromagnetic flowmeters,[28] distribution of radioisotope-labeled microspheres,[29] dye dilution,[30] and radioisotope methods.[31] Some of these findings have been corroborated for the human in studies of the human fetus itself.[32,33]

Pulmonary hemodynamics differ substantially in the prenatal as compared to the postnatal period. In fetal life the lungs have no function with regard to blood oxygenation. The fetus derives its oxygen from the placenta.

A large portion of the blood returning from the fetal placenta via the *umbilical vein* is distributed to both the right and left lobes of the liver and reaches the inferior vena cava via the hepatic veins; the remainder is shunted directly to the inferior vena cava via the *ductus venosus* (Figure 4).[26] The well-oxygenated blood mixes

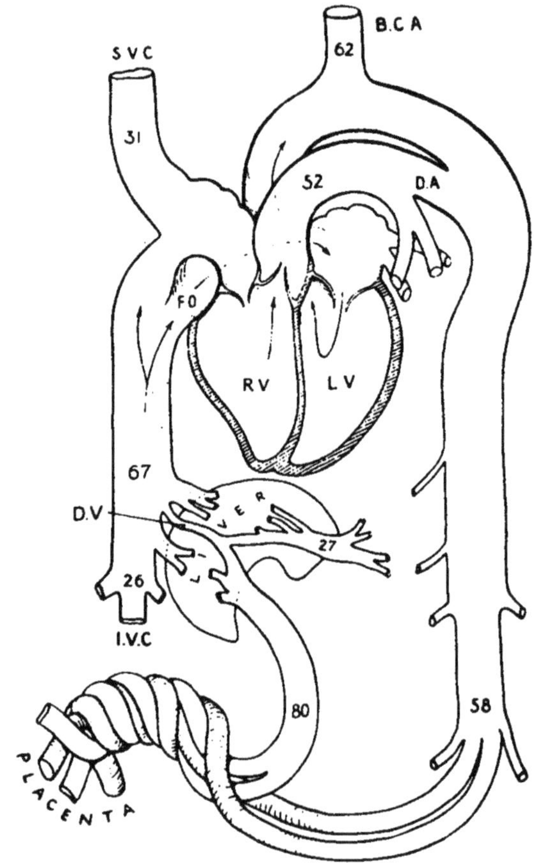

Figure 4. Diagram of circulation of mature fetal lamb. Numbers indicate approximate hemoglobin-O_2 saturation. Reproduced from Born.[26]

with the venous return from the viscera and lower body in the inferior vena cava.[34] The blood flows preferentially from the right atrium through the *foramen ovale* into the left atrium; it then enters the ascending aorta via the left ventricle to perfuse the cephalad portion of the body with the still oxygenated blood. Superior caval vein blood flows preferentially through the tricuspid valve to the right ventricle, where it is ejected into the main pulmonary artery, ductus arteriosus, and descending aorta. The *ductus arteriosus* (or ductus Botalli) connecting the main pulmonary artery with the aorta is a large channel equal in diameter to the ascending aorta. The main pulmonary artery is relatively large since two-thirds of the ventricular output is ejected into the vessel from the right ventricle. Due to the high pulmonary vascular resistance, only 10% of the combined cardiac output flows through the lungs while approximately 55% returns to the placenta and the remaining 35% perfuses the remaining body tissues.

The structures italicized above are those that mark the essential morphologic differences between fetal and adult circulations. The functional followed by the ultimate structural involution of these fetal conduits after birth, might be termed an example of "disuse atrophy."[35] For the purposes of conceptualization, their functional and structural integrity in the fetus may be attributed to the actual and potential gas exchange systems.[22]

The actual gas exchange system, that is, the umbilical-placental circulation, is a low resistance system with high blood flow. Its input and output vessels conduct about one-half of the total cardiac output, and thus are well-developed in the fetus. Cessation of flow at delivery sets the stage for the structural involution of the umbilical vessels.

The potential gas exchange system, that is, the pulmonary circulation, presents a converse picture in that it is a high-resistance (higher than the systemic arterial resistance) system with low blood flow. High resistance, low-flow pulmonary circulation is characterized by low pulmonary venous return to the left atrium and high right ventricular and right atrial pressures. The resultant pressure gradient from the right to the left atrium and the consequent flow of blood in the same direction prevents closure of the foramen ovale. At birth, pulmonary vascular resistance falls and pulmonary blood flow increases markedly with lung expansion. This phenomenon, along with decreased blood flow to the placenta, reverses the atrial pressure gradient so that the foramen ovale closes. Arterial oxygen saturation also increases so that the ductus arteriosus constricts and begins the process of involution.

Postnatal Circulation

After birth and with the initiation of pulmonary ventilation, the crucial circulatory adaptation is the rapid fall of the pulmonary vascular resistance which results in increased perfusion of the pulmonary capillaries and thus establishes alveolar-capillary gas exchange.[22] Factors thought to be responsible for the decrease in pulmonary vascular resistance include the change in arterial or alveolar Po_2, and the influence of prostaglandins that are powerful pulmonary vasodilators.[24] With the cessation of umbilical flow, right atrial pressure consequently falls and the combination of decreased pressure and resistance with the concomitant rise in left atrial pressure at first reduces and then stops flow through the foramen ovale as left atrial pressure exceeds that of the right atrium. The removal of the low-resistance, high-flow placental circuit from the systemic circulation also tends to increase left heart and systemic pressures. Constriction of the ductus arteriosus occurs at about 12 hours after birth largely in response to an increase in systemic arterial oxygen concentration. By 24 hours after birth, pulmonary artery pressure is about one-half of aortic pressure and thereafter falls more slowly to reach low adult values by 1–4 weeks after birth. A final level is reached six months after birth.

Arterial Dynamics

An understanding of the hydrodynamics of the arterial system must come not only from knowledge of the geometry but also from knowledge of the elastic properties of the arterial wall. The viscoelasticity of blood vessels has long been recognized as playing an essential role in cardiovascular behavior. In oscillatory flow, the artery must be treated as a viscoelastic tube whose diameter varies with a pulsating pressure; in addition, it will propagate pressure and flow waves, generated by the ejection of blood from the ventricle, at a certain velocity which is largely determined by the elastic properties of the arterial wall.

Capacitance. For the presegmental (or elastic arteries), the wall properties that are essential are those which lead to an estimate of vessel compliance, that is, the change in volume (V) per unit change in pressure (p) and is a measure of the stiffness or distensibility of the vascular wall.[6] Part of the energy delivered by the right ventricle in systole is used to propel the blood forward into the pulmonary artery and part of it is used to distend the major pulmonary branches. During diastole, the walls recoil and provide energy for propulsion of blood. Frequently calculated variables from which compliance can be estimated include (1) the volume distensibility coefficient (D_v), (2) the pressure-strain elastic modulus (Ep), and (3) the ratio of wall thickness (h) to vessel radius (r) and Young's modulus (E). Equations are listed as follows:

$$D_v = \Delta V / V \Delta p \quad (1)$$

$$E_p = \Delta p \cdot r / \Delta r \quad (2)$$

$$C = D_v \cdot V \quad (3)$$

$$D_v = 2/E_p; \text{ elastic, cylindrical tube of fixed length} \quad (4)$$

$$E = E_p \cdot (r/h); \text{ thin walled, cylindrical elastic tube} \quad (5)$$

$$E = 3/4 \cdot Ep \cdot (r/h); \text{ thick walled, viscoelastic tube} \quad (6)$$

$$C = [3(r/h + 1)^2 / E(2r/h + 1)] \cdot V \quad (7)$$

$$C \simeq [3r/2E \cdot h] \cdot V; r/h >> 1 \quad (8)$$

Pulse wave velocity. The pressure and flow pulse waves generated by the right ventricle contain a certain amount of information on their speed of propagation and their degree of attenuation.[36] The propagation of pressure pulse waves has been shown to be largely dependent on the extensibility of the arterial walls.[37] The Moens–Korteweg equation most clearly depicts the relationship between wall properties and pressure pulse wave velocity:[38]

$$c = (Y \cdot h/2r\rho)^{0.5}$$

where c is the wave velocity, Y is Young's modulus for the arterial wall, h is the thickness of the wall, r is the radius of the artery, and ρ is the density of blood. Bramwell and Hill modified and simplified the Korteweg equation to:

$$c = 3.57/E^{0.5}$$

where E is the distensiblity coefficient of the vessel (percentage change in volume per mm Hg).

Measurements of the pressure pulse wave velocity in humans have given figures of 1.8m/s[37] and 1.7m/s.[38] Values of 4–5 m/s[37] have been found for patients with acquired pulmonary hypertension signifying a considerable decrease in distensibility of the pulmonary arteries relative to their increase in radius. Flow pulse wave velocity values of 70 cm/s[39] have been determined. With increasing levels of pulmonary arterial pressure, the conduction time progressively diminished.

Wave reflections. In their passage through the complex branching system of the pulmonary vascular tree, the pressure and flow waves are susceptible to reflection from discontinuities at branching points, areas of change in arterial distensibility, and the high resistance arterioles.[36] When wave travel occurs in a system with closed-end type reflection sites, the reflected wave is added to incident wave to produce the measured wave. For travel in open-end reflection systems, the reflected wave is subtracted from the incident wave to produce the measured wave resulting in a damping effect. Thus, the presence of reflected waves would mean that the amplitudes of pressure and of flow waves would be increased or decreased at various places in the pulmonary arterial tree.

D. Pathophysiology

Congenital Heart Disease

Congenital heart disease is present in about eight per 1,000 of newborn infants.[40] In most instances, environmental factors interact with a genetic predisposition to produce the maldevelopment. About 8% of cases result mainly from genetic factors.[41,42] About 2% of cases are primarily the result of environmental or external factors.[43]

Congenital heart defects can be organized into those which do or do not result in cyanosis (Table 2).[44] Acyanotic defects are generally characterized by a left-to-right shunt and increased pulmonary blood flow. Left-to-right shunts may occur at

Table 1. Classification of Congenital Heart Defects Modified from Perloff et al.[45]

ACYANOTIC WITH LEFT-TO-RIGHT SHUNT
Atrial level shunt:
1. Atrial septal defect
2. Partial anomalous venous connection
3. Partial anomalous pulmonary venous connection
Ventricular level shunt:
1. Ventricular septal defect
2. Ventricular septal defect with aortic regurgitation
3. Ventricular septal defect with left ventricular to right atrial shunt
Aortic root to right heart shunt:
1. Ruptured sinus of Valsalva aneurysm
2. Coronary arteriovenous fistula
3. Anomalous origin of the left coronary artery from the pulmonary trunk
Aortopulmonary level shunt:
1. Aortopulmonary window
2. Patent ductus arteriosus
Multiple level shunts:
1. Complete common atrioventricular canal
2. Ventricular septal defect with atrial septal defect
3. Ventricular septal defect with patent ductus arteriosu
ACYANOTIC WITHOUT A SHUNT
Right heart malformations:
1. Ebstein's anomaly of the tricuspid valve
2. Pulmonic stenosis (subinfundibulur, infundibular, valvular, supravalvular)
3. Congenital pulmonary valve regurgitation
4. Idiopathic dilatation of the pulmonary trunk
Left heart malformations:
1. Congenital obstruction to left atrial inflow
2. Mitral regurgitation
3. Primary dilated endocardial fibroelastosis
CYANOTIC
Increased pulmonary blood flow:
1. Complete transposition of the great arteries
2. Double-outlet right ventricle of the Taussig-Bing type
3. Truncus arteriosus
4. Total anomalous pulmonary venous connection
5. Single ventricle without pulmonary stenosis
6. Common atrium
7. Tetralogy of Fallot with pulmonary atresia
8. Tricuspid atresia with large ventricular septal defect and no pulmonic stenosis

(*continued*)

Table 1. (Continued)

Normal or decreased pulmonary blood flow:
1. Tricuspid atresia
2. Ebstein's anomaly with right-to-left shunt
3. Pulmonary atresia with intact ventricular septum
4. Pulmonic stenosis or atresia with ventricular septal defect (Tetralogy of Fallot)
5. Pulmonic stenosis with right-to-left atrial shunt
6. Complete transposition of the great arteries with pulmonic stenosis
7. Double-outlet right ventricle with pulmonic stenosis
8. Single ventricle with pulmonic stenosis
9. Pulmonary arteriovenous fistula
10. Vena caval to left atrial communication

various levels: atrial (e.g., atrial septal defect), ventricular (e.g., ventricular septal defect), or arterial (e.g., patent ductus). Obstructive lesions without any associated shunts include pulmonary stenosis, aortic stenosis, and coarctation of aorta. Cyanosis, however, may occur in these defects in the presence of congestive heart failure, pulmonary complications, or pulmonary vascular obstructive disease.

Cyanotic defects are generally characterized by a right-to-left shunt and may be classified into two broad categories. In the first group of anomalies with intracardiac defects and obstruction to pulmonary flow, cyanosis is caused by decreased pulmonary blood flow and mixing of oxygenated and desaturated blood at the atrial and ventricular levels. In the second group, cyanosis may also be caused by the admixture of pulmonary and systemic venous returns despite normal or increased pulmonary blood flow. In most cardiac malformations classified in this group a single chamber receives the total systemic and pulmonary venous returns. The admixture lesion can occur at any cardiac level: venous—total anomalous pulmonary venous connection; atrial—single atrium; ventricular—single ventricle; and great vessel—persistent truncus arteriosus. Near uniform mixing of the two venous returns usually occurs. Complete transposition is included in this group though only partial admixture of the two venous returns occur. This incomplete mixing leads to severe hypoxia.

Hemodynamic Principles

To understand the hemodynamic consequences of congenital heart defects, a simple, physiological classification has been proposed.[45,46] It should be noted that within each category described, subgroups relating to the relative size or location of the defects are also present.

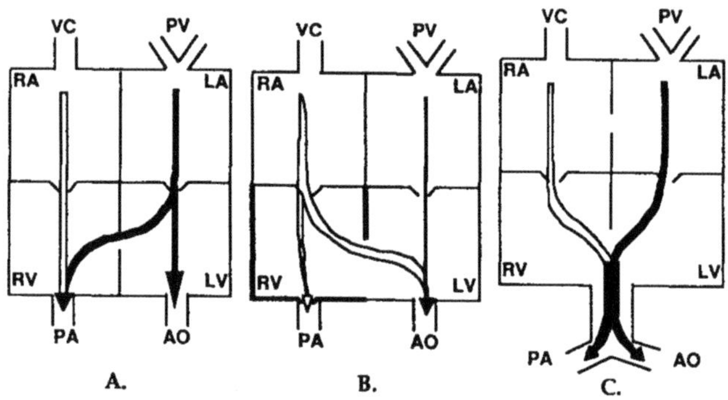

Figure 5. Flow diagrams showing the three physiological classifications (modified from Miller).[45] (**a**) Low resistance communication (example, isolated ventricular septal defect). (**b**) High resistance communication (example, tetralogy of Fallot). (**c**) Common mixing situation with low pulmonary vascular resistance (a common arterial trunk). Magnitude of blood flow is represented by the relative sizes of the arrows. Oxygenated blood is shown in black, deoxygenated blood is shown in white, and the mixture of oxygenated and deoxygenated blood is shown in grey. RA (right atrium); LA (left atrium); RV (right ventricle); LV (left ventricle); PA (pulmonary artery); AO (aorta); VC (vena cava); PV (pulmonary veins).

1. Simple communications between the right and left heart with low pulmonary vascular resistance. An example of this condition is an isolated ventricular septal defect. In Figure 5a, a large defect results in equalization of right and left ventricular pressures. Since pulmonary vascular resistance is low, a large left-to-right shunt develops as blood moves from the high resistance systemic circulation into the low resistance pulmonary circulation. The consequences of the shunt and high pulmonary blood flow are an enlarged heart and pulmonary arteries. Volume loading of the heart eventually leads to heart failure.

2. Simple communication between the right and left heart with high pulmonary vascular resistance. An example of this condition is tetralogy of Fallot (Figure 5b). This malformation has four components: ventricular septal defect, aorta overriding the ventricular septal defect, pulmonary stenosis, generally at the infundibulum, and right ventricular hypertrophy. Because of the relationship between the aorta and right ventricle, right ventricular systolic pressure is at systemic levels. The high pulmonary vascular resistance and pulmonary stenosis results in decreased pulmonary blood flow and a right-to-left shunt with arterial desaturation (cyanosis). The degree of arterial desaturation is determined by the size of the right-to-left shunt and by the amount of oxygenated blood arriving at the left heart. Increasing pulmonary blood flow via an aortopulmonary shunt (such as Blalock-Taussig beyond the site of obstruction) will increase the amount of oxygenated pulmonary venous blood and therefore reduce the degree of cyanosis.

Rudolph introduced the term "obligatory" or "nonobligatory" shunt to describe the dependance or independence of left-to-right shunts on pulmonary resistance.[47] In the case of nonobligatory shunting, the magnitude of the left-to-right shunt will be dependent on the relative difference between the pulmonary and systemic resistances. Examples of this type of shunting include an unrestricted ventricular septal defect or large ductus arteriosus. In the case of obligatory (or independent) shunting, the left-to-right shunt will always be present regardless of the pulmonary vascular resistance or impedance.

3. *Mixing Situations.* An example of a complete mixing condition is persistent truncus arteriosus, typified by a common arterial trunk (Figure 5c). With this arrangement, the low resistance pulmonary circulation will be supplied from the high resistance systemic circulation. As a result, pulmonary blood flow will be high and cardiac enlargement will develop from the increased volume load. The systemic circulation is therefore supplied with blood of less than the full oxygen saturation. Thus, the degree of arterial desaturation is inversely related to pulmonary flow. However, in the case of poor mixing such as complete transposition (in which the aorta is connected to the morphologically right and the pulmonary trunk is con-

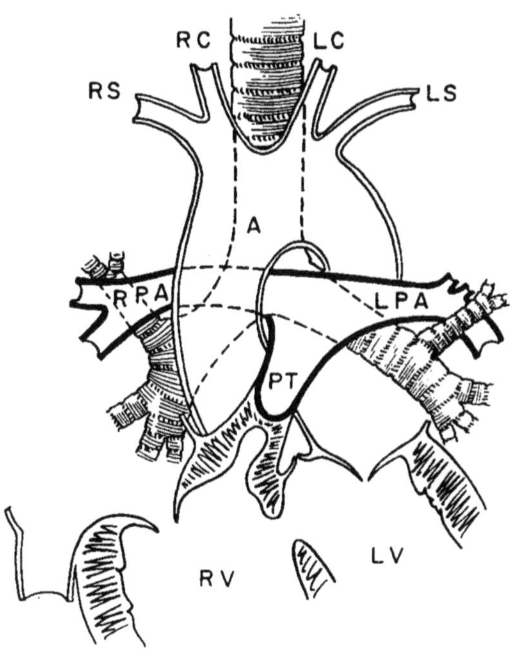

Figure 6. Atresia of the origin of the pulmonary trunk (PT) in tetralogy of Fallot (pseudotruncus arteriosus). Right pulmonary artery (RPA), left pulmonary artery (LPA), aorta (A). Reproduced from Edwards and McGoon.[50]

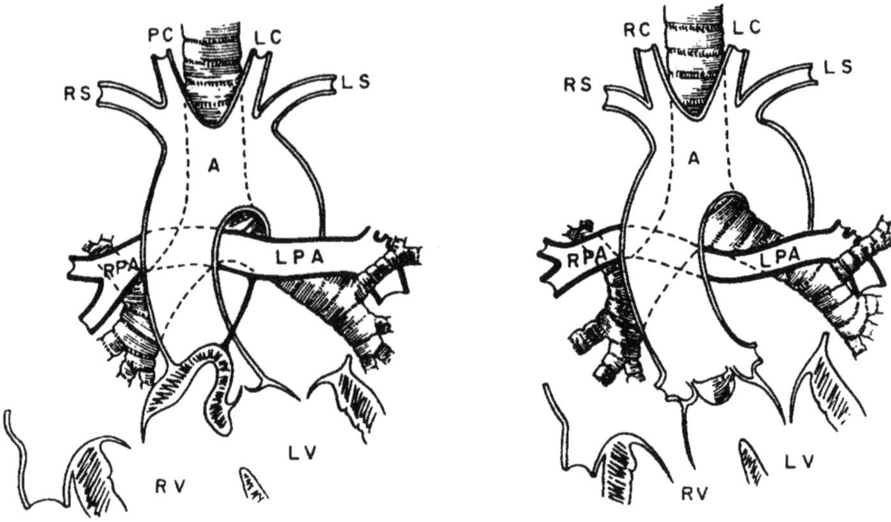

Figure 7. Atresia of the entire length of pulmonary trunk with confluence of the left (LPA) and right (RPA) pulmonary arteries. (Left) The pulmonary trunk is identifiable. (Right) The pulmonary trunk is not identifiable. Reproduced from Edwards and McGoon.[50]

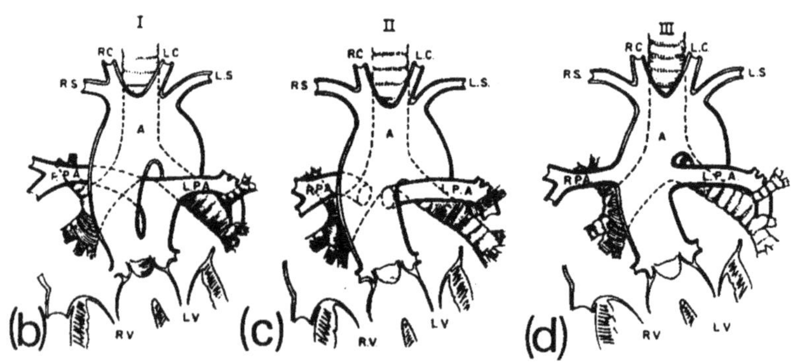

Figure 8. (b), (c), (d) Variations in the manner of origin of the pulmonary arteries in persistent truncus arteriosus. (b) Type I, in which the left and right pulmonary arteries arise from a pulmonary trunk. (c) Type II. The right and left pulmonary arteries arise independently from the dorsal aspect of the truncus arteriosus. (d) Type III. The left and right pulmonary arteries arise independently from the lateral aspects of the truncus arteriosus. Reproduced from Edwards and McGoon.[50]

nected to the morphologically left ventricles, respectively), a profound degree of arterial desaturation exists since the oxygenated blood returning to the left atrium never mixes with the largely independent systemic blood flow. Thus, palliative treatment in the form of an atrial septostomy must be performed to improve mixing.

The Pulmonary Circulation in Congenital Heart Disease

In children with congenital heart disease the pulmonary circulation may fail to develop normally even before birth.[24] Abnormalities of pulmonary vascular development which arise during fetal life may be divided into primary developmental abnormalities and secondary abnormalities which are probably due to hemodynamic changes caused by the presence of congenital heart disease. The latter therefore develop later in fetal life. Primary abnormalities may occur in children with normal hearts and are then usually associated with disordered airway development.

Among the anomalies of the pulmonary trunk are idiopathic dilatation, obstruction of the lumen, communication with the aorta, and origin of the coronary arteries from this segment of the pulmonary arterial system.[48] Idiopathic dilatation of the pulmonary trunk is characterized by dilatation of this vessel in the absence of pulmonary stenosis, of pulmonary hypertension, and of inflammatory disease of this vessel.[49] Obstruction of the pulmonary trunk may take the form of either stenosis or atresia (Figures 6 and 7).[50] Communication with the aorta may be through a window-like communication between the pulmonary trunk and the ascending aorta or through persistent truncus arteriosus when the embryonic truncus arteriosus fails to divide into the aorta and pulmonary trunk. The manner of origin of the pulmonary arteries is a basis for subdividing this condition into three basic types (Figure 8).

The pulmonary arterial branches are subject to anomalous origin, stenosis, or atresia.[48] Anomalous origin of a pulmonary arterial branch may be unilateral or bilateral. The anomalous conditions may involve origins of the pulmonary arterial branches either from the pulmonary arterial system[51] or from the aorta or from its branches[52] (Figure 9). Obstruction involving pulmonary arterial branches may take the form of local stenosis or stenoses, on the one hand, or atresia, on the other. The process of stenosis is characterized by focal obstruction or obstructions that may involve the origins of the main pulmonary arterial branches (Figure 10) and/or one or more foci in secondary and tertiary branches (Figure 11). Atresia of a pulmonary trunk may be focal or diffuse (Figure 12).

Surgical Treatment of Congenital Heart Disease

Though enormous strides have been made in cardiovascular surgical techniques over the past 20 years, a large number of challenges remain. These challenges remain because of the diversity and complexity of anatomic, hemodynamic, and

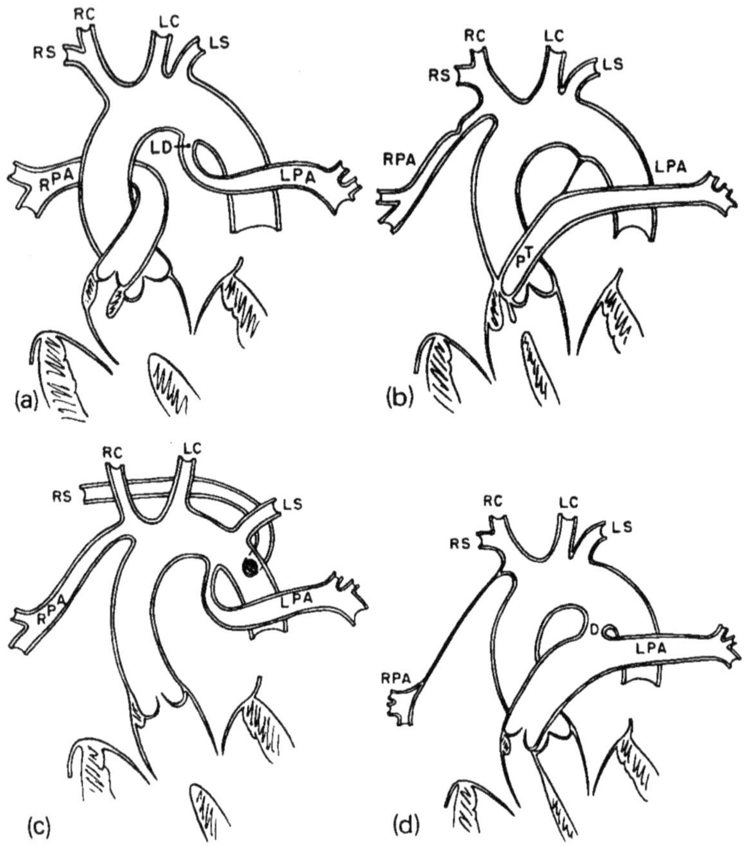

Figure 9. Examples of distal ductal origin of one or both pulmonary arteries. RPA (right pulmonary artery), LPA (left pulmonary artery), D (ductus arteriousus). (a) Left aortic arch with distal ductal origin (LD) of LPA from aortic arch. (b) Left aortic arch with distal ductal origin of the RPA from the inominate artery. (c) Solitary aortic trunk with bilateral distal ductal origin of the pulmonary arteries associated with left aortic arch. The right subclavian artery (RS) shows an aberrant origin from the aorta. (d) Distal ductal origin of RPA with atresia of vessel to the pulmonary hilus. At the latter location, the RPA is patent. From Sotomora and Edwards.[52]

electrophysiologic considerations that play major roles in the sequelae of patients with congenital heart disease before and after their operations.

Cyanosis is caused by heart defects characterized by the interruption or impairment of pulmonary blood flow, such that shunting of blood becomes necessary in order to increase pulmonary blood flow. The ideal shunt should (1) reliably increase pulmonary blood flow, (2) be relatively simple to construct, (3) provide bilateral pulmonary blood flow, and (4) facilitate pulmonary artery growth. Though early operative procedures relied on surgically created pulmonary systemic shunts,[53] extracardiac shunts using homografts and synthetic conduits[54,55] began to be used

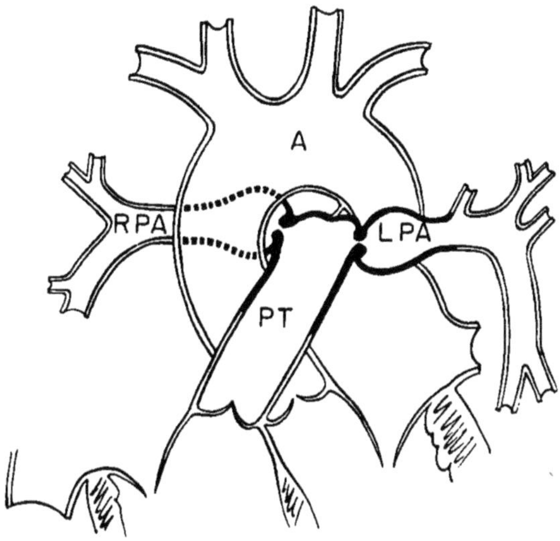

Figure 10. Diagram of localized stenosis of origin of each pulmonary arterial.[48]

Table 2. Palliative Procedures to Create New Circulatory Pathways

Objective	*Operation*	*Description*
Increase pulmonary flow	Blalock-Taussig	Subclavian artery-pulmonary artery anastomosis, end-to-side
	Blalock-Taussig, modified	Subclavian artery-pulmonary artery, interposed PTFE* graft
	Potts	Descending aorta-pulmonary artery anastomosis, side-to-side
	Waterston	Ascending aorta-pulmonary artery anastomosis, side to side
	Central	Aorta-pulmonary artery anastosis, PTFE* graft interposed
	Glenn	Superior vena cava-pulmonary artery anastomosis, side to end
	Brock	Pulmonary valvutomy and in fundibulectomy, closed
	Outflow patch	Right ventricle-pulmonary artery outflow patch
Decrease pulmonary flow	Pulmonary artery banding	Constrictive band around pulmonary artery
Improve systemic-pulmonary mixing	Rashkind	Rupture of membrane of fossa ovale by balloon catheter
	Park	Atrial septostomy using catheter blade
	Blalock-Hanlon	Atrial septostomy, closed
	Mustard or Senning (palliative)	Intra-atrial venous transposition without closure of atrial septal defect.

Note: *PTFE polytetrafluroethylene

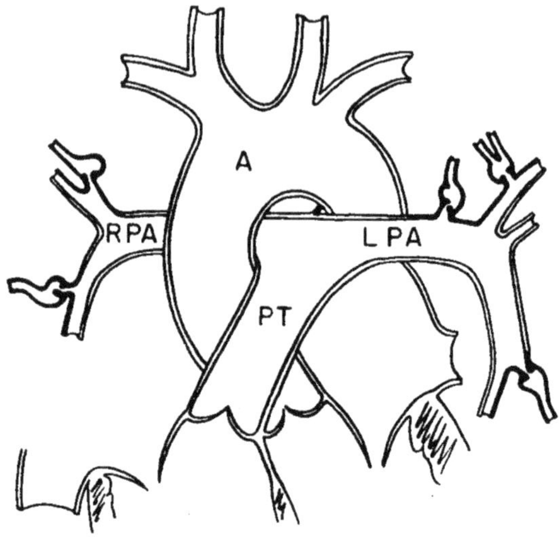

Figure 11. Diagram of multiple sites of stenosis of peripheral pulmonary arteries.[48]

for conditions when there was inadequate space within the heart for rechanneling blood (Table 2). The advent of the Fontan procedure in 1971 provided a more physiologic correction in terms of arterial oxygen saturation via a cavopulmonary (Glenn) anastomosis, a direct anastomosis between the right atrial appendage and the proximal end of the divided pulmonary artery, closure of the atrial septal defect, ligation of the main pulmonary artery, and the insertion of an aortic homograft valve into the inferior vena cava ostium (Figure 13).[56] Despite considerable progress in surgical techniques including more recent modified Fontan operations, the design requirements for an ideal shunt remain elusive: (1) synthetic conduits do not grow and frequently become obstructed,[57,58] and (2) the modified Fontan procedures will selectively benefit only those patients with reasonable anatomy of the veins and normal pulmonary vascular resistance.

Pulmonary Hypertension

Definition. Pulmonary hypertension is defined as pulmonary artery pressure above normal, a value that changes during fetal development and in the neonatal period. The age-related changes in pulmonary artery pressure are shown in Figure 14. Pulmonary hypertension is a significant risk factor in many forms of congenital heart disease. The presence of pulmonary hypertension modifies the surgical management of a child with a congenital cardiac abnormality, markedly influences the ultimate results, and poses a discouraging prognosis for life in cases when pulmonary hypertension leads to the development of irreversible pulmonary vascular disease.

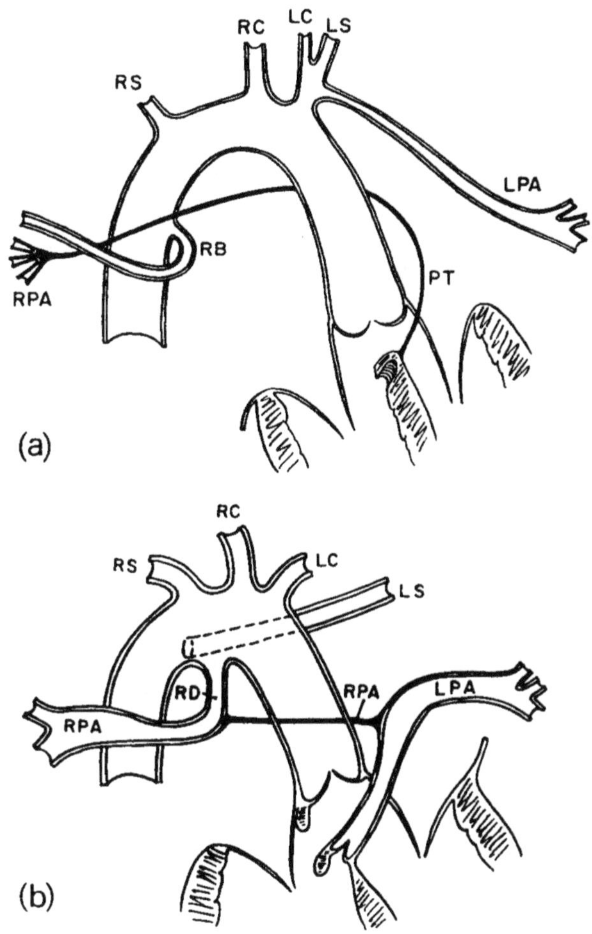

Figure 12. Atresia of major segments of pulmonary arterial system. (a) While the right pulmonary artery is patent at the hilus, the proximal segment of the right pulmonary artery and the entire pulmonary trunk are atretic. There is coexistent distal ductal origin of the left pulmonary artery from the left-sided inominate artery associated with a right aortic arch. RB = bronchial artery. (b) Atresia of the proximal segment of the right pulmonary artery, while the distal segment is patent and is supplied by a right-sided ductus arteriosus (RD) arising from a right aortic arch. The left subclavian artery (LS) is shown as an abberant origin from the aorta. Reproduced from Sotomora and Edwards.[52]

Mechanisms. In general, the principal mechanisms of pulmonary hypertension include:[59]

1. Pulmonary hypertension can be passive, that is, when right ventricular and pulmonary artery pressure is increased as a consequence of a raised "downstream"

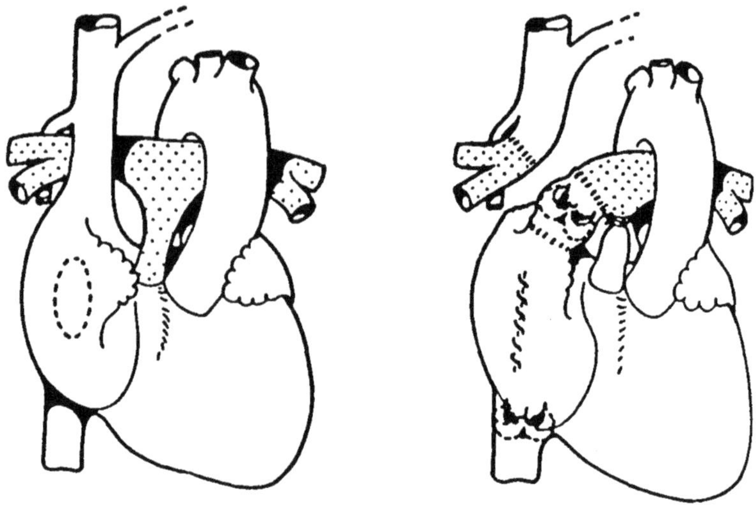

Figure 13. Fontan's original repair. Reproduced from Fontan.[56]

(pulmonary venous pressure) due to obstruction to blood flow beyond the capillary bed. The most common cause of passive pulmonary hypertension is mitral stenosis. Other causes include pulmonary venous obstruction such as total anomalous venous return. Sustained increases in pulmonary venous pressure can lead to a secondary increase in pulmonary vascular resistance due to structural changes in the walls of the intra-acinar pulmonary arteries, including medial hypertrophy and intimal thickening.

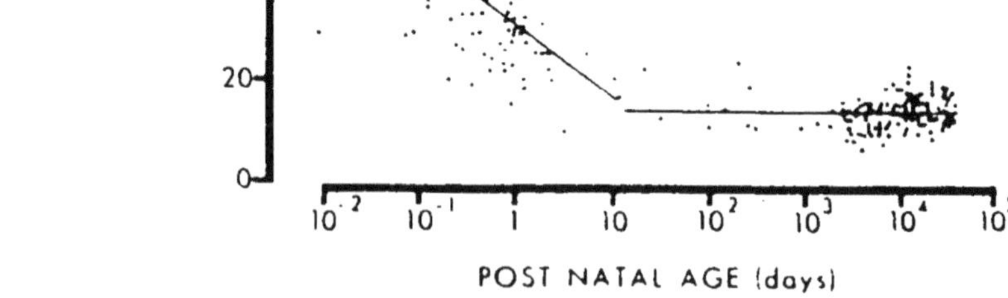

Figure 14. Relation of postnatal age to resting mean pulmonary arterial pressure in 257 measurements in normal people residing near the sea level. Reproduced from Reeves and Grover.[211]

2. Pulmonary hypertension may be due to increased pulmonary blood flow resulting from a free communication between the high resistance systemic circulation and the low resistance pulmonary circulation. It has been generally accepted that a flow of three times greater than normal must exist to increase pulmonary artery pressure. Sustained high flows result in remodeling of the pulmonary vasculature and eventually the pulmonary artery pressures may reach systemic levels.[8]

3. Pulmonary hypertension can reflect an increase in pulmonary vascular resistance. The mechanisms of an increased pulmonary vascular resistance fall into three categories:

(A) Obliterative: The loss of portions of the pulmonary vascular bed with normal function in the remaining segments. Such losses may be due to pulmonary emphysema with loss of many of the alveolar walls, one or many pulmonary parenchymal diseases that replace normal lung tissue.
(B) Obstructive: A reduction in the luminal diameter of the pulmonary vessels; this condition may be functional and reversible, or it may be an organic and generally irreversible pathologic vascular process.
(C) Hyperviscosity: An increase in blood viscosity, as in polycythemia, which causes the apparent pulmonary vascular resistance to increase. This mechanism appears to be clinically significant only when the packed cell volume reaches values greater than about 65%.

4. Pulmonary hypertension may be the result of alveolar hypoxia which is the most potent physiologic stimulus to pulmonary vasoconstriction. The mechanism by which hypoxia causes pulmonary hypertension has not yet been identified. Studies have suggested a possible mechanism involving the altered synthesis or release of chemical mediators in the lungs, such as prostaglandins which are important modulators of pulmonary vascular tone during hypoxia and during other pathophysiologic states. The role of hypoxic pulmonary vasoconstriction in the development of pulmonary hypertension is critical in those patients in whom another mechanism of pulmonary hypertension has already led to pathologic changes in the pulmonary vascular endothelium.

Pulmonary Vascular Disease

Definition. Pulmonary vascular disease refers to the pathologic changes in the vascular smooth muscle and vascular endothelium of the pulmonary vasculature caused by significant changes in pulmonary hemodynamics, as a consequence of, or in association with, congenital heart disease.[60] Pulmonary vascular disease can occur in the absence of pulmonary hypertension. However, pulmonary hypertension represents a significant risk factor for the development of pulmonary vascular disease and/or an important consequence of pulmonary vascular disease.

Pulmonary vascular disease is a primary factor governing morbidity and mortality in many congenital heart defects and is an important determinant of the timing and success of surgery.[61] The development of pulmonary vascular disease is not well understood and current measurement techniques are inadequate for assessing susceptibility to the disease, early changes of the disease, or onset of irreversibility. Pathophysiologic mechanisms remain unexplained by traditional techniques and more recent studies have focused on the molecular mechanisms for the development of pulmonary vascular disease.[62]

Classical pathology. Heath and Edwards described six grades of structural changes in pulmonary arteries associated with congenital heart defects on the basis of a study of 67 cases.[63] The disease grades were based on light microscopy analyses and classified according to a time-dependent progression, beginning with medial thickening and progressing to intimal changes. Since the last three Heath–Edward categories are morphologic variations on the theme of severe pulmonary vascular obstructive disease, other investigators have differentiated only three categories. The first recognizable (and reversible) lesion is the extension of smooth muscle into the more peripheral arteries, followed by medial (muscular) hypertrophy. These changes are followed by a reduction in the external diameter of the intra-acinar arteries that may be reversible. The more advanced, obstructive stages that are not thought to be reversible involve intimal proliferation and progressive reduction of the intravascular lumen.

Quantitative morphometrics. A number of investigators have differentiated the effects of various hemodynamic stresses on the normal growth and maturation of pulmonary vessels using quantitative morphometric studies in both experimental and clinical material.[7,17,64–68] Specifically, the size of distal arteries, the number of arteries associated with each alveolar unit, and the degree of muscularization were found to be altered by congenital heart lesions and analogous stresses. In children with congenital heart disease characterized by pulmonary hypertension, three patterns emerge that correlate with progressively worsening hemodynamic states: (1) abnormal extension of muscle into the intra-acinar arteries, (2) increased medial thickness in normally muscular arteries, and (3) reduction in arterial number and size. The sequence of changes in pulmonary vascular pathology have been found to differ among certain defects. For example, more severe, earlier intimal changes are observed in children with complete transposition and ventricular septal defect.[24]

Pulmonary input impedance. A number of investigators have explored the possibility that impedance patterns may be of value in predicting the irreversibility of pulmonary vascular disease in infants and children with pulmonary hypertension due to left-to-right shunt.[69–74] Lucas, Wilcox, and coworkers have conducted impedance studies in infants and children undergoing surgical repair of congenital heart defects.[70–74] In children with pulmonary hypertension, there are marked increases in characteristic impedance and in the frequency of the impedance minimum and phase crossover. In a study of infants with ventricular septal defects, the spectra of 25 infants (2 years or younger) were compared with the spectra of 20 children (ages 2–7 years), and the two groups were subdivided according to mean PAP: those with moderate pressure levels (35 mm Hg or less) and those with high pressure levels (at least 40 mm Hg).[74] Results indicate that though a shift in the impedance spectrum was observed in both high pressure groups, a more striking increase was observed in the older age group. This finding was consistent with morphometric results that showed the pulmonary vascular beds of children with high pulmonary artery pressures to be more impaired than those in infants with similarly high pulmonary artery pressures. Since no infant with high pulmonary artery pressure had evidence of pulmonary vascular disease, they could not determine the impedance minimum for which irreversible pulmonary vascular impairment could be assumed. Studies by Neches on 35 patients (ages 5 months to 19 years) comparing PVR and pulmonary input impedance values indicated that PVR values obtained with and without oxygen inhalation was often a poor predictor of postoperative pulmonary vascular status among patients with pulmonary hypertension.[69]

Risk factors. The primary risk factors for the development of pulmonary vascular disease as a consequence of congenital heart disease in order of increasing risk, include increased pulmonary blood flow, increased pulmonary artery pressure, and hyperviscosity.[60] The incidence of irreversible pulmonary vascular disease in defects characterized by increased pulmonary blood flow vary significantly but timing is often measured in years. Increased pulmonary blood flow occurs in the most common congenital heart defects and examples include defects of the atrial, ventricular, and atrioventricular septa and ductus arteriosus. In contrast, pulmonary hypertension is a more significant risk, with irreversible changes occurring in months to 1–2 years in age. Defects commonly associated with pulmonary hypertension include subsets of ventricular and atrioventricular septal defects and ductus arteriosus but also include defects characterized by systemic outflow tract obstruction.[75,76] In conditions with severe hyperviscosity combined with pulmonary hypertension and cyanosis, irreversible changes may be found very early, within infancy. Hyperviscosity as a consequence of polycythemia can occur in any condition characterized by a reduction in arterial oxygen tension, with common examples being tetralogy of Fallot and transposition of the great arteries (particularly in conjunction with left ventricular outflow obstruction).[77]

II. CLINICAL OBSERVATIONS

A. Introduction

Characterization of pulmonary artery hemodynamics is often necessary for diagnostic and therapeutic interventions in the clinical setting. Blood flow through the pulmonary circulation is influenced by complex anatomic, hemodynamic, and neural interactions. Hence, characterization of blood flow in the pulmonary circulation requires an adequate understanding and description of the anatomy of the blood vessels, the geometric and functional changes that occur during maturation, local and peripheral humoral influences, the interaction of the pulmonary hemodynamics with the cardiac hemodynamics, the interaction of the pulmonary hemodynamics with respiratory function, and the interaction of the pulmonary hemodynamics with autonomic and central neuroregulation. Because of the complex balance of these interactions, recognition of the dynamic nature of the physiology and pathophysiology of the pulmonary circulation is essential in the clinical setting to optimally intervene medically or surgically.

In an effort to narrow the scope of our focus, we will concentrate on the influence of pathophysiologic mechanisms induced by altered cardiopulmonary physiology on pulmonary artery hemodynamics in the child and the adult with congenital heart disease and the methodology employed to characterize pulmonary artery hemodynamics in these clinical settings. It should be emphasized, however, that this clinical description will focus on the physiologic determinants of cardiopulmonary hemodynamics, enabling the reader to apply these descriptions to altered pulmonary hemodynamics caused by any pathophysiological etiology—acquired or congenital—at any age—fetal, pediatric, or adult.

B. Clinical Indications for Medical or Surgical Intervention

The interdisciplinary approach necessitated clinically to optimally care for the child with congenital heart disease requires effective efforts to assess accurately the

Table 3. Hemodynamics in Normal Adult Man at Rest[78]

	Range of Pressure (mm Hg)			
CIRCULATION	*Systolic/diastolic*	*Mean*	*Blood Flow (1/min/m²)*	*Resistance (dyne-s/cm⁵)*
Right atrium		0–5		
Right ventricle	25–25/0–5			
Pulmonary artery	15–25/5–10	9–15		
Pulmonary capillaries		5–10		
Pulmonary veins		5–10		
Left atrium		5–10		
Left ventricle	110–130/5–10			
Aorta	110–130/70–90	85–100		
Systemic capillaries		25		
Systemic veins		5–10		
Pulmonary			3.1	
Systemic			3.1	
Pulmonary vascular				60–100
Total systemic				900–1200

Table 4. Consequences of Congenital Heart Disease on Pulmonary Hemodynamics

Alterations in flow
Increased pulmonary blood flow
Decreased pulmonary blood flow
Altered pathway for pulmonary blood flow
Alterations in pressure
Pulmonary hypertension
Alterations in structure
Pulmonary vascular disease

anatomic defect(s), to assess the physiologic consequences of the anatomic abnormalities, and to communicate these findings effectively. The consequences of altered blood flow induced by congenital heart disease and the sequels of therapeutic interventions invariably influence the pulmonary circulation (Table 4), whether by increasing pulmonary blood flow (e.g., left-to-right shunting through intracardiac septal defects), decreasing pulmonary blood flow (e.g., right heart obstructive lesions such as tetralogy of Fallot), altering the pathway of pulmonary blood flow (e.g., Fontan-Kreutzer repair), or by altering the hemodynamics to which pulmonary blood flow (e.g., pulmonary hypertension) is subjected. Successful management of children with congenital heart disease often depends on the clinician's ability to monitor pulmonary hemodynamics and to assess pulmonary vascular impairment. Critically important to an understanding of the physiological consequences of these defects is an understanding of the maturational differences incurred in cardiopulmonary function. For example, cardiac function is subject to maturational changes occurring at the cellular level in a variety of processes including maturational changes in the neurocardiac functional unit—changes in neurotransmitters, neurotransmitter content, the receptor system, innervation, the effector/transducer systems, and in the cellular components acted on by autonomic stimulation.[79] Regardless of the anatomic defects, the physiologic consequences necessitating medical and/or surgical intervention fall into three broad categories—heart failure, hypoxemia/hypoxia, and/or risk of pulmonary vascular disease (Table

Table 5. Indications for Medical and/or Surgical Intervention in the Child with Heart Disease

Heart failure
Hypoxemia/hypoxia
Risk of pulmonary vascular disease

5). All three of these physiologic consequences have the capacity to alter pulmonary artery hemodynamics and often are present concomitantly.

Heart failure is defined simply as the inability of the heart to supply an adequate cardiac output to meet the body's aerobic metabolic demands, including the metabolic demands incurred by growth; inefficiency of the heart to meet the metabolic demands can also be included in a more liberal definition of heart failure. An alteration in one or more of the physiologic determinants of ventricular function—preload, afterload, contractility, and heart rate/rhythm—can adversely affect cardiac performance beyond the compensatory mechanisms, particularly in the fetus or newborn infant where cardiac function is occurring much higher (and hence less efficiently) on the Frank-Starling curve because of maturational aspects mentioned above. As a physiologic consequence, the fetus and infant are more dependent on mechanisms to increase heart rate to increase cardiac output in response to increased metabolic demands rather than an increase in stroke volume.[80]

The etiology of hypoxemia (abnormal reduction in the arterial oxygen tension) must always be established to determine whether therapeutic intervention is immediately necessary. Hypoxia (inadequate tissue perfusion) is always a medical emergency as high morbidity and mortality are associated with uncorrected metabolic acidosis. Hypoxemia is most often associated with defects characterized by right-to-left intracardiac shunting where the effective pulmonary blood flow is reduced. Pulmonary blood flow may be entirely dependent on the patency of the arterial duct. As discussed in a previous section, the arterial duct begins to close shortly after birth, at which time the hypoxemic consequences of the ductal-dependency become manifest. Since the 1970s pharmacologic manipulation of the arterial duct to maintain patency or to reestablish patency by constant intravenous infusion of prostaglandin E_1 or E_2 has dramatically improved the care of such children by allowing for avoidance of hypoxia during transport to a center where diagnostic and therapeutic interventions can more safely take place.

Defining the pathophysiology of pulmonary vascular disease remains a fertile area for research. Many of the imaging techniques subsequently discussed are used to identify the child at risk for developing irreversible vascular changes. At present we continue to approach therapeutic interventions to eliminate the risk factor for pulmonary vascular disease in all children identified at high risk, since we know so little about the molecular basis for the etiology of the pathologic changes. The three risk factors—increased pulmonary blood flow as a consequence of left-to-right intra- or extracardiac shunting or abnormal cardiac connection (e.g., septal defect, patent arterial duct, arteriovenous fistula, transposition of the great arteries), hyperviscosity as a consequence of hypoxemia from decreased pulmonary blood flow in right heart obstructive lesions (e.g., tetralogy of Fallot, tricuspid atresia, pulmonary atresia) or hypoxemia from inadequate mixing (e.g., transposition of the great arteries), and increased pulmonary artery pressure as a consequence of increased pulmonary blood flow or increased pulmonary venous pressure—require characterization by noninvasive and invasive techniques as described below.[81]

Increased pulmonary blood flow can be distinguished physiologically using the concept of independent or obligatory flow—dependency defined relative to pulmonary vascular resistance (or impedance).[82] An example perhaps best illustrates this concept. In the child with an unrestricted ventricular septal defect, the magnitude of the left-to-right shunting, and thus pulmonary blood flow, will be dependent on the relative difference between pulmonary and systemic vascular resistances (or impedances). As physiologic influences change this relative difference, the ratio of pulmonary to systemic flow will change proportionally. Thus, this type of shunting is dependent on the status of the pulmonary vascular bed. In contrast, in the child with an atrioventricular defect with unrestricted left ventricular-right atrial shunting via the abnormal left atrioventricular valve, a significant difference in the resistances determining this flow (e.g., left ventricular systolic pressure compared to simultaneous right atrial pressure) is always present. Thus, increased flow will occur across the tricuspid and pulmonary valves, independent of the pulmonary vascular resistance. The magnitude of such a shunt will be modulated more by ventricular function. Commonly in this clinical setting, pulmonary hemodynamics are further impaired by the presence of pulmonary hypertension, increasing the burden on ventricular function and subjecting the child to higher risks for heart failure and accelerated development of pulmonary vascular disease.

C. Non-invasive Assessment of the Pulmonary Arterial Circulation—Anatomy and Hemodynamics

Physical Examination

The physiologic features associated with altered pulmonary artery hemodynamics discernible by physical examination can be generally ascribed to features associated with decreased pulmonary blood flow, increased pulmonary blood flow, and pulmonary hypertension. Cardiac situs must first be established by palpation in the child suspected of having congenital heart disease.

The child with decreased pulmonary blood flow secondary to congenital heart disease will present clinically with cyanosis. Cyanosis requires approximately five grams of circulating deoxygenated hemoglobin, thus, in children with relative anemia, cyanosis may not be as obvious as expected, even in cyanotic congenital heart disease. Despite the cyanosis, the child with congenital heart disease often appears comfortable without evidence of respiratory distress—an important distinction to differentiate hypoxemia as a consequence of a parenchemal disorder (leading to a ventilation/perfusion defect of perfused but under-ventilated portions of the lungs). The child with congenital heart disease who is cyanotic because of obstruction to blood flow to the lungs will have alterations in the second heart sound with a diminished or absent pulmonary component, as a consequence of diminished or absent flow across the pulmonary valve. Despite the most astute clinical efforts, the diagnosis of specific congenital heart defects by physical examination is often disappointing, and can only be regarded as an initial screening procedure.

The physical diagnosis of pulmonary hypertension is rarely difficult.[83] The cardiac examination predictably consists of a prominent right ventricular impulse that is either visible or easily palpable at the lower left sternal border or in the subxiphoid area (when present with normal cardiac situs), and on auscultation, a single, loud second heart sound or narrowly split second heart sound with a loud pulmonary component is present. Pulmonary systolic ejection clicks are also quite common in severe pulmonary hypertension, arising from a dilated, hypertensive proximal main pulmonary artery. Systolic murmurs at the lower left sternal border consistent with tricuspid insufficiency are sometimes present, although tricuspid insufficiency is common and commonly presents without a murmur being noted on auscultation. In severe, long-standing pulmonary hypertension, a decrescendo, high-pitched, early diastolic murmur of pulmonary insufficiency may be present along the mid-left sternal border. When pulmonary hypertension is accompanied by failure of the right ventricle, findings of systemic venous engorgement are present, including hepatosplenomegaly and peripheral edema. Abnormal v and a waves may be found in the neck veins.

Features associated with increased pulmonary artery flow are typically related to auscultatory findings associated with excessive flow crossing normal heart valves. Since the semilunar valves have approximately one-half the cross-sectional area of the atrioventricular valves, early diastolic murmurs associated with increased flow across the atrioventricular valves require more flow than the mid-systolic flow murmurs from flow across the semilunar valves. This point can be a distinguishing feature in quantifying a left-to-right shunt in the presence of normal ventricular function, since flow across the atrioventricular valves must be approximately doubled to auscultate such diastolic murmurs.

Chest Radiography

Although more sophisticated imaging modalities now exist to provide anatomic and physiologic information regarding the pulmonary circulation, chest radiography is still used routinely as a screening method to determine the status of the pulmonary vasculature, pulmonary parenchema, and cardiac situs, size, and morphology. While its role in assessing cardiopulmonary anatomy and function when compared with cross-sectional echocardiographic techniques is challenged,[84] its availability, speed, and usefulness in providing information about pulmonary features suggest that its future as an imaging modality remains quite secure.[85] Multiple textbooks are devoted to the evaluation of the chest radiograph in the child with congenital heart disease, thus, we will limit our comments to the usefulness of chest radiography in assessing pulmonary hemodynamics.

Evaluation of pulmonary hemodynamics by chest radiography includes the assessment of pulmonary ventilation and perfusion.[86] Evaluation of perfusion by chest radiography by assessing the pulmonary vasculature remains quite useful to distinguish the pathophysiology of altered pulmonary hemodynamics in the child with congenital heart disease. For example, specific diagnostic entities can be considered by evaluating the pulmonary vascularity (Fig. 15). Pulmonary vascularity on a posteroanterior chest radiograph can be assessed as normal, increased, diminished, or abnormally redistributed and each lung field must be compared with the other. For pulmonary arterial vasculature to be identified as increased by chest radiography, an increase in pulmonary blood flow of approximately 100% is required. This is quite helpful in evaluating the child with a left-to-right shunt and correlating with the physical examination. An increase in cardiac output of a similar amount (approximately 100%) is necessary to auscultate an early diastolic ventricular filling murmur across either the right or left atrioventricular valve. Diminished pulmonary vasculature typically represents obstruction of blood flow to the lungs and is an ominous radiographic finding in the newborn infant (Fig. 15). Central dilatation and peripheral pruning of pulmonary arterial vessels is noted in more advanced pulmonary vascular disease and is found with evidence for right ventricular hypertrophy as defined by retrosternal filling on the lateral chest radiograph with the cardiac silhouette. Pulmonary edema presents a more distinctive pattern of haziness in the lung fields which warrants immediate investigation as to etiology as significantly increased morbidity and mortality are associated with this finding. Specific assessment of the size of the main pulmonary artery is possible from the chest radiograph. Since the pulmonary artery is thin-walled, it dilates readily when exposed to increased flow or pressure. Dilation of the main pulmonary artery is readily visible on the chest radiograph and differentiating radiographic features are then sought to determine the physiologic etiology.

Evaluation of lung ventilation by assessment of conducting airways and lung parenchema including evaluation of lobar and lung volumes is essential to maximize information about pulmonary physiology. Evaluation of the cardiac situs and chamber enlargement by evaluation of the cardiac contour can greatly aid in the assessment of altered pulmonary hemodynamics. Since the right ventricle will be affected by altered pulmonary hemodynamics, attention to changes in shape and size must be sought. However, the sensitivity and specificity for defining changes in right ventricular function by chest radiography are less sensitive and specific than evaluation of pulmonary perfusion.

M-mode Echocardiography

The use of M-mode echocardiography to obtain information regarding pulmonary hemodynamics is used as an important adjunct to additional echocardiographic techniques. *M* (Motion)-mode echocardiography, characterized as a unidimensional imaging technique, is in fact two dimensional since the image is displayed with time on the x axis; as a consequence, cross-sectional echocardiography described below is sometimes erroneously characterized as a two dimensional

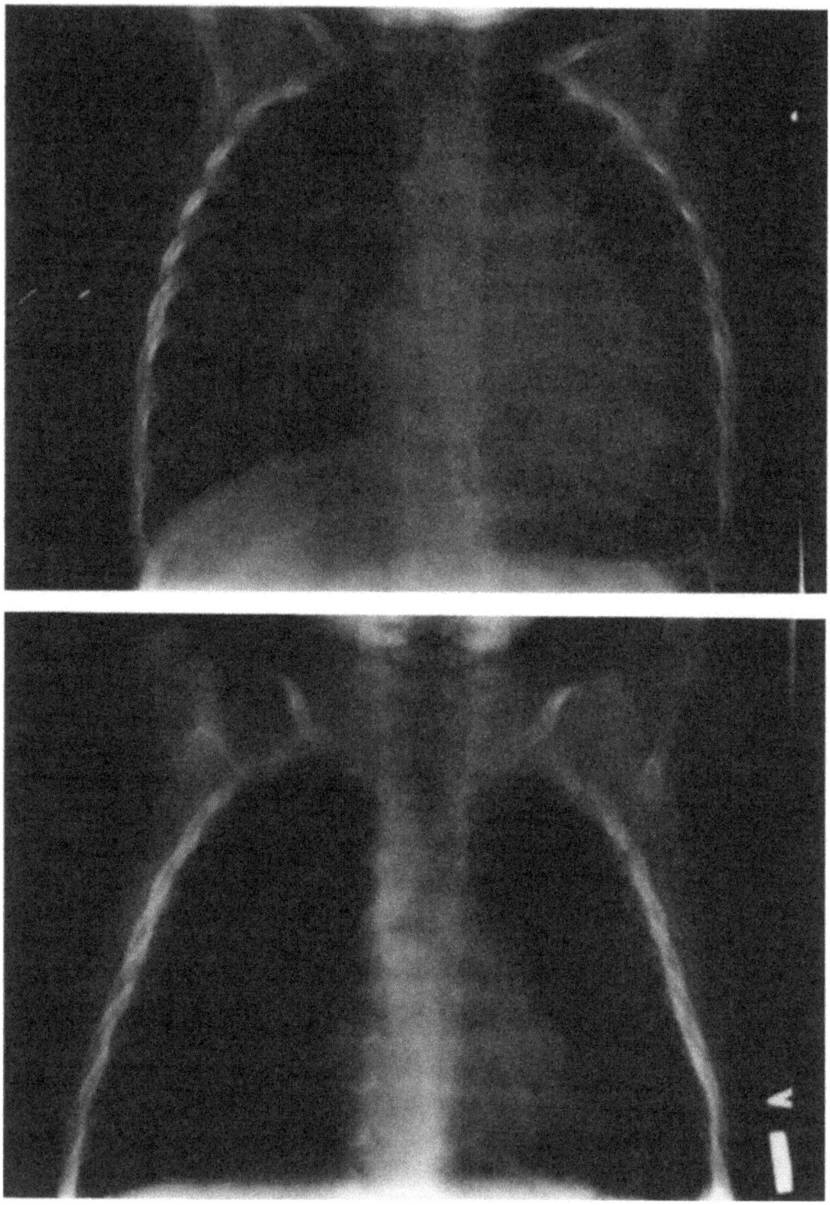

Figure 15. Chest radiographs demonstrating alterations in pulmonary vascularity as a consequence of congenital heart disease. In the chest radiograph on the top, increased pulmonary vascularity is shown from a four-month-old infant with a left-to-right shunting through a ventricular septal defect. The chest radiograph on the bottom shows decreased pulmonary vasculature from a four-month-old infant with tetralogy of Fallot with severe right-to-left ventricular shunting.

technique. M-mode examination is greatly facilitated by selecting the anatomic axis for measurement from images obtained by cross-sectional echocardiography. The principal advantage of M-mode techniques is the high repetition frequency (>1000/sec) compared to cross-sectional techniques (≤30/sec) which allows for higher time-resolution. This higher repetition frequency allows for accurate measurement of events occurring during the cardiac cycle, including movement of intracardiac structures such as valves and chamber walls.[87] Of course, ultrasound measurements in children must be referenced to appropriate norms derived from measurements accounting for growth, development, and gender. Most institutions acquire measurements following guidelines established by the American Society of Echocardiography, although even these must be modified in children to avoid the introduction of errors incurred under conditions of higher heart rates.[88]

The physiologic consequences of changes in pulmonary hemodynamics will affect the right heart structures as in enlargement of chamber size with increased flow, ventricular hypertrophy with pulmonary hypertension, and changes in flow characteristics through the right heart structures including the pulmonary artery. M-mode echocardiographic examination of pulmonary hemodynamics is principally applied to investigation of the right heart structures. Quantitative M-mode echocardiographic assessment of right heart structures can be divided into measurement of cardiac structural dimensions, right ventricular function, and valve function.[87]

Measurement of right heart structural dimensions includes the measurement of the right ventricular wall thickness, right ventricular systolic and diastolic dimensions, and right ventricular outflow tract sizes and the interventricular septal thickness and movement. Elevations in pulmonary artery pressure result in right ventricular hypertrophy definable by echocardiographic measurement of the thickness of the right ventricular anterior free wall and the ventricular septum. Increases in volume of pulmonary blood flow lead to dilatation of right heart structures reflected in echocardiographic measurement of right heart cavitary dimensions and the diameter of the right ventricular outflow tract.

Characterization of left ventricular systolic and diastolic function by M-mode echocardiographic techniques is much more developed and widely used since the shape of the left ventricle is a less complex geometric shape (assumed to be a prolate ellipse). Thus, parameters of systolic function such as ventricular volume and ejection fraction can be estimated or measured for the left ventricle, while such applications to the right ventricle are potentially less accurate. Other measurements of systolic function such as shortening fraction, peak shortening rate, velocity of circumferential fiber shortening, percent change in right ventricular anterior wall thickness, normalized peak thickening rate of the right ventricular anterior wall, and normalized peak thinning rate of the right ventricular anterior wall are independent of shape and can be measured to characterize right ventricular function. The determination of systolic time intervals in pulmonary hypertension are still used, although the sensitivity and specificity remain in debate. Systolic time intervals relate electrical to mechanical events of the heart. In the absence of an arrhythmia or electromechanical prolongation, electrical conduction through the heart varies minimally. Therefore, changes in systolic time intervals principally reflect changes in mechanical properties; in physiologic terms, changes in preload, contractility, afterload, and heart rate will affect systolic time intervals. Prolongation of the right pre-ejection period and shortening of the right ventricular ejection time has been shown to correlate with elevated pulmonary vascular resistance and pulmonary hypertension (Figure 16).[83,89–91] However, since these intervals are influenced by the physiologic determinants of ventricular function besides ventricular afterload, demonstration of corroborating findings are essential. Indices of diastolic right ventricular function obtained by M-mode echocardiographic tech-

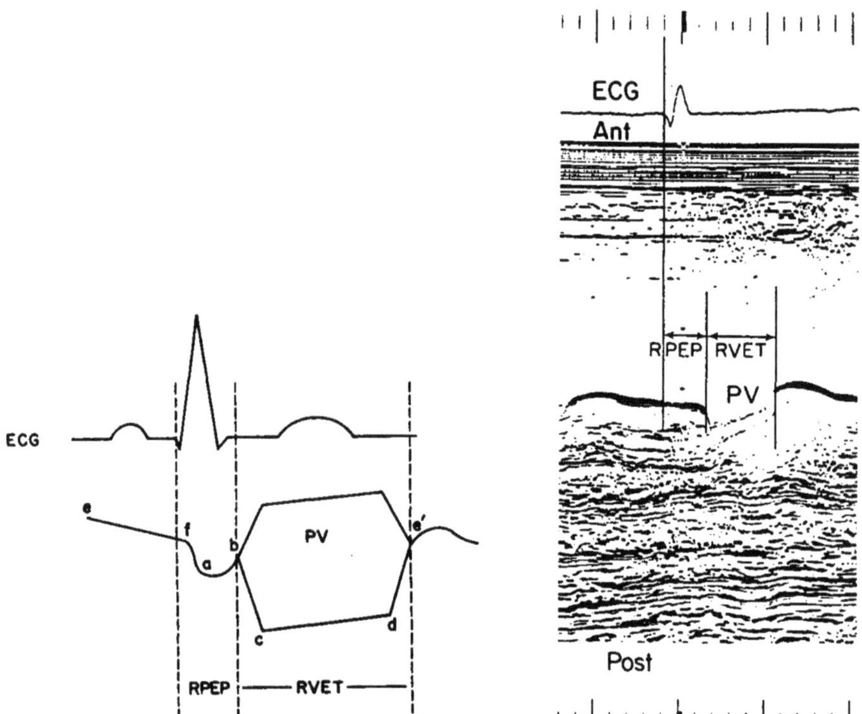

Figure 16. M-mode depiction of systolic time intervals. In the schematic drawing, the M-mode echocardiographic recording of the pulmonary valve is depicted. PV: pulmonary valve; e, é: point of valve closure; f: leaflet position in diastole prior to atrial contraction; a: dip in the tracing coinciding with atrial contraction; b: point of valve opening; c–d: maximal leaflet excursion during systole; RPEP: right pre-ejection period; RVET: right ventricular ejection time. In the second panel, derivation of RPEP and RVET is shown from a M-mode echocardiogram of the pulmonary valve at a chart speed of 100 mm/sec in an infant with pulmonary hypertension, demonstrating abnormal systolic time intervals (RPEP 80 msec; RVET 135 msec; RPEP/RVET 0.59).

niques include diastolic time intervals defining isovolumic relaxation time, relaxation time index, rate of change in diastolic dimension, normalized rate of change in diastolic dimension and others. For example, isovolumic relaxation time can be measured from M-mode echocardiographic recording of pulmonary valve closure and tricuspid valve opening. This interval has been shown to correlate with systolic pulmonary arterial pressure in children[92] and adults.[93]

Changes in the M-mode echocardiographic pattern of the pulmonary valve occur in pulmonary hypertension including partial early closure of the pulmonary valve during systole, a delayed opening of the pulmonary valve, a rapid systolic opening slope, a maximum pulmonary a wave of –2 mm, and a flat e-f slope (Figure 16).[83]

Two Dimensional Echocardiography

Cross-sectional images are generated by transducers with activated piezoelectric crystals configured for mechanical sector scanning, phased-array scanning, and linear-array scanning. In contrast to M-mode echocardiography which provides little information regarding the anatomy of the pulmonary circulation beyond the pulmonary valve, cross-sectional echocardiographic techniques for the assessment of the heart and great vessels offers the opportunity to assess anatomically the pulmonary arterial circulation beyond the primary branch pulmonary arteries and connections of the pulmonary veins to the heart. The utility of cross-sectional echocardiography to provide anatomic and functional details of cardiopulmonary function revolutionized the care of children with congenital heart disease by providing a non-invasive, bed-side, cost-effective method for the assessment of abnormal cardiopulmonary anatomy and function. Furthermore, cross-sectional scanning is superior to conventional cineangiography for demonstration of atrioventricular valve anatomy (Figure 17), which is a critical feature for determining interventional options for children whose congenital heart defects include abnormalities of atrioventricular valve anatomy.

Current techniques include the conventional transthoracic approach employing multiple echocardiographic windows, an epicardial approach to assess the adequacy of repairs during cardiac surgery, transesophageal echocardiography for diagnostic evaluation, and monitoring of ventricular function during and following cardiac surgery, and transabdominal, transthoracic approach to assess cardiac function and anatomy in the fetus. Regardless of the technique, a systematic approach is essential to optimally acquire and interpret the anatomy and function of the heart and great vessels. Anderson and colleagues[94] have long advocated a sequential approach based on describing the connections of the heart and great vessels by delineating the systemic venous connections to the heart, the atrioventricular connections and the ventriculoarterial connections. By avoiding embryologic terminology and describing the anatomy in simple descriptive form based on interconnection terminology, complete and effective communication of the disorders is accomplished (Figure 18).

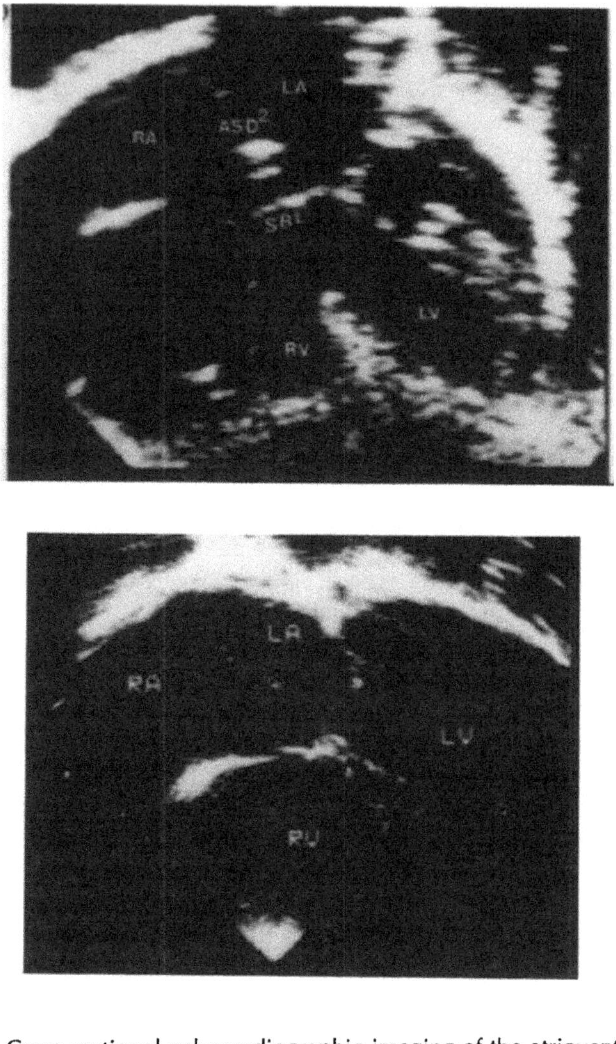

Figure 17. Cross-sectional echocardiographic imaging of the atrioventricular valve area in infants with an atrioventricular septal defect. (a) Typical appearance in the subcostal four chamber projection showing separate right and left atrioventricular valve orifices in ventricular systole. Note the abnormal coaptation of the atrioventricular valves to the crest of the ventricular septum and the deficiency above this site, producing a site for atrial shunting. (b) Similar echocardiographic projection from another child showing the common valvar orifice in which there is an extensive ventricular shunt beneath the free-floating superior leaflet. Two atrial defect are present as well. LA: left atrium; RA: right atrium; RV: right ventricle; LV: left ventricle; ASD^2: secundum atrial septal defect; SBL: superior bridging leaflet. (Reprinted with permission.[116])

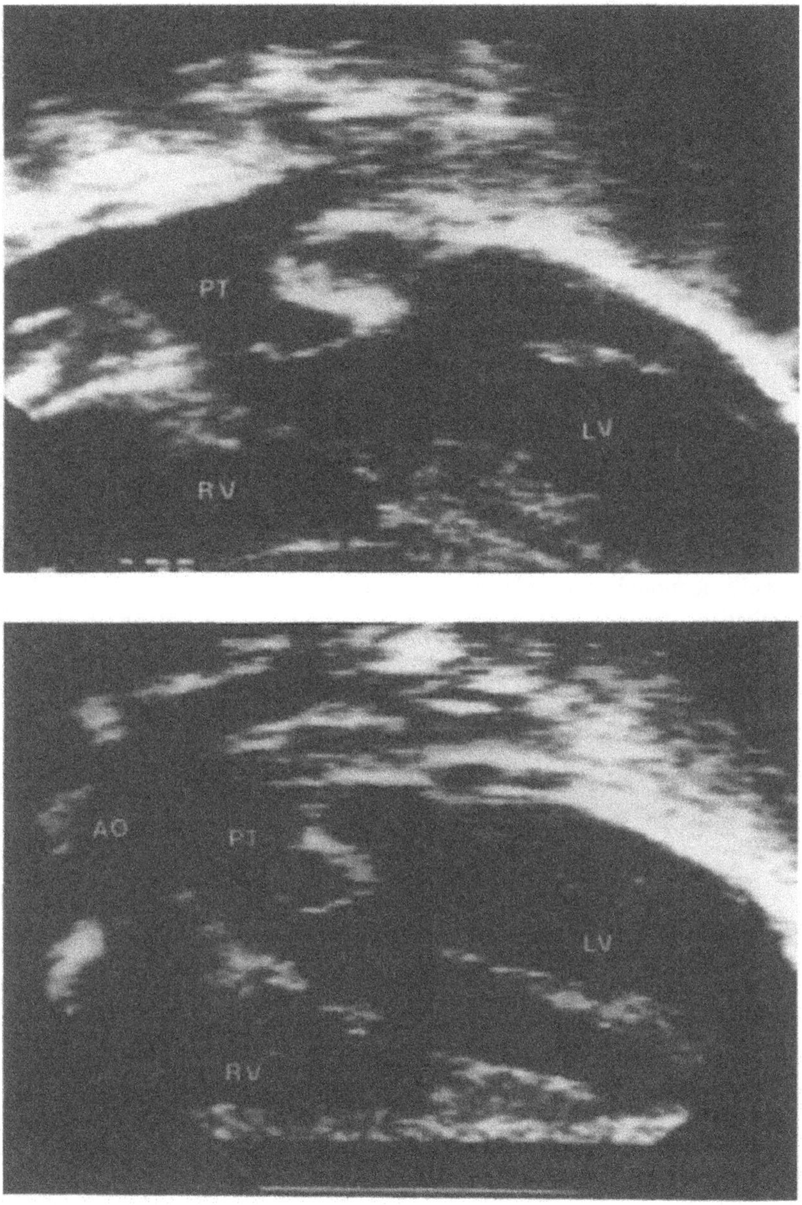

Figure 18. Subcostal paracoronal echocardiographic sections demonstrating a discordant ventriculoarterial connection with the pulmonary trunk arising from the left ventricle and bifurcating (top) and the aorta connected to the right ventricle (bottom). PT: pulmonary trunk; RV: right ventricle; LV: left ventricle; AO: aorta. (Reprinted with permission.[117])

In addition to the anatomic detail cross-sectional echocardiography provides, physiologic assessment of the right heart structures is possible by quantifying chamber sizes and volumes including calculation of ejection fraction for the right ventricle.[87,88] While such techniques are refined and routinely used for characterization of left ventricular function, the more complex geometric shape of the right ventricle has been an impediment to the assessment of right ventricular function conveniently (Figure 19). The right ventricular outflow tract with its conal anatomy (in contradistinction to the left ventricle) cannot be imaged cross-sectionally in the same plane as the cross-sectionally imaged ventricular cavity, making volume assessment include additional orthogonal scanning planes. Watanabe et al.[95] used this approach by applying Simpson's rule but excluded the entire right ventricular outflow tract area, resulting in right ventricular volumes approximately one-half of those derived from cineangiography. Silverman and Hudson[96] also applied Simpson's rule using two planes incorporating the ventricular cavity and outflow tract and obtained good correlation of right ventricular volumes and ejection fractions compared to cineangiographically derived measurements. Starling et al.[97] and Levine et al.[98] used measurements based on geometric models. The prolate ellipse model provided superior correlations.

Calculation of right atrial volumes can also be performed using uniplane and biplane techniques from standard echocardiographic approaches and the results are dependent on gender and body surface area.

Figure 19. Silicon cast of right atrium, right ventricle, and pulmonary arterial circulation obtained in situ in a lamb to demonstrate the geometry of the right ventricular outflow tract.

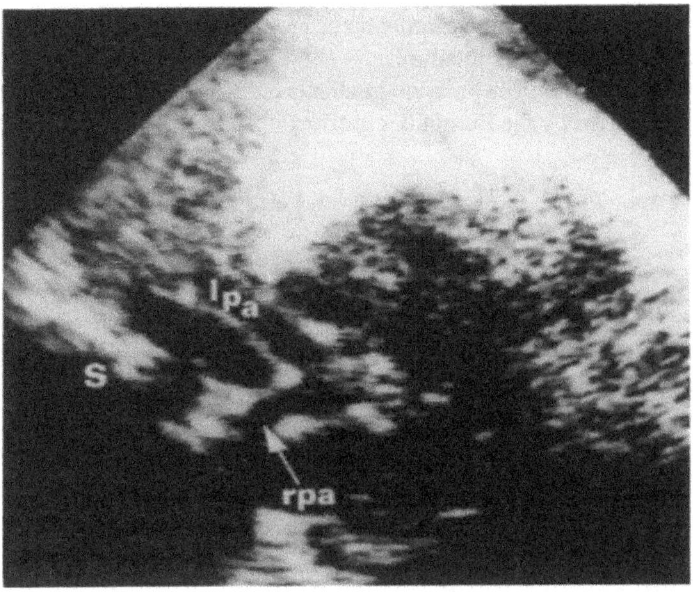

Figure 20. Demonstration of the continuity between the left and right pulmonary arteries and the lack of continuity between the pulmonary trunk and the right ventricle in a fetus with pulmonary atresia. (Reprinted with permission.[118])

Simple measurements of the diameter of the subvalvular right ventricular outflow tract, the main pulmonary artery and its primary branches are easily obtained and can be referenced to the aorta for comparison. Pulmonary artery sizes in childhood are closely related to body surface area.[87] Intracardiac shunting that increases pulmonary blood flow results in dilation of the thin-walled pulmonary vessels. Conversely, one of the major uses of cross-sectional echocardiography occurs in the setting of reduced pulmonary blood flow, such as found with right heart obstructive lesions. Characterization of the presence and size of the main pulmonary artery and the branch pulmonary arteries in these circumstances can be quite challenging, and characterizing whether these vessels if present are in direct continuity can also be difficult (Fig. 20). The latter circumstance has been greatly facilitated by incorporating color-flow Doppler mapping with the cross-sectional imaging.

Doppler Ultrasound

Doppler ultrasound techniques include pulsed-Doppler, continuous-wave Doppler, and color-flow Doppler. Quantitative Doppler echocardiography provides a powerful tool for the noninvasive assessment of pulmonary artery hemodynamics obviating the need for invasive assessment in many clinical circumstances. Doppler echocardiography can provide estimation of pressure gradients, intracardiac pressures, and pulmonary artery pressure.

The determination of peak pressure gradients utilizes the principle of convective acceleration defined by the Bernoulli equation:

$$P_1 - P_2 = \tfrac{1}{2}\rho(V_2^2 - V_1^2) + \rho \int_1^2 dV/dt(ds) + R(V)$$

where P_1 is the pressure proximal to the obstruction, P_2 the pressure distal to the obstruction, V_1 the velocity proximal to the obstruction, V_2 the velocity distal to the obstruction, ρ the mass density of blood, dV the change in velocity over the time period dt, ds the distance over which the decrease in pressure occurs, and V the velocity of blood flow. Since the inertial component is typically negligible and because V_1 is usually <1 (m/s), this formula is commonly simplified for clinical use to

$$P_1 - P_2 = 4V_2^2$$

where V_2 is in m/s, and P_1 and P_2 are in mmHg.

The acquisition of the Doppler signal includes obtaining a maximal velocity aligned as parallel as possible to the blood flow since the Doppler equation applies a cosine function to the angle of incidence. It is important to note that the Doppler-derived peak instantaneous gradient differs from the peak-to-peak pressure gradient obtained from invasive measurements, routinely exceeding the latter. Limitations of the modified Bernoulli equation are evident. Conditions that lead to relevance of the inertial component (as in long-segment obstructions) or where V_1 is >1 m/s will lead to inaccuracies. Poorly angled measurements in which maximal velocities are not obtained will lead to underestimation of such pressure gradients. Mean pressure gradients, defined as the mean of all instantaneous gradients, can be calculated by digitizing the Doppler curve and applying computer-aided calculations. These estimations of pressure gradients, when properly obtained, correlate well with catheterization-derived measurements.

In the absence of obstruction in the right ventricular outflow tract and pulmonary artery, there is little difference between systolic pulmonary artery pressure and systolic right ventricular pressure. Thus, the former can be inferred by estimating the latter. One method of assessing systolic right ventricular pressure (and hence systolic pulmonary artery pressure) is the Doppler measurement of the regurgitant jets in tricuspid insufficiency. Tricuspid insufficiency is common in children with pulmonary hypertension and can be localized by color-flow mapping from a cross-sectional echocardiographic scan. Of course, in this circumstance it is the right ventricular-right atrial gradient that is being measured (not the right ventricular systolic pressure); estimation of the right atrial pressure must be considered either by arbitrary assumption, by direct measurement when a central venous

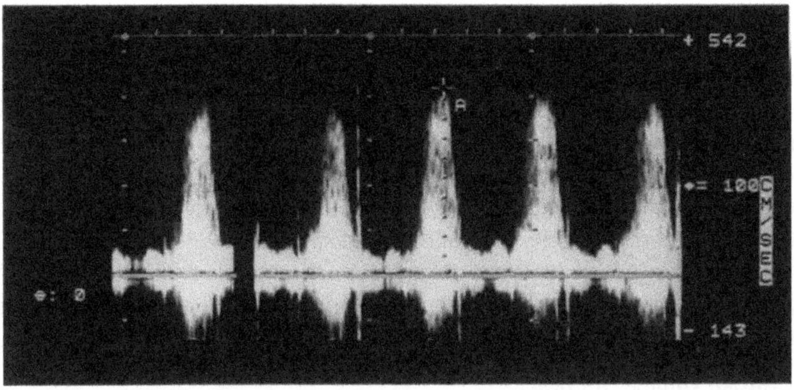

Figure 21. Demonstration of Doppler derived pulmonary artery pressure obtained from an interventricular velocity across a ventricular septal defect. Arterial blood pressure was 100/60 mm Hg. The maximal velocity across the ventricular septal defect was 4 m/sec. From the modified Bernoulli equation, the pressure gradient is 64 mm Hg. Therefore, in the absence of left and right ventricular outflow obstruction, pulmonary artery systolic pressure equals aortic systolic pressure minus the interventricular pressure gradient; thus, estimated pulmonary artery systolic pressure is 36 mm Hg.

catheter is in place, or by catheterization-derived indices and subtracted from the right ventricular-right atrial gradient to obtain the right ventricular systolic pressure.[87,99] In children with a ventricular septal defect, determination of the interventricular gradient across the defect will allow one to estimate right ventricular pressure, since the systolic left ventricular pressure can be accurately estimated by measurement of arterial blood pressure (in the absence of left ventricular outflow tract obstruction; Figure 21).

Estimation of pulmonary artery pressure (and pulmonary vascular resistance) can also be derived from characterization of Doppler velocity waveforms and measurements of Doppler time intervals. The clinical setting is routinely the child with suspected pulmonary hypertension. As mentioned previously, the right ventricular isovolumic relaxation time correlates well with pulmonary artery pressure.[92] Kitabatake et al. reported that acceleration time (time to peak velocity) of a Doppler velocity curve in the main pulmonary artery correlated inversely with pulmonary arterial pressure (PAP) and more closely with $\log_{10}$PAP.[100] Additional reports have supported these findings,[101–103] although the sensitivity and specificity have been challenged since patients with normal indices often have abnormal pulmonary artery pressures.[104] Previously discussed systolic time intervals (right ventricular ejection time, right ventricular pre-ejection time) can also be derived by Doppler techniques for estimation of pulmonary artery pressure.

Estimation of pulmonary artery volumetric flow can be calculated from cross-sectional and Doppler-derived echocardiographic measurements. Measurement of pulmonary artery flow requires measurement of the pulmonary artery diameter and mean velocity. Similarly, aortic flow can be estimated by measurement of aortic diameter and mean velocity aortic velocity. The diameters are measured using standard cross-sectional techniques and assumption is made that the vessel is circular (not true for the pulmonary artery). Velocity is measured in the center of the vessel. Use of this velocity measurement assumes a flat velocity profile and the presence of laminar flow (at times treacherous assumptions clinically). Using the following equations, respective flows in the great vessels can be obtained and referenced as a ratio of pulmonary-to-systemic flow.

Stroke volume (ml/beat) =
(mean velocity)(cross-sectional area)(R-to-R interval)/1000 ml/L

Volumetric flow (L/min) =
(mean velocity)(cross-sectional area)(60 s/min)/1000 ml/L

Satomi et al[105] have reported the clinical utility of estimating pulmonary-to-systemic flow ratio in children with left-to-right atrial shunts measuring velocities across the atrioventricular valves.

Color-flow Doppler techniques have greatly aided the clinician in identifying and characterizing blood flow in children with congenital heart disease. Evaluation of intracardiac shunts, valve insufficiency, obstructive flow, flow through hypoplastic vessels (e.g., hypoplastic pulmonary arteries), and determining cardiovascular connections has been facilitated using color-flow mapping. In this technique, computer-assigned color is used to delineate direction and velocity of flow.

Computerized Tomographic Scanning

Conventional computed tomography is not routinely used in the assessment of pathophysiologic alteration affecting pulmonary hemodynamics. In this method a moving x-ray tube moves around a patient in a circular housing in which x-ray detectors are activated. After digital processing, an anatomic slice is produced. Conventional computed tomography is better suited for imaging structures with minimal spontaneous motion, thus limiting its application to intracardiac imaging. Intrinsic diseases affecting the chest are well-suited for this mode of imaging, thereby distinguishing pathophysiological processes affecting the lung and mediastinal structures from the heart and great vessels. However, the ability of computerized tomographic scanning to provide three dimensional anatomic imaging enhances its application to the study of great vessels, particularly when coupled with contrast enhancement. Contrast enhancement is essential for maximizing vascular study, which requires intravenous access. A practical disadvantage is that the young child must be sedated adequately since the sampling rates are long compared to conventional cineangiography.

Gated computerized tomographic techniques enhance imaging of the heart at the disadvantage of significantly higher radiation exposure. The most recent advancement and refinement of so-called fast computerized tomographic scanning, cine-computerized tomographic scanning, is now commercially available and offers a method for determining accurately mass, volume, and flow characterization in anatomically normal and abnormal cardiopulmonary conditions. In this technique, an electron source produces an electron beam that is first focused then deflected magnetically onto tungsten target rings below the patient producing roentgen rays which are collected after passing through the patient.[86,106] Time-density curves obtained in electrocardiographically-synchronized flow mode provides an accurate determination of blood flow through vascular structures, including the heart. This is particularly useful in noting flow through surgical aortopulmonary anastomoses as well as intracardiac shunting through septal defect. Perhaps most important, imaging of proximal and distal pulmonary arteries can be shown with assessment of respective flow in a selected area of interest. In patients with collateral flow through surgically placed or naturally occurring aortopulmonary connections, this can be quite helpful clinically. In cine-mode, a series of scans at multiple target rings are acquired and stored digitally, allowing for subsequent sequential loop display for anatomic and quantitative assessment. Finally, in volume-mode, one target ring is scanned and the table is moved over this ring.

Cine-computerized tomographic scanning must offer distinct advantages for imaging of pathophysiologic processes affecting pulmonary hemodynamics over echocardiographic techniques to employ its use clinically. Imaging of the distal pulmonary arteries is an example of the superiority of cine-computerized tomographic scanning.[107] Examination of intracardiac tumors and thrombus is particularly well-suited for cine-computerized tomographic scanning since accurate assessment of mass can be obtained. Despite the developing capability of this modality, its use in the assessment of pulmonary hemodynamics compared to other techniques as discussed in this section remains limited in most centers caring for children with congenital heart disease. The rapid development and application of magnetic resonance imaging has further reduced the application of computerized tomographic scanning in the assessment of congenital heart disease and its consequences.

Radionuclide Imaging

Radionuclide techniques have been used for years to provide information regarding cardiopulmonary function. These techniques are safe, noninvasive, and can be used to acquire functional information during resting or stress conditions. First-pass radionuclide angiocardiogram can be used to detect and quantify both left-to-right and right-to-left intracardiac shunting. A bolus of a radionuclide is injected via a peripheral vein while a gamma camera images the heart and lungs; time activity curves can then be generated in addition to the sequential imaging. With left-to-right shunts a recirculation peak is seen that is proportional to the magnitude of the shunt. With right-to-left intracardiac shunting an early systemic peak is seen before complete circulation through the lungs. This early peak is also proportional to the size of the shunting. Limitations are encountered when bidirectional shunting is present. Gated equilibrium nuclide angiography is a second method used where red blood cells are labeled and the data acquisition is synchronized to the electrocardiogram to obtain a cardiac blood pool angiogram. Both of these techniques can be used to assess ventricular function including global and regional wall motion, ejection fraction, regurgitant fraction, and cardiac output. Areas of interest which include the left ventricle and right ventricle include overlapping structures which must be eliminated to achieve accurate assessment of ventricular function; this is usually more of a problem for the right ventricle. Hurwitz et al. published normal values for ejection fraction in children in 1984.[108]

One important application of these techniques in assessing cardiopulmonary hemodynamics is to quantify lung perfusion. As we have discussed previously, altered pulmonary blood flow almost always accompanies the child with congenital heart disease. Radionuclide scanning can be used to determine accurately the distribution of lung perfusion, which is extremely valuable information in the child in whom the distribution of lung perfusion may be altered as in branch pulmonary artery stenosis or in the child with an aortopulmonary palliative shunt.

Magnetic Resonance Imaging

Magnetic resonance imaging provides a safe, noninvasive method to image the heart and lungs. Resonance contrast between blood and blood vessel (including heart) walls provides an adequate interface for imaging by this method. The principal impediment to its use in the study of children with altered pulmonary hemodynamics initially was movement artifact created by the cardiac cycle and respiratory cycle. Appropriate synchronization mechanisms triggering on the electrocardiogram have allowed for the development of imaging to determine anatomic detail and physiologic function. It is beyond the scope of this text to describe in detail the different techniques available. The acquisition phase of conventional magnetic resonance imaging takes place by scanning transectional slices which can be viewed individually and sequentially in a loop format; this method requires considerable time (minutes) for acquisition. Newer dynamic cardiac magnetic resonance imaging techniques can provide transectional images with a single transectional slice performed in 20–60 ms depending on the format and subsequently displayed in a cine-loop format.[109] Images are acquired in multiple orthogonal planes for subsequent anatomic and physiologic interpretation.

Cardiac function can be studied by magnetic resonance imaging by measurement of ventricular volumes with derivative calculations such as ejection fraction for both ventricles. Blood flow through a great vessel can be calculated in the child providing a method for determining pulmonary and systemic blood flow with extracardiac or intracardiac shunting. Assessment of segmental ventricular movement provides information regarding segmental function. Wall thickness can be measured. Muscle metabolism can be studied using magnetic resonance spectroscopy by identifying resonance peaks of the phosphate energy chain. Thus, evaluation of the right heart structures can be performed which provides a method for evaluating cardiac function in the child with altered pulmonary hemodynamics.

Morphologic assessment of anatomic detail provided by magnetic resonance imaging is best illustrated by examples derived from studies of children with altered pulmonary hemodynamics secondary to congenital heart disease as illustrated in Figure 22.

Magnetic resonance imaging is a superb modality for assessing the pulmonary vasculature in children with congenital heart disease, particularly in those children in whom conventional techniques are poorly suited. The child with obstruction to pulmonary blood flow in whom a question arises about the continuity of the hypoplastic pulmonary arteries poorly imaged by cross-sectional and color-flow Doppler echocardiography is well-suited for evaluation by magnetic resonance imaging.

Despite the ability of this mode of imaging to provide anatomic and functional detail, its use must be assessed in the context of cost. It is expensive. Justification and risk/benefit analysis of information obtained from this technology must be compared with that available from less expensive and more convenient access provided by other imaging techniques such as cross-sectional echocardiography.

D. Invasive Assessment of the Pulmonary Arterial Circulation—Anatomy and Hemodynamics

Digital Subtraction Radiography

Digital subtraction radiography provides an alternative approach for imaging the cardiovascular system and is beginning to be used for acquiring physiologic information that will prove useful for the study of pulmonary hemodynamics. Although originally promoted as an almost non-invasive method requiring only a peripheral intravenous access, delivery of the contrast centrally is required for optimal imaging. Thus, its use is supplanting conventional cineangiography during a conventional cardiac catheterization procedure. Used in this manner, the physiologic pressure data are obtained by direct measurement. One advantage of this technique is its relative cost to x-ray cineangiography, and in some centers digital subtraction radiography has replaced conventional cineangiography. Since multiple small injections can be used, elimination of biplane equipment has been proposed, although the suitability of this approach is debated. The use of earlier equipment was hampered by the lack of sufficiently sophisticated software to achieve adequate imaging of the moving heart and great vessels; this deficiency was overcome by supplementing the masting techniques with so-called time interval-difference im-

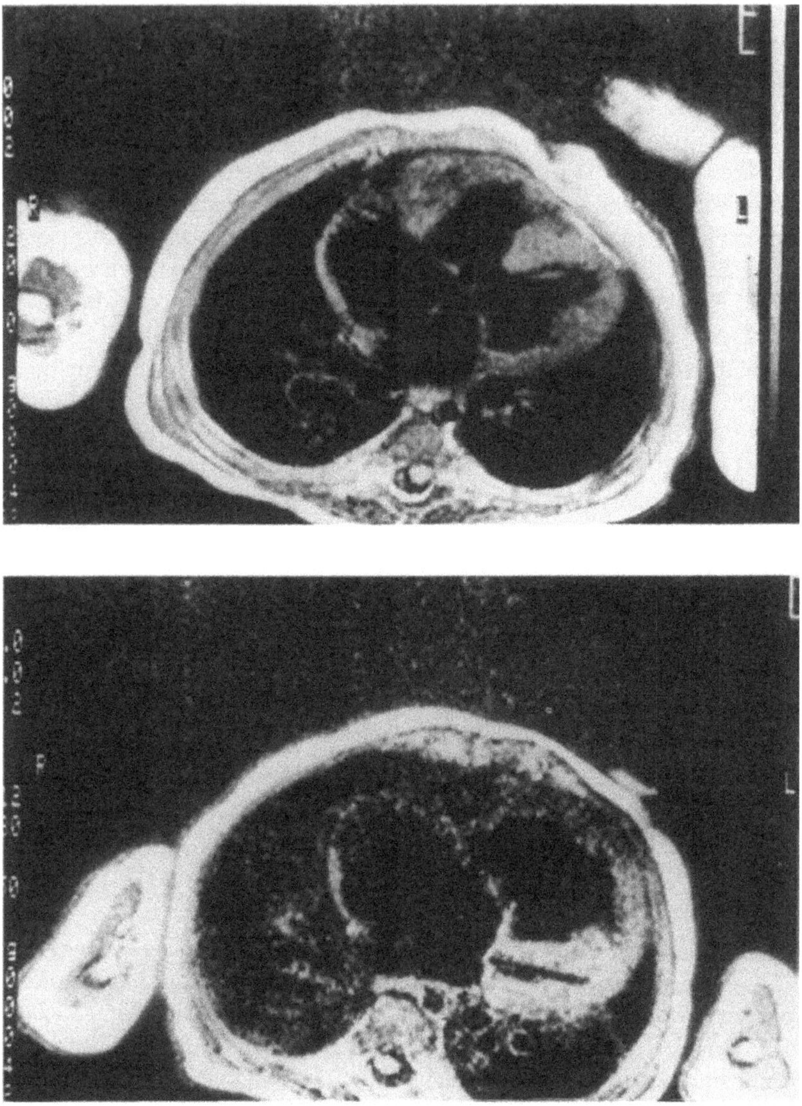

(continued)

Figure 22. Magnetic resonance imaging in congenital heart disease. (a) Resonance imaging in "four chamber" projection to show the extent of separation between the bridging leaflets and the ventricular septum; compare with Figure 2b. (b) Resonance imaging showing an atrioventricular septal defect with marked right ventricular dominance. (c) Resonance imaging in coronal and sagittal projections from a child after a Fontan procedure showing the anastomosis (*) between the right atrium (RA) and the pulmonary artery (PA). (Reprinted with permission.[116,119])

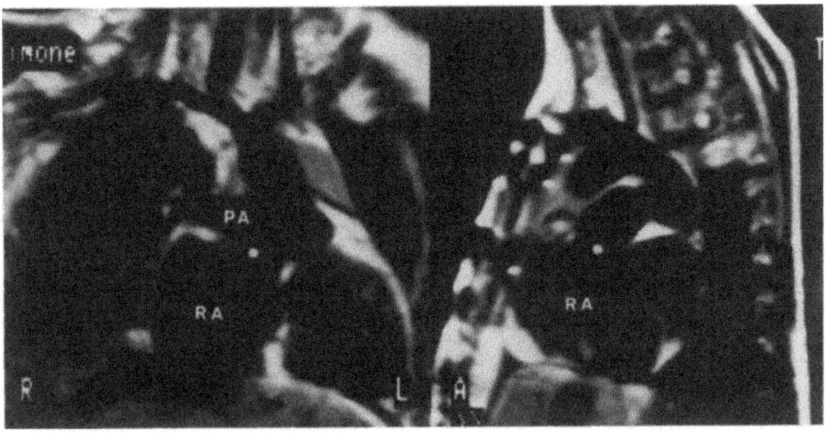

Figure 22. (Continued)

aging.[110] Current equipment allows for image-acquisition superior to conventional cineangiographic techniques. Current techniques are recognized to be at least comparable in image quality as conventional cineangiography while offering advantages for the pediatric patient—reduction in dose of radiation, reduction in dose of contrast agent, and the potential reduction in dose-dependent complications (e.g., renal toxicity, acute osmolality shifts), and potential reduction in vascular injury since smaller catheters can be used for contrast delivery.

As mentioned above, application to assessment of altered pulmonary hemodynamics in the child with congenital heart disease usually occurs in the context of acquiring pressure and flow data by direct measurement. Quantifying systemic and pulmonary flows and cardiac chamber volumes, ejection fractions, regurgitant fractions can now be calculated with appropriate software allowing for adjunctive information for the investigation of altered pulmonary hemodynamics in which the physiologic consequences are altered pulmonary flow or chamber volumes.

Cardiac Catheterization

Cardiac catheterization remains a powerful clinical tool for the accurate assessment of cardiac and pulmonary vascular anatomy and pulmonary hemodynamics, despite the maturation and application of cross-sectional and Doppler ultrasound techniques. Clinically the pendulum has probably shifted to an under-utilization of this technique. Hemodynamic information obtained by direct measurement is the standard by which all other techniques are compared.

The usual approach to gain vascular access is via transcutaneous inguinal access through the femoral vein and/or artery. Occasionally a cut-down approach must be used or an alternative site such as brachial, axillary, or subclavian areas selected.

Once catheter access is gained to the heart, systemic veins and great vessels, oximetric and pressure data are obtained sequentially and cineangiography then performed.

Oximetry data are obtained by the measurement of oxygen saturation from blood samples obtained from the various sequential catheter positions. This is a fast, accurate method for detection and assessment of extra- and intracardiac shunts in the child with congenital heart disease and is used as discussed below in the estimation of cardiac output.

Determination of blood flow during a cardiac catheterization is an important part of characterizing pulmonary hemodynamics. This process takes on an added complexity in children with congenital heart disease because the extracardiac and intracardiac shunts can necessitate the determination of both pulmonary and systemic blood flows. The common invasive means for determining flow are derivations of the indicator dilution principle, with the most direct example being the green dye curves.[80,111] However, both the Fick method and thermodilution techniques use the principle of indicator dilution with oxygen and heat being the respective indicators. The Fick method, dating back to its description in 1870,[112] is commonly used to calculate pulmonary and systemic blood flow in children undergoing cardiac catheterization. The Fick method assumes that a steady-state equilibrium exists between oxygen uptake by the body and oxygen extraction from inhaled air. It is based on the equation

$$\text{Cardiac output (L/min/m}^2) = \frac{\text{oxygen consumption (ml/min/m}^2)}{\text{arteriovenous oxygen content difference (ml/L)}}.$$

The limiting feature of the Fick method is obtaining an accurate measurement of oxygen consumption. This is particularly true in newborn infants. However, the Fick method is useful for determining relative flows such as the pulmonary to systemic flow ratio.

Calculation of pulmonary blood flow (Qp) using the Fick method is dependent on measuring the arteriovenous oxygen content difference across the pulmonary circulation—from the pulmonary vein (PV) and pulmonary artery (PA)—and across the systemic circulation to measure systemic blood flow (Qs)—as illustrated by the following equations.

$$Qp = \frac{\text{oxygen consumption}}{\text{PV O}_2\text{ content–PA O}_2\text{ content}}$$

$$Qs = \frac{\text{oxygen consumption}}{\text{PV O}_2\text{ content–PA O}_2\text{ content}}.$$

In the denominator, blood samples obtained to calculate the arteriovenous oxygen content must be mixed adequately, which poses problems in the child with intra-cardiac or extracardiac shunting. For example, obtaining a mixed systemic venous sample is not possible in the child with a left-to-right shunt through an atrial septal defect as the caval venous blood is mixed with contributions from the shunt. In the presence of shunts in the heart and great vessels, Qs can be calculated using the Fick method, knowing the mixed venous (MV) and systemic arterial (SA) saturations.

$$Qs = \frac{\text{oxygen consumption}}{SA\,O_2\text{ content}-MV\,O_2\text{ content}}$$

$$Q_{L\to R}\text{shunt} = Qp - Qs.$$

Effective pulmonary blood flow (Qep) is a useful clinical concept and is defined as the volume of systemic venous blood reaching the lungs for oxygenation; it can be calculated using oximetric data including mixed venous saturations (MV).

$$Qep = \frac{\text{oxygen consumption}}{PV\,O_2\text{ content}-MV\,O_2\text{ content}}.$$

Thus, right-to-left shunting and left-to-right shunting can be calculated.

$$Q_{L\to R}\text{ shunt} = Qp - Qep$$

$$Q_{R\to L}\text{ shunt} = Qs - Qep.$$

Thermodilution techniques are used commonly in the cardiac catheterization laboratory and the intensive care unit to determine cardiac output, since multiple measurements are possible; this technique is particularly suitable for physiologic assessment of pharmacologic intervention. As stated above, temperature change is the indicator whereby iced saline is injected through a catheter with a proximal port and the fall in temperature measured at a distal thermistor, based on a mathematical relationship derived from the Stewart-Hamilton equation.[80,113,114]

$$\text{Cardiac output} = \frac{V_i(T_B - T_i)S_iC_i60\text{k}}{S_BC_B\int_0^{\infty} DT_B(t)dt}$$

V = volume
i = injectate
B = blood
T = temperature
S = specific gravity
C = specific heat
t = time
k = correction factor for heat transfer from injection site to proximal port.

Multiple-port, balloon-tipped catheters with thermistors for calculation of cardiac output are commonly used with proximal ports used for monitoring central venous pressure when the catheter is placed in a branch pulmonary artery. With the balloon inflated, a pulmonary artery wedge pressure can be obtained for calculation of pulmonary vascular resistance (see below).

Other methods for determining cardiac output include the indicator dilution technique using indocyanine green, which is still used in some centers. Blood flow is proportional to the change in concentration of the dye measured distally. Catheter-tipped velocimeters or electromagnetic flow probes are used in conjunction with cross-sectional echocardiographic or angiographic measurement of vessel diameters to calculate blood flow at a specific site, usually the pulmonary artery or aorta.

In clinical practice pressure data are obtained during cardiac catheterization using fluid-filled catheters attached to a strain-gauge type external transducer via a small volume displacement manifold.[115] Even in newborn infants, newer, commercially available, thin-walled catheters allow for adequate frequency response characteristics to record high quality pressure measurements. Occasionally, catheter-tip microtransducers are necessary to obtain high-fidelity measurements for calculation of cardiac mechanics.[114] Pressures are obtained in every cardiac chamber accessible. Access to the left ventricle includes the transvenous approach across the patent oval fossa with advancement through the left atrioventricular valve or retrogradely from the femoral artery across the aortic valve.

Evaluation of the right atrial pressure includes measurement of mean pressure (older children, 2–6 mm Hg; infants, 0–4 mm Hg) and assessment of the *a*, *c*, and *v* waves, corresponding to atrial systole, ventricular systole, and atrioventricular filling, respectively.[111] In the right atrium, the *a* wave is typically the dominant wave in contrast to the left atrium where the *v* wave typically dominates. In the spontaneously breathing child, intrathoracic caval and right atrial pressures fall with inspiration and rise with expiration. By convention, intracardiac pressures are reported at end-expiration unless otherwise noted. In children with pulmonary hypertension, the *v* wave is often elevated indicating diastolic changes in the compliance of the right ventricle. Tricuspid insufficiency obviously will alter the right atrial pressure tracing, usually elevating mean right atrial pressure and producing a prominent *c* wave that may blend into the *v* wave.

The configuration of right ventricular pressure is similar to left ventricular pressure with peak systolic pressure reaching 15–25 mm Hg in normal older children. End-diastolic pressure is usually measured at the point prior to the rapid ventricular upstroke. If right ventricular pressure is elevated, measurement in the subvalvar right ventricular outflow area as well as the body of the right ventricle is necessary to determine if subvalvar obstruction is present. Typically, withdrawal pressures from the pulmonary artery to right ventricle are obtained to evaluate for obstruction at supravalvular, valvular, and subvalvular sites.

Pulmonary artery systolic pressure is usually similar to the right ventricular systolic pressure; diastolic pressure in the pulmonary artery in older children is approximately 8–12 mm Hg, with mean pulmonary artery pressure below 20 mm Hg considered normal, although such pressures are usually much lower, in the range of 10–16 mm Hg. But in the newborn infant, branch pulmonary artery systolic pressures are usually lower, representing the disproportionate fetal size of the branch vessels compared to the main pulmonary artery. Elevation in pulmonary vascular resistance usually results in elevations in pulmonary artery pressures. However, in conditions of low pulmonary blood flow, irreversible pulmonary vascular disease can be present without elevation in pressure and calculation of pulmonary vascular resistance can erroneously provide a falsely low value that may not be appreciated until therapeutic intervention improves pulmonary blood flow. When pulmonary hypertension is present, vasodilators are administered to assess pulmonary artery pressure and pulmonary vascular resistance as described below. Pulmonary wedge pressure is obtained by advancing an end-hole catheter distally into both branch pulmonary arteries until a wedge position is reached or by occlusion of the distal pulmonary artery by a balloon-tipped catheter. Venous waveforms are thus obtained which correlate with left atrial pressures. Since occlusion of the vessel occurs, there is no flow, thus no resistance and pressure should mirror left atrial pressure; in practice a damped signal is usually obtained with identifiable *a* and *v* waves.

Calculation of vascular resistance continues to be used as an important determinant for characterizing cardiopulmonary hemodynamics. The concept of vascular resistance is of course based on the Poiseuille equation,

$$Q = \frac{\Delta P \pi r^4}{8 \eta l}$$

where Q is defined as flow, ΔP the difference between inflow and outflow pressure, r the radius of the tube, l the length of the tube, and h the viscosity of the fluid. Hydraulic resistance, defined as the ratio of mean pressure drop to flow, based analogously to Ohm's law, allows resistance (R) to be expressed

$$R = \frac{\Delta P}{Q} = \frac{8 \eta l}{\pi r^4}.$$

Viscosity and length are assumed to be constant, the former a debatable assumption in newborn infants with high hematocrits, however, since the radius component is a power function, one can see that vascular changes induced by pulmonary vascular disease that compromises intravascular lumen will disproportionately influence this calculation. Furthermore, the Poiseuille equation was derived from steady-state flow through non-compliant cylindrical tubes; clearly these conditions are violated in the compliant, pulsatile pulmonary circulation subjected to turbulent flow in stenotic conditions. Although such limitations exist, the clinical use of vascular resistance for defining pathophysiologic states of altered pulmonary hemodynamics remains an integral part of clinical decision making. Pulmonary vascular resistance is defined as the difference in mean pressure across the pulmonary bed (mean pulmonary artery pressure—mean left atrial pressure) divided by pulmonary blood flow. Such an equation yields resistance units expressed as mm Hg·L or can be converted to metric resistance units expressed in dynes·sec·cm^{-5} by multiplying the former by 80. Access to the left atrium for direct pressure measurement is possible in children through a patent ovale fossa. However, when this is closed the mean left atrial pressure must be estimated by measuring the left ventricular end-diastolic pressure or the pulmonary arterial wedge pressure. Occasionally, access to the pulmonary artery is impossible and estimates of the mean pulmonary artery pressure are derived from pulmonary vein wedge pressures. The interpretation of pulmonary vascular resistance must take into account the normal maturational differences discussed in a previous section. The normal pulmonary vascular resistance in children is 1–3 units/m^2, however, in the newborn infant, pulmonary vascular resistance is elevated (8–10 units/m^2)[111] and falls to adult levels over approximately 4–6 weeks, unless altered by physiologic changes induced by congenital heart disease. Frequently, pulmonary vascular resistance is expressed as a percentage or fraction of systemic vascular resistance.

In the child with congenital heart disease complicated by elevated pulmonary vascular resistance, vasodilators are often administered in an effort to determine the "reactivity" of the pulmonary vascular bed to aid in the assessment of defining risks for surgical intervention. Vasodilators such as supplemental oxygen, tolazoline, prostaglandin E_1, and prostacyclin I_2 are administered after basal measurements and the pressure/flow measurements repeated. If supplemental oxygen is administered, dissolved oxygen must be included in calculation oxygen content for flow calculations. Of course, measurement of pulmonary vascular resistance in the cardiac catheterization is influenced by many variables. In the child in whom it is necessary to evaluate more fully the response to vasodilators, a balloon-tipped, end-hole thermodilution catheter can be placed after a more complete diagnostic study and the child returned to an intensive care unit for continuous monitoring of the pulmonary artery pressure and repeated measurements of pulmonary vascular resistance obtained. Lack of reactivity can imply irreversible pulmonary vascular disease and additional evaluations then considered for assessment of risks for therapeutic interventions, such as pulmonary artery wedge angiography or open lung biopsy.

Angiography

Pulmonary angiography allows assessment of pulmonary arterial anatomy and assessment of the pulmonary venous connections (Figure 23). The consequences of altered pulmonary hemodynamics on pulmonary arterial morphology can then be examined. Unlike acquired lesions, access to the pulmonary circulation can be difficult and at times impossible in children with congenital heart disease (e.g., pulmonary atresia). In such circumstances, access through alternative pathways for

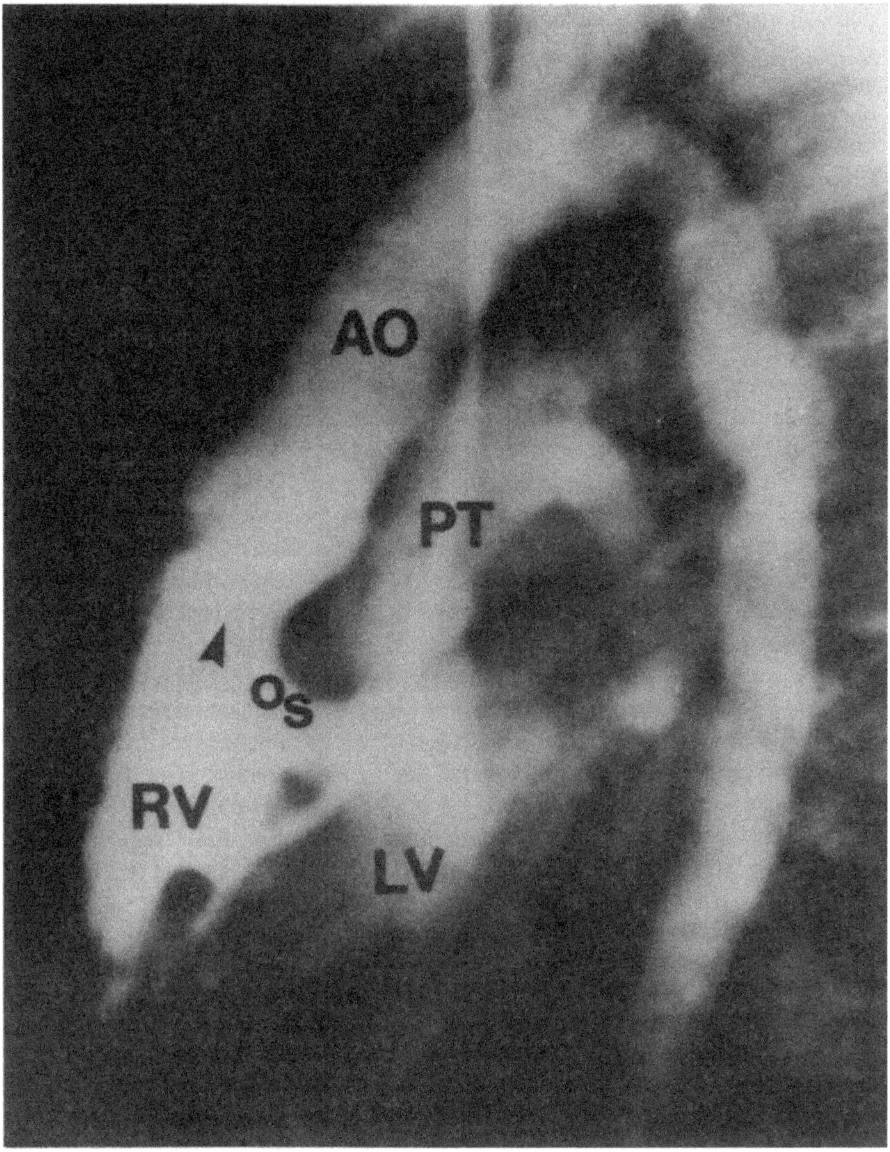

Figure 23. Right ventriculogram performed in long axial projection from an infant with transposition of the great arteries and an outlet ventricular septal defect. Note the subaortic infundibulum (arrow). AO: aorta; RV: right ventricle; LV: left ventricle; PT: pulmonary trunk; OS: outlet septum. (Reprinted with permission.[117])

blood to the lungs can be used, including aortopulmonary collaterals, an arterial duct, and previously placed surgical aortopulmonary shunts. Occasionally, retrograde filling of the pulmonary arteries via a retrograde pulmonary wedge angiogram is helpful when the size and continuity of the branch and main pulmonary arteries is in question.

Assessment of ventricular function by calculating chamber volume and dimension provides calculation of derivative indices such as ejection fraction and ventricular wall mass incorporating digitized angiograms is routinely used in adults and can be applied well to the pediatric patient.

Therapeutic Cardiac Catheterization Techniques

The advent of non-surgical, transcatheter interventional techniques to physiologically palliate or correct alterations in cardiopulmonary function secondary to congenital heart disease has greatly obviated the need to subject children to cardiac surgery for a variety of defects as listed in Table 6. Although these techniques are not employed to assess pulmonary hemodynamics, the reader is well advised to be aware of the dramatic changes brought about these techniques in the management of children with altered pulmonary hemodynamics. The application of these techniques often follows the diagnostic portion of a cardiac catheterization procedure and is thus linked closely clinically.

Early procedures included the balloon septostomy of the atrial septum in the child dependent on a communication in the atrial septum for adequate effective pulmonary blood flow mixing (e.g., transposition of the great vessels) or for lesions in which the cardiac output must traverse such a communication (e.g., tricuspid atresia). The balloon dilation of obstructive lesions such as pulmonic stenosis and the interruption of left-to-right shunting as in closure of an atrial septal defect or arterial duct is now routine management of such problems. An area of increasing

Table 6. Specific Congenital Heart Defects with Altered Pulmonary Hemodynamics for which Successful Transcatheter Intervention is Used

Atrial septostomy—balloon or blade
Transcatheter deliver of a closure device
Atrial septal defect—oval fossa type
Ventricular septal defect—perimembranous and muscular
Atrial septal fenestration following Fontan procedure
Aortopulmonary collaterals
Balloon dilation
Pulmonic stenosis—subvalvular, valvular, supravalvular, branch
Aortic stenosis—subvalvular, valvular
Mitral stenosis—acquired and congenital
Systemic and pulmonary venous obstruction—congenital, post-surgical
Laser or radiofrequency "valvotomy"
Pulmonary atresia

sophistication is the dilation and maintenance of stenotic pulmonary arteries by stents. Furthermore, occlusion techniques for intrapulmonary vessels by coils and other hardware allow successful obliteration of undesirable pulmonary shunting.

It is now necessary to evaluate post-procedure pulmonary hemodynamics by the same techniques that we have been discussing to determine the efficacy of such procedures invasively and noninvasively.

III. ANIMAL STUDIES

A. Introduction

Though advances in intraluminal catheter[120] and Doppler ultrasound techniques[121–123] allow clinicians to measure many key parameters related to the human pulmonary circulation, animal studies are still needed to explore and explain the fluid dynamics of the pulmonary outflow track. The main pulmonary artery and its bifurcation present a unique and challenging juncture[124–127]—a short, curved, viscoelastic, somewhat elliptical, tapered tube that divides into two vessels with a branch: parent cross-sectional area ratio that is routinely <1:1, in contrast to the 1.19–1:26:1 ratio considered the norm for arterial dichotomously branching sites.[23] Thus, the velocity patterns are complex and difficult to assess: encircling flow probes that constrict or bind the vessel tend to alter the flow field being studied[128]; pulsed Doppler ultrasound techniques that should enable the measurement of velocity profiles from an external position are difficult to implement *in vivo* given the oblique axis requirements and the vessel's curvature, taper, short length, and so on. Some of these difficulties may be overcome with the advancement of MRI velocity measurement techniques.[16] However, due to costs and physical constraints, MRI will continue to have limitations as a tool for assessing blood flow in animals on a routine/chronic basis.

The preponderance of animal studies have been performed in dogs, however, pertinent results have also been obtained in rabbits, cats, lambs, swine, cows, and horses. The results of these investigations can be reviewed in many excellent books that focus on or encompass the pulmonary circulation.[1,2,129–142] Though much has been accomplished to increase our understanding of the hemodynamics of the pulmonary circulation, definitive answers to numerous important questions are still unknown: how flow and velocity waveform shapes can be used to estimate pulmonary pressures and resistances; how to use velocity and dimension measurements to make reliable estimates of cardiac output; how to make allowances for normal-for-age variations in hemodynamics and geometry; how to assess, interpret, and manipulate the coupling properties of the right ventricle-lungs connection; and how to optimize surgical outcomes of interventions undertaken when normal coupling mechanisms are missing. Answers to the above are predicated on our understanding of the hemodynamic conditions expected, the relationships between velocity patterns and the underlying hemodynamics, and the concept of input impedance.

B. Hemodynamics

The normal pulmonary circulations of mammals have many consistent anatomic and functional features.[143] In addition to the geometric description given above, features that are particularly pertinent to studies of pulmonary artery flow dynamics in contrast to their systemic counterparts areas follow:[1] major artery walls are thinner and more distensible, pressures and resistances are approximately one-sixth systemic levels, pulse pressures are lower, and the blood flow waveform, which has much of its energy in its first harmonic, is rounder and more symmetric.

Also, some hemodynamic variables are also relatively constant in animals of different sizes, (notably pulmonary blood pressures and blood velocity).[23] Normal adult mean pulmonary artery pressures are approximately 15 mmHg with a similar pulse pressure of approximately 15 mmHg. Mean adult pulmonary blood velocity is approximately 20 cm/s. Other variables, such as cardiac output, peripheral vascular resistance, pulse wave velocity, and heart rate do vary with size and/or species, but a universally accepted indexing equation has not been derived. Equations that work well at the intraspecies level, for example, on the basis of weight or body surface area, fail at the interspecies level.[143–145] Those equations that work well at the interspecies level tend to have characteristics that are not desirable at the intraspecies level.

Changes with Age

As a result of the frequency and significance of pulmonary vascular changes in patients with congenital heart disease, the ability to assess pulmonary hemody-

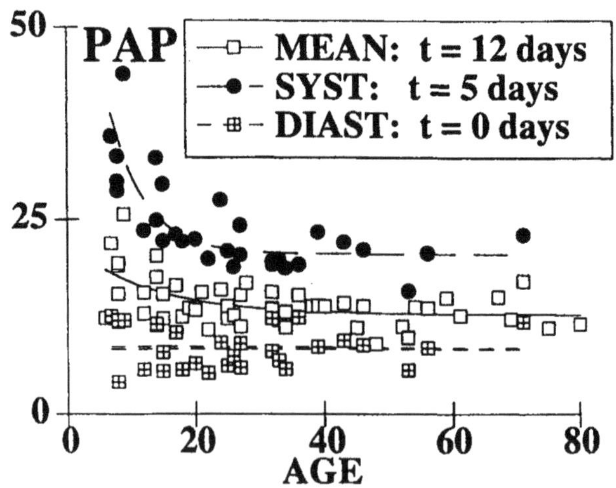

Figure 24. Time constants for mean, systolic, and diastolic pulmonary artery pressure with age.

namics is of particular importance to pediatric clinicians.[146,147] Proper diagnosis and treatment often depend on knowing what is "normal for age," and few systems undergo as dramatic a change with birth and infancy as the pulmonary circulation:[8,67,148] pressures and resistances decrease rapidly, vessels proliferate, and vessel walls thin and change composition.

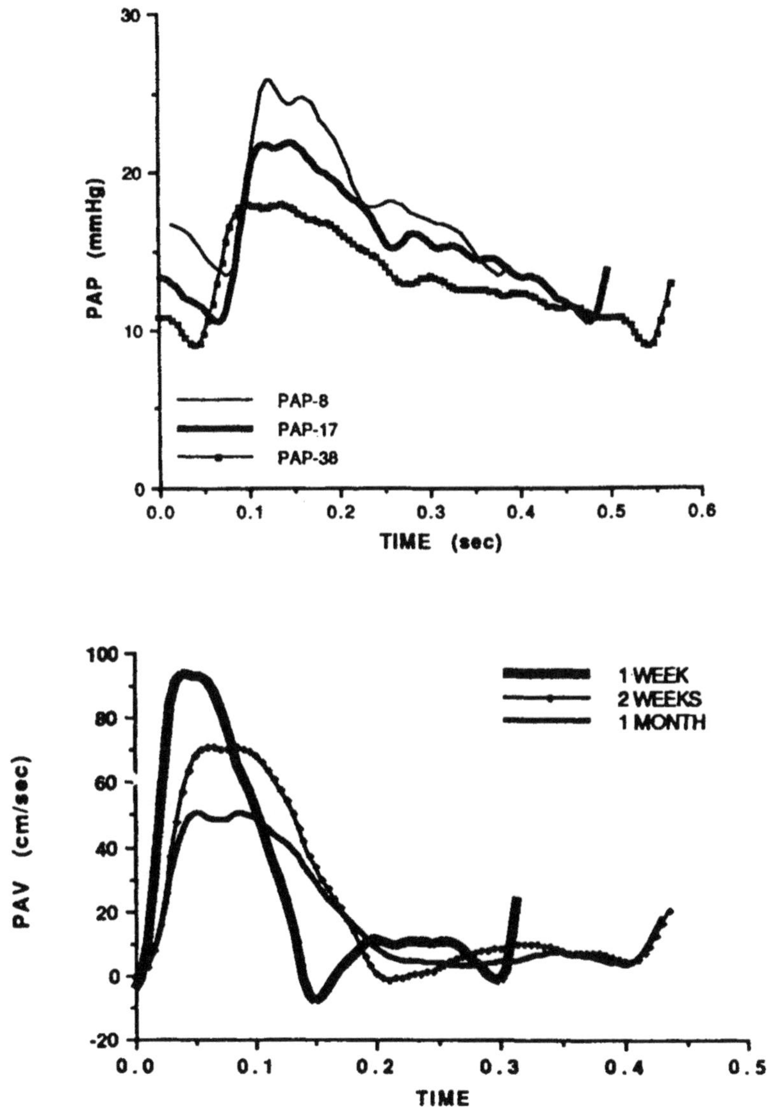

Figure 25. Representative pressure (upper) and velocity (lower) waveforms at one week, two weeks, and one month of age.

Given size constraints required for appropriate instrumentation, the infant animal model of choice has been the newborn lamb. Results obtained in alert, chronically instrumented animals[149,150] between the ages of three days and three months, indicate that the preponderance of pressure and compliance changes are 98% (3 time constants) complete by the end of the first month (Fig. 24). Systolic pressures fell from a projected 87 mm Hg to 20 mm Hg, mean pressure fell from 22 mm Hg to 13 mm Hg, and diastolic pressure remained relatively unchanged at approximately 8 mmHg. (Since much of the fall in pulmonary arterial pressure in newborn lambs is reported to be complete within 1–2 hours of birth, these intercept values may be underestimated.)[147,148] Comparable equations for velocity indicated that peak velocity fell from up to 145 cm/s to approximately 60 cm/s and that mean velocity fell from 40 cm/s to the expected 20 cm/s. Representative pressure and velocity waveforms at one week, two weeks, and one month are shown in Figure 25. The decrease in compliance/increase in elastance was indicated by a decrease in the pressure-strain elastic modulus ($E_p = \Delta P \cdot r_0/\Delta r$)) from an estimate of 750 mm Hg at birth to 96 mm Hg. Variations in pulsatile changes in diameter measured with micromanometers and the related pressure area loops are illustrated in Figures 26 and 27.

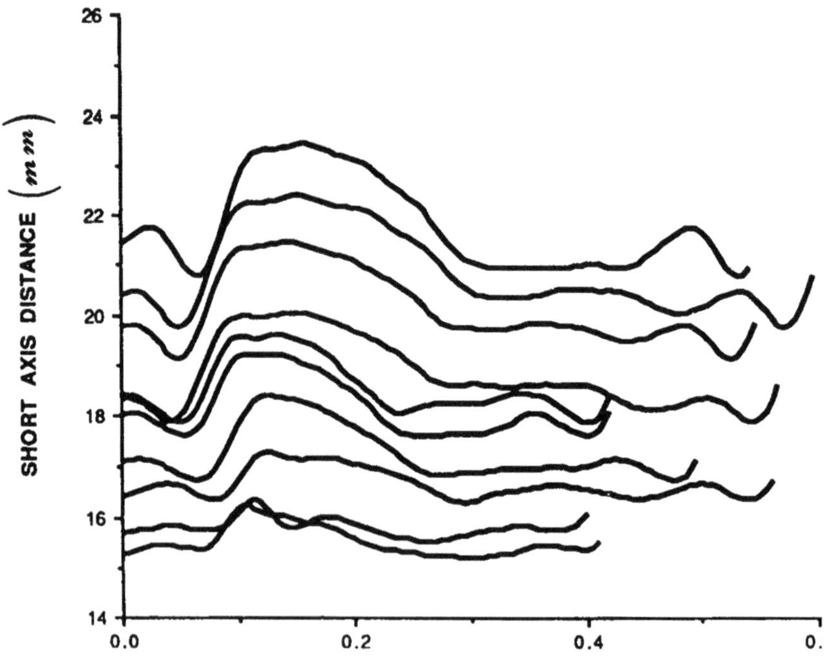

Figure 26. Phasic dimension changes with age. Dimensions always increased with increasing age. From bottom to top, ages were 6, 8,12, 17, 25, 28, 38, 59, 67, and 80 days.

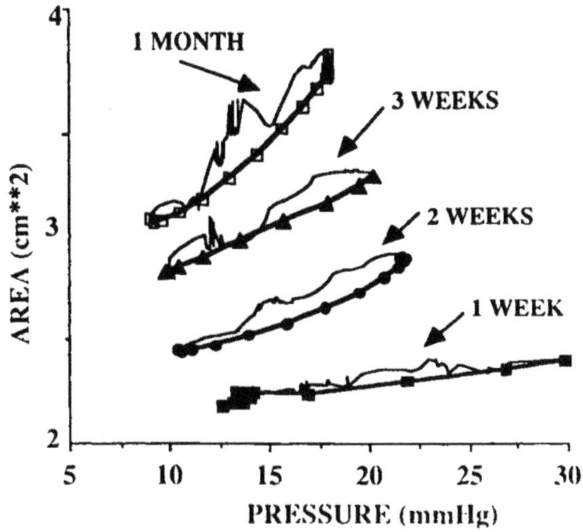

Figure 27. Representative pressure/area loops at four different ages. Symbols designate systolic rise time: solid squares = eight days, circles = 17 days, triangles = 25 days, and open squares = 38 days. (Reproduced from Lucas et al.[149])

In a comparable terminal hemodynamic studies in piglets between the ages of one hour to two weeks,[151] the regression in mean PA pressure went from a high of 30 mm Hg (at 12 hours) to a mean of 16 mmHg. Estimated changes in E_p, based on pulse wave velocity measurements, decreased from 334 to 78 mm Hg. *In vivo* and post-mortem studies in man indicate that the rapid regression of wall thickness at birth is followed by gradual stiffening with an estimated increase in E_p of 1.9 mm Hg/year between the ages of one and 50 years,[152] a finding that has not yet, to the author's knowledge, been substantiated in animals models.

Changes with Disease

The need for improved techniques for monitoring pulmonary hemodynamics and assessing pulmonary vascular impairment in patients with congenital or acquired lesions has prompted investigations of relationships between abnormal pulmonary circulations and pulmonary artery blood velocity patterns that can be observed noninvasively with pulsed Doppler ultrasound. Features associated with pulmonary hypertension in humans include decreased rise time (time from onset of systole to peak velocity)[55,100,102,104,154–158] and a velocity waveform with a triangular or skewed shape.[159]

Typical normotensive compared to hypertensive waveform shapes are illustrated in Figure 28. Data on the left were obtained from a one-month-old lamb with normal pulmonary pressures and resistances, while data on the right were obtained from a one-month-old lamb in whom pulmonary vascular injury was initiated via a

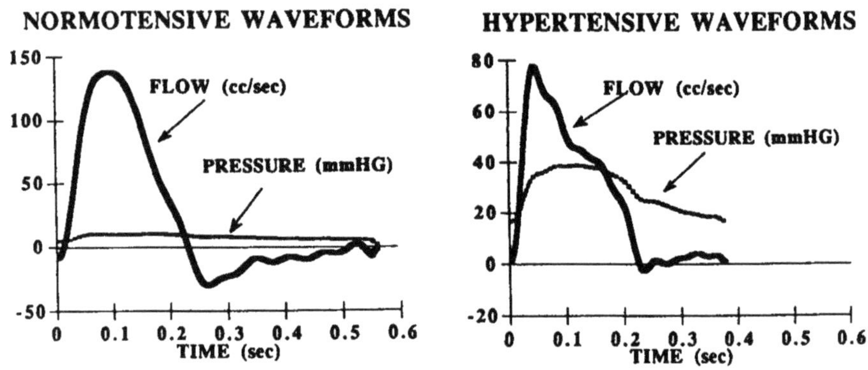

Figure 28. Typical pressure and flow waveforms from a one-month-old lamb with normal PAP and PVR (left) and from a one-month-old lamb with pulmonary hypertension (right) induced by monocrotaline pyrrole injection at 3–5 days of age. Note the triangular shape of the hypertensive flow waveform.

monocrotaline pyrrole injection at 3–5 days of age.[160] While the triangular shape associated with pulmonary hypertension is unmistakable,[161] no quantitative techniques derived for estimating pulmonary pressure and flow solely from features of pulmonary velocity waveforms have proven to be sufficiently reliable, to be widely adopted in clinical practice. Suggested reasons for failure include: (1) irregularity in pulmonary velocity profiles (discussed in detail below) such that indices vary depending on the position of the sampling site in the main pulmonary artery or the right ventricular outflow tract[162–166]; (2) disagreement as to whether and how to correct for variations in heart rate;[167,168] and (3) incomplete understanding of differences in responses due to acute compared to chronically induced hypertension.

One explanation for the change in waveform shape, sometimes referred to as mid-systolic notching, is based on a consideration of reflected waves, as illustrated in Figure 29. The measured pressure and flow waveforms shown in Figure 28 were separated into projected forward and reverse components using the method of Westerhof et al.[169,170] Under normal conditions, the wave returns in late systole-to-diastole, creating a wide incisura and augmenting diastolic pressure similar to the action of a counterpulsating ventricular assist device. A decrease in compliance leading to increased pulse wave velocity *and/or* the initiation or accentuation of a more proximal reflection site would effect an earlier return of the reflected wave. With the earlier return, the mid-to-late systole pressure waveform is augmented and the flow waveform is diminished, creating the skewed/notched velocity waveform.

Furuno et al.[171] showed that dogs with acute pulmonary hypertension induced by constriction of both branch arteries displayed a similar mid-systolic deceleration of the pulmonary flow wave, which they likewise attributed to reflection. Results from

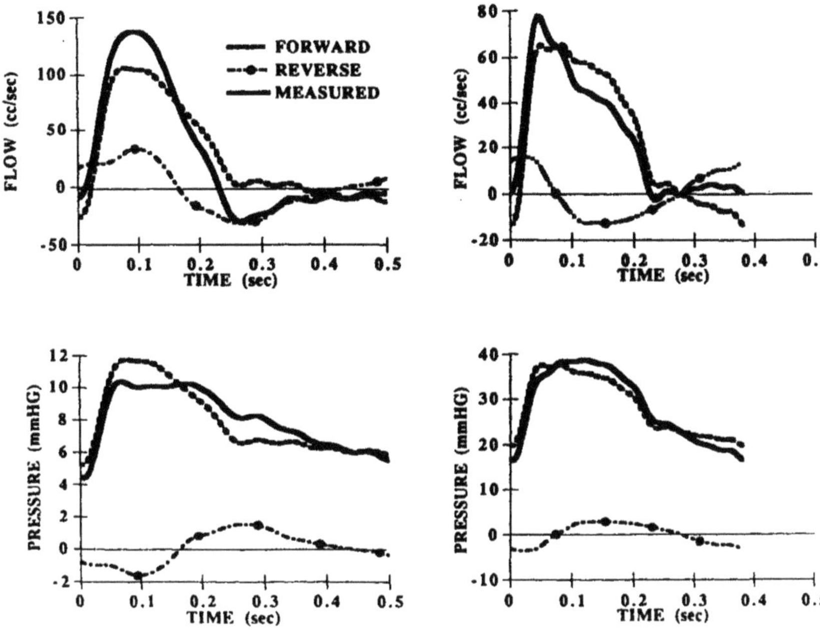

Figure 29. Separation of measured pressure and flow waveforms from Figure 28 into forward and reverse components. Normal waveforms are shown on the left and hypertensive waveforms are shown on the right. Under normal conditions, the wave returns in late systole-to-diastolie, creating a wide incisure and augments diastolic pressure. Under hypertensive conditions, the earlier return of the reflected wave augments the mid-to-late systole pressure waveform and the flow waveform is diminished, creating the skewed or notched velocity waveform.

studies in which hypertension was created by embolization vary from no evidence of waveform alteration[171,172] to marked changes[173] in waveform shape. Variation in findings can probably be attributed to variations in the size and quantity of emboli produced, as an increase in peripheral resistance alone does not guarantee the necessary increase in compliance or the development of a pertinent reflection site.

Other explanations of decreasing velocity/flow waveforms in concert with an increasing intraluminal pressure have been hypothesized. For example, in similar studies in dogs in which acute pulmonary hypertension was also created by pulmonary artery constriction, Tahara et al.[174] were able to relate the transient decrease in pulmonary artery flow in mid-systole to a transient reversal of the pulmonary artery-right ventricular pressure gradient.[133] Thus, the dip may be attributed to a decrease in the driving pressure gradient that may or may not be reflection related. Another possibility, which has not been substantiated, is a dynamic intrabeat change in downstream events, for example, a pressure varying pulmonary vascular compliance.[175]

A second feature of velocity waveforms that some investigators have found useful in assessing pulmonary hypertension is the degree and duration of backflow during late systole and early diastole.[176–180] However, examination of velocity profiles in normal animals often shows flow reversals of the magnitude associated with pulmonary hypertension in humans, which will be discussed in more detail below. A significant backflow near the posterior main pulmonary artery wall of humans with normal pulmonary circulations has been noted.[161] Current investigations imply, however, that the relationship between pulmonary pressures and backflows is much more complex than originally anticipated. The emerging scenario that is imaging is that the backflow present under conditions of normal and severe pulmonary hypertension disappears under conditions of moderately elevated pressures.[181]

Velocity Profiles

As indicated above, the difficulty in reliably relating velocity waveform shapes to pulmonary hemodynamics has been attributed to a dependence of waveform shape on intravessel position, that is, the existence of an irregular velocity profile. Hemodynamic theory predicts that the entrance profile to the major vessels—aorta and main pulmonary artery—should be flat;[1,23] early hot-film anemometry stud-

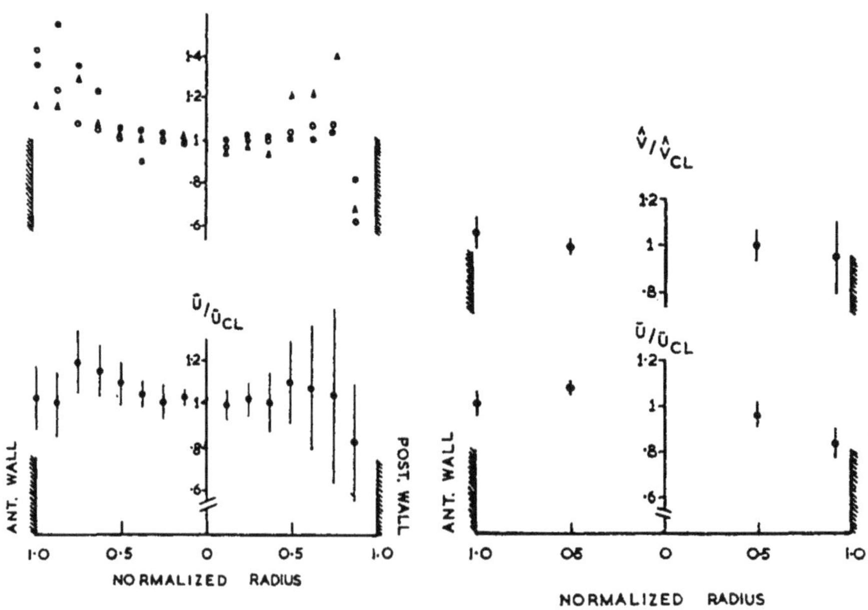

Figure 30. Velocity profiles across the transverse axis of the main pulmonary artery of dogs (left) and humans (right) obtained with a thin anemometer. (Reproduced from Rueben et al.[26])

ies[182] in which the main pulmonary artery was encircled by a supporting cuff indicated that this was true. Reuben et al.[26] studied five dogs and five patients and concluded that the mean velocity profiles were approximately flat (Figure 30); Paulsen[183] studied five dogs and concluded that the mean velocity profile was "centrally flat with a slight deviation with the highest velocity nearer to the posterior wall." Measurements were only made during systole since bi-directional flow can not be readily distinguished via this technique.

In studies in dogs and lambs using an intraluminal pulsed Doppler needle probe capable of measuring bi-directional flow and examining the entire cardiac cycle,[184] the mean profile along an anterior-posterior axis was consistently skewed toward the posterior wall (Figure 31).[163] The skew could be attributed to the large flow

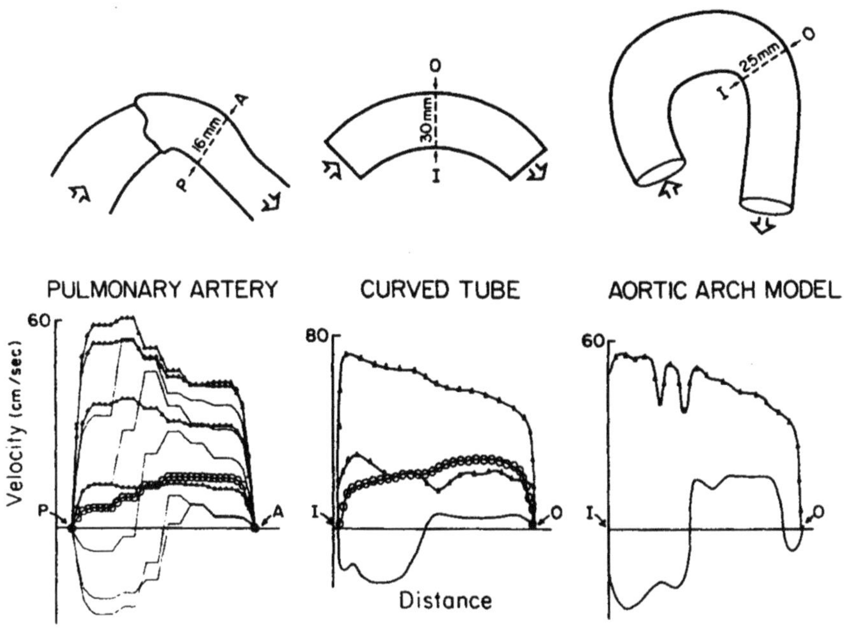

Figure 31. Relationship between pulmonary artery velocity profiles observed *in vivo* and those observed *in vitro* in studies of pulsatile flow in curved tubes. Figures for the *in vivo* pulmonary artery profiles were obtained from lamb studies while figures for the curved tube and aortic arch model were adapted from data published by Chandran et al.[3] and Yearwood and Chandran,[4] respectively. The anterior wall (A) of the pulmonary artery corresponds to the outer wall (O) of the tubes and the posterior wall (P) corresponds to the inner wall of the tubes (I). Schematics show the plane and axis over which the profile was measured. Instantaneous profiles during early systole (*triangles*) and late sytole plus early diastole (*line*) are shown. Mean velocity profiles (*circles*) were not published for the aortic arch model. (Reproduced from Lucas et al.[163])

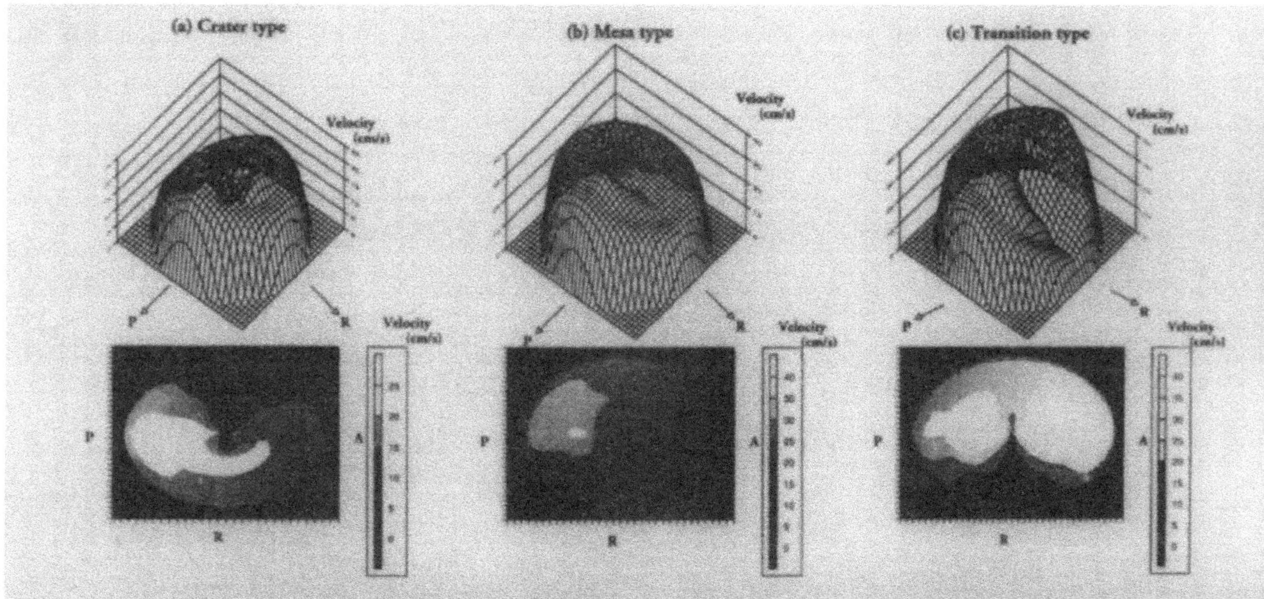

Figure 32. Analysis of three dimensional velocity profiles projected from velocity measurements made along the right-left and the anterior-posterior axes of the main pulmonary artery in lambs with normal hemodynamics and monocrotaline pyrrole induced pulmonary hypertension. Three profile configurations were identified. (Reproduced from Katayama et al.[185])

reversal region observed during late systole and diastole along the posterior wall, a pattern that was remarkably consistent with the patterns observed in the mid-arch region of a simple curved tube[3] and in the region of greatest curvature in a model made from a cast of an adult human aorta.[4]

In a more detailed analysis of three dimensional profiles, projected from velocity measurements made along the right-left and the anterior-posterior axes in lambs with normal hemodynamics and monocrotaline pyrrole induced pulmonary hypertension, the highest forward as well as reverse velocities were found consistently near the posterior wall.[185] All of these findings are predicted by fluid dynamic theory; under conditions of pulsatile flow, the highest velocities going into a curve will be along the wall with the smallest radius of curvature.[1] Three profile configurations, designated as "crater," "mesa," and "transition," were identified (Fig. 32). Locations of the triangular shaped/mid-systolic notched waveforms in hypertensive animals were examined and were found to vary with intraluminal position.

The most detailed examination of a pulmonary velocity profile was recently reported for the porcine pulmonary trunk.[186] Hot-film anemometry was used to measure forward systolic velocity at 41 locations in 90 kg pigs. The pattern observed was described as a counterclockwise rotation during early systole, that subsequently turned clockwise and ultimately counterclockwise to become flat during the early and late deceleration phases. Highest velocities were consistently found near the anterior rather than posterior wall, as observed in dogs and lambs.

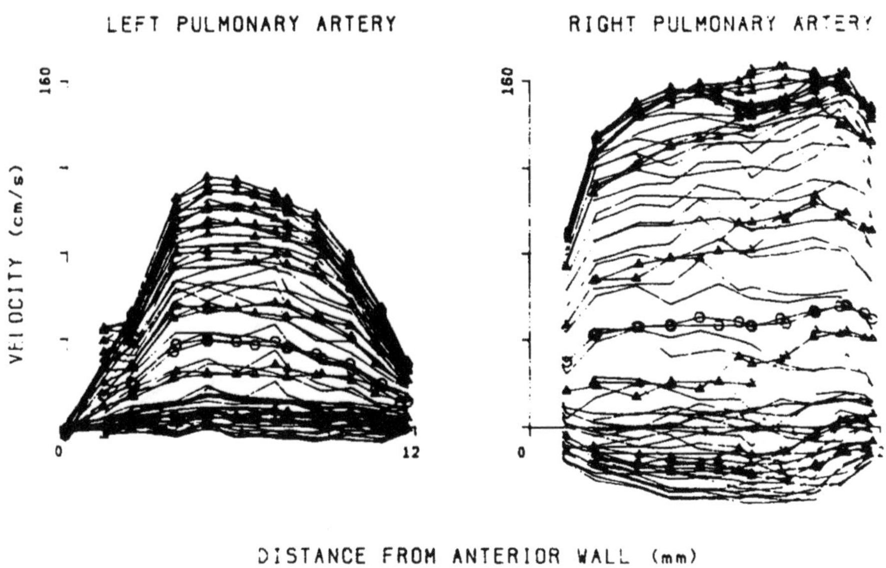

Figure 33. Instantaneous velocity profiles for the left pulmonary artery (left) and right pulmonary artery (right). Abcissa values are given as distance from the posterior wall as recorded from range-age.

Mean velocity varied markedly at different sampling sites with a difference between the maximum mean and the "mean of the means," which was considered to be the best indicator of bulk flow, differing by an average of 30%. Again, the method did not allow assessment of flow reversal patterns.

While little is known about the velocity profile in the main pulmonary artery, even less information is available about the profile in the branches. Some data have been collected in branch arteries of open-chest lambs using 20 MHz pulsed Doppler crystals mounted approximately 1 cm from the bifurcation and oriented anterosuperiorly.[187] Profiles were obtained by range-gating the pulsed Doppler signal across each of the vessels from the anterior to the posterior wall. Profiles obtained were relatively flat with similar mean velocities (Figure 33). The most consistent difference observed was a marked increase in flow reversal in the right compared to left pulmonary artery during diastole. Profiles were repeated during 7.5% O_2 inspiration. Maximum flow reversal in both branches correlated well with pulmonary vascular resistance (Figure 34) but not with mean pulmonary artery pressure.[188] Branch velocity profiles were likewise obtained in the lambs in which chronic hypertension had been initiated by monocrotaline. Flow reversal was still

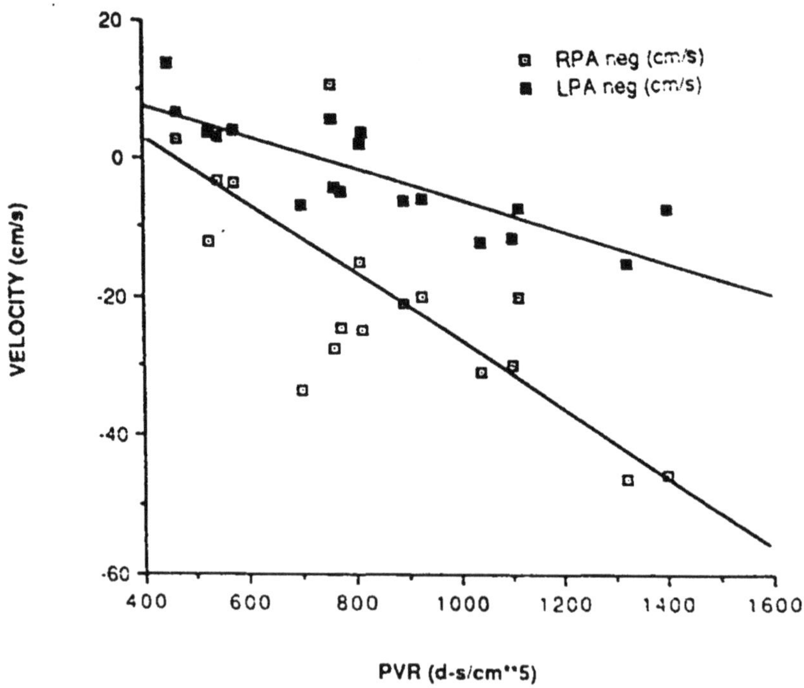

Figure 34. Relationship between PVR and maximum flow reversal in the branch pulmonary arteries.

more marked in the right than in the left pulmonary artery. The correlation between a shortened acceleration time and pulmonary artery pressure was observable in the left pulmonary artery waveforms, but not in the right.

C. Input Impedance

The pulmonary input impedance spectrum serves two major purposes: (1) to represent the load the pulmonary vascular bed places on the right ventricle,[189] and (2) to describe the response of the pulmonary vascular bed to altered hemodynamic states and to pharmacological and surgical interventions.[62] Configurations observed have been interpreted via a variety of mathematical models,[7] ranging from simple lumped parameter systems to elaborate systems attempting to incorporate individual vessels across the entire vascular bed.

The shape of the spectrum is quite simple and is remarkably similar across species, including man.[1,23] Modulus values fall from a DC term that is related to pulmonary vascular resistance to a minimum in the 2–5 Hz range, followed by a maximum at twice the minimum frequency, followed by a fairly constant value that is often termed "the characteristic impedance" or impedance in the absence of reflections. Phase values at low frequencies are negative as compliance dominates the system and flow leads pressure, then become zero at the first modulus minimum value, and then fluctuate around the zero line. The frequency of the first minimum modulus value in conjunction with an estimate of pulse wave velocity (c_o) is used to estimate the distance to a major reflection site (distance = wavelength/4 = $c_o/4f_{min}$). Reflections at the main pulmonary artery bifurcation area are difficult to detect via this mechanism as the energy level at the high frequencies that represent the quarter wavelength distance to this site generally have signal to noise ratios that do not enable accurate impedance estimation.

The compliant proximal pulmonary arteries, as indicated by a characteristic impedance value that is approximately 25–30% of the resistance value, effectively decouple the right ventricle from the high resistance downstream vessels, conserving the power required to pump blood through the pulmonary vascular bed in a pulsatile manner. Under normal conditions, the right heart acts as a flow pump capable of doubling flow rates without markedly increasing pulmonary artery pressures. Theoretical analyses indicate that the "matching" observed minimizes power requirements.[189] Investigators have also shown that the right ventricle is very sensitive to changes in afterload and makes adaptations that have been described as a transition from a flow to a pressure pump. For most purposes, a modified Windkessel representation of the input impedance spectrum is adequate for representing this right ventricular afterload. As indicated in Figure 35 in which a Windkessel has been fit to the spectrum derived mathematically for the input impedance spectrum of a cat, a spectrum derived from the morphological and morphometric data compiled by Fung and coworkers that represents the first complete representation of a mammalian pulmonary vascular tree.[190,191]

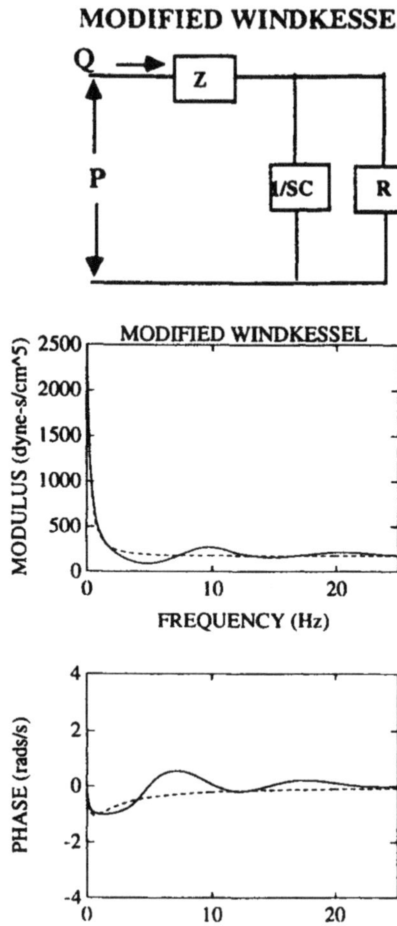

Figure 35. The modified Windkessel model (top) was used to fit input impedance spectra derived using Womersley's constrained tube equations for a 24-generation model of the pulmonary vascular bed of the cat (solid line in middle and bottom figures). The best fit impedance spectra are shown in the dotted lines.

Like the shape of the velocity waveform, the shape of the impedance spectrum has also been extensively studied as a tool for assessing the functional status of the entire pulmonary vascular bed on the basis of measurements made at the entrance. Undoubtedly, the major features of the impedance spectrum at frequencies slightly higher than the DC component (>0.1 Hz) are dominated by the characteristics of the proximal arterial bed. However, significant changes downstream are often reflected in the impedance spectrum via an adaptation of the arterial bed. For example, increases in pulmonary pressure via constriction of small vessels, that is,

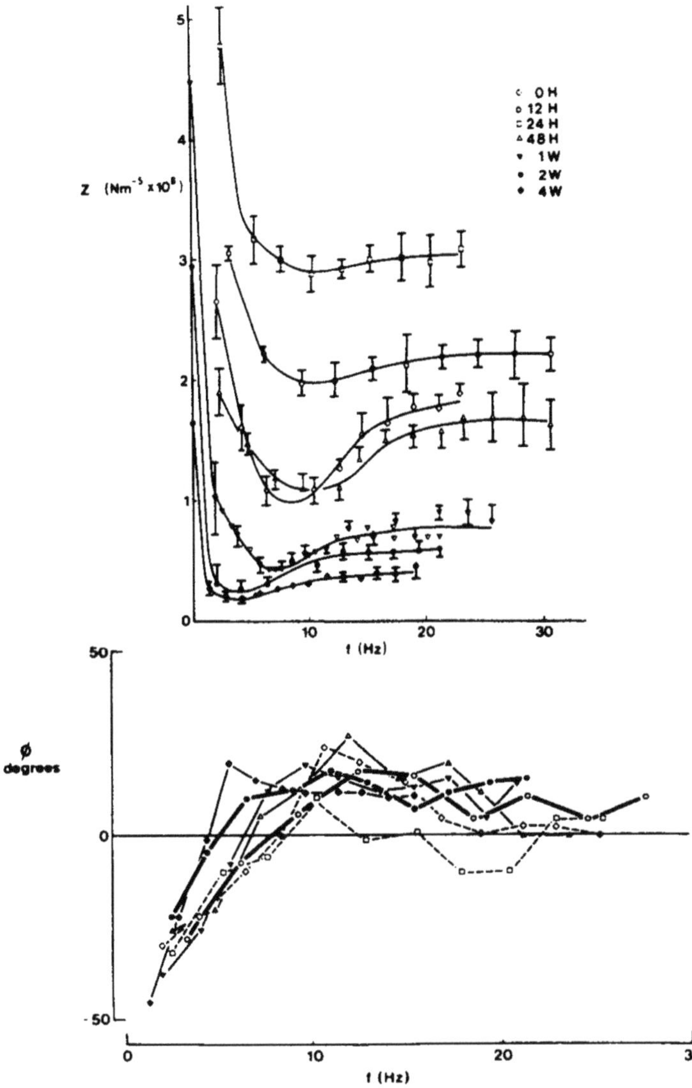

Figure 36. Changes in pulmonary vascular impedance modulus and phase with growth in a series of pigs reported by Greenwald et al.[151] 0 H (hour) pigs were studied at birth. Four W pigs were studied at weeks after an average weight gain from 1.45 to 10.8 kg. Other studies were conducted at intermediate periods (12 hours to two weeks). The modulus units are newton meters; the phase units are degrees.

hypoxia or bead embolization, or an elevation in left atrial pressure, will have a minimal immediate effect, however, the spectra will change with time as pulmonary arterial walls adapt to the increased wall tension.

As a mechanism that should be viewed as determining the pulsatile relationship between pressure and flow, with both variables being highly age-dependent in newborns, the input impedance spectrum is likewise highly age-dependent. Spectra calculated for the piglets ranging from one hour to four weeks in age,[151] demonstrated the expected decrease in resistance, characteristic impedance, and shift to the left in the modulus minimum as arterial walls thin and grow. Pulse wave velocity decreases and the major reflection site projected to be at the end of the arterial tree increases in distance (Figure36). These findings are consistent with our studies in lambs that indicate that the impedance spectra achieve a relatively steady state level by the end of one week (Figure 37).

Decreases in compliance and increases in pulse wave velocity as a result of chronic hypertension have the effect of increasing characteristic impedance and shifting the spectrum to the right, as indicated in Figure 38 by composite spectra computed for groups of normotensive and hypertensive lambs, whose representative hemodynamic waveforms were illustrated in Figures 28 and 29 above. The hypertensive spectrum suggests markedly stiffer arteries and the addition of a more proximal reflection site. The inflection around 10 Hz is reasonably related to the site that corresponds with 3–4 Hz in the controls. However, continued negative phase values up to 20 Hz are consistent with a prominent, more proximal site. Composite input impedance to branch vessels showed the same shift for hypertensive animals;[192] however, there was less evidence of reflections and the modulus values could be readily fit to the modified "Windkessel" model (Figure 39). The shift observed in the hypertensive spectrum is comparable to that observed in children with ventricular septal defects and is evidence of irreversible pulmonary vascular obstructive disease.[77]

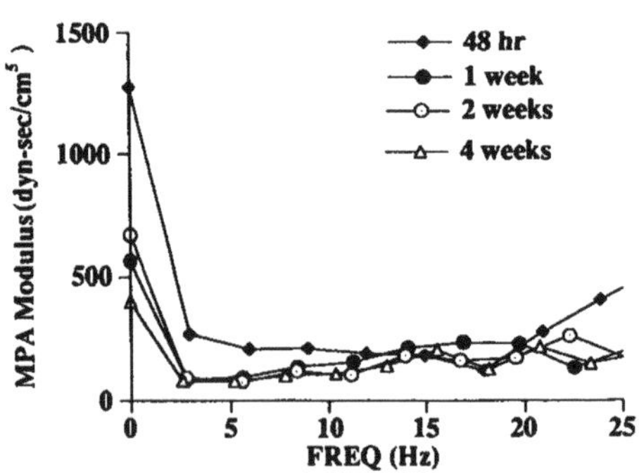

Figure 37. Changes in pulmonary input impedance modulus spectra with maturation in lambs ranging in age from 48 hours after birth to four weeks.

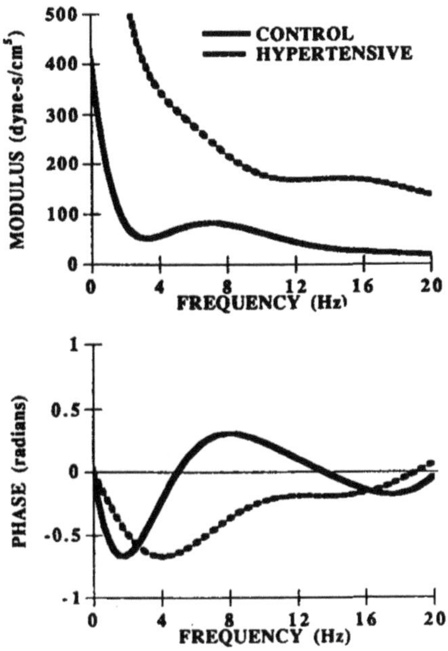

Figure 38. Composite pulmonary input impedance modulus spectra for normotensive and hypertensive lambs.

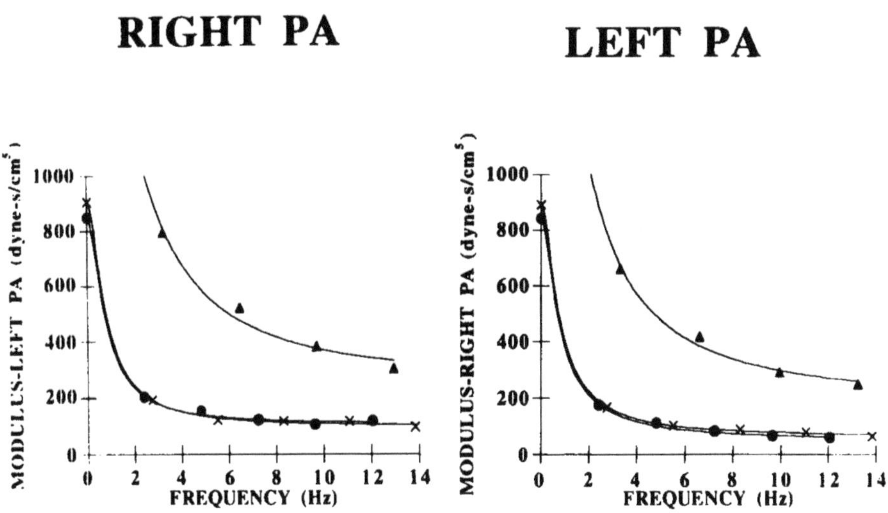

Figure 39. Composite input impedance modulus spectra for the branch pulmonary vessels fitted to the modified Windkessel model. Triangles represent data from hypertensive lambs.

The responses of the spectrum to numerous interventions have been well documented:[62,189,193–197] vasoconstriction with serotonin, sympathetic nerve stimulation, nonadrenergic vasoconstriction, exercise, hemorrhage, right-left shunts, acutely and chronically elevated left atrial pressure, inspiratory versus expiratory periods, acute and long-term pneumonectomy, and so forth. Of particular importance to investigators and clinicians hoping to assist an impaired circulation, however, are those interventions or insights that can be used to decrease resistance to pulsatile flow and/or maximize right ventricular efficiency if it is an operational component of the loop. Examples include interventions that may increase vessel compliance, decrease characteristic impedance, and decrease pulse wave velocity. Heart rates may be controlled to meet minimal outflow resistance and optimal settings could be determined for synchronizing respirators and pacemakers. Implanted extracardiac conduits could be designed to "impedance" match their surroundings. The possibilities to be explored seem unlimited!

IV. *IN VITRO* MODEL STUDIES

A. Introduction

Because of the technical difficulties in obtaining sufficiently accurate and well-resolved hemodynamic data *in vivo*, most studies directed toward an understanding of the arterial hemodynamics have been carried out *in vitro* or by computer simulation. The characteristics of the pulmonary artery hemodynamics have, however, not been studied by computer simulation due to its complicated arterial geometry. On the other hand, *in vitro* studies of the pulmonary artery hemodynamics have been performed only in a few groups.[5,52,198–204]

B. Flow Visualization and Pressure Measurement Studies

Sung[201] conducted flow visualization in an adult-sized pulmonary artery model with a "normal" pulmonic valve (5.35 cm^2 in orifice area). The study was performed using a 7 mW He-Ne laser as the light source. A schematic diagram of the experimental set-up is shown in Figure 40. The laser beam was first converted into a sheet of light by directing it through a glass rod, 3 mm in diameter. The sheet of laser light was then applied to illuminate the flow fields in the test model. Polystyrene particles (Amberlite ion exchange resin, Rohm and Hass Co., Philadelphia, Pennsylvania), 100 mm in diameter, were suspended in the fluid as tracers. The nomenclature used to describe various regions in the test model is illustrated in Figure 41.

During the acceleration phase, flow that emerged from the valve was evenly distributed (see Figure 42). A tiny region of flow separation was observed immediately distal to the valve near the edge of the bottom leaflet. As flow traveled downstream in the main pulmonary artery and impinged on the apex of the

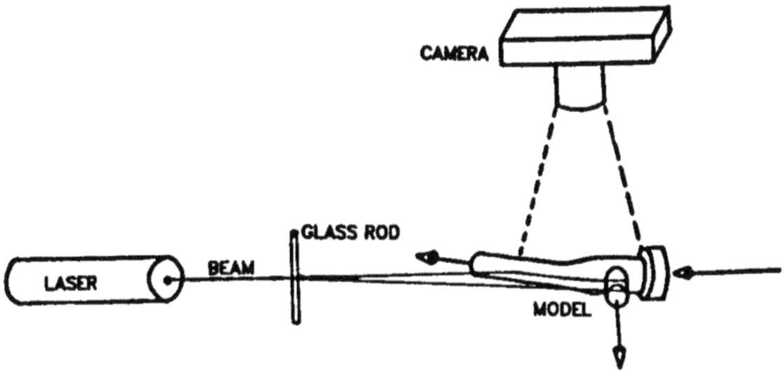

Figure 40. Schematic drawing of the flow visualization set-up.

bifurcation, it broke up into two streams. The two streams then flowed along the inner walls of the left and right pulmonary arteries, respectively. The flow in the left pulmonary artery was evenly distributed. No regions of flow separation or secondary flow were observed. In the right pulmonary artery, a small region of low-axial-velocity was observed at its origin near the outer wall. The flow recovered immediately downstream of the low-axial-velocity region.

At peak systole (Figure 43), the instantaneous flow rate increased and a broad central flow was seen in the main pulmonary artery. The tiny region of flow

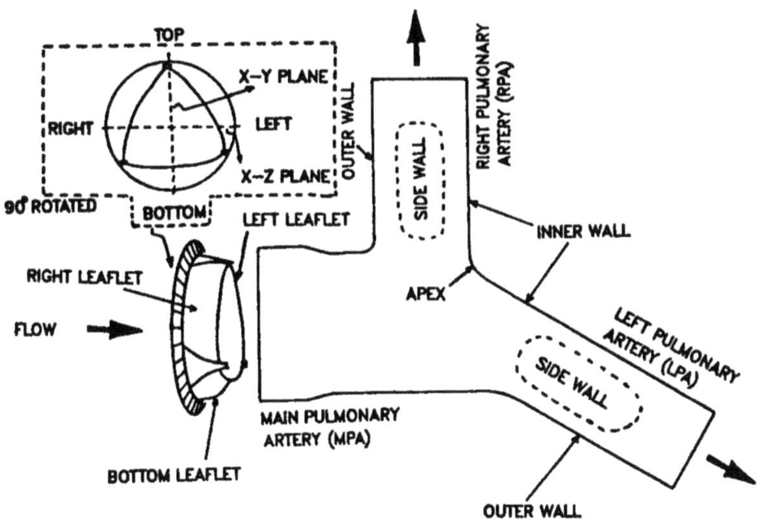

Figure 41. Nomenclature of regions in the test model.

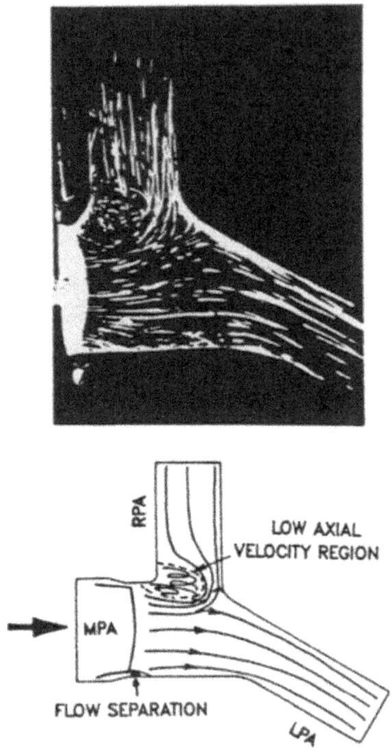

Figure 42. Flow visualization results for the normal valve case during the acceleration phase.

separation observed distal to the valve during the acceleration phase was still present. In the left pulmonary artery, the flow stream originating from the apex of the bifurcation started rotating. This was the origin of the secondary flow in the left pulmonary artery. The flow became evenly distributed after a short distance. In the right pulmonary artery, the streamlines near the main pulmonary artery deviated and appeared to flow around a region of low-axial-velocity. Distal to the low-axial-velocity region, the flow became evenly distributed due to its strong secondary flow.

During the deceleration phase (Figure 44), in the main pulmonary artery, the broad central flow seen at peak systole decreased in intensity. In the left pulmonary artery, the origin of the secondary flow was located slightly farther downstream than that observed at peak systole. In addition, the low-axial-velocity region in the right pulmonary artery appeared to be smaller than at peak systole.

To inspect the effects of various degrees of pulmonic valvular stenosis on the flow fields in the main, left, and right pulmonary arteries, Philpot[198] performed flow visualization and pressure measurements in the same model. The valves used in the study were a mildly stenotic valve, a moderately stenotic valve, and a severely

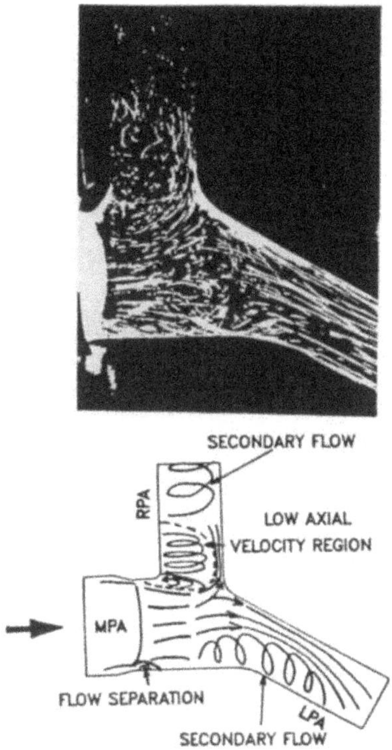

Figure 43. Flow visualization results for the normal valve case at peak systole.

stenotic valve. The orifice areas of the mildly, moderately, and severely stenotic valves were 1.0, 3.0, and 0.5 cm^2, respectively.

The results of the flow visualization showed a jet-like flow exiting the stenotic valve, which first passed the main pulmonary artery, then bypassed the right pulmonary artery, and finally hit the inner wall of the left pulmonary artery. As the valve became more stenotic, the jet-like flow in the main pulmonary artery became narrower and hit the inner wall of the left pulmonary artery farther downstream from the bifurcation apex (see Figure 45). In addition, it was noticed that the intensities of the secondary flows in both the left and right pulmonary arteries appeared to be stronger as the valve became more stenotic.

In the same study performed by Philpot, the pressure measurements revealed that the pressure drops in the main and right pulmonary arteries were greater than those recorded in the left pulmonary artery. Also, larger variations in pressure were monitored across the left pulmonary artery as compared to the right pulmonary artery.

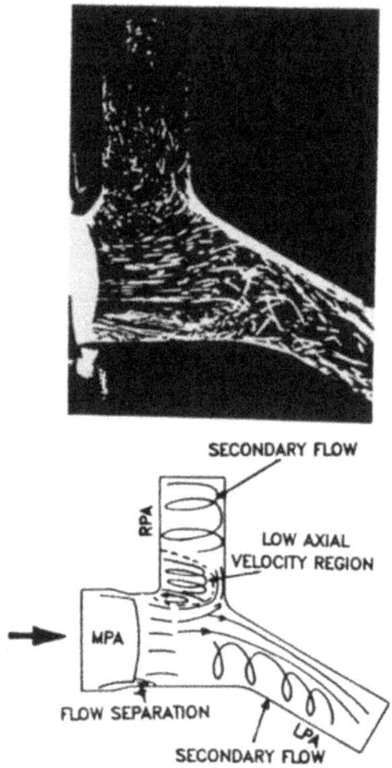

Figure 44. Flow visualization results for the normal valve case during the deceleration phase.

To assess the effects of varying radii of curvature of the right ventricular outflow tract and main pulmonary artery on the flow fields in the main, right, and left pulmonary arteries, Lynch[5] conducted flow visualization in a one-month lamb pulmonary artery model. Three models were studied; one with no curvature, one with the correct anatomic curvature, and one with an over accentuated curvature on the right ventricular outflow tract and main pulmonary artery. All other geometric parameters were held constant. Pulsatile flow visualization studies were conducted at nine flow conditions; heart rates of 70, 100, and 140 beats/min; and cardiac outputs of 1.2, 2.5; and 3.5 liters/min with corresponding mean pulmonary pressures of 10, 20, and 30 mmHg.

Changes were observed in the pulmonary artery hemodynamics as the curvature of the outflow tract, heart rate, and mean pulmonary pressure were varied. An increase in vessel curvature correlated with an increase in the disturbed nature of the overall flow field. The radial nature of the flow field became more intense and the secondary flow phenomena became more prominent as vessel curvature in-

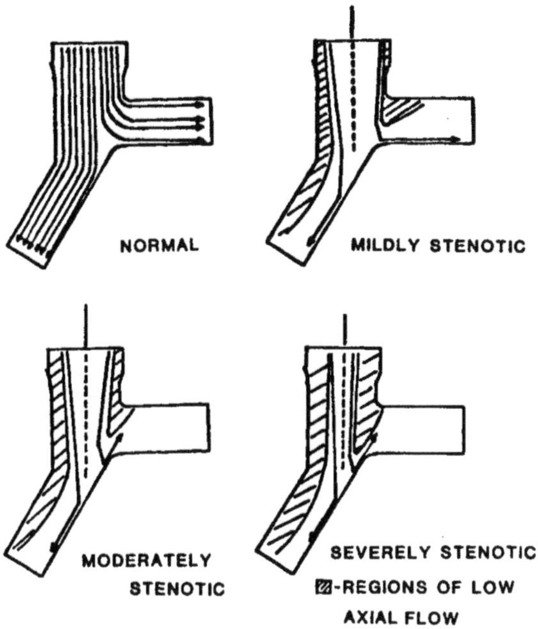

Figure 45. Schematic of the flow patterns observed at peak systole with varying degrees of stenosis.

creased (Figure 46). In addition to increasing the radial nature of the flow field, an increase in main pulmonary artery and right ventricular outflow tract curvature resulted in flow separation regions which formed faster and grew larger (Figures 47 and 48).

Heart rate had a marked effect on the pulmonary flow field. As the heart rate increased, the separation regions in all three vessels grew faster and appeared earlier in the cardiac cycle. In addition, as the heart rate increased, the size of the separation region decreased (Figure 49). Furthermore, it was noticed that the initiation of flow reversal was affected by the heart rate; flow reversal occurred later in the cardiac cycle at lower heart rates.

The pulmonary artery pressure affected the pulmonary artery flow field in two ways. First, the magnitude of reverse flow increased with an increase in pulmonary pressure. At the higher pressure, the magnitude of the reverse flow velocity was greater than at the lower pressure, as evidenced by the longer streaklines and greater concentration of particles at the higher pressure. Second, as pulmonary pressure increased, fewer coherent structures such as well-defined vortices were formed. The vortices observed at the lower pressure were more defined than those seen at the higher pressure.

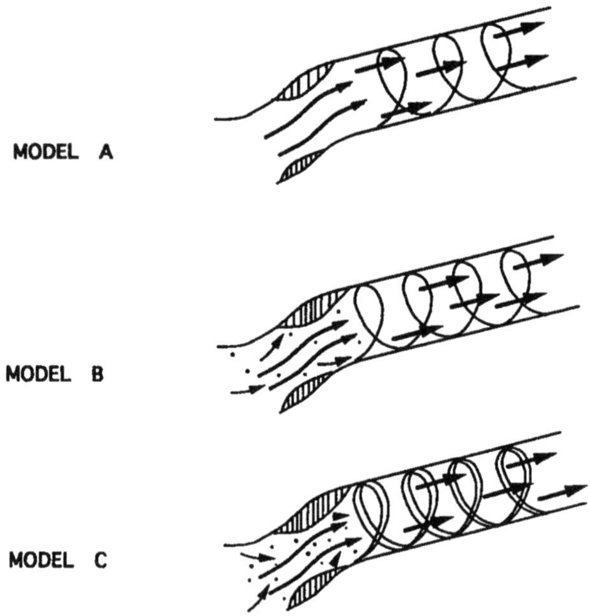

Figure 46. Schematics of the flow fields in the LPA of Models A, B, and C as viewed from the right-left plane, soon after peak systole. The flow conditions were 3.5 l/min, 70 bpm, and 30 mmHg. An increase in RVOT and MPA curvature correlates with an increase in the secondary nature of the flow field.

Low et al. conducted a flow study on the Fontan surgical connections under steady flow conditions.[204] The Fontan operations are performed as treatment for congenital heart defects such as tricuspid atresia and univentricular heart. The surgical procedure involves bypassing the right ventricle as reported by Fontan and Baudet,[52] and there are presently two main configurations; atriopulmonary and cavopulmonary connections. In the atriopulmonary connection, the right atrium is connected to the pulmonary artery either by a tube graft or by direct anastomosis. In the cavopulmonary connection, the superior vena cava is connected directly to the pulmonary artery and so too is the inferior vena cava which is being extended by an intra-atrial channel. In both connections, the absence of right-ventricle pressure results in very little pressure to drive the blood flow. In Low's study, the comparative pressure measurements on the two connections were determined.

It was seen that the pressure-drop coefficients were generally lower in the cavopulmonary connection. In this configuration, the pressure drop occurred mainly at the junction. In the atriopulmonary connection, there were additional

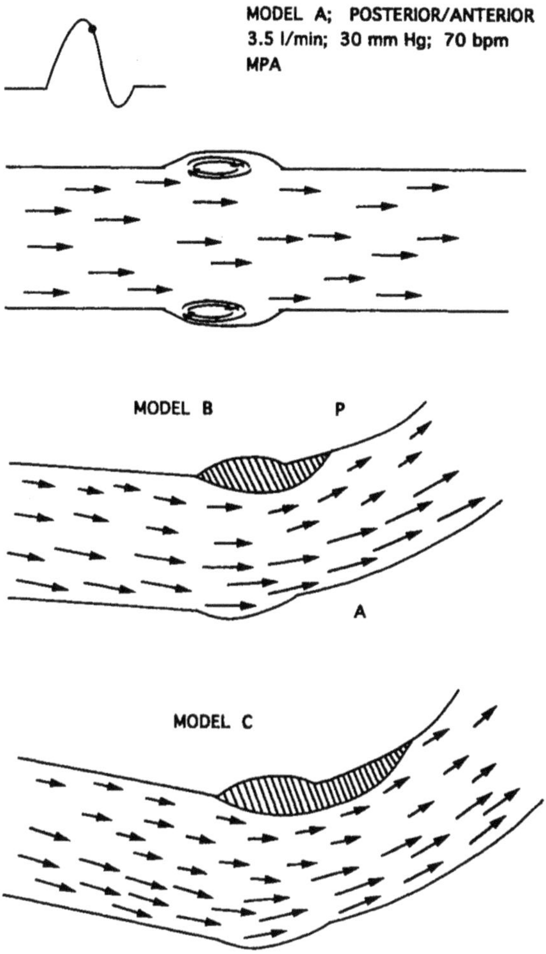

Figure 47. Schematics of the flow fields in the MPA of Models A, B, and C as viewed from the posterior-anterior plane, at the beginning of the deceleration phase. The flow conditions were 3.5 l/min, 70 bpm, and 30 mmHg. An increase in vessel curvature from Model A to Model C caused: (1) the flow to become skewed towards the anterior vessel wall; and (2) the flow separation region in the posterior sinus to increase in size and extend farther downstream.

pressure drops due to the sudden flow—area enlargement and contraction at the atrium. There was also higher junction losses due to the higher velocity in the connecting duct from the atrium to the artery. Also, the pressure loss coefficient of the atriopulmonary connection was 15–60% higher than the cavopulmonary. This might be related to the larger number of disturbed flow regions in the atriopulmon-

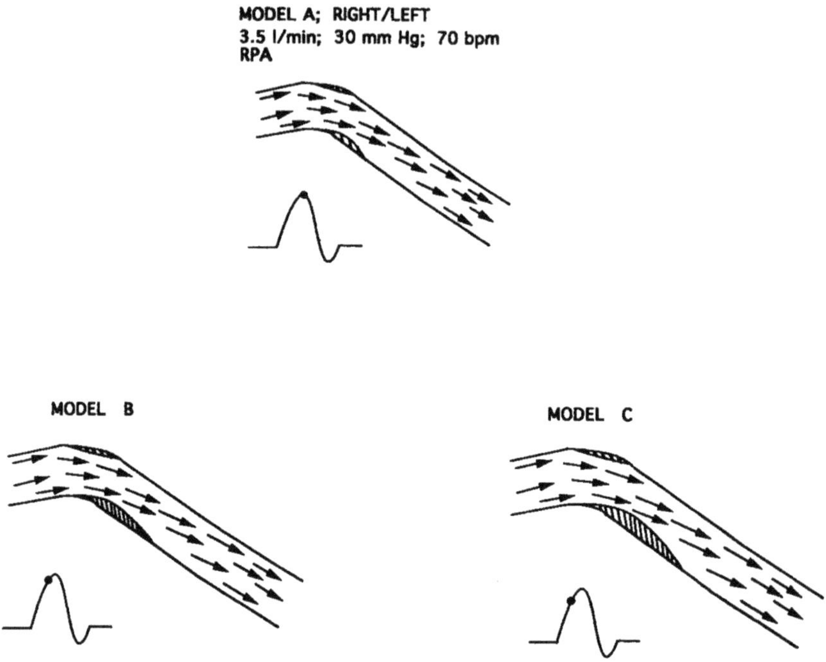

Figure 48. Schematics of the flow fields in the RPA of Models A, B, and C as viewed from the right to left plane showing the onset of flow separation. The flow conditions were 3.5 l/min, 70 bpm, and 30 mmHg. Flow separation appeared earlier in the cardiac cycle in Model C, the model with greatest curvature. As the model became more curved, the separation regions extended farther downstream.

ary model, which caused large turbulent and separation losses. The results of Low's study were consistent with a flow visualization study performed by de Leval.[205]

C. Velocity and Turbulence Mapping

Although flow visualization is useful in qualitatively describing flow fields, quantitative data such as velocity and turbulence intensity are also essential in gaining a thorough understanding of the pulmonary artery hemodynamics. For instance, while flow visualization illustrated regions of flow separation and/or secondary flow in the pulmonary artery, the exact extent and magnitude of these regions were still not known. In addition, the flow visualization technique used in the aforementioned studies had a disadvantage in that it did not show the nature of the secondary flow perpendicular to the plane of illumination. To obtain a three-dimensional view of the flow fields in the pulmonary artery with various degrees of valvular stenosis, Sung conducted a laser-Doppler velocity and turbulence mapping

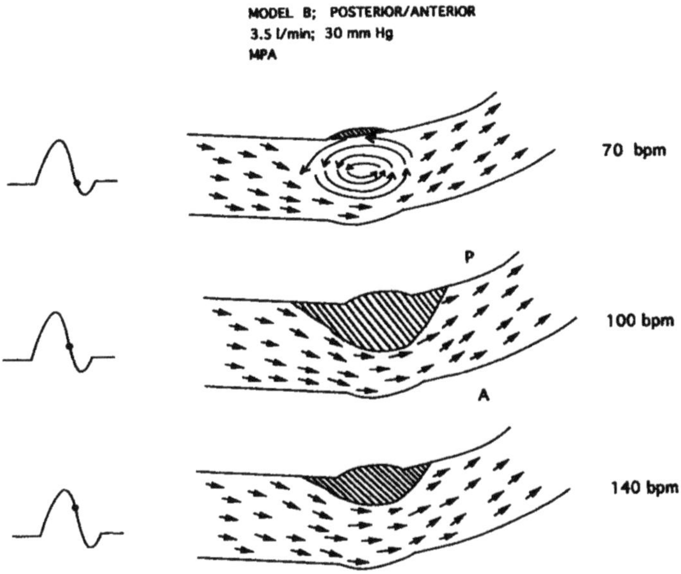

Figure 49. Schematics of the flow fields in the MPA of Model B at 70,100, and 140 bpm as viewed from the posterior-anterior plane at the point of maximum separation. Flow conditions were 3.5 l/min and 30 mmHg. As the heart rate increased, the size of the separation region in the posterior sinus decreased. Also, an increase in heart rate caused a decrease in the stability of the flow field and the formation of coherent structures.

in two orthogonal planes in the same model used for the flow visualization study.[201–203] The same "normal," mildly, moderately, and severely stenotic valves as used in the flow visualization were employed in the study. A schematic diagram of the pulmonary artery model together with the locations of the laser-Doppler velocity measurement is shown in Figure 50.

For the case of the "normal" valve, in the main pulmonary artery, a broad central flow was observed throughout the entire systole. The maximum axial velocity was measured at peak systole. In the left pulmonary artery, the axial velocities were approximately evenly distributed in the perpendicular plane. However, in the bifurcation plane, they were slightly skewed toward the inner wall at peak systole and during the deceleration phase. In the right pulmonary artery, the axial velocity in the perpendicular plane has a very marked M-shaped profile during the peak systole and deceleration phases. In the bifurcation plane, higher axial velocities were observed along the inner wall, while lower axial velocities were observed along the outer wall and in the center. Overall, relatively low levels of turbulence were observed in all the branches during systole. The maximum turbulence inten-

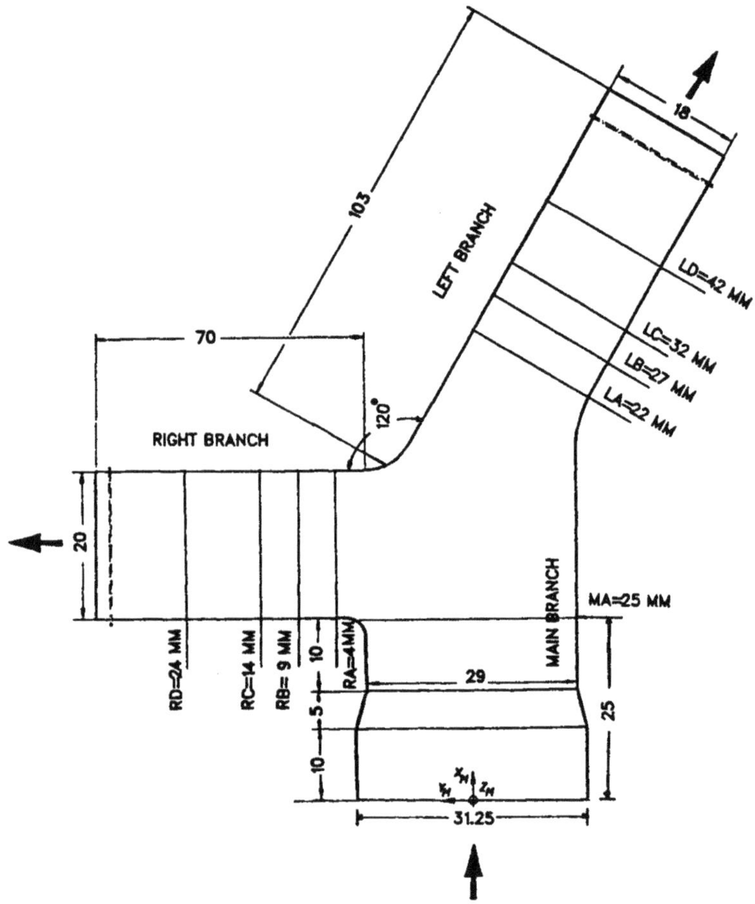

Figure 50. Schematic of adult pulmonary artery model, showing the location and planes of the velocity profile measurements.

sity measured was at the boundary of the broad central flow in the main pulmonary artery at peak systole.

Changes in flow fields in the pulmonary artery were observed as the degree of valvular stenosis increased. In the main pulmonary artery, a jet-like flow was observed distal to the stenotic valve. A higher degree of valvular stenosis corresponded to a higher velocity jet. In addition, as the degree of stenosis increased, the jet-like flow became narrower and produced larger regions of flow separation and/or stagnation surrounding the jet-like flow (Figure 50).

The turbulence intensity of the jet-like flow grew as the degree of valvular stenosis increased (Figure 51). Additionally, it was noticed that relatively high

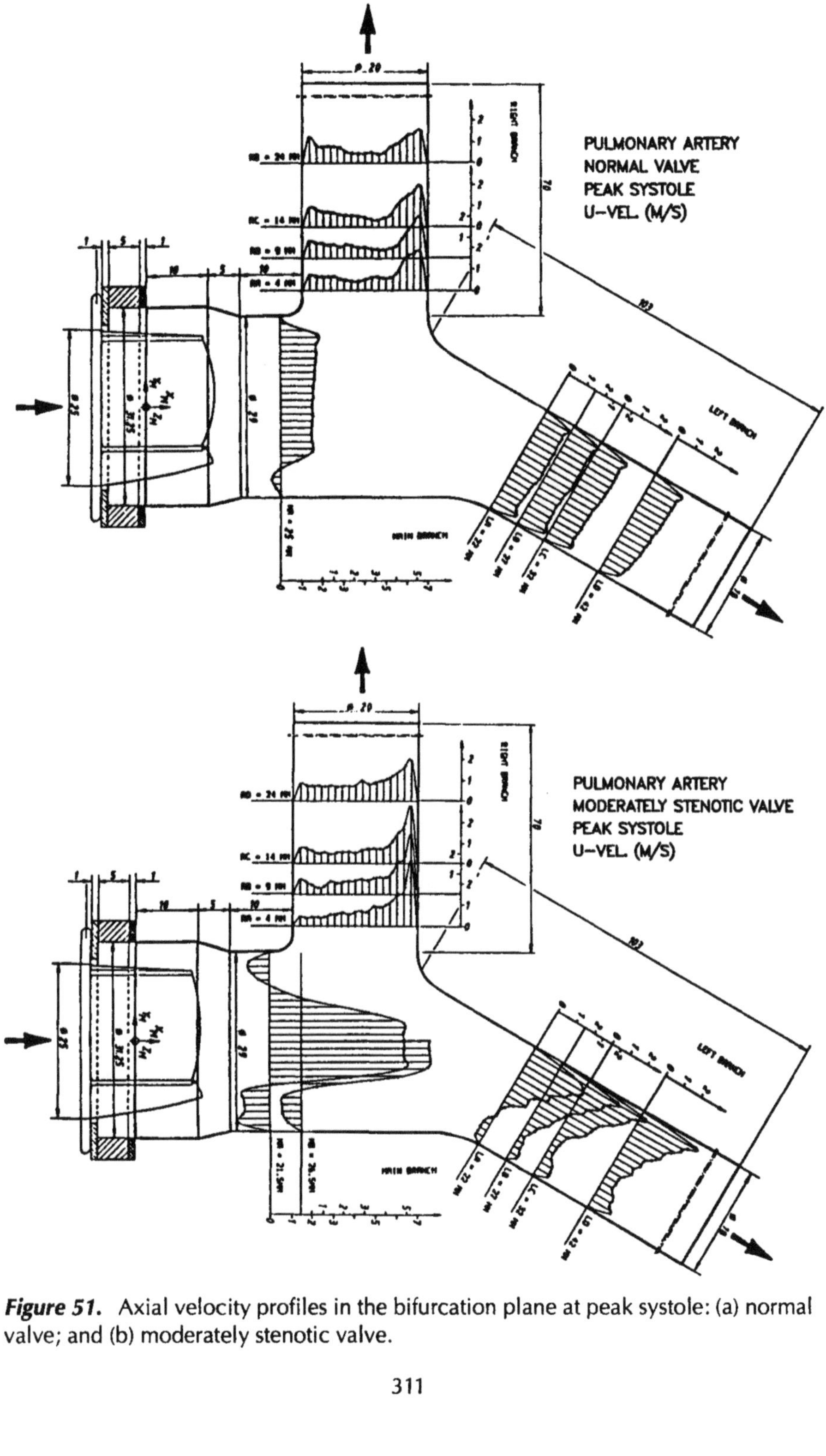

Figure 51. Axial velocity profiles in the bifurcation plane at peak systole: (a) normal valve; and (b) moderately stenotic valve.

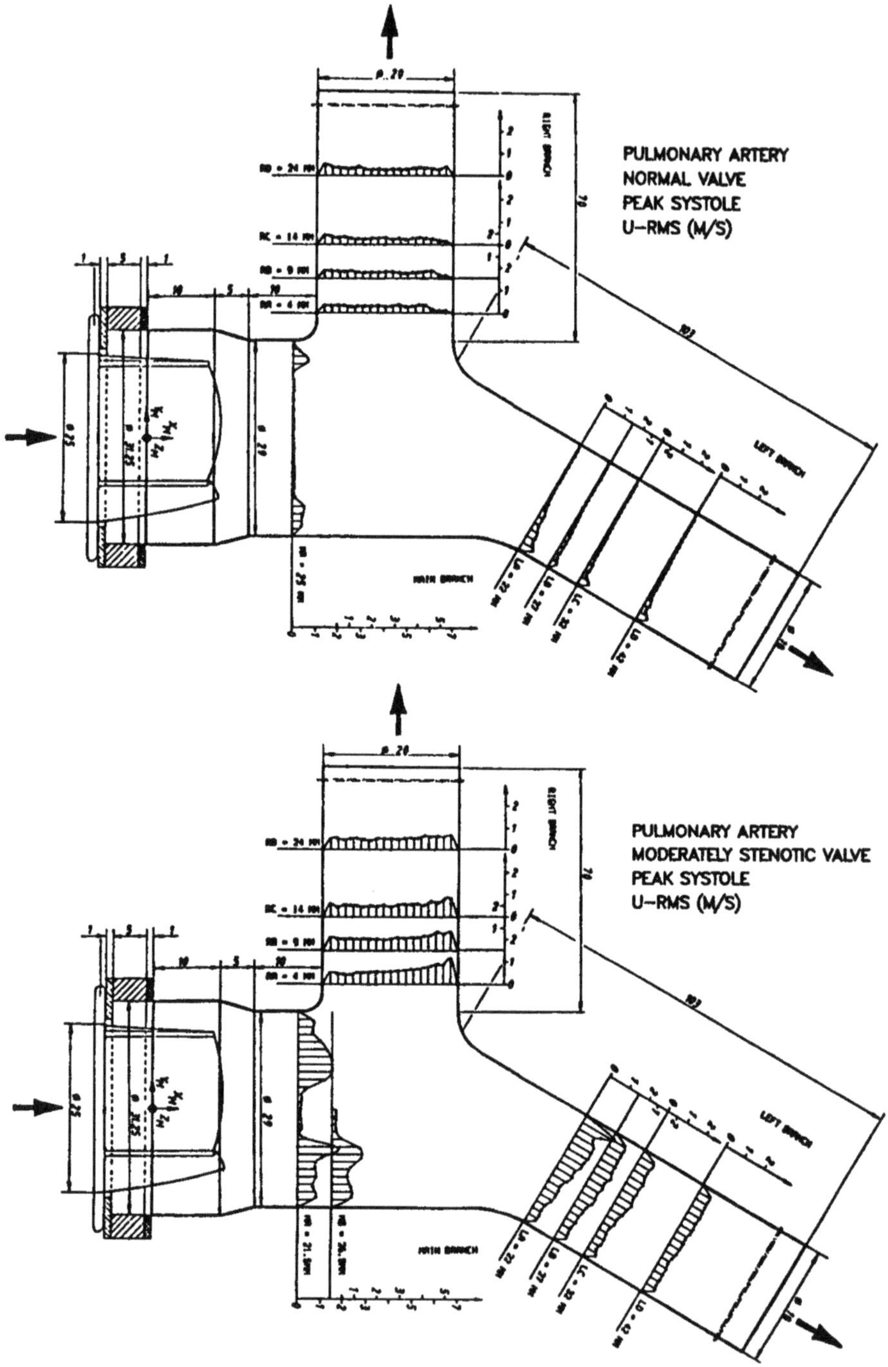

Figure 52. RMS axial velocity fluctuation profiles in the bifurcation plane at peak systole: (a) normal valve; and (b) moderately stenotic valve.

turbulence intensity was observed at the boundary of the jet-like flow for all the cases studied. Of note is that the turbulence intensity at location MB was slightly higher than that measured at location MA in all the cases studied. Also, the turbulence intensity obtained during the peak systole and deceleration phases was higher than during the acceleration phase. These two phenomena indicated that turbulence might develop gradually in the poststenotic flow field, since it took time for some unstable modes of motion in the flow field to grow into turbulence. Therefore, the maximum turbulence intensity did not occur immediately distal to the stenotic valve (i.e., location MA), and the flow field during the acceleration phase was more stable than those observed during the peak systole and deceleration phases. These results seemed to confirm the conclusion drawn by Nerem and Seed in an *in vivo* study of the aortic flow disturbances.[206]

Cineangiograms in several previous publications show that the right pulmonary artery branches at a right angle form the main pulmonary artery, whereas the left pulmonary artery, originating more distally than the right pulmonary artery, is a direct continuation of the main pulmonary artery.[207–209] This particular anatomy of the pulmonary artery played a significant role in the flow velocity patterns in both the left and right pulmonary arteries, especially as the degree of valvular stenosis was increased.

As shown in the flow visualization study, the jet-like flow existing from the stenotic valve first traveled through the main pulmonary artery and bypassed the origin of the right pulmonary artery. It then hit the inner wall of the left pulmonary artery and broke up into two jet streams; left-pulmonary-artery (LPA) jet stream and right-pulmonary-artery (RPA) jet stream. For the case of the mildly stenotic valve, the maximum axial velocity measured in the LPA jet stream was near the inner wall at location LA. The LPA jet stream then diverged as it traveled downstream from locations LA to LD. This implied that the LPA jet stream, originating from the main pulmonary artery, hit the inner wall of the left pulmonary artery near location LA. For the moderately stenotic valve, the LPA jet stream converged slightly between locations LA and LB, and then began to diverge between locations LB and LD. The maximum axial velocity measured was at the inner wall near location LB. This suggested that the LPA jet stream hit the inner wall of the left pulmonary artery near location LB. For the case of the severely stenotic valve, the LPA jet stream first converged slightly as it traveled from locations LA to LC, with the highest measured axial velocity at location LC, and then diverged from locations LC to LD. This indicated that the LPA jet stream hit the inner wall of the left pulmonary artery near location LC. From these results, it can be inferred that with increasing degree of valvular stenosis, the LPA jet stream hit farther downstream in the left pulmonary artery. This phenomenon which resulted in more blood flow being sent into the left lung than into the right lung was also observed in an *in vivo* study conducted by Chen et al.[207]

As described above, the distributions of the flow in the main and left pulmonary arteries were strongly dependent on the degree of valvular stenosis. However, the flow fields in the right pulmonary artery did not change significantly with varying degrees of valvular stenosis. This suggested that the flow distribution across the right pulmonary artery was mainly determined by the geometry of the bifurcation. This was because the jet-like flow observed in the main pulmonary artery bypassed the origin of the right pulmonary artery and broke into two jet streams downstream of the bifurcation apex in the left pulmonary artery. Therefore, a larger portion of the momentum possessed by the jet-like flow in the main pulmonary artery was transferred to the LPA jet stream rather than to the RPA jet stream. Since the momentum of the RPA jet stream did not increase much with increasing degree of valvular stenosis, the maximum axial velocity measured in the RPA jet stream was near location RA for all the cases studied. The turbulent intensity in the right pulmonary artery, however, increased slowly with the degree of valvular stenosis.

D. Secondary Flow

The secondary flow patterns in the left and right pulmonary arteries changed with time during the cardiac cycle for all the cases studied. For the case of the "normal" valve, as observed in the flow visualization study, no secondary flow was present in the left pulmonary artery during the acceleration phase. At peak systole and during the deceleration phase a pair of counter-rotating secondary flows were observed in the left pulmonary artery. The two counter-rotating secondary flows were relatively symmetric about the bifurcation plane. Similar patterns were noticed in the right pulmonary artery: no secondary flow was observed during the acceleration phase, while a pair of counter-rotating secondary flows were present during the peak systole and deceleration phases. However, as observed in the flow visualization study, the strength of the secondary flows in the right pulmonary artery was stronger than that observed in the left pulmonary artery.

The secondary flow patterns in the left and right pulmonary arteries started to change as the valve became stenotic. For the case of the mildly stenotic valve, as seen in the "normal" valve case, a pair of counter-rotating secondary flows were observed at peak systole and during the deceleration phase. However, the two counter-rotating secondary flows were no longer symmetric about the bifurcation plane. It was observed that the secondary flow along the top side wall was stronger than the one along the bottom side wall in both the left and right pulmonary arteries. Furthermore, the strength of the secondary flows in both branches increased as compared to that observed with the normal valve. However, the percentage increment in the left pulmonary artery was greater than in the right pulmonary artery, even though the strength of the secondary flows in the right pulmonary artery was still stronger than in the left pulmonary artery.

The secondary flow patterns with the moderately and severely stenotic valves were similar to those observed with the mildly stenotic valve. The strength of the secondary flows, however, increased as the degree of valvular stenosis increased. Again, similar to the findings with the mildly stenotic valve, the increment of the strength of the secondary flows in the left pulmonary artery was greater than in the right pulmonary artery with both the moderately and severely stenotic valves.

REFERENCES

1. Nichols, W.W.; O'Rourke, M.F. *McDonald's Blood Flow in Arteries*. Lea & Febiger, Philadelphia, 1990.
2. Pedley, T.J. *The Fluid Mechanics of Large Blood Vessels*. Cambridge University Press, Cambridge, 1980.
3. Chandran, K.B.; Yearwood, T.L.; Wieting, D.W. An experimental study of pulsatile flow in a curved tube. *J. Biomech.* **1979**, *12*, 793–805.
4. Yearwood, T.L.; Chandran, K.B. Physiological pulsatile flow experiments in a model of the human aortic arch. *J. Biomech.* **1984**, *15*, 683–704.
5. Lynch, P.G.; Saylor, A.; Ha, B.; Lucas, C.; Henry, G.W.; Ferreiro, J.I.; Yoganathan, A.P. The effects of curvature on fluid flow fields in pulmonary artery models: Flow visualization studies. *J. Biomech. Eng.* **1993**, *115*, 97–103.
6. Wagenvoort, C.A.; Wagenvoort, N. Pulmonary vascular bed: Normal anatomy and response to diseases. In: *Pulmonary Vascular Diseases* (Moser, K.M., Ed.). Marcel Dekker, New York, 1979, pp. 1–109.
7. Lucas, C.L. Fluid mechanics of the pulmonary circulation. *CRC Crit. Rev. Biomed. Eng.* **1984**, *10*, 317–392.
8. Davies, G.; Reid, L. Growth of the alveoli and pulmonary arteries in childhood. *Thorax* **1970**, *25*, 669.
9. Kay, J.M. The pulmonary vasculature and experimental pulmonary hypertension in animals. In: *The Pulmonary Circulation in Health and Disease* (Will, J.M.; Dawdon, C.A.; Weir, E.K.; Buckner, C.K., Eds.). Academic Press, Orlando, 1987.
10. Langston, C.; Holder, P. Pulmonary vascular changes in infants and children. In: *Pulmonary Circulation in Health and Disease* (Will, J.A.; Dawson, C.A.; Weir, E.K.; Buckner, C.K., Eds.). Academic Press, Orlando, 1987, pp. 57–74.
11. Heath, D.; Wood, E.H.; DuShane, J.W.; Edwards, J.E. The structure of the pulmonary trunk at different ages and in cases of pulmonary hypertension and pulmonary stenosis. *J. Pathol. Bact.* **1959**, *77*, 443–456.
12. Ferencz, C.; Libi-Sylora, M.; Greco, J. Age-related characteristics of human pulmonary arterial tree circulation. *Circulation* **1967**, *34(II)*, 107.
13. Wagenvoort, C.A.; Neufield, H.N.; Edwards, J.E. The structure of the pulmonary arterial tree in fetal and early postnatal life. *Lab. Invest.* **1961**, *10*, 751–762.
14. Kuriyama, K.; Gamsu, G.; Stern, R.G.; Cann, C.E.; Herfkers, R.T.; Brundage, B.H. CT-determined pulmonary artery diameters in predicting pulmonary hypertension. *Invest. Radiol.* **1984**, *19*, 16–22.
15. Wu, X.; Latson, L.A.; Driscoll, D.J.; Ensing, G.J.; Ritman, E.L. Dynamic three-dimensional anatomy of pulmonary arteries in pigs with aorto-pulmonary artery shunts. *Am. J. Physiol. Imaging* **1987**, *2*, 169–175.
16. Bogren, H.G.; Klipstein, R.H.; Mohaiddin, R.H.; et al. Pulmonary artery distensibility and blood flow patterns: A magnetic resonance study with pulmonary arterial hypertension. *Am. Heart J.* **1989**, *118*, 990–999.
17. Gefter, W.B.; Hatabu, H.; Dinsmore, B.J.; Axel, L.; Palevsky, H.I.; Reichek, N.; Schiebler, M.L.; Kressel, H.Y. Pulmonary vascular cine MR imaging: A noninvasive approach to dynamic imaging of the pulmonary circulation. *Radiology* **1990**, *176*, 761–770.
18. Rudolph, A.M. *Congenital Diseases of the Heart*. Yearbook Medical Publishers, Chicago, 1974, pp. 29–48.

19. Attinger, E.O. Pressure transmission in pulmonary arteries related to frequency and geometry. *Circ. Res.* **1963**, *12*, 623.
20. O'Rourke, M.F.; Milnor, W.R. Relation between differential pressure and flow in the pulmonary artery of the dog. *Cardiovasc. Res.* **1971**, *5*, 558–565.
22. Mitzner, W.; Chang, H.K. Hemodynamics of the pulmonary circulation. In: *Respiratory Physiology* (Chang, H.K.; Paiva, M., Eds.). Marcel Dekker, New York, 1989, pp. 561–631.
23. Milnor, W.R. *Hemodynamics*. Williams & Wilkins, Baltimore, 1989.
24. Caro, C.G.; Harrison, G.K. Observations on pulse wave velocity and pulsatile blood pressure in the human pulmonary circulation. *Clin. Sci.* **1962**, *23*, 317–329.
25. Milnor, W.R.; Conti, C.R.; Lewis, K.B.; O'Rourke, M.F. Pulmonary arterial pulse wave velocity and impedance in man. *Circ. Res.* **1969**, *25*, 637–649.
26. Reuben, S.R.; Swadling, J.P.; Lee, G.J. Velocity profiles in the main pulmonary artery of dogs and man, measured with a thin-film resistance anemometer. *Circ. Res.* **1970**, *27*, 995–1001.
27. Peltonen, T.; Hirvonen, L. Experimental studies on fetal and neonatal circulation. *Acta Paed. Scan.* **1965**, *54*(Suppl. 161).
28. Born, G.V.R.; Dawes, G.S.; Mott, J.C.; Widicombe, J.G. Changes in the heart and lung at birth. *Cold Spring Harbour Symp. Quant. Biol.* **1954**, *19*, 102.
29. Dawes, G.S.; Mott, J.C.; Widdicombe, J.G. The fetal circulation in the lamb. *J. Physiol.* **1954**, *126*, 563.
30. Assali, N.S.; Morris, J.A. Maternal and fetal circulations and their relationships. *Obs. Gyn. Surv.* **1964**, *19*, 923.
31. Rudolph, A.M.; Heymann, M.A. Circulatory changes during growth in fetal lamb. *Circ. Res.* **1970**, *26*, 289.
32. Mahon, W.A.; Goodwin, J.W.; Paul, W.M. Measurement of individual ventricular outputs in the fetal lamb by an indicator dilution technique. *Circ. Res.* **1966**, *19*, 191.
33. Zweizig, H.Z.; Kuhl, D.E.; Katz, R.; Polgar. Distribution of pulmonary blood flow in fetal and newborn lambs. *Resp. Physiol.* **1970**, *8*, 160.
34. Lind, J.; Weglius, C. Human fetal circulation: Changes in the cardiovascular system at birth and disturbances in the post-natal closure of the foramen ovale and ductus arteriosus. *Cold Spring Harbour Symp. Quant. Biolog.* **1954**, *19*, 109.
35. Lind, J.; Stern, L.; Weglius, C., Eds. *Human Fetal and Neonatal Circulation*. Charles C. Thomas, Springfield, 1964.
36. Dawes, G.S. *Fetal and Neonatal Physiology*. Yearbook Medical Publishers, Chicago, 1968.
37. Scarpelli, E.M. *Pulmonary Physiology of the Fetus, Newborn, and Child*. Lea & Febiger, Philadelphia, 1975, pp. 116–139.
38. Dexter, L. Pulmonary vascular disease in acquired heart disease. In: *Pulmonary Vascular Diseases* (Moser, K.M., Ed.). Marcel Dekker, New York, 1979, pp. 427–488.
39. Haworth, S.G. Pulmonary vasculature. In: *Pediatric Cardiology* (Anderson, R.H.; Macartney, F.J.; Shinebourne, E.A.; Tynan, M., Eds.). Churchill Livingstone, London, 1987, pp. 123–157.
40. Hoffman, J.I.E. Incidence, mortality, natural history. In: *Pediatric Cardiology* (Anderson, R.H.; Macartney, F.J.; Shinebourne, E.A.; Tynan, M., Eds.). Churchill Livingstone, London, 1987, pp. 3–14.
41. Rose, V.; Gold, R.J.M.; Lindsay, G.; et al. A possible increase in the incidence of congenital heart defects among the offspring of affected parents. *J. Am. Coll. Cardiol.* **1985**, *6*, 376.
42. Allan, L.D.; Crawford, D.C.; Chita, S.K.; et al. Familial recurrence of congenital heart disease in a prospective group of mothers refereed for fetal echocardiography. *Am. J. Cardiol.* **1986**, *58*, 334.
43. Zierler, S. Maternal drugs and congital heart disease. *Obstet. Gynecol.* **1985**, *65*, 155.
44. Edwards, J.E. Congenital pulmonary vascular disorders. In: *Pulmonary Vascular Diseases* (Moser, K.M., Ed.). Marcel Dekker, New York, 1979, pp. 527–572.
45. Greene, D.G.; Baldwin, E. deF.; Baldwin, J.S.; Himmelstein, A.; et al. Pure congenital pulmonary stenosis and idiopathic congenital dilatation of the pulmonary artery. *Am. J. Med.* **1949**, *6*, 24–40.

46. Edwards, J.E.; McGoon, D.C. Basence of anatomic origin from heart of pulmonary arterial supply. *Circulation* **1973**, *47*, 393–398. Perloff, J.K. *The Clinical Recognition of Congenital Heart Disease*. WB Saunders Co., Philadelphia, 1987, pp. 4–5.
47. Jue, K.L.; Lockman, L.A.; Edwards, J.E. Anomalous origins of pulmonary arteries from pulmonary trunk ("crossed pulmonary arteries"). Observation in a case with 18 trisomy syndrome. *Am. Heart J.* **1966**, *71*, 807–812.
48. Sotomora, R.F.; Edwards, J.E. Anatomic identification of the so-called absent pulmonary artery. *Circulation* **1978**, *57*, 624–633.
49. Blalock, A.; Taussig, H.B. Surgical treatment of malformations of the heart in which there is pulmonary stenosis or pulmonary atresia. *JAMA* **1945**, 128–189.
50. Rastelli, G.C.; Dugley, P.A.; Davis, G.D.; Kirklin, J.W. Surgical repair for pulmonary valve atresia with coronary-pulmonary artery fistula: Report of a case. *Mayo Clin. Proc.* **1965**, *40*, 521–527.
51. Ross, D.N.; Somerville, J. Correction of pulmonary atresia with a homograft aortic valve. *Lancet* **1966**, *2*, 1446–1447.
52. Fontan, F.; Baudet, E. Surgical repair of tricuspid atresia. *Thorax* **1971**, *26*, 240–248.
53. Agarwal, K.C.; Edwards, W.D.; Feldt, R.H.; Danielson, G.K.; Puga, F.J.; McGoon, D.C. Clinicopathological correlates of obstructed right-sided porcine-valved extracardiac conduits. *J. Thorac. Cardiovasc. Surg.* **1981**, *81*, 591–601.
54. Agarwal, K.C.; Edwards, W.D.; Feldt, R.H.; Danielson, G.K.; Puga, F.J.; McGoon, D.C. Pathogenesis of nonobstructive fibrous peels in right-sided porcine-valved extracardiac conduits. *J. Thorac. Cardiovasc. Surg.* **1982**, *83*, 584–589.
55. Kosturakis, D.; Goldberg, S.J.; Allen, H.D.; Loeber, C. Doppler echocardiographic prediction of pulmonary arterial hypertension in congenital heart disease. *Am. J. Cardiol.* **1984**, *53*, 1110–1115.
56. Cloez, J.L.; Schmidt, K.G.; Birk, E.; et al. Determination of pulmonary to systemic blood flow ratio in children by a simplified Doppler echocardiographic method. *J. Am. Coll. Cardiol.* **1988**, *11*, 825.
57. Morimota, K.; Matsuzaki, M.; Tohma, Y.; et al. Diagnosis and quantitative evaluation of secundum-type atrial septal defect by transesophageal Doppler echocardiography. *Am. J. Cardiol.* **1990**, *66*, 85.
58. Ritter, S.B. Transesophageal real-time echocardiography infants and children with congenital heart disease. *J. Am. Coll. Cardiol.* **1990**, *18*, 569.
59. Kersting-Sommerhoff, B.A.; Diethelm, L.; Stanger, P.; et al. Evaluation of complex congenital ventricular anomalies with magnetic resonance imaging. *Am. Heart J.* **1990**, *120*, 133.
60. Hoffman, J.I.E.; Heymann, M.A. Pulmonary arterial hypertension secondary to congenital heart disease. In: *Pulmonary Hypertension* (Weir, E.K.; Reeves, J.T., Eds.). Futura Publishing Co., Mount Kisco, 1984, pp. 73–114.
61. West, J.B. *Pulmonary Pathophysiology*. Williams & Wilkins, Baltimore, 1987, p. 131.
62. Hopkins, R.A. The pathophysiology of pulmonary vascular disease associated with congenital cardiac lesions. In: *Ventricular Vascular Coupling* (Yin, C.P., Ed.). Springer-Verlag, New York, 1987, pp. 42–78.
63. Rabinovitch, M. Problems of pulmonary hypertension in children with congenital cardiac defects. *Chest* **1988**, *93*, 119S–125S.
64. Haworth, S.G. Pathophysiological and metabolic manifestations of pulmonary vascular disease in children. *Herz.* **1992**, *17*, 254–261.
65. Heath, D.; Edwards, J.E. The pathology of hypertensive pulmonary vascular disease: A description of six grades of structural changes in the pulmonary arteries with specific reference to congenital cardiac septal defect. *Circulation* **1958**, *18*, 533–547.
66. Hislop, A.; Haworth, S.G.; Shinebourne, E.A.; Reid, L. Quantitative structural analysis of pulmonary vessels in isolated ventricular septal defect. *Br. Heart J.* **1975**, *37*, 1014–1021.
67. Reid, L.M. The pulmonary circulation: Remodeling in growth and disease. *Am. Rev. Resp. Dis.* **1979**, *119*, 531.

68. Haworth, S.G. Normal structural and functional adaptation to extrauterine life. *J. Pediatr.* **1981**, *98*, 911–918.
69. Rabinovitch, M.; Herrera-Deloen, V.; Casteneda, A.R.; Reid, L. Growth and development of the pulmonary vascular bed in patients with tetralogy of Fallot with or without pulmonary atresia. *Circulation* **1981**, *64*, 1234–1249.
70. Rabinovitch, M.; Reid, L.M. Quantitative structural analysis of the pulmonary vascular bed in congenital heart defects. In: *Pediatric Cardiovascular Disease* (Engle, M.A., Ed.). FA Davis Co., Philadelphia, 1981, pp. 149–169.
71. Haworth, S.; Hislop, A. Pulmonary vascular development: Normal values of the peripheral vasculature. *Am. J. Cardiol.* **1983**, *52*, 579–583.
72. Rabinovitch, M.; Keane, J.F.; Norwood, W.I.; Casteneda, A.R.; Reid, L. Vascular structure in lung tissue obtained at biopsy correlated with pulmonary hemodynamic findings after repair of congenital heart defects. *Circulation* **1984** *69*, 655–667.
73. Neches, W.H.; Park, S.C.; Mathews, R.A.; et al. Pulmonary artery impedance in evaluation of pulmonary vascular disease. *Circulation* **1980**, *62(III)*, 330.
74. Lucas, C.L.; Wilcox, B.R.; Coulter, N.A. Contrasting pulmonary vascular response to closure of atrial and ventricular septal defects in children. *Circulation* **1973**, *48*, 121.
75. Lucas, C.L.; Wilcox, B.R.; Coulter Jr., N.A.; Pulmonary vascular response to atrial septal defect closure in children. *J. Surg. Res.* **1975**, *18*, 571.
76. Lucas, C.L.; Wilcox, B.R.; Coulter, N.A. Contrasting pulmonary blood flow profiles in children with atrial and ventricular septal defects. *Cardiovasc. Res.* **1976**, *10*, 1–12.
77. Wilcox, B.R.; Lucas, C.L. Pulmonary input impedance in children with left-right shunt, *J. Surg. Res.* **1980**, *29*, 40.
78. Lucas, C.L.; Radke, N.F.; Wilcox, B.R.; et al. Maturation of pulmonary input impedance spectrum in infants and children with ventricular septal defect. *Am. J. Cardiol.* **1986**, *57*, 821–827.
79. Long, W.A.; Henry, G.W. Autonomic and central neuroregulation of fetal cardiovascular function. In: *Fetal and Neonatal Physiology* (Polin, R.A.; Fox, W.W., Eds.). WB Saunders, Philadelphia, 1991, pp. 629–645.
80. Denfield, S.; Henry, G.W. Postoperative cardiac intensive care. In: *Neonatal and Fetal Cardiology* (Long, W.A., Ed.). WB Saunders, Philadelphia, 1990, pp. 812–829.
81. Henry, G.W.; Long, W.A.; Lucas, C.L. New directions in the investigation of pulmonary hypertension in children. In: *Pulmonary Heart Disease in the Young* (Harned Jr., H.S., Ed.). Little, Brown, Boston, 1990, pp. 379–398.
82. Rudolph, A.M. *Congenital Diseases of the Heart.* Year Book Medical Publishers, Chicago, 1974, pp. 47–48.
83. Henry, G.W. Noninvasive assessment of cardiac function and pulmonary hypertension in persistent pulmonary hypertension of the newborn. *Clin. Perinatol.* **1984**, *3*, 627–640.
84. Tynan, M. Pediatric cardiology—then and now. *Cardiol. Young* **1981**, *1*, 3–10.
85. Chrispin, A.; Baker, E.J. Imaging the heart and chest. In: *Pediatric Cardiology* (Anderson, R.H.; Macartney, F.J.; Shinbourne, E.A.; Tynan, M. Eds.). Churchill Livingstone, London, 1987, pp. 201–233.
86. Delany, D.J. Imaging of the right heart. In: *Pulmonary Heart Disease in the Young* (Harned Jr., H.S., Ed.). Little, Brown, Boston, 1990.
87. Snider, A.B.; Serwer, G.A. *Echocardiography in Pediatric Heart Disease.* Year Book Medical Publishers, Chicago, 1990, pp. 78–133.
88. Silverman, N.H. *Pediatric Echocardiography.* Williams & Wilkins, 1993, pp. 37–38.
89. Valdes-Cruz, L.M.; Dudell, G.G.; Ferrara, A. Utility of M-mode echocardiography for early identification of infants with persistent pulmonary hypertension of the newborn. *Pediatrics* **1981**, *68*, 515–525.
90. Johnson, G.L.; Cunningham, M.D.; Desai, N.S.; Cottrill, C.M.; Noonan. J.A.. Echocardiography in hypoxemic neonatal pulmonary disease. *J. Pediatr.* **1980**, *96*, 716–720.

91. Riggs, T.; Hirschfield, S.; Fanaroff, A.; Liebman, J.; Fletcher, B.; Meyer, R. Persistence of fetal circulation syndrome: An echocardiographic study. *J. Pediatr.* **1977**, *91*, 626–631.
92. Hatle, L.; Angelsen, B.A.J.; Tromsdal, A. Noninvasive estimation of pulmonary artery systolic pressure with Doppler ultrasound. *Br. Heart J.* **1981**, *45*, 157–165.
93. Stevenson, J.G.; Kawabori, I.; Guntheroth, W.G. Noninvasive estimation of peak pulmonary artery pressure by M-mode echocardiography. *J. Am. Coll. Cardiol.* **1984**, *4*, 1021–1027.
94. Anderson, R.H.; Macartney, F.J.; Shinbourne, E.A.; Tynan, M., Eds. *Pediatric Cardiology*. Churchill Livingstone, London, 1987, pp. 65–82.
95. Watanabe, T.; Katsume, H.; Matsukubo, H.; Furukawa, K.; Ijichi H. Estimation of right ventricular volume with two-dimensional echocardiography. *Am. J. Cardiol.* **1982**, *49*, 1946–1953.
96. Silverman, N.H.; Hudson, S. Evaluation of right ventricular volume and ejection fraction in children by two-dimensional echocardiography. *Pediatr. Cardiol.* **1983**, *4*, 197–203.
97. Starling, M.R.; Crawford, M.H.; Sorensen, S.G.; O'Rourke, R.A. A new two-dimensional echocardiographic technique for evaluation right ventricular size and performance in patients with obstructive lung disease. *Circulation* **1982**, *66*, 612–620.
98. Levine, R.A.; Gibson, T.C.; Aretz, T.; Gillam, L.D.; Guyer, D.E.; King, M.E.; Weyman. A.E. Echocardiographic measurements of right ventricular volume. *Circulation* **1984**, *69*, 497–505.
99. Berger, M.; Haimowitz, A.; Van Tosh, A.; Berdoff, R.L.; Goldberg, E. Quantitative assessment of pulmonary hypertension in patients with tricuspid regurgitation using continuous wave Doppler ultrasound. *J. Am. Coll. Cardiol.* **1985**, *6*, 359–365.
100. Kitabatake, A.; Inoue, M.; Asao, M.; Masuyama, T.; Tanouchi, J.; Morita, T.; Mishima, M.; Uematsu, M.; Shimazu. T.; Hori, M.; Abe, H. Noninvasive evaluation of pulmonary hypertension by a pulsed Doppler technique. *Circulation* **1983**, *68*, 302–309.
101. Martin-Duran, R.; Larman, M.; Trugeda, A.; Vazquez De Prada, J.A.; Ruano, J.; Torres, A.; Figueroa, A.; Pajaron, A.; Nistal, F. Comparison of Doppler-determined elevated pulmonary arterial pressure with pressure measured at cardiac catheterization. *Am. J. Cardiol.* **1986**, *57*, 859–863.
102. Marchandise, B.; De Bruyne, B.; Delaunois, L.; Kremer, R. Noninvasive prediction of pulmonary hypertension in chronic obstructive pulmonary disease by Doppler echocardiography. *Chest* **1987**, *91*, 361–365.
103. Isobe, M.; Yazaki, Y.; Takaku, F.; Koizumi, K.; Hara, K.; Tsuneyoshi, H.; Yamaguchi, T.; Machii, K. Prediction of pulmonary arterial pressure in adults by pulsed Doppler echocardiography. *Am. J. Cardiol.* **1986**, *57*, 316–321.
104. Matsuda, M.; Sekiguchi, T.; Sugishita, Y.; Kuwako, K.; Iida, K.; Ito, I. Reliability of non-invasive estimates of pulmonary hypertension by pulsed Doppler echocardiography. *Br. Heart J.* **1986**, *56*, 158–164.
105. Satomi, G.; Nakazawa, M.; Momma, K.; Kikuchi, N.; Sasaki, S. A new method for calculating pulmonary to systemic flow ratio in patients with atrial septal defect using multigated Doppler echocardiography. *Med. Biol. Eng. Comp.* **1991**, *29(Suppl. 1)*, 248.
106. Toombs, B.D. Computed tomography. In: *The Science and Practice of Pediatric Cardiology* (Garson Jr., A.; Bricker, J.T.; McNamara, D.G., Eds.). Lea & Febiger, Philadelphia, 1990, pp. 880–904.
107. Eldridge, W.J.; Flicker, S. Evaluation of congenital heart disease using cine-CT. Am *J. Card. Imag.* **1987**, *1*, 38–50.
108. Hurwitz, R.A.; Kurec, A.; Treves, S. Right ventricular and left ventricular ejection fractions in pediatric patients with normal hearts. First pass radionuclide angiocardiography. *Am. Heart J.* **1984**, *107*, 726–732.
109. Vick, G.W., III; Rokey, R.; Johnston, D.L. Nuclear magnetic resonance and positron emission tomography—clinical aspects. In: *The Science and Practice of Pediatric Cardiology* (Garson Jr., A.; Bricker, J.T.; McNamara, D.G., Eds.). Lea & Febiger, Philadelphia, 1990, pp. 852–879.

110. Pinsky, W.W.; Jones, D.W. Digital substraction angiography. In: *The Science and Practice of Pediatric Cardiology* (Garson Jr., A.; Bricker, J.T.; McNamara, D.G., Eds.). Lea & Febiger, Philadelphia, 1990, 907–912.
111. Rudolph, A.M. *Congenital Diseases of the Heart*. Year Book Medical Publishers, Chicago, 1974, pp. 49–167.
112. Fick, A. *Uber die Messung des Blutquantums in den Herzventrikeln*. Sitz der Physik Med Ges Wurzburg, 1897, p. 16.
113. Stewart, G.N. Researches on the circulation time and on the influences which affect it. IV. The output of the heart. *J. Physiol.* **1897**, *22*, 159.
114. Kinsman, J.M.; Moore, J.W.; Hamilton, W.F. Studies on the circulation: I. Injection method. Physical and mathematical considerations. *Am. J. Physiol.* **1929**, *89*, 322–330.
115. Grossman, W. *Cardiac Catheterization and Angiography*. Lea & Febiger, Philadelphia, 1980, pp. 103–123.
116. Anderson, R.H.; Baker, E.J.; Ho, S.H.; Rigby, M.L.; Ebels, T. The morphology and diagnosis of atrioventricular septal defect. *Cardiol. Young* **1991**, *1*, 290–305.
117. Rigby, M.L.; Chan, K-Y. The diagnostic evaluation of patients with complete transposition. *Cardiol. Young* **1991**, *1*, 26–40.
118. Allan, L.D.; Cook, A. Pulmonary atresia with intact ventricular septum in the fetus. *Cardiol. Young* **1992**, *2*, 367–376.
119. Stern, H.; Bauer, R.; Schroter, G.; Sauer, U.; Emmrich, P.; Bühlmeyer, K. Magnetic resonance imaging after modified Fontan procedure. *Cardiol. Young* **1992**, *2*, 158–167.
120. Mills, C.J.; Gabe, I.T.; Gault, J.H.; Mason, D.T.; Ross Jr., J.; Braunwald, E.; Shillingford, J.P. Pressure-flow relationships and vascular impedance in man. *Cardiovasc. Res.* **1970**, *4*, 405.
121. Sahn, D.J.; Valdes Cruz, L.M. New advances in two dimensional Doppler echocardiography. *Prog. Cardiovasc. Dis.* **1986**, *28(5)*, 367–382.
122. Sanders, S.P.; Yeager, S.; Williams, R.G. Measurement of systemic and pulmonary blood flow and Qp/Qs ratio using Doppler and two-dimensional echocardiography. *Am. J. Cardiol.* **1983**, *51*, 952–956.
123. Goldberg, S.J.; Sahn, D.J.; Allen, H.D.; Valdes-Cruz, L.M.; Hoenecke, H.; Carnahan, Y. Evaluation of pulmonary and systemic blood flow by two-dimensional echocardiography using fast Fourier transform spectral analysis. *Am. J. Cardiol.* **1982**, *50*, 1394–1400.
124. Melbin, J.; Brown, D.J.; Noordergraaf, A. Pulmonary blood flow and large vessel function. *Proc. 4th AVMF* **1986**, 40–46.
125. Melbin, J.; Gopalakrishna, R.; Noordergraff, A. Three dimensional laminar flow in distorting, axisymmetric, axially varying vessels. *Bull. Math. Biol.* **1975**, *37*, 489–504.
126. Melbin, J.; Gopalakrishna, R. Flow and distortion phenomena in vessel models for the pulmonary trunk. In: *Cardiovascular System Dynamics* (Baan, J.; Noordergraaf, A.; Raines, J., Eds.). MIT Press, Cambridge, MA, 1978, pp. 309–317.
127. Fry, D.L.; Patel, D.J.; de Freitas, F.M. Relationship of geometry to certain aspects of hydrodynamics in larger pulmonary arteries. *J. Appl. Phys.* **1962**, *17*, 492–496.
128. Grant, B.J.; Paradowski, L.J.; Fitzpatrick, J.M. Effect of perivascular electromagnetic flow probes on pulmonary hemodynamics. *J. Appl. Physiol.* **1988**, *65(4)*, 1885–1890.
129. Bergel, D.H. *Cardiovascular Fluid Dynamics*, Vol. 2. Academic Press, London, 1972.
130. Caro, C.G.; Pedley, T.J.; Schroter, R.C.; Seed, W.A. *The Mechanics of the Circulation*. Oxford University Press, Oxford, 1978.
131. Dinnar U. *Cardiovascular Fluid Dynamics*. CRC Press, Boca Raton, 1981.
132. Fung. Y.C. *Biomechanics: Motion, Flow, Stress, and Growth*. Springer-Verlag, New York, 1990.
133. Harned, H.S. Jr. *Pulmonary Heart Disease in the Young*. Little, Brown, Boston, 1990, pp. 379–398.
134. Harris, P.; Heath, D. *The Human Pulmonary Circulation: Its Form and Function in Health and Disease*, 2nd ed. Churchill Livingston, New York, 1977, p. 392.

135. Hwang, N.H.C.; Normann, N.A. *Cardiovascular Flow Dynamics and Measurements*. University Park Press, Baltimore, 1977, p. 665.
136. Chang, H.K. Mitchner and Chang book.
137. Noordergraaf, A. *Circulatory System Dynamics*. Academic Press, New York, 1978.
138. Bauer, R.D.; Busse, R. *The Arterial System: Dynamics, Control Theory and Regulation*. Springer-Verlag, Berlin, 1978.
139. Weir, E.K.; Reeves, J.T. *Pulmonary Hypertension*. Futura Publishing Company, Mount Kisco, 1984.
140. West, J.B., Ed. *Bioengineering Aspects of the Lung*. Marcel Dekker, New York, 1977.
141. Will, J.A.; Dawson, C.A.; Weir, E.K.; Buckner, C.K. *The Pulmonary Circulation in Health and Disease*. Academic Press, New York, 1987.
142. Yin, C.P. *Ventricular Vascular Coupling*. Springer-Verlag, New York, 1987.
143. Dawson, T.H. *Engineering Design of the Cardiovascular System of Mammals*. Prentice Hall, Englewood Cliffs, NJ, 1991.
144. Ferenza, C. Pulmonary arterial design in mammals: morphologic variation and physiologic constancy. *Johns Hopkins Med. J.* **1969**, *125*, 207.
145. Liu, Z.R.; Yin, F.C. Normalization of hemodynamic parameters: application to vascular resistance and impedance. *Am. J. Physiol. Apr.* **1987**, *252(4 Pt 2)*, R710–709.
146. Heymann, M.A.; Hoffman, J.I.E. Persistent pulmonary hypertension secondary to congenital heart disease. In: *Pulmonary Hypertension* (Weir, E.K.; Reeves, J.T., Eds.). Futura Publishing Company, Mount Kisco, 1984, pp. 73–114.
147. Heymann, M.A.; Hoffman, J.I.E. Persistent pulmonary hypertension syndromes in the newborn. In: *Pulmonary Hypertension* (Weir, E.K.; Reeves, J.T., Eds.). Futura Publishing Company, Mount Kisco, 1984, pp. 45–71.
148. Rudolph, A.M.; Auld, P.A.M.; Golinko, R.J.; Paul, M.H. Pulmonary vascular adjustments in the neonatal period. *Pediatrics* **1961**, *28*, 28–34.
149. Lucas, C.L.; Henry, G.W.; Frantz, E.; Ha, B.; Ferreiro, J.I.; Wilcox, B.R. The role of maturation in pulmonary vascular dynamics. In: *Biofluid Mechanics III* (Schneck, D.J.; Lucas, C.L., Eds.). New York University Press, New York, 1990, 23–32.
150. Lucas, C.L.; Henry, G.W.; Frantz, E.G.; et al. Ultrasonic assessment of pulmonary hemodynamics in maturing chronically instrumented lambs with normal and abnormal pulmonary circulations. In: *Physiol Fluid Dynamics AIII* (Swamy N.V.C.; Singh, M., Eds.). Narosa Publishing House, 1992, pp. 38–43.
151. Greenwald, S.E.; Johnson, R.J.; Haworth, S.G. Pulmonary vascular input impedance in the newborn and infant pig. *Cardiovasc. Res.* **1985**, *18*, 44–50.
152. Gozna, D.E.; Marble, A.E.; Shaw, A.; Holland, J.G. Age-related changes in the mechanics of the aorta and pulmonary artery of man. *J. Appl. Physiol.* **1974**, *36*, 407.
153. Masuyama, T.; Kodama, K.; Kitabatake, A.; Sato, H.; Nanto, S.; Inoue, M. Continuous-wave Doppler echocardiographic detection of pulmonary regurgitation and its application to noninvasive estimation of pulmonary artery pressure. *Circulation* **1986**, *3*, 484–492.
154. Dabestani, A.; Mahan, G.; Gardin, J.M.; Takenaka, K.; Burn, C.; Allfie, A.; Henry, W.L. Evaluation of pulmonary artery pressure and resistance by pulsed Doppler echocardiography. *Am. J. Cardiol.* **1987**, *59*, 662–668.
155. Friedman, D.M.; Bierman. F.Z.; Barst, R. Gated pulsed Doppler evaluation of idiopathic pulmonary artery hypertension in children. *Am. J. Cardiol.* **1986**, *58*, 369–370.
156. Graettinger, W.F.; Greene, E.R.; Voyles, W.F. Doppler predictions of pulmonary artery pressure, flow and resistance in adults. *Am. Heart. J.* **1987**, *113*, 1426–1437.
157. Serwer, G.A.; Cougle, A.G.; Eckerd, J.M.; Armstrong, B.E. Factors affecting use of the Dopper-determined time from flow onset to maximal pulmonary artery velocity for measurement of pulmonary artery pressure in children. *Am. J. Cardiol.* **1986**, *58*, 352–356.

158. Stevenson, J.G. Comparison of several noninvasive methods for estimation of pulmonary artery pressure. *J. Am. Soc. Echocardiol.* **1989**, *2*, 157–171.
159. Robinson, P.J.; Macartney, F.J.; Wyse, R.K.H. Non-invasive diagnosis of pulmonary hypertension. *Internat. J. Cardiol.* **1985, 1986**, *11*, 253–259.
160. Lucas, C.L.; Henry, G.W.; Ha, B.; Ferreiro, J.I.; Frantz, E.G.; Wilcox, B.R. Characterization of pulmonary artery blood velocity patterns in lambs. In: *Biofluid Mechanics: Blood Flow in Large Vessels* (Liepsch, D.W., Ed.). Springer-Verlag, Berlin, 1990, pp. 171–184.
161. Hatle, L.; Angelsen. *Doppler Ultrasound in Cardiology: Physical Principles and Clinical Applications.* Lea & Febiger, Philadelphia, 1985, pp. 257–264.
162. Lucas, C.L.; Henry, G.W.; Keagy, B.A.; Criado, E.; Ferreiro, J.I.; Ha, B.; Baudino, M.D.; Wilcox, B.R. Continuous measurement of pulmonary blood flow using a retractable pulsed Doppler probe. *J. Ultrasound Med.* **1987**, *6*, 181–190.
163. Lucas, C.L.; Henry, G.W.; Ferreiro, J.I.; Ha, B.; Keagy, B.A.; Wilcox, B.R. Pulmonary blood velocity profile variability in open-chest dogs: influence of acutely altered hemodynamic states on profiles and influence of profiles on the accuracy of techniques for cardiac output determination. *Heart Vessels* **1988**, *4*, 65–78.
164. Panidis, I.P.; Ross, J.; Mintz, G.S. Effect of sampling site on assessment of pulmonary artery blood flow by Doppler echocardiography. *Am. J. Cardiol.* **1986**, *58*, 1145–1147.
165. Henry, G.W.; Johnson, T.A.; Ferreiro, J.I.; Hsiao, H.S.; Lucas, C.L.; Keagy, B.A.; Lores, M.E.; Wilcox, B.R. The velocity profile in the main pulmonary artery of a canine model. *Cardiovasc. Res.* **1984**, *18*, 620–625.
166. Frantz, E.G.; Henry, G.W.; Lucas, C.L.; Keagy, B.A.; Lores, M.E.; Criado. E.; Ferriero, J.I.; Wilcox, B.R. Characteristics of blood flow velocity in the hypertensive canine pulmonary artery. *Ultrasound Med. & Biol.* **1986**, *12*, 379–385.
167. Gardin, J.M.; Sato, D.A.; Rohan, M.K.; Shu, V.W.; Allfie, A.; Gardin, S.K.; Henry, W.L. Effect of acute changes in heart rate on Doppler pulmonary artery acceleration time in a porcine model. *Chest Nov.* **1988**, *94(5)*, 994–997.
168. Mallery, J.A.; Gardin, J.M.; King, S.W.; Ey, S.; Henry, W.L. Effects of heart rate and pulmonary artery pressure on Doppler pulmonary artery acceleration time in experimental acute pulmonary hypertension. *Chest Aug.* **1991**, *100(2)*, 470–473.
169. Westerhof, N.; van dem Bos, C.G.; Laxmiranarayan, S. Arterial reflection. In: *The Arterial System: Dynamics, Control Theory and Regulation* (Bauer, R.D.; Busse, R., Eds.). Springer-Verlag, Berlin, 1978.
170. Westerhof, N.; Sipkema, P.; van dem Bos, G.C.; Elzinga, G. Forward and backward waves in the arterial system. *Cardiovasc. Res.* **1972**, *6*, 648–56.
171. Furuno, Y.; Nagamoto, Y.; Fujita, M.; Kaku, T.; Sakurai, S.; Kuroiwa, A. Reflection as a cause of mid-systolic deceleration of pulmonary flow wave in dogs with acute pulmonary hypertension: Comparison of pulmonary artery constriction with pulmonary embolisation. *Cardiovasc. Res.* **1991**, *25*, 118–124.
172. Calvin, Jr., J.E.; Baer, R.W.; Glantz, S.A. Pulmonary artery constriction produces a greater right ventricular dynamic afterload than lung microvascular injury in the open chest dog. *Circ. Res.* **1985**, *56(1)*, 40–56.
173. Henry, G.W.; Katayama, H.; Lores, M.E.; Lucas, C.L.; Ferreiro, J.I. Intraluminal pulsed Doppler evaluation of the pulmonary artery velocity time curve in a canine model of acute pulmonary hypertension. *Chest* **1991**, *100*, 474–479.
174. Tahara, M.; Tanaka, H.; Nakao, S.; et al. Hemodynamic determinants of pulmonary valve motion during systole in experimental pulmonary hypertension. *Circulation* **1981**, *64*, 1249–1255.
175. Turkevich, D.; Groves, B.M.; Micco, A.; Trapp, J.A.; Reeves, J.T. Early partial systolic closure of the pulmonic valve relates to severity of pulmonary hypertension. *Am. Heart J.* **1988**, *115*, 409–418.

176. Redel, D.A.; Fehske, W. Diagnosis and follow-up of congenital heart disease in children with the use of two-dimensional Doppler echocardiography. *Ultrasound Med. Biol.* **1984**, *10*, 249–258.
177. Yock, P.G.; Naasz, C.; Schmittger, I.; Popp, R.L. Doppler tricuspid and pulmonic regurgitation in normals: Is it real? *Circulation* **1984**, *70*, II–40.
178. Okamoto, M.; Miyatake, K.; Kinoshita, N.; Sakakibara, H.; Nimura, Y. Analysis of blood flow in pulmonary hypertension with the pulsed Doppler flowmeter combined with cross sectional echocardiography. *Br. Heart J.* **1984**, *51*, 407–415.
179. Takao, S.; Miyatake, K.; Izumi, S.; Kinoshita, N.; Sakakibara, H.; Nimura, Y. Physiological pulmonary regurgitation detected by the Doppler technique and its differential diagnosis. *J. Am. Coll. Cardiol.* **1985**, *5*, 499.
180. Zeiher, A.; Bonzel, T.; Hollschlager, H.; Just, H. Noninvasive evaluation of pulmonary hypertension by quantitative contrast m-mode echocardiography. *Eur. Heart J.* **1984**, *62(Suppl. 1)*.
181. Mattos, S.S.; Severi, R.; Cavalacanti, C.V.; Fonte Freire, M.; Brindero Filho, D. Recognition of normal flow patterns in children: A Doppler echocardiographic study. In: *Physiol Fluid Dynamics AIII* (Swamy, N.V.C.; Singh, M., Eds.). Narosa Publishing House, 1992, pp. 64–72.
182. Schultz, D.L.; Tunstall Pedoe, D.S.; Lee, G.J.; Gunning, A.J.; Bellhouse, B.J. Velocity distribution and transition in the arterial system. In: *CIBA Foundation Symposium on Circulatory and Respiratory Mass Transport* (Wolstenholme, G.E.W.; Knight, J., Eds.). Churchill Livingstone, London, 1969, pp. 172–199.
183. Paulsen, P.K. The hot-film anemometer--a method for blood velocity determination. *Eur. Surg. Res.* **1980**, *12*, 149–158.
184. Hartley, C.J.; Cole, J.S.; Hartley, C.J. An ultrasonic pulsed Doppler system for measuring blood flow in small vessels. *J. Appl. Physiol.* **1974**, *37*, 626–629.
185. Katayama, H.; Henry, G.W.; Lucas, C.L.; Ha, B.; Ferreiro, J.I.; Frantz, E.G.; Krzeski, R. Three dimensional visualization of pulmonary blood flow velocity profiles in lambs. *Jpn. Heart J.* **1992**, *33(1)*, 95–111.
186. Somod, L.; Hasenkam, J.M.; Kim, W.Y.; Nygaard, H.; Paulsen, P.K. Three dimensional visualisation of velocity profiles in the normal porcine pulmonary trunk. *Cardiovasc. Res.* **1993**, *27(2)*, 291–295.
187. Henry, G.W.; Lucas, C.L.; Ha, B.; Ferreiro, J.I.; Wilcox, B.R. The velocity profile in the branch pulmonary arteries of a lamb model. In: *Proceedings 7th International Conference Workshop Cardiovascular System Dynamics Society*. Zuoz, Switzerland, August 1986.
188. Lucas, C.L.; Henry, G.W.; Ha, B.; Ferreiro, J.I.; Wilcox, B.R. Relationship between flow reversal and pulmonary artery pressure and resistance in branch pulmonary arteries of lambs. *Automedica* **1987**, *9*, 9 (abstract).
189. Piene, H. Matching between right ventricle and pulmonary bed. In: *Ventricular Vascular Coupling* (Yin, C.P., Ed.). Springer-Verlag, New York, 1987, 180–202.
190. Fung, Y.C.; Yen, M.R.; Sobin, S.S. Models of the pulmonary hemodynamics based on anatomic and functional data. In: *The Pulmonary Circulation in Health and Disease* (Will, J.A.; Dawson, C.A.; Weir, E.K.; Buckner, C.K., Eds.). Academic Press, Orlando, 1987, pp. 199–214.
191. Zhuang, F.Y.; Fung, Y.C.; Yen, R.T. Analysis of blood flow in cat's lung with detailed anatomical and elasticity data. *J. Appl. Physiol.* **1983**, 1341–1348.
192. Lucas, C.L.; Henry, G.W.; Ferreiro, J.I.; Ha, B.; Wilcox, B.R. Comparison of input impedance spectra in branch pulmonary arteries of a lamb model of pulmonary hypertension. *Am. Rev. Resp. Dis.* **1987**, *135*, A514.
193. Meyers, C.H.; Purut, C.M.; DAmico, T.A.; Smith, P.K.; Sabiston Jr., D.C.; Van Trigt, P. Pulmonary arterial impedance after single lung transplantation. *J. Surg. Res.* **1992**, *52(5)*, 459–465.
194. Zuckerman, B.D.; Orton, E.C.; Stenmark, K.R.; Trapp, J.A.; Murphy, J.R.; Coffeen, P.R.; Reeves, J.T. Alteration of the pulsatile load in the high altitude calf model of pulmonary hypertension. *J. Appl. Physiol.* **1991**, *70(2)*, 859–868.

195. Fitzpatrick, J.M.; Grant, B.J. Effects of pulmonary vascular obstruction on right ventricular afterload. *Am. Rev. Respir. Dis.* **1990**, *141(4 Pt 1)*, 944–952.
196. Crouch, J.D.; Lucas, C.L.; Keagy, B.A.; Wilcox, B.R.; Ha, B. The acute effects of pneumonectomy on pulmonary vascular impedance in the dog. *Ann. Thorac. Surg.* **1987**, *43(6)*, 613–616.
197. Grant, B.J.; Fitzpatrick, J.M.; Lieber, B.B. Time-varying pulmonary arterial compliance. *J. Appl. Physiol.* **1991**, *70(2)*, 575–583.
198. Philpot, E.F.; Yoganathan, A.P.; Woo, Y.R.; Sung, H.W.; Franch, R.H.; Sahn, D.J.; Valdes-Cruz, L.M. *In vitro* pulsatile flow visualization studies in a pulmonary artery model, *J. Biomech. Eng.* **1985**, *107*, 368–375.
199. Yoganathan, A.P.; Ball, J.; Woo, Y.R.; Philpot, E.F.; Sung, H.W.; Franch, R.H.; Sahn, D.J. Steady flow velocity measurements in a pulmonary artery model with varying degrees of pulmonic stenosis, *J. Biomech.* **1986**, *19*, 129–146.
200. Gardin, J.M.; Sung, H.W.; Yoganathan, A.P.; Ball, J.; McMillian, S.T.; Henry, W.L. Doppler flow velocity mapping in an *in vitro* model of the normal pulmonary artery, *J. Am. Coll. Cardiol.* **1988**, *12*, 1366–1376.
201. Sung, H.W.; Yoganathan, A.P. Axial flow velocity patterns in a normal human pulmonary artery model: pulsatile *in vitro* studies. *J. Biomech.* **1990**, *23*, 201–214.
202. Sung, H.W.; Philpot, E.F.; Nanda, N.C.; Yoganathan, A.P. Axial flow velocity patterns in a pulmonary artery model with varying degrees of valvular pulmonic stenosis: pulsatile *in vitro* studies. *J. Biomech.* **1990**, *23*, 563–578.
203. Sung, H.W.; Yoganathan, A.P. Secondary flow velocity patterns in a pulmonary artery model with varying degrees of valvular pulmonic stenosis: pulsatile *in vitro* studies. *J. Biomech. Eng.* **1990**, *112*, 88–92.
204. Low, H.T.; Chew, Y.T.; Lee, C.N. Flow studies on the Fontan surgical connections. In: *Proceedings of the 7th International Conference on Biomedical Engineering*. Singapore, 1992, pp. 514–516.
205. de Leval, M.R.; Kilner, P.; Gewillig, M.; Bull, C. Total cavopulmonary connection: A logical alternative to atriopulmonary connection for complex Fontan operations. *J. Thorac. Cardiovasc. Surg.* **1988**, *96*, 682–695.
206. Nerem, R.M.; Seed, W.A. An *in vivo* study of aortic flow disturbances. *Cardiovasc. Res.* **1972**, *6*, 1–14.
207. Chen, J.I.T.; Robinson, A.E.; Goodrich, J.K.; Lester, R.G. Uneven distribution of pulmonary blood flow between left and right lungs in isolated pulmonary stenosis. *Am. J. Roentgenology* **1969**, *107*, 343–350.
208. Gay, Jr., B.B.; Franch, R.H. Pulsations in the pulmonary arteries as observed with roentgenoscopic image amplification. *Am. J. Roentgenology* **1960**, *83*, 335–344.
209. Muster, A.J.; van Grondelle, A; Paul, M.H. Unequal pressures in the central pulmonary arterial branches in patients with pulmonary stenosis. *Pediatr. Cardiol.* **1982**, *2*, 7–14.
210. Davies, G.; Reid, L. Growth of the aveoli and pulmonary arteries in childhood. *Thorax* **1970**, *25*, 669.
211. Reeves, J.T.; Grover, R.F. High altitude pulmonary hypertension and pulmonary edema. *Prog. Cardiol.* **1975**, *4*, 99.

IN VIVO HEMODYNAMIC MEASUREMENTS

J. Michael Hasenkam

I. INTRODUCTION

Hemodynamic measurements can be used for monitoring cardiovascular reactions to external or internal actions. The measurements can also have a purpose of their own and be used as tools for research-oriented studies to gain deeper insight into the dynamics of blood flow.

Advances in Hemodynamics and Hemorheology,
Volume 1, pages 325–372.

ISBN: 1-55938-634-7.

Understanding of blood flow in humans has been a challenge throughout the history and has always been appreciated as a condition for life. Leonardo da Vinci made very detailed studies of the heart and was the first to describe the heart as the most powerful muscle in the body, composed of four chambers. He also described the autonomous function of the atria as being separated from the ventricles by valves, and he described the function of the semilunar aortic and pulmonary valves. He even proposed to make *in vitro* flow visualization studies of bovine aortic valves mounted in glass tubes and using seeds as tracer.[1] This idea was based on his excellent understanding of the closing mechanism of the valve leaflets, promoted by vortices in the sinuses of valsalvae (see Figure 1).

New aspects of hemodynamics were not conceived until Harvey in the sixteenth century described the dynamics of the circulation; and later Bernoulli (1700–1782) used blood pressure as a central parameter in his law on energy conservation.[2] His

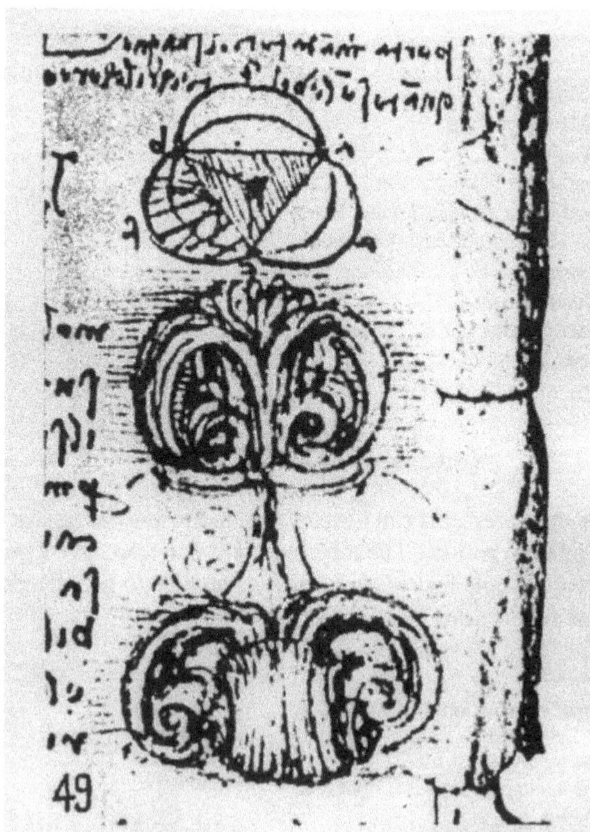

Figure 1. Sketches from Leonardo da Vinci's studies on the closing mechanism of the semilunar valves promoted by eddies in the sinuses of Valsalva.

contributions to the understanding of the circulation were supplemented by Thomas Young (1773–1829) who described the modulus of elasticity and derived a premature expression for pressure loss in tubes.

Based on Young and Bernoullis' theories, Poiseuilles (1799–1869) extended the knowledge about the circulation by conceiving "Poiseuilles law."

$$Q = (\pi * R^4 * \Delta p)/8 * v * L \tag{1}$$

Q = flow; R = tube radius; Δp = pressure difference; v = dynamic viscosity; L = tube length.

The conditions for Poiseuilles law are that the fluid is Newtonian, the flow is laminar (non-turbulent), and the tube (vessel) is rigid. Obviously, none of these conditions are fulfilled in the cardiovascular system but in spite of that Poiseuilles law has a good deal of applicability to *in vivo* hemodynamic studies.

Poiseuilles was also the first to use the mercury manometer to measure blood pressure and accordingly he may be responsible for the general use of the unit "mm Hg" for blood pressure measurement. Actually Stephen Hales was the first to describe a method for blood pressure measurement in 1726 (about the same time that Bernoulli published his mathematical solutions to the conservation of energy). One might speculate to what extent the advancements of hemodynamics would have been accelerated if the concepts of these two had been combined to commence hemodynamic investigations! The history of advances in hemodynamics progressed slowly but steadily until the 1930s when the technical advances in all other scientific fields expanded research possibilities. In 1936 Kolin[3] was the first to introduce the electromagnetic flowmeter for blood flow measurements. Also in 1936 Machella[4] introduced the hot-wire anemometer for blood velocity measurements. In the 1950s the ultrasound method for cardiac imaging was introduced by Satumora[5] among others.

With the availability of these and even more advanced methods for hemodynamic studies the progress in this scientific field accelerated considerably. Today a variety of measuring techniques and models are used all over the world to contribute to the ever increasing understanding of the dynamics of the circulation.

II. ANIMAL STUDIES

Animal models, like *in vitro* models, have the quality of being manipulable but at the same time closely comparable to humans. They constitute a valuable bridge for transfer of methods and hypotheses from the *in vitro* laboratory models to humans.

Small animals like mice, rats, or rabbits are seldom used for cardiovascular studies, although some hemodynamic measuring techniques can be used to evaluate the influence of any stimulation to the animal. In this section only larger animals will be mentioned.

For cardiovascular studies in the heart and great vessels there are certain animals which are used more than others. Monkeys obviously have the highest degree of human compatibility in respect to anatomy and physiology. These animals are, however, seldom used, due to limited availability and ethical concerns and because they are also costly to procure and care for.

Dogs have a fairly high degree of human compatibility with respect to the cardiovascular system and they have been the most frequently used laboratory animal for blood flow and blood velocity measurements. Dogs, like monkeys and cats, are in many countries psychologically considered more closely associated with people, and special laws and regulations have been made to regulate the use of these animals for research purposes.

The coronary arteries show obvious differences in anatomy compared with humans.[6] Both the course of the vessels and the vast amount of collaterals are deviations from human coronary anatomy. Consequently, dogs are not appropriate animals for coronary artery studies when human similarity is a demand.

The canine arterial tree has a high degree of human similarity. The ascending aorta is well defined,[7] in contrast to cloven-footed animals; the aortic arch has approximately the same course as in humans,[8,9] but unlike humans, the abdominal aorta terminates as a trifurcation.[10] Dogs have an obvious advantage in terms of handling; they are not as sensitive to surgical procedures as, for example, are pigs; they are easy to anaesthetize and the postoperative course is uneventful for most surgical procedures. For special purposes other species play a dominating role. Sheep and lambs are used for biological heart valve studies[11] and for pulmonary artery blood velocity studies (e.g., Katayama et al.[12]).

Heart valve prostheses are often studied in calves (e.g., ref. 13). They have the quality of being large animals at a young age and they provide the opportunity to study large valve sizes in growing animals. These qualities are attractive from both a surgical and a methodological point of view since the implanted valve can be studied under gradually changing hemodynamic conditions as the animal grows. As indicated above calves do not have a well defined ascending aorta, since it bifurcates immediately after the aortic valve.

Like calves, pigs are not pet animals and therefore they pose fewer ethical problems with respect to research use. Consequently, they are used as laboratory animals with increasing frequency. From a cardiovascular research point of view pigs have a high degree of human similarity. The coronary artery anatomy is very close to that of humans, and there is virtually no coronary collateral circulation.[6,14] They have a short, but well defined ascending aorta,[15] and the aortic arch is comparable to humans apart from the brachiocephalic vessels, of which pigs only have two. The descending aorta (thoracic and abdominal) generally shows the same anatomical features as in humans, apart from the most distal part which terminates as a trifurcation, as in most other large animals. The pulmonary artery hemodynamics have also been studied with the aim of transferring data to human physiology.[16]

Beyond the above-mentioned laboratory animals, other species have been introduced in hemodynamic research. Cats were used by Matre and Segadal[17] for ascending aorta blood flow measurements. Heart valve implantations have been made in goats[18] and coronary blood velocity measurements have been performed in horses.[19]

A common feature of animal models is the biological complexity which can be considered as both an advantage and a disadvantage, depending on which studies are being conducted. The biological reactions (e.g., vascular compliance, blood pressure fluctuations, heart rate variability) are by definition already present and do not have to be simulated. This quality may be desirable when a physiological reaction to a certain stimulus is wanted (e.g., increased cardiac contractility, heart rate and peripheral vascular resistance after injection of adrenalin). It might, however, also disturb the validity of the research model if stable hemodynamic conditions are required for acquisition of reproducible measurements. Animal models also provide all the modulating mechanisms which are characteristic for biological systems. Therefore, vascular reactivity, autoregulation of tissue flow and so on do not have to be simulated as it has to be in the *in vitro* environment. Consequently, *in vivo* models have to be used with their inherent complexity of reactions, and these reactions have to be interpreted. As in every other research area, models have to be selected carefully according to the kind of information which is searched for.

III. HUMAN MEASUREMENTS

Healthy volunteers and patients are obviously the ultimate targets for studies of cardiovascular physiology and pathophysiology in humans.

Evidently, such studies must be conducted with optimal focus on safety and with as little discomfort as possible. This implies that the measuring techniques must be primarily non-invasive, and at least not involve any health hazards to the person being measured on. A general issue in human measurements is also the fact that data acquisition time is normally much more limited than in animal studies. Last but not least: invasive methods used in patients must apply transducers, cables, and so forth which are sterilizable, and the demands for patient safety are much higher than for non-invasive techniques.

The ethical aspects in human hemodynamic studies are associated with a number of rules and laws. First of all the Helsinki II declaration[20] must be respected in all scientific studies involving human beings or living human tissue. The rules of this declaration are applied worldwide, and basically no scientific data can be published without proper justification that the data have been acquired under respect of the declaration.

Many countries have established local or national scientific ethical committees which supervise that scientific studies in humans are conducted only if the participating volunteers or patients have been sufficiently informed and if they have given oral and written consent. The formal requirements for conducting studies in humans are therefore generally more complex than for animal studies. In spite of all the cumbersome work which has to be done prior to human studies they are evidently extremely important to conduct as a final investigation before clinical routine measurements are introduced or before consequences are taken in terms of patient treatment.

IV. METHODS

Relevant aspects of data analysis and interpretation are discussed elsewhere in this book. In this part only measuring techniques applicable to *in vivo* investigation will be presented and discussed.

The measurable parameters in cardiovascular research which will be described in this chapter are based on measurement of the following qualities: Blood flow, blood velocity, blood pressure, and sound and vibration. All parameters are physically interrelated since the flowing blood can be described in terms of its volume flow measured in 1 min^{-1}, its velocity measured in m s^{-1}, and the pressure inside the blood vessel measured in kPa or in mm Hg. Blood flow disturbances in terms of turbulence which causes pressure fluctuations can be measured as precordial murmurs and quantitated in units of Pa, ms^{-1}, or ms^{-2}, (acceleration). All these basic measures can be used to calculate energy dissipation, energy loss, degree of blood flow disturbance, compliance/elasticity, shear stresses, wall stress, and so forth. In this way a variety of biomechanic and fluid-/hemodynamic aspects can be interpreted by quantitative measures.

A. Blood Flow Measurements

Blood flow is a gross parameter which, throughout the years, has caused a good deal of methodological difficulties since all measuring techniques presented hitherto have been associated with considerable inaccuracy and several sources of errors when applied *in vivo*. Blood flow is suitable as a reference value to describe cardiac output or to measure organ blood supply by measurement on end arteries (e.g., renal arteries), or to measure venous blood flow as a summation of different inputs (e.g., blood flow in the coronary sinus).

The oldest method for cardiac output measurement *in vivo* was introduced by Adolph Fick in 1870. He showed that flow in a given period of time is equal to the amount of substance entering the stream of flow divided by the difference between the concentrations of substance before and after the point of entry. According to MacDonald's[21] the principle was first applied *in vivo* by Grehart and Quinquart in 1886 to measure cardiac output in dogs. The substance normally used for calculating cardiac output by the Fick Principle is oxygen and the procedure is schematically shown in Figure 2. The oxygen uptake is measured by having the patient or laboratory animal breathe through a mask to measure the amount of oxygen

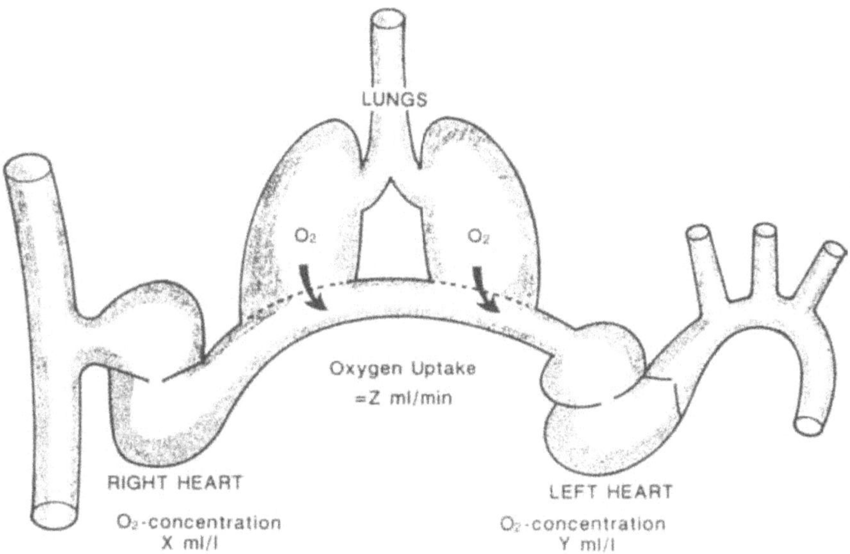

Figure 2. An illustration of Fick's principle for cardiac output measurement.

$$\text{CO [l/min]} = \frac{O_2\text{consumption[ml/min]}}{\text{Arterial } O_2\text{concentration[ml/l]-venous } O_2\text{concentration[ml/l]}}$$

extracted from the gas supply. Simultaneous blood samples are taken preferentially from the pulmonary artery and from a systemic artery. Cardiac output is then measured as:

$$\text{CO[l/min]} = \frac{\textit{Pulmonary } O_2\textit{-uptake}[ml/\text{min}]}{\textit{arterial } O_2\textit{-concentration}[ml/l]\textit{-veneous } O_2\textit{-concentration}[ml/l]}. \quad (2)$$

The Fick principle is considered the "gold-standard" for cardiac output measurement and has been used as reference for a high number of cardiac output measuring techniques which have been published in the literature. Evidently, the method is cumbersome, pulmonary artery catheterization is necessary and it does not provide instantaneous measures of cardiac output. However, during cardiac catheterization procedures when a pulmonary artery catheter is easy to position, application of the Fick principle is justified since the patient is normally in a steady state during the measurement period.

Another "traditional" method for cardiac output measurement is the indicator dilution technique. It is based on a continuous measurement of the concentration of an indicator in the bloodstream downstream of an injection site where the indicator ideally is injected as a bolus. The method has a widespread clinical application in terms of the thermodilution technique first introduced for human use by Ganz et al. in 1971.[22] The technique utilizes cold saline as indicator. A four-lumen catheter is advanced to the pulmonary artery, guided by continuous pressure recordings at the tip of the catheter and aided by a saline-filled balloon at the catheter's distal part. Rapid injection of cold saline gushing out through a side-hole positioned in the right atrium provide the possibility to monitor the temperature change by a thermistor positioned at the pulmonary artery part of the catheter. The principle of measurement is illustrated in Figure 3. By integrating the area underneath the temperature curve, a measure for cardiac output can be computed. The

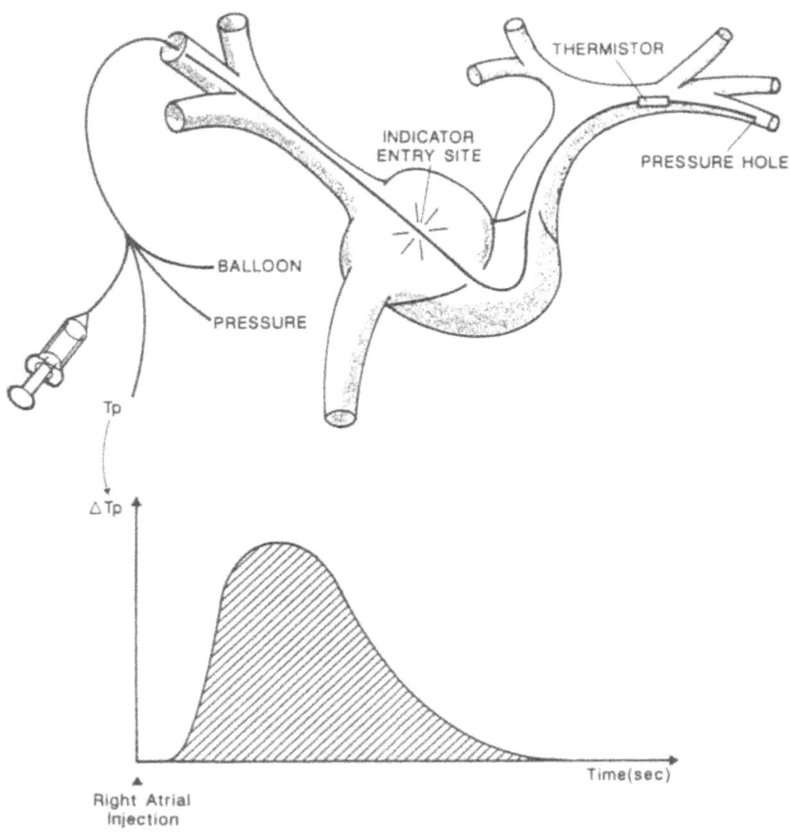

Figure 3. The thermodilution principle. A four lumen catheter is advanced from a large vein to the pulmonary artery, thereby, positioning the injection site at the right atrium level. A rapidly injected volume of cold isotonic saline is mixed with the blood and the temperature change is traced by the thermistor located at the distal end of the catheter. The area under the curve is directly convertable to a measure for cardiac output.

formula which is used for the calculation of cardiac output is the modified Stewart-Hamilton formula:

$$CO = C_t \frac{V_i(T_b - T_i)}{\int_0^\infty \Delta T_b\,(t)\,dt} \tag{3}$$

where V_i = injected volume of cold saline (normally 5–15 ml), T_b and T_i is the temperature of blood and injectate respectively. ΔT_b is the instantaneous change in blood temperature measured in the pulmonary artery. C_T is a constant which comprises blood and saline density and specific heat. In most applications it also comprise compensation for heat loss in the catheter, myocardial tissue and the fact that indicator injection is not administered as a bolus. The calculation of cardiac output by this method is normally computerized and thus provide the possibility for display of cardiac output shortly after saline injection.

The thermodilution technique is probably the most extensively evaluated method for cardiac output measurement (e.g., ref. 23). When the technique is optimized the accuracy is in the range of 10%, but in general clinical use the value is probably more realistically in the range of 20%.

The advantage of the thermodilution technique is its applicability in critically ill patients as the apparatus available today is simple and user-friendly. It does not need patient cooperation and it is generally considered a safe method for critically ill patients. The disadvantages are the invasive nature of the technique which limits the applicability of the technique to patients who are already in interventional therapy or under invasive diagnostic evaluation. Injection of cold saline cannot be made uncritically, especially not in children or in patients with renal failure.

A variant of the bolus injection thermodilution technique as described above is the constant infusion thermodilution technique. An example of the use of this technique is coronary flow measurements by coronary sinus catheterization. A constant infusion of saline is applied at the distal end of the catheter and the temperature is monitored by a thermistor placed proximal to the infusion site. The temperature is a result of various thermic characteristics and the flow which passes the catheters. This technique has a narrow range of applications limited to flow in smaller vessels.

The above methods are the "true" flow measuring techniques. Other methods are based on the measurement of a mean blood velocity in the vessel cross sectional area and multiplication of the vessel cross sectional area at the site of the measurement according to the formula:

$$Q = V * A \tag{4}$$

where Q = flow (m^3/min), V = mean blood velocity (m s^{-1}), and A = cross sectional area (m^2).

One of these techniques is the **electromagnetic flowmeter**. It is based on Faraday's law which states that if a wire is moved in a magnetic field, a current is induced in the wire. The voltage (U) created in the wire depend on the magnetic field strength (B), the length of the wire (L) and its velocity (v). The parameters are inter-related as follows:

$$U = B * L * v. \tag{5}$$

Since blood is a conductor of electricity, its velocity can be measured by placing a magnetic field around the blood vessel and measure the induced electromagnetic force. The principle is schematized in Figure 4. There are, however, some basic assumptions which have to be fulfilled: (1) The magnetic field has to be uniform, (2) the conductor (the blood vessel flow axis) has to be orthogonal to the direction of the magnetic field, and (3) the velocity profile has to be axisymmetric. There must be no interfering electrical signals near the measurement site.

By multiplying the mean blood velocity with the vessel cross sectional area, blood flow can be estimated and formula (5) can be rewritten as:

$$Q = \frac{U*A}{B*D} \tag{6}$$

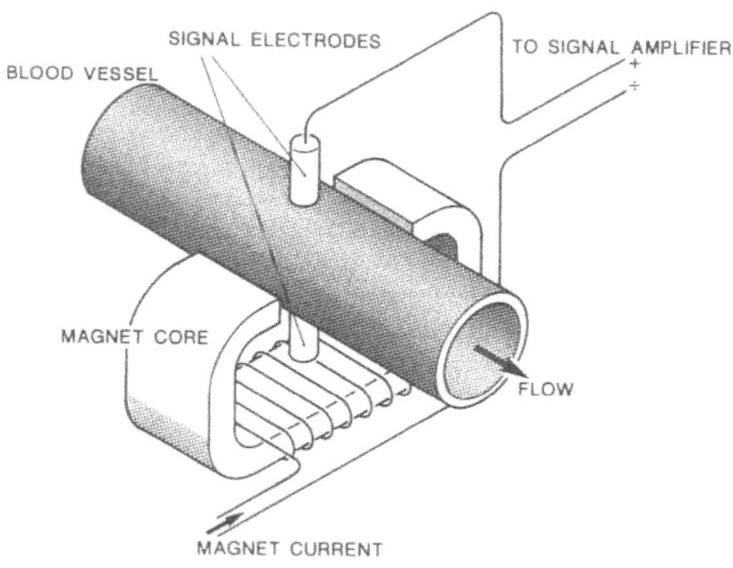

Figure 4. Diagrammatic illustration of the electromagnetic flowmeter principle of measurement. When blood flows in the vessel through the magnetic field an electromagnetic force is induced and detected by signal electrodes. The amplified signal can be displayed on an oscilloscope, tape recorder, or merely on a meter as an indication of instantaneous blood flow.

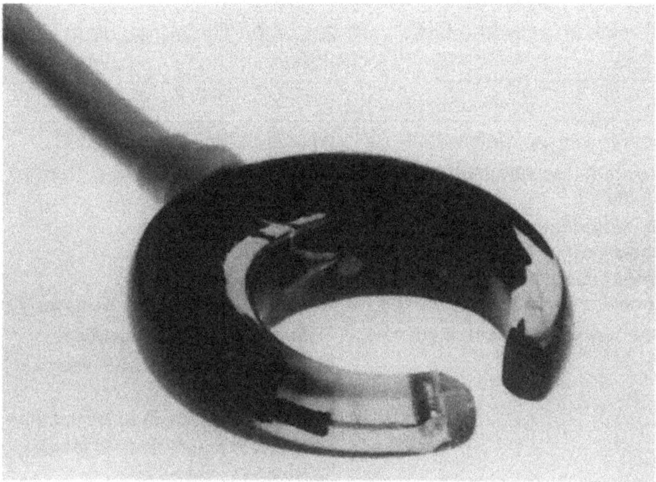

Figure 5. Photograph of an electromagnetic flow probe to be used *in vivo*—if applicable also as an implanted device around a vessel for chronic measurements. The probe shown here is produced by Vingmed.

where Q = blood flow, U = the measured voltage, A = vessel cross sectional area, B = the magnetic field strength, and D = the vessel diameter.

The basic assumptions are evidently not obtainable in practical use. The most prominent difficulty in electromagnetic flow measurement is the electrical noise from the environment. To reduce this problem the transducer is normally operated with a cyclic changing excitation current on the magnet. Normally the magnetic excitation voltage is activated by a square wave function so that detection of induced current is gated. This implies a limited frequency response in the electromagnetic flowmeter technique. The -3dB cut-off frequency is normally in the range of 30–50 Hz, which is sufficient for laminar flow but insufficient for detection of turbulent flow.

The transducers which are operated with the electromagnetic flowmeter may have different designs for perivascular, in-line or even intraluminal application. The probes are sterilizable and can, therefore, be used intra-operatively or be implanted in laboratory animals for chronic measurements. An example is shown in Figure 5.

Some advantages and disadvantages of the electromagnetic principle of measurement are listed in Table 1. Operating electromagnetic flowmeters *in vivo* is a cumbersome task; since base-line adjustment should be performed by mechanical occlusion on the vessel which is studied (although electronic zero-adjustment by some manufactures is claimed to be sufficient). Gain-adjustment should be made by manual injection of a fixed volume of blood for a short period of time where all other blood flow through the transducer is eliminated. Consequently, these two procedures can potentially disturb the organ blood flow which is being studied. An

Table 1. Advantages and Disadvantages of the Electromagnetic Flowmeter Technique

Advantages	*Disadvantages*
Applicable both *in vitro* and *in vivo*	Frequent calibration is needed
Instantaneous blood flow registration (flow curve available)	Zero calibration must be performed in situ by vessel occlusion
Fairly reliable (reproducibility, in laboratory, *can* be accurate within 5%)	Sensitive to: Hematocrit
Perivascular (non-invasive in respect to the flowing blood)	Velocity profile Electrical noise (ECG, measuring equipment)
Implantable (allows repeated long-term measurements)	Tight probe-vessel fit is critical
	Vessel wall thickness and electrical conductance
	Reduced accuracy in turbulent flow
	Sensitive to magnetic field interference
	Sensitive to temperature
	Require dissection of the target vessel

essential detail in operating electromagnetic flowmeters is securing a good contact to the vessel wall. This is obtained by using a probe slightly smaller than the outer vessel diameter and by lubricating or even immersing the measuring site in saline.

As mentioned in Table 1 there are several other technical sources of errors which have to be addressed in these measurements. A comprehensive overview of electromagnetic measurements is presented in "MacDonald's Bloodflow in arteries."[21] Because of the inherent inaccuracy and the many sources of errors, the electromagnetic flowmeter seems to be gradually replacing the **transit time flowmeter**. Like the electromagnetic flowmeter, transit time flow probes are available for measurements on blood vessels with a diameter of 1–35 mm and the probes are sterilizable, which means that intraoperative and chronic measurements with implanted probes are feasible.[24] A commercially available transit time flowmeter probe is shown as an example in Figure 6.

The transit time technique is basically a pulsed ultrasound technique where a burst of ultrasound is transmitted from two piezo-electric crystals which act as transmitter and receiver in an alternating fashion (shown schematically in Figure 7). A master oscillator is used for time base and a transit burst of ultrasound with the master oscillator frequency is emitted from the crystal which first acts as a transmitter. The time delay between the reflected signal detected by the receiving crystal and the master oscillator signal is stored. After a short period when all echoes have died out, the roles of the two transducers are reversed. The delay is detected

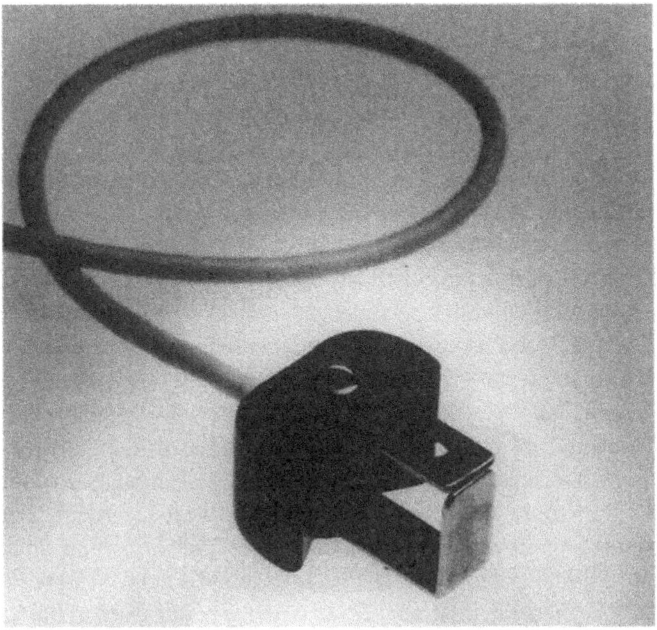

Figure 6. Photograph of a transit time flowmeter probe which is designed to be mounted perivascularly. Two piezoelectric ultrasound crystals are mounted in the upper plastic coated part of the transducer and the lower part acts as a reflector in order to reflect the ultrasound beams to the two piezoelectric crystals which acts as transmitter and receiver in an alternating fashion. The transducer shown here is manufactured by Transonics Inc.

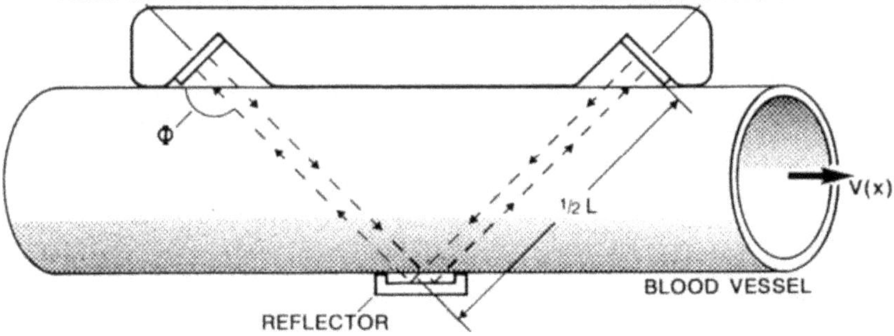

Figure 7. Illustration of the transit time principle of measurement for blood flow measurement. Two piezoelectric ultrasound crystals are mounted in a plastic cuff and "points" at an acoustic mirror on the opposite site of the vessel. The two crystals are emitting ultrasound bursts simultaneously and subsequently they act as receivers. The transit time difference for the ultrasound bursts are proportional to the blood velocity and thereby blood flow.

after a new transmission in the reversed setting is stored. The difference between the two stored values is related to the volumetric flow signal. This sequence of measurements is repeated every millisecond.

An electronic zero-flow reference signal is generated by subtracting two consecutive upstream phase measurements, rather than an upstream and a downstream measurement. This eliminates the need for clamping the vessel to establish zero-flow baseline (which is necessary for most electromagnetic flowmeters). A crystal-controlled internal time-delay is generated to provide a scale factor reference signal, allowing the flow-induced transit-time delay to display the volumetric flow directly in l/min. As long as good acoustic contact is secured, accurate flow estimates are made for vessel diameters between 0.5 and 1.0 times the probe size.

The transit time method is rather accurate with a deviation of approximately 5% from the actual flow rate.[25] It is essentially independent of the velocity profile as long as it is fairly axi-symmetric. Another feature is that angulation of the probe is not critical for the accuracy, and the susceptibility to electric noise is less than that of the electromagnetic flowmeter. Turbulence poses a methodological problem for the transit time flowmeter since the transit of ultrasound is delayed more in turbulent flow than in a complementary laminar flow. With the renaissance of the transit time flowmeter technique during the last 10 years, aided by the design of new implantable probes and versatile electronic equipment, it seems justified to expect this technique to gradually replace the dominating role hitherto possessed by electromagnetic flowmetry. Especially for clinical measurements it is obviously advantageous to use a method which requires a minimum of calibration and only limited operator experience.

In our research group we have exchanged the electromagnetic flowmeter technique with the transit time flowmeter both for animal experimental and clinical flow measurements. Advantages and disadvantages of the technique are listed in Table 2.

Cardiac output measurement can also be made with this technique, provided that the ascending aorta or pulmonary trunk is surgically accessible. If cardiac output

Table 2. Advantages and Disadvantages of the Transit Time Flowmeter Technique

Advantages	*Disadvantages*
Applicable *in vitro* and *in vivo*	Require dissection of the target vessel
Implantable	Unable to measure accurately in turbulent flow
No calibration is required	
No zero-adjustment by vessel clamping	
Probe size is not critical	
Insensitive to probe angulation	
Accurate (within 5% *in vivo*)	
Instantaneous blood flow measurement (flow curve available)	

measurements are otherwise requested we apply the thermo-dilution technique for intermittent measurements.

Evidently, the ultimate choice of measuring technique depends on access to a suitable measuring site, the demands for accuracy, operator experience, and availability of equipment. Apart from the blood flow measuring techniques mentioned here it should be underlined that several other means of estimating blood flow have been reported in the literature. These methods are based on blood velocity measurements (described below) comprising mean velocity and cross sectional area measurements or estimations (e.g., refs. 26, 27, 28, and 29). Others estimate cardiac output by impedance cardiography[30] or pulse pressure contour analysis (e.g., ref. 31), but none of these methods have gained widespread use.

B. Blood Velocity Measurements

Measuring of blood velocity provides the opportunity to calculate volume flow, and point blood velocity measurements enables detailed description of velocity fields. As for the volume flow rate techniques each method has its advantages and disadvantages depending on measuring conditions, accessibility, and which qualities are requested.

The **hot-film anemometer (HFA)** can be used for point blood velocity measurements by means of a thin film mounted on a quartz cone. The quartz cone which is normally approximately 1 mm in diameter is mounted on a needle which can be inserted into the vessel lumen via a small punctured hole. A schematic drawing of the probe is shown in Figure 8.

The HFA operating principle is that the hot-film is heated up about 5°C above the surrounding blood temperature. The passing blood tends to cool off the film by convection, and the power required to maintain film temperature is an expression of fluid velocity at the measuring point. The film power is, via a negative feedback loop based on a Wheatstone bridge, regulated sensitively enough to allow a high frequency response of the HFA measuring equipment. This means that when the laminar underlying blood velocity is sufficiently high, the overlying turbulent fluctuations can be registered accurately. Paulsen et al.[32] found that the -3 dB cut-off frequency for underlying laminar velocities of 50 cm/s and 10 cm/s were 500 Hz and 100 Hz, respectively.

The relation between the HFA output voltage and fluid velocity is non-linear. Therefore, linearization of the analog HFA-signal has to be performed either by analog[33] or digital technique.[34] The voltage–fluid velocity relation has to be determined individually for each probe. Normally, a known fluid velocity is applied to the probe by means of a turn-table (see Figure 9). The calibration procedure should also comprise other measures such as viscosity (hematocrit) and fluid temperature. Apart from the calibration procedure there are some sources of errors which have to be taken into account. These include: fibrin accumulation on the hot-film, temperature changes in the blood, viscosity changes (hemodilution due

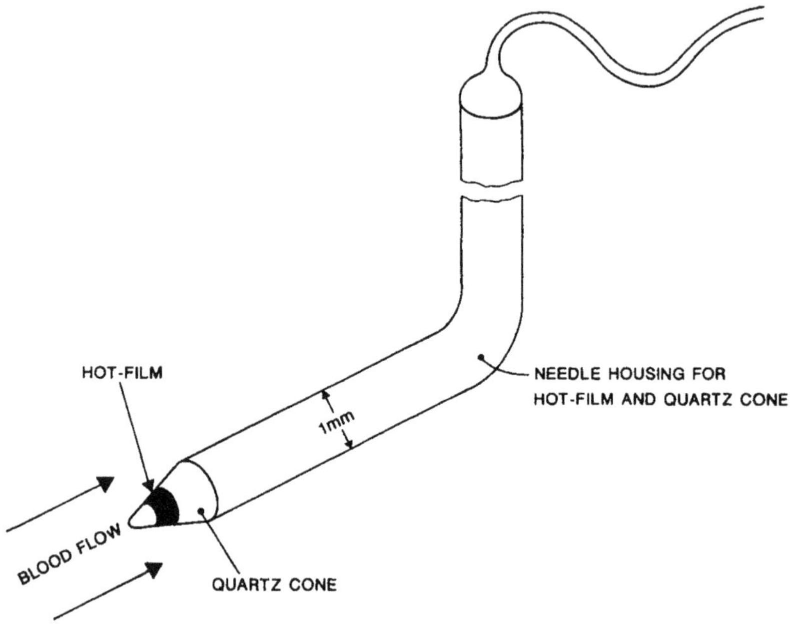

Figure 8. Schematic representation of the hot-film anemometer probe. The thin film is deposited on a quartz cone which is mounted at the tip of an L-shaped stainless steel needle. The tip of the hot-film anemometer is introduced in the vessel lumen where the hot-film can register the velocity of flowing blood around the tip of the probe but not the direction of blood velocity.

to intravenous infusions) and angulation in relation to the flow axis. The HFA probe is sensitive to flow around the hot-film but it is unable to detect the direction of the flow; furthermore, calibration is normally only performed for one direction in relation to the probe. The velocity signal from a measurement in the ascending aorta illustrates some of the features of the HFA technique (Figure 10). In the systolic deceleration phase turbulent velocity fluctuations are seen, and the output voltage of the HFA does not reach zero-level at any time since there is always some fluid motion around the hot-film. The advantages and disadvantages of HFA are summarized in Table 3.

Since the HFA probe has to be inserted into the vessel lumen there are some potential electrical hazards which have to be addressed; this is obviously particularly important when human measurements are conducted. A patient safety unit (and as a surplus also a probe protection unit) has been designed earlier by our group.[35]

Intraluminal positioning of needle probes has been accomplished in various ways depending on the measuring task. Ling et al.[36] used a micrometer device to position the probe in 1–2 mm increments along a vessel diameter. Amyot et al.[37] also

Figure 9. An example of a turn-table used for calibration of hot-film anemometer probes. By determining the distance from the probe tip to the center point of the turnplate and its rotational speed, it is possible to calculate the fluid velocity, applied on the hot-film.

included a rigid collar clamp to restrict the vessel from expanding during systole. The method which enables the most extensive mapping of velocity fields in large vessels was designed by Paulsen and Gormsen.[38] The device, which is shown in Figure 11, makes it possible, from an extravasal approach, to position the probe tip at 41 measuring points distributed evenly in the vessel's cross sectional area in a

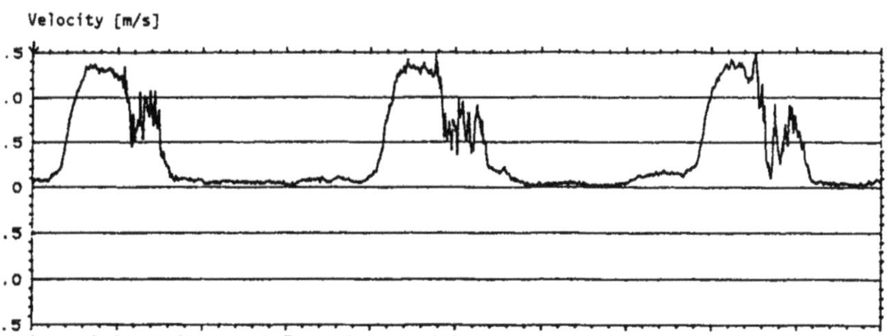

Figure 10. The output signal from the hot-film anemometer with the hot-film probe positioned centrally in the ascending aortic lumen of a pig. Note the turbulent velocity fluctuations in the systolic deceleration phase, and that the velocity tracing does not cross the zero-line at any time.

Table 3. Advantages and Disadvantages of the Hot-film Anemometry Technique

Advantages	*Disadvantages*
Applicable *in vitro* and *in vivo*	Invasive
High upper frequency limit	Direction insensitive
High spatial resolution	Sensitive to:
High signal/noise ratio	viscosity (hematocrit)
	temperature change
	dirt (fibrin) accumulation
	angulation of the probe
	Cumbersome calibration and measurement procedure
	Patient safety considerations due to electrical hazards

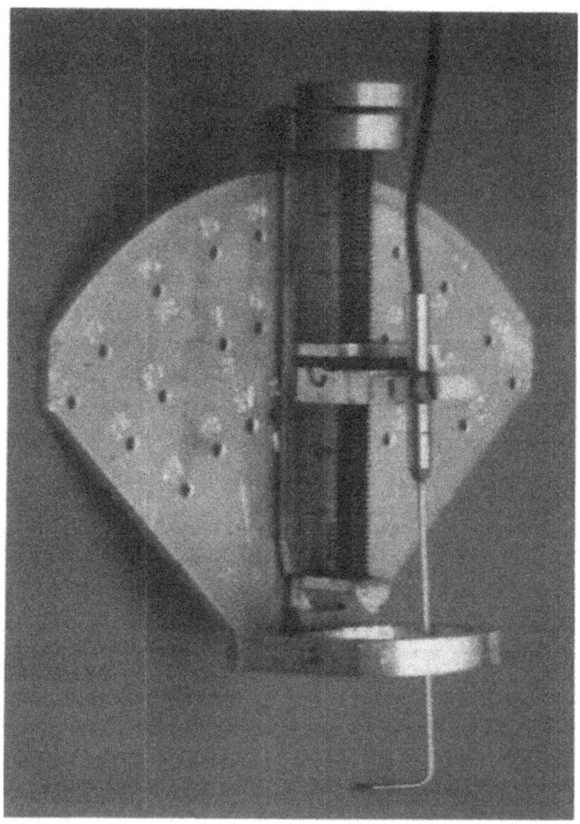

Figure 11. The hot-film anemometer probe positioning device designed by Paulsen and Gormsen. The device enables positioning of the probe tip at 41 measuring points distributed evenly in the vessel's cross sectional area.

plane perpendicular to the vessel axis. Thus, it is feasible to make fairly detailed mappings of velocity fields with three dimensional visualization techniques (see later in this chapter). Measurements in 41 measuring points is appropriate in vessels with a diameter exceeding 12 mm, since the spatial resolution of the HFA probe is 1–2 mm. For the large vessels, like the ascending aorta or the pulmonary trunk, the accuracy of probe positioning is compromised by the pulsatile vascular movements due to the function of the heart and lungs. These movements imply that the vessel is not at a constant position, and maintaining the measuring point is comparable to aiming at a moving target. Consequently, the probe positioning device has to be held manually in these applications,[39,40] but despite this source of inaccuracy, the data provided by this technique have a high degree of conformity with *in vitro* data acquired with HFA.[15]

HFA of different designs have been used for a variety of applications. Seed and Wood[41] evaluated different designs of needle probes for cardiovascular studies and pointed out some weak and strong points in different ways of mounting hot-films. Some researchers have used catheter mounted probes (e.g., ref. 42), others have used flush-mounted hot-films for wall shear stress measurements (e.g., ref. 43). The needle probes have been manufactured in a variety of designs since they are mostly custom-defined. Description of these different probe designs is beyond the scope of this presentation.

The **Doppler ultrasound** technique has gained increasing popularity throughout the recent years, both for research and routine clinical purposes. The most attractive features of the ultrasound technique is the non-invasive nature of measurements and the fact that data acquisition can be made with little or no discomfort for the person (or animal). Furthermore, there are no known hazards to the patient.[44] Consequently, ultrasound studies can be conducted frequently and for extended periods. Ultrasound can be used for continuous measurements of point velocities of blood either percutaneously, intra-operatively, intravascularly, or from an esophageal aspect. In the same procedure, or even simultaneously, a dynamic two dimensional image of the anatomical structures can be made. Thus, it is possible to inter-relate flow phenomena to the cyclic changes in geometry in, for example, intracardiac flow and flow patterns around heart valves and in the great vessels. A detailed description of the imaging part of ultrasonography is beyond the scope of this chapter which will focus on blood velocity measuring techniques and its potentials. The measuring technique is based on the Doppler principle which states that when an ultrasound wavefront is reflected from a moving object its frequency is shifted according to the object's velocity. The relation between the ultrasound frequency shift (f_d), object velocity (v), emitted frequency (f_0), and the angle of insonation (Φ) in relation to the direction of the moving object is:

$$f_d = \frac{2v * f_0}{c} * \cos \Phi \tag{7}$$

where c = the ultrasound velocity in tissue (1540 m/s).

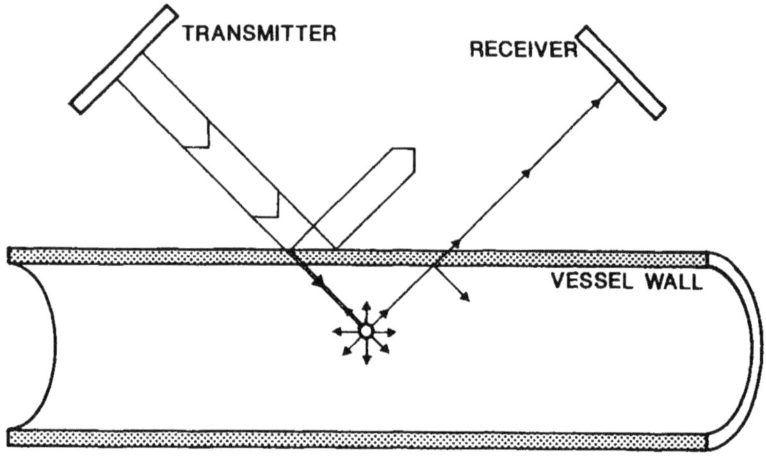

Figure 12. The scattering effect of an ultrasound wave emitted from a transmitter at a certain angle and sent through a vessel wall. Only a minor fraction of the ultrasound wave remains to be transmitted through the vessel lumen due to reflection at the vessel wall. Again only a minor portion of the signal is returned in the direction of the receiving crystal and the second passage through the vessel wall attenuates the signal once more.

There are two main operational principles of Doppler ultrasound for blood velocity measurements namely: continuous wave and pulsed wave technique. In the *continuous wave* technique one piezoelectric transducer crystal is emitting ultrasound continuously and another crystal physically separated from the emitter is constantly receiving the reflected, frequency shifted ultrasound as illustrated in Figure 12. With this technique a Doppler shifted ultrasound wave can be received from anywhere within the cross field of the ultrasound beams, which means that the frequency spectrum of the Doppler shifted signal is very broad banded and that no range discrimination can be performed. The clinical value of the continuous wave technique is the ability to detect the highest and lowest velocities within a beam, but it is unable to determine the exact distance from the transducer. The range limitation of continuous wave Doppler ultrasound is determined by the signal/noise ratio as indicated in Figure 12.

For practical experimental and clinical purposes continuous wave can be used for estimation of the highest blood velocity (a jet) within the insonated area but not the location; consequently spatial velocity mapping cannot be performed as with the HFA-technique.

The *pulsed wave* principle is illustrated in Figure 13. The pulsed Doppler technique uses a single piezoelectric crystal which is operated as emitter and receiver in an alternating fashion. Bursts of ultrasound with 5–20 cycles are insonated into the tissue of interest. The crystal is then operated in receiver mode,

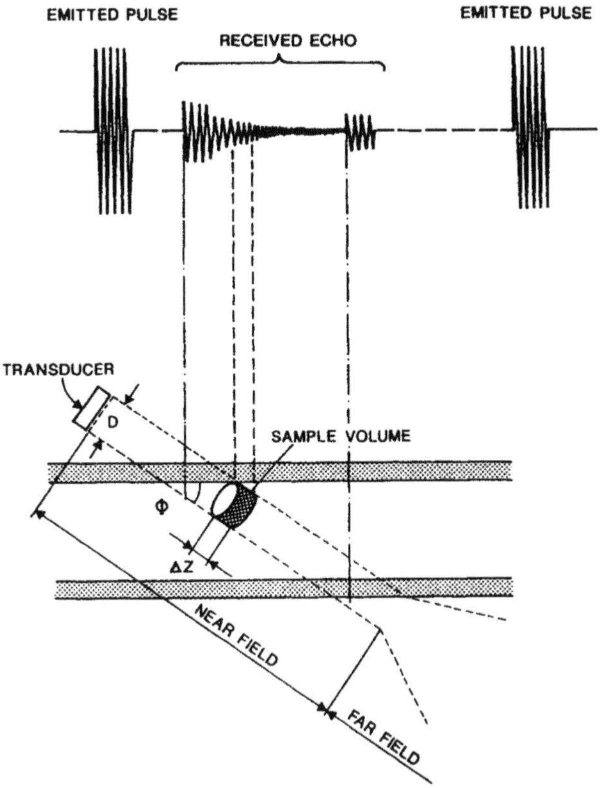

Figure 13. The operational principle of the pulsed Doppler ultrasound technique. A pulse of ultrasound energy is transmitted through the vessel wall at an angle (Φ), and by range-gating all signals received are neglected except for a small time interval which represents the sample volume. For further details please refer to the text.

and a predetermined number of milliseconds after the burst-emission, reflections are registered within a time gate. The gate delay determines the depth (location) of the sample volume and the gate width determines its size in the beam direction. The width of the piezoelectric crystal determines the width of the sample volume. The transducer has an operating range within the near field (see Figure 13). This range depends on the transducer frequency and diameter. Echoes received from the far field have an energy which does not justify reliable signal processing for velocity estimations. These range-velocity limitations will be described later.

Positioning of the sample volume is performed by adjusting the range gating function of the velocimeter and in duplex scanners, which comprise both imaging and velocity measuring modalities, the sample volume position can be verified visually on the screen.

By using multiple range gates each with its own mean frequency estimator blood velocity measurements can be made simultaneously, in multiple sample volumes across a vessel in the direction of insonation (multigate technique).

The electronic part of a continuous wave equipment consists of a small piezo-electric crystal which is excited at its resonance frequency (f_0) by a master oscillator. Ultrasound waves are then emitted from the crystal in the measurement direction. When the ultrasound is reflected or scattered from the tissue and blood cells toward the receiving crystal (see Figure 12) with a slightly changed frequency, it is capable of exciting the receiver crystal mechanically. This excitation of the crystal gives rise to a high-frequency electric signal which is fed into a receiving amplifier and subsequently demodulated by comparing this signal with the master oscillator frequency. The output of the demodulator is the frequency shifted Doppler signal (f_d) which is sent through a zero crossing detector or subjected to Fast Fourier Transform to calculate the velocity.

As indicated in the basic Doppler formula, angulation of the transducer in relation to the flow axis has an impact on the measured frequency shift and consequently also on the measured blood velocity. For practical clinical purposes, angulation below 30° is of minor importance but angulation of more than 60° has a significant impact, since not only is it important to determine the angle of insonation accurately but small changes in this angulation changes the velocity signal output significantly. Moreover, changing directions of blood velocity obviously changes the insonation angle, in other words, vortices and changing jet orientations also causes ambiguity.

It is important to acknowledge that the emitted ultrasound wave has an amplitude which is far higher than the reflected ultrasound amplitude (usually in the range of 80 dB). This is because the received echoes are the product of scattering in contrast to ideal reflection (illustrated in Figure 12). Therefore, even slight reduction in the emitted amplitude might have a significant impact on the quality of the received echo and therefore the inherent low signal/noise ratio may be more aggravated. When two media with different acoustic impedance are in close proximity then the reflection is even more accentuated. This is particularly relevant if the tissues contain air or hard tissue, bone, or foreign materials. This can pose practical problems in measuring through vascular grafts, in areas with atherosclerosis or in infected tissue with gas. In cardiac ultrasonography ribs and lung tissue are compromising access to unlimited insonation of the heart and great vessels.

Transducer contact to the surface where the ultrasound has to go through is an important aspect to appreciate. Good transducer coupling is normally secured by ultrasonic gel on the transducer head. Since the acoustic impedance in soft tissue and water is approximately identical, water is a suitable contact medium which is often used intraoperatively as stand-off and contact medium with good results.

A special technique for visualizing velocity fields is the color Doppler technique. This modality is basically an expansion of a multigated pulsed Doppler technique where multiple sample volumes are range-gated along a single ultrasound beam. By sweeping this beam over the scanning plane within a confined angle, multiple sample volumes can be spread over a relatively large area. For interpretation of this vast amount of data from a scanning plane the velocity direction and size are color coded; normally so that flow toward the transducer is presented in red, flow away from the transducer is encoded in blue, and flow 90° in relation to the scanning plane is black. Flow with multiple directions are encoded in various ways depending on the ultrasound equipment; often in yellow and/or green colors. The velocity is also semiquantified by the intensity of the color. Some ultrasound manufacturers even offer facilities for extracting data from each individual sample volume for computerized individual quantitation of blood velocity.

There are obviously limitations with the color Doppler technique since the beams of ultrasound are insonating at different angles with an inaccuracy of velocity quantitation and in extreme cases also ambiguity in determining the direction of flow. The distance between the ultrasound beams increases with increasing distance from the transducer. This increasing gap between the sample volumes is masked by special signal processing algorithms specific for each ultrasound equipment. Consequently, the reliability of data in the most distant part of the insonated area is diminishing gradually. The principles of color Doppler ultrasound is described in a very interpretable and illustrative approach by Kisslo et al.[45] and general Doppler ultrasound principles are described in more detail by for example, Hatle and Angelsen[46] and Atkinson and Woodcock.[47]

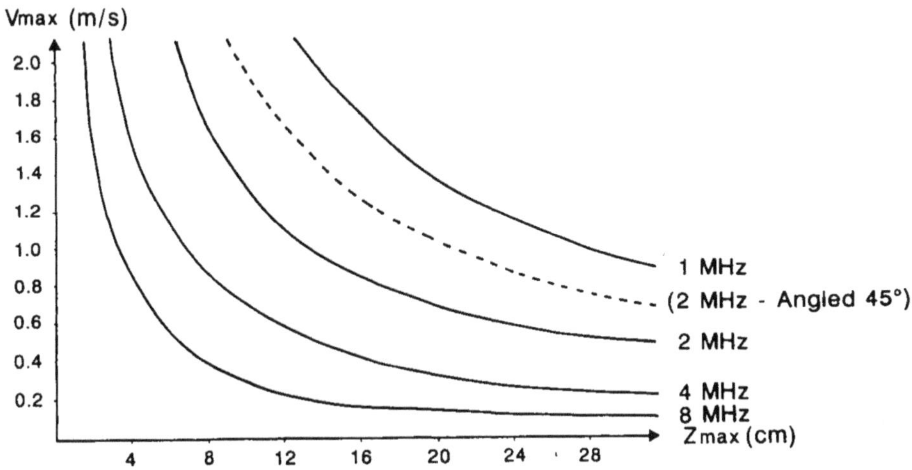

Figure 14. The relation between the maximum measuring depth of an ultrasound transducer and the corresponding maximum velocity which can be measured in that depth. If data are obtained above these curves range-velocity ambiguities will occur. The figure at each curve indicates the type of ultrasound transducer by its emitting frequency. The dotted line indicates the range-velocity relationship for a 2 MHz transducer with an angle of insonation at 45°.

For the understanding of "the bottom-line statements" given here it is also necessary to acknowledge the range-velocity limitations of pulsed Doppler ultrasound which represent a constant factor: when the working range of the transducer is increased the maximum velocity, which can be measured, is decreased and vice versa. This relation, which is characteristic for each transducer type, is illustrated graphically in Figure 14. It underscores the fact that the near field of the transducer is not a uniform power field. In a simple way the relation illustrates, that if blood velocities up to 1.0 m/s should be measured with an 8 MHz transducer then the distance to the most distant sample volume should not exceed 3.5 cm. Accordingly, the choice of Doppler transducer depends on which qualities has the highest priority in each particular study.

For transcutaneous precordial measurements the most appropriate transducer frequency is typically in the range of 2–3.5 MHz as a compromise of sufficient penetration and accuracy of velocity measurements. When the distance from the transducer to the target is shorter, (e.g., in children, or when only anterior structures are of interest) higher transducer frequency can be selected to improve the accuracy

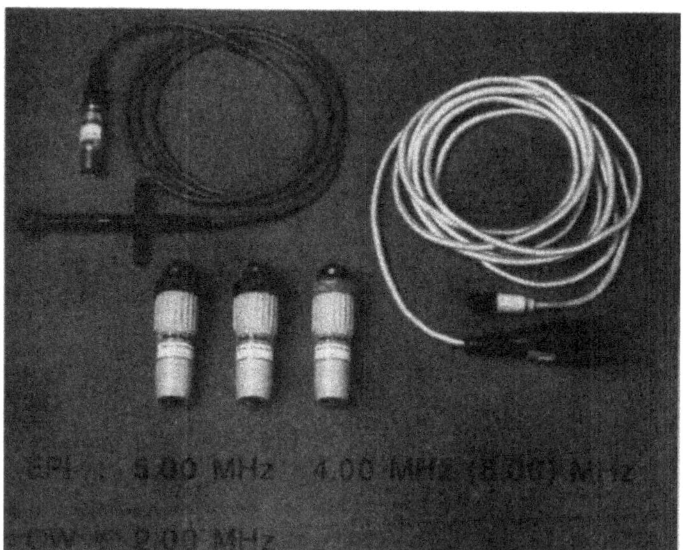

Figure 15. Some examples of ultrasound transducers (all are produced by Vingmed Sound, Norway). Top left is a 10 MHz single Doppler transducer applicable for continuous wave and pulsed Doppler measurements. In the bottom are three conventional mechanical phased array transducers capable of being operated in all modalities each with its individual frequency. At the right is an example of a miniaturized mechanical phased array transducer which is designed for intraoperative use where access is limited, for example, at the lateral and posterior aspect of the heart during cardiac surgery.

of blood velocity measurements. The transducers used for this purpose have a physical size and are designed to be hand-held (see Figure 15). For intraoperative application it is seldom a problem to get close enough to the target; high-frequency transducers can be used here. These are normally in the range of 5–10 MHz. However, another problem may arise in these measurements since the near field does not start at the front end of the transducer but a small distance away (depending on the type of transducer). Therefore, measurements close to the contact surface is sometimes only possible when some stand-off material which has the same ultrasound characteristics as the target is interposed. Special sterilizable stand-off materials are commercially available for intraoperative use but sterile saline can often provide equally good results. (An example of a transducer designed for intraoperative use is shown in Figure 15). Cardiovascular flow measurements can also be performed from an esophageal approach using a transducer mounted on a conventional gastroscope. In this way the distance to the posterior aspect of the heart is short, and the entire thoracic aorta can usually be visualized and subjected to blood velocity measurements with a 5 MHz transducer. Transesophageal ultrasonography is normally performed as duplex scanning (both imaging and Doppler modalities). It *can* be performed in cooperative patients or volunteers without any sedation, but tranquilizers are normally administered for such an investigation.

Pure Doppler transducers are applicable as well, and these are manufactured in a variety of designs defined by the producing company or custom-designed. An example of a commercially available Doppler transducer is also shown in Figure 15.

High frequency Doppler transducers applied for blood velocity measurements with a high degree of accuracy within a small sample volume often necessitate intraluminal intrusion of the probe. This is most often embodied as a needle probe comparable to the hot-film anemometer needle probe. Typically a 20 MHz crystal is used to "point" at a sample volume a few millimeters upstream of the probe. This implies that although the method is invasive the probe does not disturb the measuring area. Like the HFA probe it is possible to position this intraluminal probe by extravasal adjustment. This application is normally used for detailed dynamic visualization of velocity profiles in large vessels (e.g., ref. 48).

Within the last five years catheter-mounted ultrasound transducers have been introduced to make intraluminal 2-D echo studies and Doppler measurements. The transducer frequency which is used for these studies are in the range of 15–25 MHz. The catheters can be manufactured to very small sizes (3 French) and can be manipulated into human coronary arteries. The technique is used clinically to visualize atherosclerotic plaques prior to and after percutaneous transluminal angioplastic (PTCA) procedures in order to verify the nature of the stenosis and to detect possible significant dissection in the vessel wall. The Doppler ultrasound modality can be used to quantitate the hemodynamic impact of a coronary artery stenosis and to evaluate the efficiency of, for example, PTCA. The accuracy of these measurements is difficult to evaluate since these techniques are very new and still lack a significant part of methodological evaluation but the technique itself does seem to imply interesting potentials. One of the major concerns which have to be addressed is the problem of orientation since it is very difficult, without concomitant X-ray angiography, to know exactly where in the arterial tree the transducer is positioned.

Another approach to blood velocity measurements has been adventitial implantation of miniaturized Doppler probes as an implantable, extractable transducer.[27] This type of Doppler probe has obvious advantages in opening the possibility for chronic measurements postoperatively in experimental animals and even in patients. It has also been applied with multigate technique for continuous estimation of cardiac output in patients after open heart surgery.[49]

Transducer technology is a specialty of its own and detailed descriptions of different transducer designs and operational principles can be found in the literature (e.g., refs. 46, 47). There are three main principles of transducer operation: phased array, linear array, and mechanical. The phased array type utilizes excitation of a matrix of small crystals either simultaneously or slightly delayed in relation to each other depending on the operational characteristics. There are no moving elements in this transducer, the size is fairly small and the versatility is better than its two counterparts. On the other hand, cross talk between the different crystals is a frequent problem, and the spatial resolution in the distal part of the near field is reduced compared to the mechanical transducer.

The linear array transducer has a line of crystals which emit ultrasound in parallel lines, and in this way avoids the problem of increasing distance between the ultrasound wavefronts with increasing distance from the transducer. It is more voluminous than its two counterparts and therefore potentially faces a problem with insonating through windows especially in transthoracic applications.

The mechanical transducer operates with moving parts and therefore faces problems with durability and it has an audible function. For the intravascular probes it is particularly difficult to rotate the crystal by a rotatable shaft within these very thin catheters but it has proven technically feasible. Like the phased array transducer it has the advantage of a small emission and receiving point, and since it physically scans through a plane it does not leave any areas un-investigated, regardless of the distance from the transducer.

Like most other aspects in the field of ultrasonography, selection of an appropriate transducer is a compromise between different demands to the measurement. Pulsed Doppler ultrasound is measured at a certain angle in respect to the vessel's main flow axis. This means that the sample volumes, which are confined to the ultrasound beam area, are positioned at an oblique line compared with the main flow axis. In long tubes this is not an issue of major concern but in areas with bifurcations, curvatures, or downstream of vessel/valvular stenoses or prosthetic heart valves it is essential for data interpretation that the plane of measurement is orthogonal to the main flow axis. A measuring scheme like that described in the

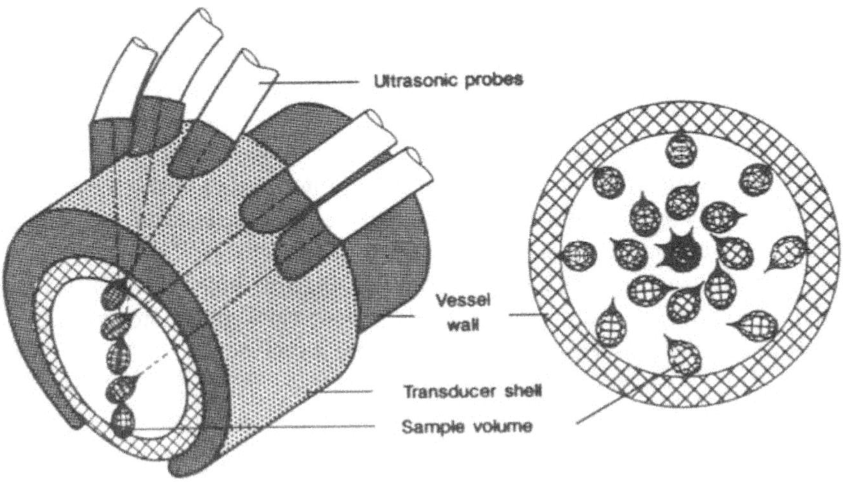

Figure 16. Schematic illustration of our own custom designed 5-element 10 MHz ultrasound transducer. The individual ultrasound probe is positioned in a hole in the perivascular PVC-shell and "points" at a sample volume on a diameter perpendicular to the main flow axis. By rotating the PVC-shell on the vessel different diameters can be subjected to point blood velocity measurement, and in this way blood velocity mapping can be performed in the entire cross-sectional area.

hot-film anemometry section is not feasible without accurate positioning of the Doppler ultrasound probe at different insonation locations.

We have addressed this problem in our group by designing and manufacturing our own Doppler ultrasound probes. Five individually range-gated probes have been mounted in a PVC-shell with holes to fit the probes (Figue 16). The shell secures alignment with the vessel's long axis; it keeps the probes in a stable and reproducible position and the direction of the holes enable measurements of blood velocities in a plane orthogonal to the flow axis. The insonation angle of the five probes ranges from 30° to 65°, each probe is operated individually with appropriate angle correction of the velocity signal. The sample volumes are positioned evenly along a diameter in the vessel, and by rotating the PVC shell on the vessel every diameter can be insonated, and the entire cross sectional area of a vessel can be subjected to blood velocity mapping in a plane orthogonal to the flow axis. The PVC-shells are made in different sizes in order to fit a range of vessel diameters from 12–40 mm.

Doppler ultrasound is normally limited to measurement of laminar velocities since commercially available ultrasound velocimeters have a low frequency response (-3dB below approximately 50 Hz). Quantitation of turbulence with the use of Doppler ultrasound has been attempted by several authors (e.g., refs. 50, 51, 52, and 53). There are several approaches to solve the problem of inherent low frequency response of Doppler ultrasound. The signal processing technique can be modified at the level where the emitted and received signals (quadrature signals) are compared to compute a velocity signal. The most frequent techniques are the zero crossing technique and the Fast Fourier Transform; these can be modified in various ways for expanding the frequency response. Nygaard et al.[53] evaluated the zero crossing technique comprising a modified filter setting in the demodulator section. In this way the -3dB cut-off frequency could be increased from 30 Hz to 240 Hz. They also made an overview of various other signal processing techniques which are designed for a reliable high-frequency registration of blood flow disturbances. Another approach for using Doppler ultrasound to quantitate turbulence is the use of spectral broadening index (SBI) which is defined as:

$$SBI = 1 - \frac{F_{\text{mean}}}{F_{\text{max}}} \tag{8}$$

where F_{mean} and F_{max} are the systolic mean and maximum frequencies, respectively.[54] This method is obviously based on a reliable determination of the mean and maximum frequency in the Doppler shifted signal. Since this determination is associated with considerable inaccuracy also the SBI is associated with sources of error. Sillesen[52] used SBI for estimating the degree of carotid artery stenosis and found a reasonable correlation with angiography and transstenotic pressure loss.

Apart from improved signal processing and the use of better parameters for estimating the degree of turbulence, further advances can be achieved in transducer technology since the size of the sample volume is critical for the accuracy of determining fluctuating velocity components. In a large sample volume the small-scale fast fluctuating velocity components are blurred by the many other low frequency large-scale velocity components which are also detected in the area of observation. Transducers can be designed to focus the ultrasound beam and thereby diminish the sample volume to less than the width of the piezoelectrical crystal. This beam focusing can be performed electronically by the phased array transducers and physically in the mechanical transducers. The focusing principle is described in further details by, for example, Feigenbaum.[55]

To summarize the application of Doppler ultrasound technique for quantitative measurement of turbulence it is obviously a method with a significant potential, first of all because it is non-invasive in nature. It is probably a matter of further technical, electronic, and mathematical achievement to improve the ability to quantitate turbulence in a reliable fashion. At this point, the necessity of a great deal of operator experience seems inevitable for *in vivo*, and in particular, patient studies of turbulence. Overall the echo-Doppler technique has gained extremely widespread use for cardiac and vascular applications both in scientific and clinical applications. It is still fairly expensive equipment to procure but obviously not as expensive as magnetic resonance scanners. Due to the other advantages mentioned in Table 4 and the continuing technical advances, measuring techniques based on

Table 4. Advantages and Disadvantages of the Pulsed Doppler Ultrasound Technique

Advantages	*Disadvantages*
Non-invasive	Low spatial resolution
No calibration	Low signal/noise ratio
Accurate	Range-velocity limitation (pulsed wave)
Combination with two dimensional echo-cardiography is feasible	Spectral broadening due to: 1. Transit time distortion
Implantable in certain applications	2. Disturbed or turbulent flow
	3. External noise
	Operator experience essential

ultrasound seem to be destined to become the overall tool for hemodynamic evaluation within the following years.

The **magnetic resonance** technique is increasingly applied for blood velocity measurement since it is purely non-invasive and seems to be very accurate in both imaging and velocity measurements. Operation of this very advanced equipment requires a vast amount of technical skills and does not seem to be a method for general clinical or research application within the next few years. The MR-technique for blood velocity measurement is described in further detail in another chapter of this book.

C. Blood Pressure Measurements

As mentioned in the introduction blood pressure is a very old hemodynamic parameter which has been a reference measure for several decades. Clinically it is measured by a **sphygmomanometer** on virtually every patient admitted to the hospital. The blood pressure measuring technique is known by every young medical student and is performed by wrapping a sphygmomanometer cuff around the upper arm. By a hand bulb the cuff is inflated while the cuff pressure is monitored on a mercury manometer. The pressure is increased well above the expected systolic blood pressure and a stethoscope is positioned over the brachial artery distal to the cuff. The cuff pressure is slowly lowered while both looking at the manometer and listening for the sound of turbulent blood flow in the brachial artery. The pressure where the first murmur is heard is registered as the systolic pressure and while still relieving the cuff pressure the diastolic blood pressure is registered at the level where the audible sound of blood flow in the artery disappears or changes its character markedly. Evidently the pressure level where the blood flow is totally obstructed is a measure for the systolic pressure but the diastolic pressure is more difficult to interpret since the sound of flowing blood does not always disappear, and the point where it changes its character is often difficult to determine accurately.

Blood pressure measurement with this technique may therefore vary considerably depending on the "operator." The blood pressure technique has an obvious advantage in being non-invasive, harmless, (provided that the cuff pressure is not raised too much), and can be repeated frequently. It provides a good measure for *changes* in blood pressure in a particular individual. The technique can be automated by a machine which can inflate, measure the cuff pressure, and register systolic/diastolic and mean blood pressure. This feature is often utilized by anesthesiologists while monitoring anesthetized patients or patients in intensive care. The disadvantages of the sphygmomanometer method is its inherent discontinuity. The quantitative accuracy might be questionable since the cuff pressure might not reflect the intra-arterial pressure due to the inter-individual variability in tissue characteristics of the upper arm (adipose, muscular, and skeleton tissue). Furthermore, a dependent stiffening of the arterial wall changes its resistance to external pressure. The disappearance of the turbulent murmur which reflects the diastolic pressure depends on the blood flow through the artery and obviously also on the auditory ability of the "operator." Advantages and disadvantages of the sphygmomanometer technique are listed in Table 5. The sphygmomanometer blood pressure technique should be selected for studies where monitoring of changes in blood pressure is more important than the absolute value and in cases where invasive measurements cannot be justified for ethical or practical reasons.

For more detailed blood pressure measurements invasive techniques most often have to be applied; and by that more knowledge also has to be gained about the quality of blood pressure which is to be measured and the capability of the equipment. More basic theory and general pitfalls in blood pressure measurement and interpretation can be found in text books on these subjects (e.g., ref. 21).

Measuring blood pressure with **fluid filled catheters** is an invasive technique which requires intravascular access either by puncture or dissection of the vessel. A manometer line is inserted by Seldinger technique (needle puncture, guide wire insertion followed by protrusion of a catheter over the guide wire). Once the

Table 5. Advantages and Disadvantages of the Sphygmomanometer Method for Blood Pressure Measurements

Advantages	*Disadvantages*
Non-invasive	Discontinuous
Reliable for intra-individual monitoring	Inaccurate
Harmless	"Operator"-dependent
No discomfort	
Can be repeated frequently	
Inexpensive	
Can be automatized	

catheter is in its intravasal position, it is coupled to a manometer line and through that to the pressure transducer (see Figure 17). Interposed are two three way stopcocks which enable zero adjustment and exposure of a calibration pressure. Isotonic saline is slowly and constantly infused in the manometer and pressure catheter to avoid intrusion of blood into the catheter and thereby plugging it with coagulated blood. The blood pressure is transmitted through the pressure catheter and manometer line to the pressure transducer which registers the pressure by a strain gauge technique. Deflection of the membrane is reflected by an electric signal which is amplified and verified on a paper recorder, monitor, or other recording media. The distance from the measuring site in the vessel to the pressure transducer is often fairly long (approximately 1 m). The medium through which the pressure wave is transmitted is obviously critical for the fidelity of the waveform which is recorded by the transducer. The catheter must be stiff enough to avoid dampening of the pressure amplitude and the fluid in the catheter should ideally be incompressible. It is therefore essential to avoid air microbubbles in the pressure line. Likewise, resonance phenomena should be avoided since excitation with pressure fluctuations at a frequency similar to the system's natural frequency can cause oscillations and thereby artifacts in the pressure recording. These and other sources of errors are described by, for example, Falsetti et al.[56] The frequency response of fluid filled catheter systems as described above are in the range of 20–50 Hz depending on the environment in which they are operated. The limitation of the frequency response is normally not the pressure transducer itself, but the catheter and the fluid within it. Therefore, evaluation of the frequency response of these kinds of systems should comprise the entire measuring set-up. This can be done by using the sinusoidal frequency response to the step response techniques. These techniques are described in detail by, for example, Nichols and O'Rourke.[21] In addition to the amplitude response, phase distortion of the system also has to be characterized, especially when timing of blood pressure in relation to other biological signals is an essential parameter. When a thorough characterization of frequency response and phase distortion has been performed it is also important to consider whether electronic compensation for insufficient performance should be applied. The technique for doing this is described by, for example, Falsetti et al.[56] Another way of improving the frequency response is to reduce the distance through which the pressure wave has to pass by shortening or even avoiding the manometer line (Figure 17). A possible artifact is the so-called "catheter whip distortion." Whenever a liquid filled catheter is struck, a transient pressure is created by the acceleration. If such a transient pressure wave coincides with the system's natural frequency, pressure fluctuations will be superimposed on the pressure tracing. In the cardiac cavities and the large vessels, sudden pressure and flow changes can cause the pressure catheter to collide with cardiovascular structures leading to catheter whip distortion. This artifact is normally easy to depict from the pressure tracings from the characteristic high frequency oscillations superimposed on the peak systolic pressure waveform. Dynamic pressure artifacts can arise when high blood velocities

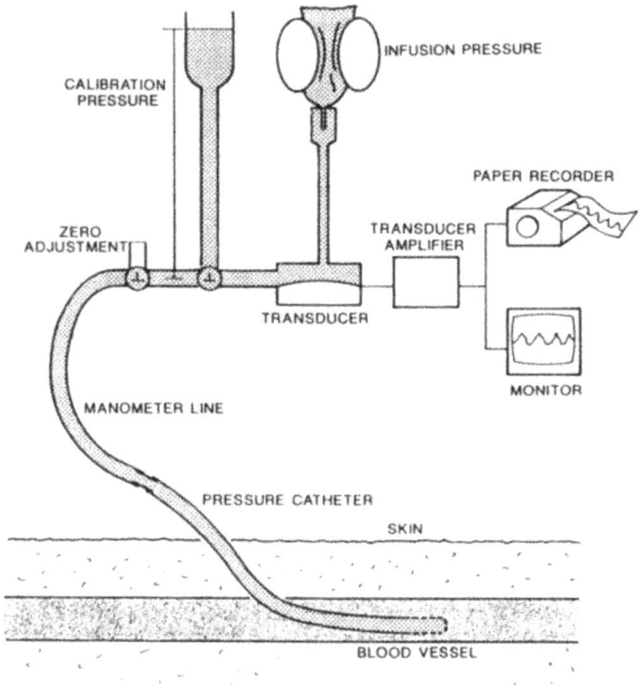

Figure 17. Diagrammatic representation of the fluid filled catheter technique for intravascular blood pressure measurement. The catheter is inserted transcutaneously into the vessel lumen and connected to an external pressure transducer via a manometer line. The pressure sensor converts pressure to an electric signal which can be amplified and displayed. Using three way stopcocks, zero adjustment and predetermined calibration pressure can be applied. Isotonic saline is slowly and constantly infused into the catheter.

are directed toward the inlet orifice of the pressure catheter. Consequently, the recording of the kinetic energy is converted to pressure energy giving rise to elevated intravascular pressure. The dynamic pressure artifacts are most likely to occur in areas with high blood velocities, that is, in large arteries and downstream of vascular stenoses.

Fluid filled catheters are routinely used clinically for cardiac catheterizations with the aim of measuring intracavitary and vascular blood pressures. Pressure losses ("pressure gradients") across valvular or vascular stenoses are used clinically as an indication of the hemodynamic resistance of an obstruction. The hemodynamic resistance (R) is defined as:

$$R = \frac{\Delta p}{Q} \tag{9}$$

Table 6. Advantages and Disadvantages of Fluid Filled Catheters for Blood Pressure Measurements

Advantages	*Disadvantages*
Continuous	Invasive
Reproduceable	Low upper frequency limit
Stable	Possible artefacts due to:
Applicable for chronic measurements	overshoot
Inexpensive (compared with the tip transducer technique)	dampening
	catheter whip distortion
	tapping on manometer lines
	Calibration is necessary
	Phase distortion

where Δp is the trans-stenotic pressure loss and Q is the blood flow through the stenosis. Accordingly, Δp cannot be interpreted as an expression of the severity of a stenosis without taking the blood flow into consideration.[57] Ideally the blood pressure on each side of the stenosis should be measured simultaneously and measured by a differential pressure transducer to avoid calibration inaccuracies.[58] Advantages and disadvantages of the fluid filled catheter technique are listed in Table 6.

As indicated above, the frequency response could be improved by shortening the distance between the strain gauge transducer and the tip of the catheter. In the **catheter-tip transducer** full advantage of this approach has been taken. The strain gauge sensor is mounted at the distal end of a conventional catheter so that pressure measurement can be made directly at the site of interest (see Figure 18).

The size of these strain gauge mounted catheters can be reduced to 2 French (< 1 mm in diameter) and obviously provide the opportunity to make accurate intravascular blood pressure measurements even in small vessels.

As indicated above, a short distance from the measuring site to the strain gauge sensor implies diminished phase delay. The low damping mass which the pressure wave must pass has a major impact on the frequency response which is flat up to 10–20 kHz depending on the transducer design. Thus, pressure recordings can be made with high fidelity even throughout the audio frequency range (see Figure 19). This feature has been used for measuring small pressure fluctuations in turbulent flow which can also be heard as murmurs (e.g., ref. 59). This technique also allows the presence of turbulence to be identified since high-frequency pressure fluctuations has been seen to coincide with high velocity fluctuations registered with a hot-film anemometer.[60] The main advantage of the catheter-tip transducer is its high frequency response. This virtue is of main importance when dp/dt (pressure change as a function of time) measurements are requested since a rapid rise-time in the

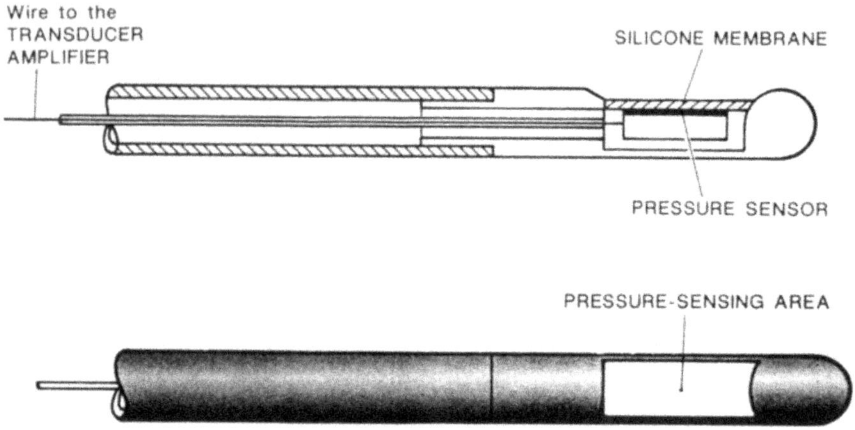

Figure 18. The microtip catheter principle. A strain gauge sensing element is mounted in a rigid frame at the tip of a conventional intravascular catheter. The catheter can be maneuvered intravascularly like any other catheter, although in this case it is not possible to introduce a guide wire.

systolic acceleration phase has to be traced accurately in order not to misinterpret an otherwise well-contracting myocardium as failing.

Recordings with a catheter-tip transducer can be justified even for normal blood pressure measurements since at very high pulse rates (> 300 min^{-1}) fluid filled catheters are often unable to reflect the rapid rise of the systolic blood pressure. These high pulse rates are frequently found in small laboratory animals. The advantages and disadvantages of the tip-transducer are listed in Table 7.

Various design modifications of the tip-transducer catheter is offered by various manufactures: (1) Double transducers which can be used to measure transvalvular or transstenotic pressure difference, and (2) combined pressure and blood velocity measuring catheters comprising a Doppler ultrasound transducer. The advantage of

Table 7. Advantages and Disadvantages of Tip Transducers for Invasive Blood Pressure Measurements

Advantages	*Disadvantages*
Continuous	Potential electric hazards
Reproducible	Invasive
Stable	Expensive (compared with fluid filled catheter technique)
No calibration is required	
High upper frequency limit	Fragile
Applicable for chronic measurements	
High precision	

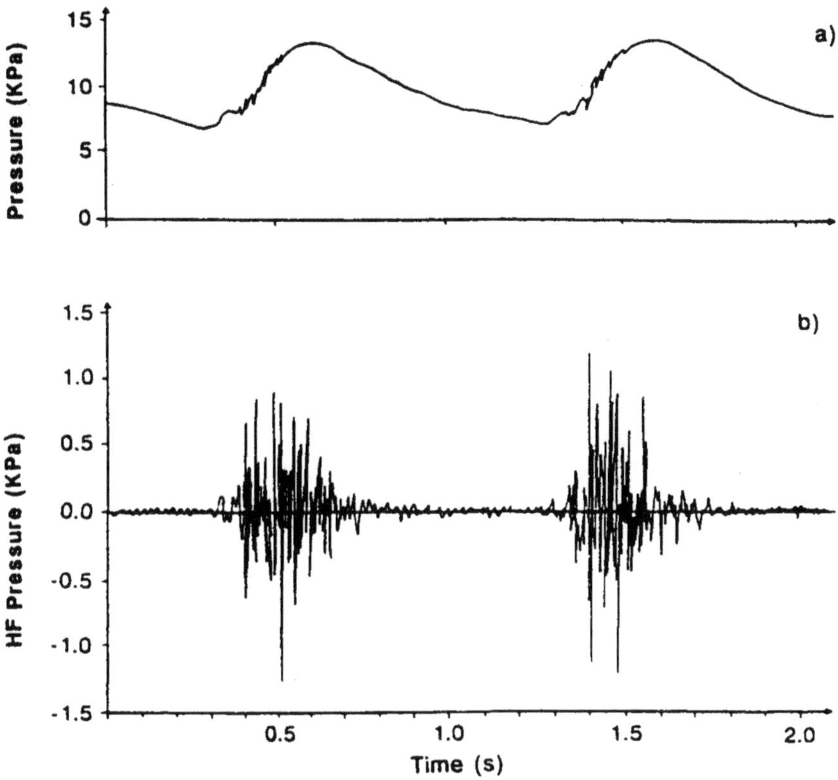

Figure 19. (a) Blood pressure signals recorded with a Millar® tip transducer catheter positioned in the ascending aorta of a patient with aortic valve stenosis. (b) The same signal as in (a) after 20 Hz high-pass filtering and amplification. Note the high-frequency systolic pressure fluctuations. HF = high frequency.

such combined catheters is of course the small size of a duplex measuring system and the possibility to make simultaneous pressure and flow measurements. But exact positioning of a catheter in the cross sectional area is not feasible in large vessels or in cardiac cavities.

From a methodological point of view a trend toward the application of purely non-invasive measuring techniques must be a goal for all scientists within hemodynamic studies.

The **stethoscope** has been the first choice of method for hemodynamic evaluation for more than two centuries. It provides the opportunity to detect flow disturbances in the cardiovascular system because turbulence causes vibrations in the surrounding tissue and from here to the body surface where it is heard as murmurs. Due to the periodical nature of murmurs arising from blood flow disturbances it is generally possible to diagnose the origin of the murmur fairly reliably. The major drawback of investigations made by a stethoscope is the inability to verify the results and consequently, the inability to provide quantitative data with this method. As an attempt to communicate the quality of cardiovascular murmurs by objective means **phonocardiography** was introduced in the 1950s. The murmurs are registered by a contact microphone and an amplifier, and displayed on an oscilloscope or a paper ink writer. The advantage of this technique is that the contact microphone has a linear frequency response in the range of 10–800 Hz, which is certainly not the case for the stethoscopes used clinically. Furthermore, the same data can be subjected to judgment by several doctors and data storage provides the possibility to compare the results from one investigation to another with long time intervals. Due to the inability to provide quantitative data the use of phonocardiography has gradually diminished through the years since the introduction of new investigational methods has reduced phonocardiography to a screening method rather than a diagnostic tool. In this role, it does not provide any significant advantages over the stethoscope which is always available for clinicians.

With the significant improvements in signal processing techniques it has recently been feasible to quantitate phonocardiographic signals.[61,62,63] We have used a 150 ms time window to extract the systolic murmur. The mean energy density spectrum of isolated murmur signals from 20 heart cycles are subjected to Fast Fourier Transform to display the mean energy density spectrum as illustrated in Figure 20. The ratio between the energy in the high-frequency range and the energy in the entire frequency range

$$\left[\frac{E_2}{E_1 + E_2}\right] \tag{10}$$

in Figure 20 reflect the degree of flow disturbance and by that the severity of a stenosis. Since this method is based on spectral analysis of precordial vibration signals this expanded phonocardiographic technique is called **Spectral Vibro Cardiography (SVCG)**. The rationale for utilizing SVCG for quantitation of the degree of flow disturbance (turbulence) is the Strouhal relation:

$$St = \frac{f_0 d}{U} \tag{11}$$

where St = Strouhal number, f_0 = corner frequency of the generated pressure fluctuations, d = orifice diameter, and U = jet velocity. This relation is described in further detail (e.g., refs. 64, 65).

Previous authors have assumed that the attenuation of the vibration waveform as it passes from the turbulent zone to the precordium does not exhibit interindividual variation. When recordings of high-frequency pressure fluctuations in the ascending aorta in a patient with aortic valve stenosis are stored simultaneously with the precordial vibration signal then this assumption proves invalid (see Figure 21). The transthoracic transfer function has been estimated in humans[66,63] and discloses a

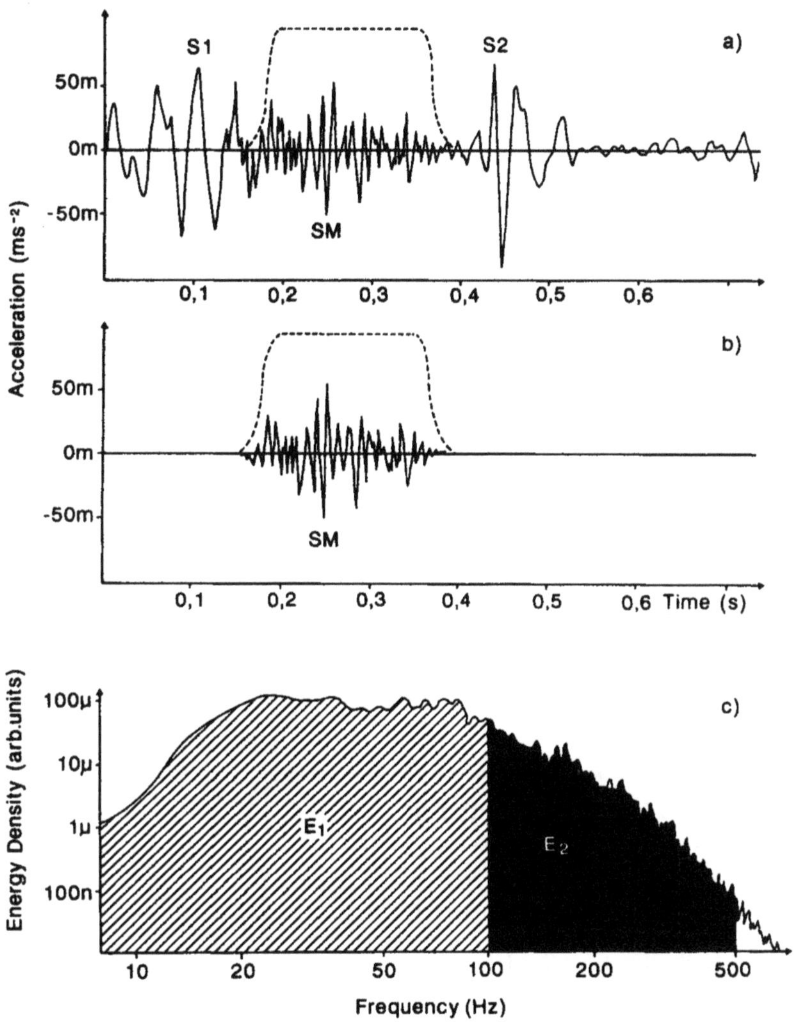

Figure 20. (a) Precordial vibration signal displaying one cardiac cycle with first and second heart sound indicated by S1 and S2. A smooth-edged 150 ms time window is used to extract the systolic murmur (SM). (b) The isolated systolic murmur signal. (c) The mean energy density spectrum of the systolic murmur from at least 20 heart cycles. The energy in the high-frequency range compared with that in the entire frequency range reflects the degree of flow disturbance and is thus a measure for the severity of flow obstruction.

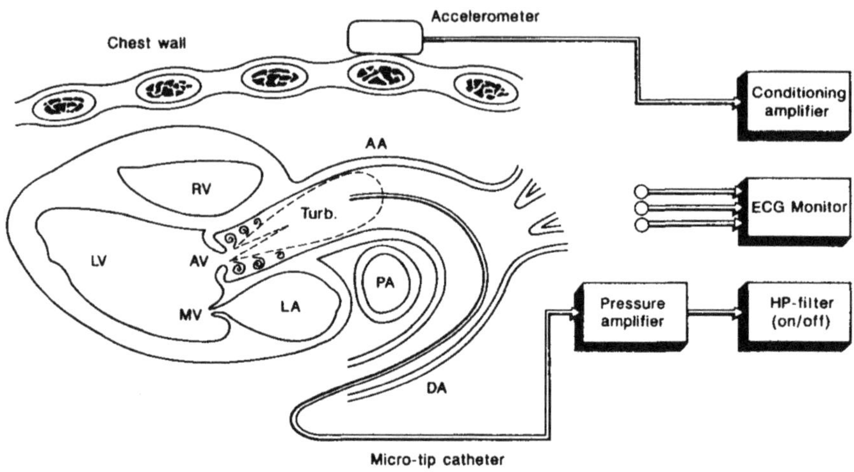

Figure 21. The attenuation of vibration wavefronts through the chest wall can be quantitated by simultaneous recordings of intra-aortic measurement of high-frequency pressure fluctuations downstream of an aortic valve stenosis, and precordial vibrations registered by an accelerometer. The two signals are synchronized with surface ECG.

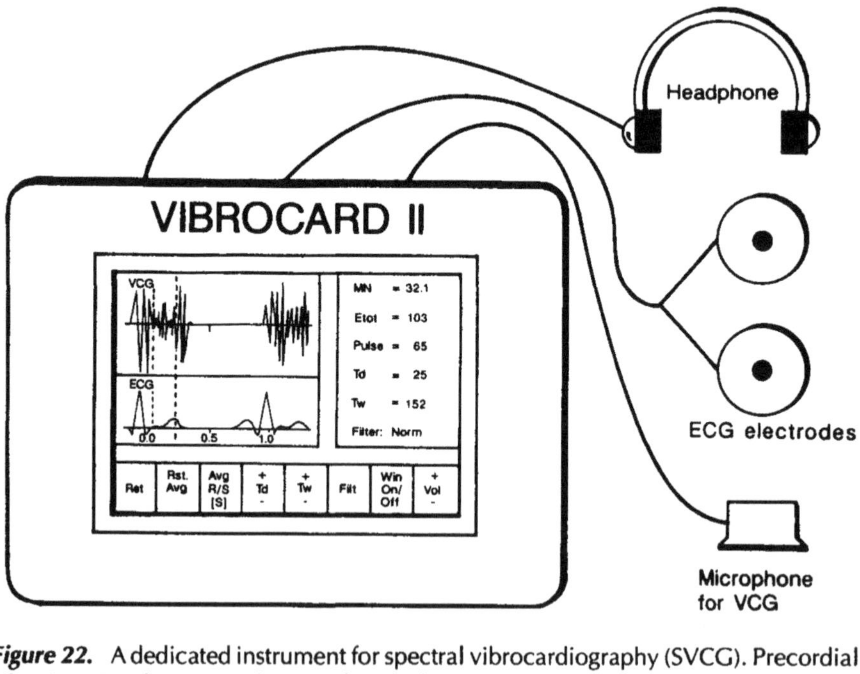

Figure 22. A dedicated instrument for spectral vibrocardiography (SVCG). Precordial vibration signals are synchronized with the ECG, and the murmur is extracted for computation of the spectral energy distribution to display energy spectra, time signals, and energy ratio. Construction by Bang & Olufsen Technology A/S, Denmark.

frequency-dependent attenuation with individual baseline attenuation; no further attenuation occurs up to approximately 25 Hz and beyond this "corner" frequency an individually increasing attenuation depends on the patient's bodily stature. The individual attenuation of vibrations can be compensated for. By application of a dedicated vibration transducer and high-speed signal processing all signal energy ratios and derived parameters can be displayed on-line (Figure 22).

The SVCG comprising individual compensation for thoracic wall attenuation is a fairly new technique which still needs further validation, but so far it has provided promising results in estimating the aortic transvalvular pressure loss in patients with aortic valve stenoses. The technique is designed for clinical use but it is also applicable in animal experimental research as a non-invasive tool for chronic experiments.

V. DATA PRESENTATION

Blood flow and blood pressure can be presented in terms of a signal tracing to show the temporal variation in amplitude and waveform. They are gross hemodynamic measures which serve as reference for other measures and/or as monitoring parameters. Both values can be subjected to further computation to provide a variety of derived parameters.

Graphical data presentation cannot be described comprehensively in this chapter but some general principles will be given for illustration. Almost every research group has its own standards for layout, image processing, and notation. This non-uniformity in data presentation may sometimes hamper a thorough comparison of data between the centers. It is caused by individual attitude to presentation technique whether the emphasis is put on simplicity in graphics with the potential risk of not being sufficiently comprehensive to provide a good overview of data. On the other hand advanced graphic illustrations which comprise large amounts of information may blur the details.

Signal tracing gives an idea of the quality of the raw data and is often convenient background information for interpretation of the signal processing. An example of signal tracings and processed signals are shown in Figure 23. The instantaneous blood velocity signal is obtained in the central part of the ascending aortic lumen downstream of a prosthetic aortic valve implanted in a pig. The instantaneous blood velocity signal ($U(t)$) can be shown to be the sum of two velocity components: The turbulent velocity component ($u(t)$) and the laminar (underlying) velocity component $\langle u(t) \rangle$

$$U(t) = \langle u(t) \rangle + u'(t). \tag{12}$$

The illustration of these three signal components are illustrated in Figure 23. The instantaneous velocity waveform is shown as the laminar (dense line) plus the overlying fluctuating velocity component. The turbulent velocity component is

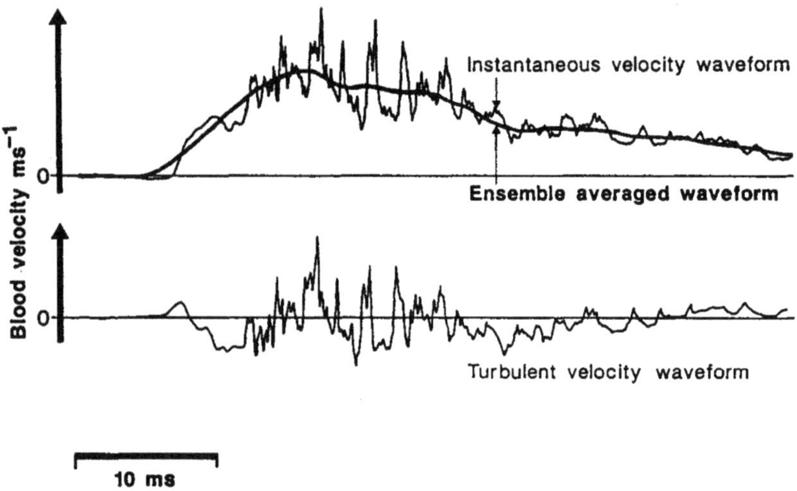

Figure 23. Example of velocity tracings showing the instantaneous, laminar (ensemble averaged), and turbulent velocity component.

extracted by ensemble averaging of the instantaneous velocity signal to obtain the laminar (or deterministic) velocity component. Subsequently, the laminar velocity component is subtracted from the instantaneous velocity to give the turbulent (or non-deterministic) signal component. This form of data presentation illustrates both the raw data and some of the signal processed data before it enters specialized further processing (velocity profiles or turbulence analysis).

The principles for raw data presentation illustrated here by blood velocity signals can obviously be applied to other types of hemodynamic signals such as blood pressure, vibration, and blood flow.

Blood velocity is mostly measured within small confined areas in order to provide detailed mappings of velocity fields. These can be visualized as the velocity at specific times in the cardiac cycle along a diameter as shown in Figure 24a. This is the traditional simple way of data presentation with no graphical manipulation. It has the drawback of not visualizing the entire flow field and for *in vivo* measurements the lack of temporal development might also deprive one of important information. Therefore, Figure 24b yields more information about this aspect by including time as a graphical third dimension. The price for this feature is that some of the data is hidden behind the highest data points in the plot. To compensate for this it is possible in most computer software to display the profile from different aspects.

As for the pure diametral plot shown in Figure 24a this visualization technique does not provide information of the flow field outside of the diameter. By sacrificing the temporal information and emphasizing the spatial distribution of velocities,

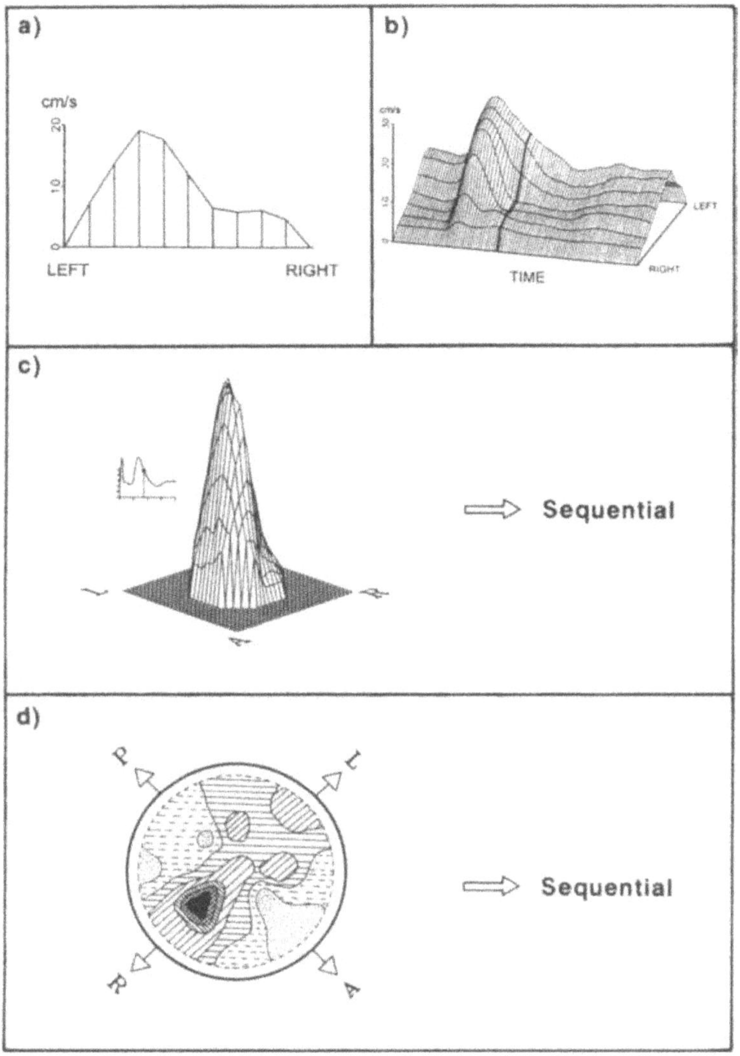

Figure 24. Some different ways of visualizing velocity fields. (a) Point blood velocities presented as mean cardiac cycle velocity along an anterior-posterior diameter of the abdominal aorta. (b) The same anterior-posterior diametrial velocity profile as in (a) plotted with computerized three dimensional drawing technique to comprise the temporal velocity profile development. (c) The same velocity measurements as in (a) but obtained from the entire vessel cross sectional area visualized three dimensionally with orientation in relation to the vessel wall indicated at the foot-plate and the blood velocity is indicated by the height of the profile. Sequential plots of this profile at different times in the cardiac cycle can illustrate the temporal development of the velocity field. (d) The spatial distribution of velocities across the cross sectional area can also be visualized in gray-scales or in colors and again temporal development can be visualized by sequential plots.

another type of three dimensional visualization can be made as shown in Figure 24c. Here the blood velocities have been measured in small confined areas in the entire cross sectional area and the data points are displayed as the height in the column where the foot plate illustrates the measuring plane, and orientation in relation to the vessel wall is given by letters. This type of data presentation is obviously rather comprehensive in its amount of data, and in order to avoid a confusing, complex mosaic contour on the top of the profile it is customary to apply some graphical "smoothing" (e.g., ref. 7). This smoothing gives an aesthetic image of the flow field, but there is a risk of overdoing and consequently suppressing velocity data, for example, a very localized jet will be dampened considerably because the surrounding data points represent low or even oscillating velocities. As for the three dimensional visualization shown in Figure 24b, there is also a problem with hidden lines and data points.

The spatial distribution of blood velocities in the entire cross-sectional area can also be visualized as a contour plot (like in geographic maps) or as grey scale (or color) plots. A grey scale plot is shown in Figure 24d as an example. The advantage of this type of velocity mapping is obviously its ability to show the entire cross-sectional area in one figure and also that no graphical smoothing is used. Since every grey scale represents an interval of velocity values the resolution of the display depends on the difference between the highest and lowest data value in the flow field; very skewed profiles offer a low resolution while fairly flat profiles yield a fine resolution of velocity distribution. In this type of plot there are no problems with hidden lines or data points. The two latter types of graphical display of velocity distribution in the cross-sectional area lack the feature of visualizing the temporal development of the velocity profile throughout the cardiac cycle. This can be compensated for by sequential plotting throughout the cardiac cycle. The ultimate utilization of this technique is plotting of multiple velocity profiles throughout a cardiac cycle and then photographing each plot several times to obtain an animation of the velocity profile development. This animation technique provides an excellent interpretation of the vast amount of complex data.

As stated above, these presentation techniques are illustrated by blood velocity data but can be applied to other hemodynamic measures as well, especially visualization of the spatial and temporal distribution of turbulence.

In both scientific and clinical settings it is of great importance to visualize both anatomy/geometry and flow patterns simultaneously in order to evaluate the interaction between anatomic structures and the flow conditions. These features are available in color-echo-Doppler ultrasound equipment and in MR-scanners. In the ultrasound setting the anatomic structures are visualized by a B-mode display and the colors representing the direction and magnitude of blood velocities are superimposed on the 2-D image. Thus, it is possible to have this information on-line, which is obviously a major advantage if other measures are to be taken simultaneously or if some kind of manipulation with the area of interest is performed. The limitations of this data presentation technique are basically those mentioned under

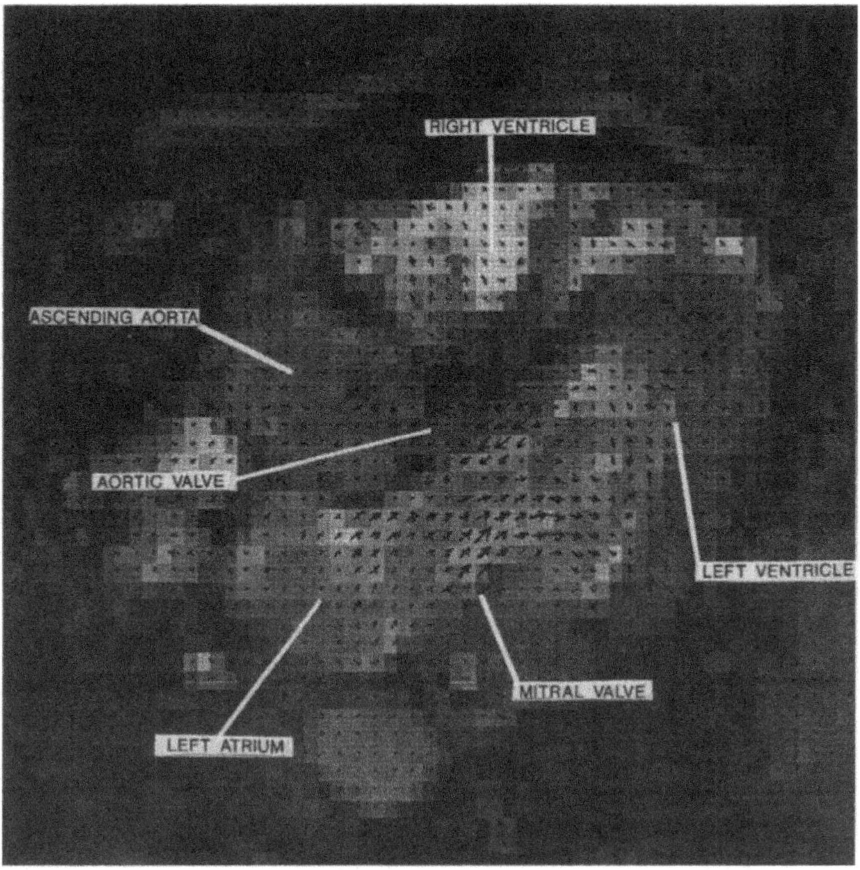

Figure 25. MR-scanning image with blood velocity data (vectors) superimposed. The anatomical structures are indicated in the figure and blood velocities are shown as vectors where each arrow represents the resulting vector of three velocity directions. The overall interpretation of velocity data discloses a large vortex in the left ventricle and high blood velocities across the mitral valve in diastole.

the description of color-Doppler technique, and that the study area is confined to the insonation plane. The color Doppler display is not illustrated here. MR-scanning provides an opportunity to depict, simultaneously, velocities from three orthogonal directions within a voxel. This obviously adds another dimension to the complexity of hemodynamic data and increases the difficulty in comprising many data in a single plot. The problems have been addressed hitherto by computation of vector plots, which imply that the resultant vector in a predefined number of sample volumes is calculated from the three velocity directions. These resultant vectors in the entire field of view are then superimposed on the MR-scanning image to provide the possibility to correlate velocity field data with anatomic data. An example of such a vector plot and anatomic image is shown in Figure 25.

As for the other cross sectional area visualizations mentioned above, sequential plots throughout the cardiac cycle can be obtained for a more detailed overview. Likewise, the sequential plots can be used for animation in order to give a good overview of the blood velocity data and the corresponding dynamic changes in anatomy.

VI. FUTURE PERSPECTIVES

In the field of hemodynamic research there is a trend toward increasingly advanced measuring techniques. The traditional sharp separation between *in vitro* and *in vivo* research is likely to diminish[15] since some measuring techniques can be applied both *in vitro* and *in vivo*. The fruitful dialog between *in vitro* and *in vivo* researchers implies that these different scientific approaches can be fully utilized: some questions can be more thoroughly answered by detailed, time consuming *in vitro* studies when the model can be verified by supporting *in vivo* data; other questions can be answered more meaningfully by animal or human studies when basic fluid mechanical knowledge has been gained from *in vitro* studies. It is important that researchers from these two specialized fields experience the work in laboratories from the opposite field to appreciate the experimental conditions of the different models. This important interchange of especially young researchers is important to strengthen the trend of "breaking walls" which has begun. This will improve the quality of the process from *in vitro* to animal experimental and finally human measurements.

For measuring techniques which are meant to be used for clinical purposes there is a trend toward increasingly non-invasive methods such as echo-Doppler cardiography, MR-scanning, and vibrocardiography. Therefore, large amounts of research activity is spent on these techniques in order to enhance their impact on hemodynamic research. This activity concerns both method improvements and application for new purposes.

The invasive techniques such as transit time flowmeters and intraluminal ultrasonography will continue to be important research tools since they can be applied in animal experimental research and to monitor catheter invasive procedures in clinical applications. The older quantitative invasive techniques such as hot-film anemometry and electromagnetic flowmetry will probably gradually be replaced by the above-mentioned newer techniques, since they are invasive and not as accurate as their modern counterparts.

With the rapidly improving computer technology on-line data presentation is feasible for increasingly complicated signal processing. Thus, on-line display of velocity profiles, turbulence levels, and an array of derived hemodynamic parameters is possible, often along with cardiovascular imaging capabilities.

The vision of bringing a patient with cardiac or vascular symptoms to a powerful investigative equipment and shortly after having a full diagnostic elucidation with no discomfort of the patient is no longer unrealistic science fiction.

ACKNOWLEDGMENTS

The presentation of the scientific work in this chapter is based on several years of cooperation and friendship with my colleagues Professor Peter K. Paulsen, M.D. D.M. Sc., Professor Hans Nygaard, ME; research fellow Erik Morre Pedersen, M.D., and recently other members of the Hemodynamic Research Group at Skejby Sygehus, Aarhus University Hospital.

Our scientific activities have been possible due to economic support from a variety of private foundations—The Danish Heart Foundation in particular. Furthermore, we have enjoyed willing support from The Institute of Experimental Clinical Research, Aarhus University and Skejby Sygehus, Aarhus County, Denmark.

REFERENCES

1. McMurrich, J.P. Leonardo da Vinci—the Anatomist. Williams & Wilkins, Baltimore, 1930.
2. Hopkins, R.W. Presidential address: Energy, poise and resilience- Daniel Bernoulli, Thomas Young, J.L.M. Poiseuille and F.A. Simeone. *J. Vasc. Surg.* **1991**, *13(5)*, 777–784.
3. Kolin, A. An electromagnetic flowmeter. Principle of the method and its application to bloodflow measurements. *Proc. Soc. Exp. Biol.* **1936**, 35–53.
4. Machella, T.E. The velocity of blood flow in arteries in animals. *Am. J. Physiol.* **1936**, *115*, 632–644.
5. Satumora S. Ultrasonic Doppler method for the inspection of cardiac functions. *J. Acoust. Soc. Am.* **1957**, *29*, 1181–1185.
6. Christensen, G.C.; Campeti, F.L. Anatomic and functional studies of the coronary circulation in the dog and pig. *Am. J. Vet. Res.* **1959**, 18–26.
7. Paulsen, P.K.; Hasenkam, J.M. Three-dimensional visualization of velocity profiles in the ascending aorta in dogs, measured with a hot-film anemometer. *J. Biomech.* **1983**, *16(3)*, 201–210.
8. Sisson & Grossman's: *The Anatomy of the Domestic Animals,* Vol. 5. WB Saunders, Philadelphia, London, Toronto, 1975.
9. Seed, W.A.; Wood, N.B. Velocity patterns in the aorta. *Cardiovasc. Res.* **1971**, *5*, 319–330.
10. Hutchison, K.J.; Karpinski, E. *In vivo* demonstration of flow and turbulence downstream of graded stenoses in canine arteries. *J. Biomech.* **1985**, *18(4)*, 285–296.
11. Barnhart, G.R.; Jones, M.; Ishihara, T.; Chavez, A.M.; Rose, D.M.; Ferrans, V.J. Failure of porcine aortic and bovine pericardial prosthetic valves: an experimental investigation in young sheep. *Circulation* **1982**, *66(suppl. I)*, I150–I153.
12. Katayama, H.; Henry, G.W.; Lucas, C.L.; Ha, B.; Ferreiro, J.I.; Frantz, E.G.; Krzeski, R. Three-dimensional visualization of pulmonary blood flow velocity profiles in lambs. *Jpn. Heart J.* **1992**, *33(1)*, 95–111.
13. Jansen, J.; Reul, H. A synthetic three-leaflet valve. *J. Med. Enging. & Tech.* **1992**, *16(1)*, 27–33.
14. Horneffer, P.J.; Healy, B.; Gott, V.L.; Gardner, T.J. The rapid evolution of amyocardial infarction in an end-artery coronary preparation. *Circulation* **1987**, *76(suppl. V)*, V39–V42.
15. Hasenkam, J.M. Studies of velocity fields and turbulence downstream of aortic valve prostheses *in vitro* and *in vivo*. *Dan. Med. Bull.* **1990**, *37(3)*, 235–249.
16. Sømod, L.; Hasenkam, J.M.; Kim, W.Y.; Nygaard, H.; Paulsen, P.K. Three-dimensional visualisation of velocity profiles in the normal porcine pulmonary trunk. *Cardiovasc. Res.* **1993**, *27*, 291–295.

17. Matre, K.; Segadal, L. Simultaneous measurement of velocity, diameter, flow and pressure in the ascending aorta of cats. Evaluation of an extractable Doppler ultrasound probe for continuous monitoring of aortic blood flow. *Cardiovasc. Res.* **1988**, *22*, 855–863.
18. Björk, V.O.; Sternlieb, J. Artificial heart valve testing in goats. *Scand. J. Thor. Cardiovasc. Surg.* **1986**, *20*, 97–102.
19. Nerem, R.M.; Rumberger, J.A.; Gross, D.R.; Hamlin, R.L.; Geiger, G.L. Hot-film anemometer velocity measurements of arterial blood flow in horses. *Circ. Res.* **1974**, *34*, 193–203.
20. Declaration of Helsinki II.
21. *McDonald's Blood Flow in Arteries: Theoretical, Experimental and Clinical Principles*, 3rd ed. (Nichols, W.W.; O'Rourke, M.F., Eds.). Edward Arnold, London, 1990.
22. Ganz, W.; Donoso, R.; Marcus, H.S.; Forrester, J.S.; Swan, H.J.C. A new technique for measurement of cardiac output by thermodilution in man. *Am. J. Cardiol.* **1971**, *27*, 392–396.
23. Sørensen, M.B.; Bille-Brahe, N.E.; Engell, H.C. Cardiac output measurement by thermal dilution. Reproducibility and comparison with the dye-dilution technique. *Ann. Surg.* **1976**, *183(1)*, 67–72.
24. Barnes, R.J.; Comline, R.S.; Dobson, A.; Drost, C.J. An implantable transit-time ultrasonic blood flowmeter. *J. Physiol.* **1983**, *35*, 2–3 P.
25. Burton, R.G.; Gorewit, R.C. Ultrasonic flowmeter—uses wide-beam transit time technique. *Medical Electronics* **1984**, *86(2)*, 68–73.
26. Paulsen, P.K.; Andersen, M. Continuous registration of blood velocity and cardiac output with a hot-film anemometer probe, mounted on a Swan-Ganz thermodilution catheter. *Eur. Surg. Res.* **1981**, *13*, 376–386.
27. Matre, K.; Segadal, L.; Engedal, H. Continuous measurement of aortic blood velocity, after cardiac surgery, by means of an extractable Doppler ultrasound probe. *J. Biomech. Enging.* **1985**, *7*, 84–88.
28. Samstad, S.O.; Torp, H.G.; Linker, D.T.; Rossvoll, O.; Skjaerpe, T.; Johansen, E.; Kristoffersen, K.; Angelsen, B.A.; Hatle, L. Cross-sectional early mitral flow velocity profiles from colour Doppler. *Br. Heart J.* **1989**, *62*, 177–184.
29. Kim, W.Y.; Pedersen, E.M.; Sømod, L.; Hasenkam, J.M.; Nygaard, H.; Paulsen, P.K. Cardiac output measurement with a perivascular multielement Doppler transducer. *Proceedings from VI Mediterranean Conference on Medical and Biological Engineering (Medicon '92) Capri* **1992**, 661–664.
30. Boer, P.; Roos, J.C.; Geyskes, G.G.; Dorhout Mees, E.J. Measurement of cardiac output by impedance cardiography under various conditions. *Am. J. Physiol.: Heart and Circulation Physiol* **1979**, *6(4)*, H491–H496.
31. English, J.B.; Hodges, M.R.; Sentker, C.; Johansen, R.; Stanley, T.H. Comparison of aortic pulse-wave contour analysis and thermodilution methods of measuring cardiac output during anesthesia in the dog. *Anesthesiology* **1980**, *52*, 56–61.
32. Paulsen, P.K.; Hasenkam, J.M.; Nygaard, H.; Gormsen, J. Analysis of the dynamic properties of a hot-film anemometer system for blood velocity measurements in humans. *Med. & Biol. Eng. & Comput.* **1987**, *25*, 195–200.
33. Paulsen, P.K. The hot-film anemometer—a method for blood velocity determination. I. *In vitro* comparison with the electromagnetic blood flowmeter. *Eur. Surg. Res.* **1980**, *12*, 140–148.
34. Hasenkam, J.M.; Westphal, D.; Reul, H.; Gormsen, J.; Giersiepen, M.; Stødkilde-Jørgensen, H.; Paulsen, P.K. Three-dimensional visualization of axial velocity profiles downstream of six different mechanical aortic valve prostheses, measured with a hot-film anemometer in a steady state flow model. *J. Biomech.* **1987**, *20(4)*, 353–364.
35. Paulsen, P.K.; Nissen, T. Patient safety unit for a hot-film anemometer, used for blood velocity determination in humans. *Med. & Biol. Eng. & Comput.* **1982**, *20*, 625–627.
36. Ling, S.C.; Atabek, H.B.; Fry, D.L.; Patel, D.J.; Janicki, J.S. Application of heated-film velocity and shear probes to hemodynamic studies. *Circ. Res.* **1968**, *23*, 789–801.
37. Amyot, J.W.; Francis, G.P.; Kiser, K.M.; Falsetti, H.L. Measurement of sequential velocity development in the aorta. *Am. Soc. Mech. Eng.* **1970**, *paper 70 - DA/BHF/13*, 1–5.

38. Paulsen, P.K.; Gormsen, J. A device for easy and exact intraluminal positioning of velocity probes in great vessels. *Clin. Phys. Physiol. Meas.* **1981**, *2(1)*, 53–56.
39. Paulsen, P.K.; Hasenkam, J.M.; Stødkilde-Jørgensen, H.; Albrechtsen, O. Three-dimensional visualization of velocity profiles in the ascending aorta in humans. A comparative study among normal aortic valves, St. Jude Medical and Starr-Edwards Silastic Ball valves. *Int. J. Artif. Organs* **1988**, *11(4)*, 277–292.
40. Hasenkam, J.M.; Pedersen, E.M.; Østergaard, J.H.; Nygaard, H.; Paulsen, P.K.; Johannsen, G.; Schurizek. B.A. Velocity fields and turbulent stresses downstream of biological and mechanical aortic valve prostheses implanted in pigs. *Cardiovasc. Res.* **1988**, *22(7)*, 472–483.
41. Seed, W.A.; Wood, N.B. Development and evaluation of a hot-film velocity probe for cardiovascular studies. *Cardiovasc. Res.* **1970**, *4*, 253–263.
42. Stein, P.D.; Sabbah, H.N. Turbulent blood flow in the ascending aorta of humans with normal and diseased aortic valves. *Circ. Res.* **1976**, *39(1)*, 58–65.
43. Tillmann, W.; Schlieper, H. Modification of miniature flush surface wall shear probes for biomedical use. *J. Phys. E: Sci. Instrum.* **1979**, *12*, 371–372.
44. Kossoff, G.; Barnett, S.B. (Eds.). Report of first symposium on safety and standardization of ultrasound in obstetrics. *Ultrasound Med. Biol.* **1986**, *12(9)*, 675–724.
45. Kisslo, J.; Adams, D.B.; Belkin, R.N. Doppler color flow imaging. Churchill Livingstone, New York, Edinburgh, London, Melbourne, 1988.
46. Hatle, L.; Angelsen, B. *Doppler Ultrasound in Cardiology. Physical Principles and Clinical Applications*, 2nd ed. Lea & Fibiger, Philadelphia, 1985.
47. Atkinson, P.; Woodcock, J.P. *Doppler Ultrasound and Its Use in Clinical Measurement.* Academic Press, London, 1982.
48. Lucas, C.L.; Henry, G.W.; Ferreiro, J.I.; Ha, B.; Keagy, B.A.; Wilcox, B.R. Pulmonary blood velocity profile variability in open-chest dogs: Influence of acutely altered hemodynamic states on profiles, and influence of profiles on the accuracy of techniques for cardiac output determination. *Heart Vessels* **1988**, *4*, 65–78.
49. Svennevig, J.L.; Grip, A.; Lindberg, H.; Geiran, O.; Hall, K.V. Continuous monitoring of cardiac output postoperatively using an implantable Doppler probe. *Scand. J. Thor. Cardiovasc. Surg.* **1986**, *20*, 145–149.
50. Giddens, D.P.; Khalifa, A.M.A. Turbulence measurements with pulsed Doppler ultrasound employing a frequency tracking method. *Ultrasound in Med. & Biol.* **1982**, *8(4)*, 427–437.
51. Garbini, J.L.; Forster, F.K.; Jørgensen, J.E. Measurement of fluid turbulence based on pulsed ultrasound techniques. Part 1. *J. Fluid Mech.* **1982**, *118*, 445–470.
52. Sillesen, H. Diagnosis and hemodynamic evaluation of internal carotid artery stenoses by Doppler ultrasound. Ph.D. Thesis, University of Copenhagen, Denmark, 1990.
53. Nygaard, H.; Hasenkam, J.M.; Pedersen, E.M.; Kim, W.Y.; Paulsen, P.K. A new perivascular multielement pulsed Doppler ultrasound system for *in vivo* studies of velocity fields and turbulent stresses in large vessels. *Med. in Biol. Eng. & Comput.* **1994**, *32*, 55–62.
54. Kassam, M.S.; Cobbold, R.S.C.; Johnston, K.W.; Graham, C.M. Method for estimating the Doppler mean velocity waveform. *Ultrasound in Med. Biol.* **1982**, *8(5)*, 537–544.
55. *Feigenbaum H. Echocardiography*, 4th ed. Lea & Fibiger, Philadelphia, 1986.
56. Falsetti, H.L.; Mates, R.E.; Carroll, R.J.; Gupta, R.L.; Bell, A.C. Analysis and correction of pressure wave distortion in fluid-filled catheter systems. *Circulation* **1974**, *49*, 165–172.
57. Ford, L.E.; Feldman, T.; Chiu, Y.C.; Caroll, J.D. Hemodynamic resistance as a measure of functional impairment in aortic valvular stenosis. *Circ. Res.* **1990**, *66*, 1–7.
58. Brogan, W.C.; Lange, R.A.; Hillis, L.D. Accuracy of various methods of measuring the transvalvular pressure gradient in aortic stenosis. *Am. Heart J.* **1992**, *123*, 948–953.
59. Sabbah, H.N.; Stein, P.D. Investigation of the theory and mechanism of the origin of the second heart sound. *Circ. Res.* **1976**, *39(6)*, 874–882.

60. Sabbah, H.N.; Blick, E.F.; Stein, P.D. High-frequency pressure fluctuations: Their significance in the documentation of turbulent blood flow. *Cathet. Cardiovasc. Diagn.* **1977**, *3*, 375–384.
61. Johnson, G.R.; Myers, G.S.; Lees, R.S. Evaluation of aortic stenosis by spectral analysis of the murmur. *J. Am. Coll. Cardiol.* **1985**, *6*, 55–63.
62. Donnerstein, R.L. Continuous spectral analysis of heart murmurs for evaluating stenotic cardiac lesions. *Am. J. Cardiol.* **1989**, *64*, 625–630.
63. Nygaard, H.; Thuesen, L.; Hasenkam, J.M.; Pedersen, E.M.; Paulsen, P.K. Assessing the severity of aortic value stenosis by spectral analysis of cardiac murmurs (spectral vibrocardiography) I: Technical aspects. *J. Heart Valve Dis.* **1993**, *2*, 454–467.
64. Tobin, R.J.; Chang, I-D. Wall pressure spectra scaling downstream of stenoses in steady tube flow. *J. Biomech.* **1976**, *9*, 633–640.
65. Lees, R.J.; Dewey, Jr., C.F. Phonoangiography: A new non-invasive diagnostic method for studying arterial disease. *Proc. Natl. Acad. Sci. USA* **1970** *67*, 935–942.
66. Meno, F.; Reddy, P.S.; Bernardi, L. Heart sound propagation in the human thorax. *Clin. Phys. Physiol. Meas.* **1985**, *6(2)*, 123–129.

HEMODYNAMICS OF VASCULAR PROSTHESES

T. V. How, R. A. Black, and P. E. Hughes

Advances in Hemodynamics and Hemorheology,
Volume 1, pages 373–423.

ISBN: 1-55938-634-7.

I. INTRODUCTION

Prosthetic vascular grafts are used routinely for the reconstruction of diseased arteries. Their function is either to replace a dilated or ruptured portion of a major artery, or to bypass an occluded segment of artery thereby restoring the blood supply to distal ischaemic tissues. Ideally, this function should be maintained throughout the lifetime of the patient, since the replacement of a failed graft is associated with poorer outcome[1,2] and reoperation leads to disappointment and distress for the patient. Large-diameter textile fabric grafts made of synthetic materials such as polyethylene terepththalate (Dacron®) perform well when they are used to replace the large major arteries of the body. This is not the case when these prostheses are scaled down to a diameter of less than about 6 mm for use in bypass procedures involving the peripheral or coronary arteries, for example. The patient's own saphenous vein is the graft of choice for surgery to the arteries of the lower limbs (i.e., the femoral, popliteal, and tibial artery segments). In a small, but significant proportion of patients, however, the saphenous vein is not suitable and expanded polytetrafluoroethylene (ePTFE) prostheses are considered by most vascular surgeons to be the best alternative. Despite their poor long-term performance in comparison with saphenous vein grafts, ePTFE grafts are often favored for their durability vis-à-vis their resistance to degradation and aneurysm formation. Nevertheless, a number of other small diameter prostheses are currently under development.[3] Most of these are fabricated from polyurethanes because this polymer exhibits lower thrombogenicity and higher compliance than either Dacron or ePTFE. Because of the potential risk of degradation, however, polyurethanes will not become wholly acceptable until attempts now being made to synthesize more biostable polyurethanes prove to be successful.[4,5]

The performance of vascular prostheses is dependent on many factors: these may be divided into (a) graft and material-related factors, such as the surface characteristics, the physical, mechanical, and hemodynamic properties; and (b) host-related factors, such as the hematological properties of the patient and the conditions of the inflow and outflow vessels.[6] The microstructure of the prosthesis is considered to be particularly important as it plays a critical role in the healing process. The intima of natural blood vessels is covered with an endothelium which is antithrombogenic. Although a neointima forms on the luminal surface of vascular prostheses, this is composed mainly of compacted fibrin and is devoid of endothelial cells except at the anastomoses. Nonetheless, the neointima can prevent thrombosis, at least in the large diameter prostheses, since it shields the flowing blood from the thrombogenic base-material. A microporous prosthesis having inter-communicating pores that extend through the wall is desirable, as this promotes the formation of a firmly adherent neointima.

Small-diameter prostheses fail most commonly by thrombotic occlusion. Early failures, within the first few weeks of the operation, are attributed to the thrombogenicity of the graft surface, inadequate blood flow, the hematological status of the patient or errors of surgical technique. Prostheses also fail months or years following implantation secondary to intimal thickening, usually at the distal anastomoses,[7] and/or progression of the underlying atheromatous disease in the recipient artery. Progression of these disease processes leads to a gradual reduction in blood flow and, eventually, to graft occlusion by thrombus. The localized intimal thickening caused by proliferation of cells at the anastomoses, specifically smooth muscle cells, should be differentiated from that which is caused by the deposition of an acellular matrix throughout the luminal surface of the prosthesis. Although more diffuse than intimal hyperplasia, the acellular matrix also accumulates and, in this respect, the internal diameter of the prosthesis may be critical: the smaller the diameter is at the time of implantation, the greater will be the effect of any intimal thickening in terms of its resistance to blood flow.

The causes of anastomotic intimal hyperplasia are not fully understood, although several factors have been implicated including the surface characteristics, the mismatch in the mechanical properties of the prosthesis to those of the recipient artery[8,9] and abrupt changes in compliance at the anastomoses caused by suturing techniques.[10,11] It is widely accepted that hemodynamic factors are also involved, although the actual mechanisms are still unclear. There is evidence that intimal hyperplasia and thrombosis tend to occur in regions where the wall shear rates are low. The presence of turbulence and vortices can be detrimental to the cellular elements of blood and endothelial cells on vascular walls. Recent research shows that endothelial cells growing to the luminal surface of vascular grafts can produce mitogens that are similar to platelet-derived growth factors (PDGF),[12,13] and that these mitogens can stimulate sub-endothelial smooth muscle cell proliferation (intimal hyperplasia). It has also been suggested that flow disturbances present at anastomoses may stimulate endothelial cells to release these PDGF-like growth factors.

The purpose of this chapter is to review the factors involved in the failure of bypass grafts and more specifically to examine the influence of hemodynamics in graft thrombosis and development of intimal hyperplasia. The importance of hemodynamics in the design of new—and the improvement of existing—small-diameter vascular grafts and the assessment of their performance *in vitro* and *in vivo* is also discussed.

II. BIOLOGICAL RESPONSES TO HEMODYNAMIC FORCES

A. The Hemodynamic Environment

Blood flow in the body is very complex owing to the high degree of geometric complexity of the cardiovascular system, with its multiple bends, branches, and abrupt changes in cross-sectional area and the unique mechanical properties of its constituent blood vessels. Add to this the pulsatile nature of the flow and the "anomalous" viscous properties of whole blood itself and it becomes clear why the hemodynamic environment is often extremely difficult to characterize adequately. This is especially true in the vicinity of an arterial reconstruction where flow separation can create regions of disturbed and stagnant flow in close proximity. As a result, the cells within the flowing blood—and those which line the vessel wall—are subjected to a wide range of fluctuating shear stresses. Highly-disturbed flow can be detrimental to blood cells causing platelet activation, or even hemolysis. The release of cell contents can trigger the cascade of biochemical events that leads to the formation of emboli and mural thrombus. On the other hand, deposition of platelet aggregates and cell debris onto a surface is more likely to occur in regions of flow recirculation and stagnation. It is at these sites, where the residence time of fluid is excessive, that coagulation factors and mitogenic substances can accumulate to levels capable of inducing the long-term structural changes which are characteristic of diseased blood vessels and which eventually impair the function of the surgically-altered circulation.

B. Hemostasis and Thrombosis in Vascular Prostheses

Blood coagulation is a complex cascade of chemical interactions that leads to the formation of a fibrin clot, the structure in which the various cellular elements of blood become trapped. One of the key events that marks the onset of thrombosis is platelet adhesion, which is consolidated by subsequent platelet activation and aggregation. Platelets do not ordinarily interact with the intact vessel wall, but they do so when the endothelial cell lining has been damaged and its basement membrane and underlying collagen fibers have been exposed.[14] Once activated, platelets undergo a series of characteristic shape changes that are accompanied by the secretion of the contents of internal structures, the organelles known as granules (the platelet release reaction[15]). Through a process of positive feedback, the liberation of ADP, ATP, and platelet-specific proteins from a relatively small number of platelets activates a far greater number of platelets which are then able to adhere to each other (cohesion) and thus to platelets attached to the damaged portion of the vessel wall.[16] Although the interaction between some of these platelets and the surface will be reversible, they soon adhere to the surface in sufficient numbers to form a seal that promptly arrests the flow of blood. Thereafter, the coagulation process continues with the polymerization of fibrinogen into a network of fibrin strands that reinforces the original platelet plug and into which other blood cells, including erythrocytes, become trapped. The process is essentially a self-limiting one: once fibrinogen has been converted into fibrin the surface induces no further activation to the blood flowing across it. In addition, the production of prostaglandins (PGI_2) by adjoining endothelial cells prevents the spread of thrombus beyond the injured surface by quenching any further platelet activation and coagulation. The structure progressively stabilizes into a smooth, glossy covering that is known as *red thrombus* owing to its substantial red blood cell content. After a period, the thrombus diminishes in size and is eventually digested by enzymatic processes (fibrinolysis) once the underlying healing process is complete.

Thrombosis is a major cause of failure of vascular prostheses. Early failures have been attributed to technical errors and the initial thrombogenicity of the polymeric graft material. In this context, degranulation of platelets that come into contact with adhesive proteins immobilized on the artificial surface liberates platelet agonists into the adjacent bloodstream[17] which, in turn, prompts further platelet activity. In this way, the exposed surface rapidly becomes covered in platelet aggregates. Since the artificial surface of a vascular prosthesis rarely becomes covered in a viable monolayer of endothelial cells in humans, the process of platelet adhesion and activation, which is normally self-limiting in other parts of the circulation, is known to continue for many years following implantation and the surface never becomes fully stable.

The weight of clinical evidence leaves little doubt that the nature of the blood flow in the vicinity of such a reactive biomaterial surface plays a key role in determining the outcome of bypass procedures. The lower patency rate noted with vein grafts and prostheses that cross the knee, for example, has been attributed to the lower rates of flow through these grafts. Sauvage et al.[18] studied the relationship between blood flow rate and the formation of thrombus in arterial prostheses and proposed a minimum flow velocity, or "thrombotic threshold velocity" (TTV), below which the prosthesis failed by thrombotic occlusion; the TTV for a given material was defined as the time-average velocity for which at least 50% of the graft was covered in thrombus deposits. In contrast, at flow velocities above the threshold value for a particular material or surface, the luminal surface of the prosthesis was largely free of thrombus through the action of hemodynamic shear forces. The role of wall shear rate or, more accurately, shear *stress* in the formation of mural thrombus is demonstrated by the higher threshold velocity of crimped versus non-crimped Dacron material. These findings have been confirmed *in vitro* by Hashimoto et al.[19] who measured the effect of shear rate on the quantity of thrombus accumulating on various artificial surfaces with a cone-in-plate viscometer. Their experiments, which showed that the mass of thrombus that formed on each surface was less at higher rates of shear, lead these authors to conclude that shear forces on an artificial surface act to inhibit the formation of blood clots. Under laminar flow conditions, the rate of platelet adhesion depends on the diffusion-controlled arrival of platelets at the surface at low wall shear rates ($<600 s^{-1}$). As the rate of shear increases, the process becomes reaction-rate controlled and the number of platelets coming into contact with the surface exceeds the available binding sites on that surface. At greater shear rates still, there is a leveling-off, followed by a reduction in the numbers of adherent cells, owing to the reduced residence time of cells adjacent to and through the removal of platelet aggregates from the surface under the action of shear stresses developed at the blood-material interface.

Rheological Aspects of Thrombus Formation

Whole blood is a suspension of cells of various sizes and concentration. By far the largest are the erythrocytes, bi-concave discoid-shaped cells that are 8–10 μm in diameter and 1–3 μm thick. Since these cells can outnumber platelets by 20:1 and thus occupy a far greater volume, it is the behavior of these cells that determines the viscous properties of whole blood.[20] Thus, the flow of blood differs from that of an ideal, homogenous, incompressible fluid with constant viscosity flowing steadily through a rigid tube of uniform cross-section. In reality, blood flows through compliant vessels and is itself a complex suspension of cells and proteinaceous substances, the viscosity of which depends greatly on hematocrit and the local flow rate. The distribution of different cell types and cell aggregates across a blood vessel results in a velocity profile that is not parabolic, like its idealized counterpart, but one that is more blunt.[21] The velocity gradient, or rate of shear, at the blood-surface interface is therefore greater. The tendency for red blood cell aggregates to occupy the core of the flow under laminar flow conditions causes platelets to be displaced to the margins of the flow. In other words, there is an increase in platelet diffusivity toward the vessel wall several orders of magnitude greater than that attributed to Brownian motion alone.[22] This marked increase in diffusivity leads to an increase in interactions between platelets and the vessel wall in the event of damage to the vessel wall.[23–25] Platelet diffusivity is modified, in turn, by factors which affect interactions between blood cells, such as hematocrit, red cell size, and deformability and shear rate.[26,27] In addition to their physical presence, red cells contribute to the process of thrombus formation through the release of substantial amounts of ADP on hemolysis at high hematocrits or by shear-induced means.[28]

The process by which platelets become attached to an artificial surface (adhesion) and to each other (cohesion) in the first place is dependent upon the interaction between platelet membrane receptors, plasma glycoproteins, and appropriate surface-bound adhesive proteins.[29] The sequence of events on initial exposure to flowing blood begins with adsorption of plasma proteins, such as fibrinogen, onto the artificial surface, followed by platelet deposition; platelet activation is required before cohesion of fluid-phase platelets and platelet aggregates can take place. The main participants in the above processes are the platelet membrane glycoproteins Ib (GPIb) and IIb-IIIa complex (GPIIb-IIIa), and the adhesive ligands von Willebrand factor (vWF) and fibrinogen in plasma. Glycoproteins GPIb and GPIIb-IIIa provide two ways in which unstimulated platelets can adhere to a thrombogenic surface that presents fibrinogen (fibrin) and/or vWF. While the interaction between GPIIb-IIa on inactive platelets and fibrinogen is irreversible, interaction of GPIb with immobilized vWF is reversible at this stage: platelets must become fully activated before GPIIb-IIIa will bind with surface ligands other than fibrinogen (e.g., fibronectin, vWF, vitronectin), allowing adherent platelets to spread and adhesive proteins to bridge the gap between these platelets and those in circulation.[30] The resulting thrombus is composed predominantly of compacted fibrin and platelets.

The sequence of events that leads to the aggregation of platelets under laminar flow conditions is now well understood.[31,32] Exposure of platelets to shear stresses in excess of 60–80 dynes cm^{-2} results in vWF binding to GPIb and an increase in intracellular calcium ion concentration of 2–3-fold.[32] Having been so activated, platelet GPIIb-IIIa is then able to bind with soluble adhesive proteins which effectively bridge the gap between platelets. Under static conditions, adhesive molecules other than vWF are involved in aggregation of activated platelets. Studies in flowing blood, however, have shown that platelet adhesion and shear-induced aggregation depend on vWF being present, especially at high rates of shear.[33,34] Similarly, thrombus formation on type I collagen is independent of vWF at shear rates below 800 s^{-1}, but becomes largely dependent on vWF at greater rates of shear.[35] Studies of this kind suggest that the multimeric nature of vWF makes the binding of platelet glycoproteins with this molecule particularly resistant to hydrodynamic shear forces.[36,37] This type of research may yield future therapies for those undergoing vascular surgery. The work of Hanson et al., for example, showed that the administration of anti-GPIIb-IIIa monoclonal antibodies was successful in blocking platelet aggregation and thereby preventing thrombotic occlusion of prosthetic grafts implanted in a baboon model.[38] Alternative approaches involve the passivation of biomaterial surfaces prior to implantation with heparin-like substances, or with proteins that would compete with adhesive plasma proteins such as fibrinogen in order to reduce platelet–surface interactions.

Flow Disturbances and Turbulent Blood Flow

Under steady, laminar flow conditions, a fluid moves in well-defined streamlines and the transfer of momentum across the streamlines takes place only by diffusion; any disturbances to the flow are rapidly damped out through viscous dissipation. Increasing the flow velocity above a certain critical value decreases the ability of viscosity to damp out these flow disturbances and the flow becomes disordered. The governing parameter that determines whether the flow is laminar or turbulent is the Reynolds number, Re, which is the ratio of inertial forces to viscous forces for a given flow. For fully-developed pipe flow, the transition from laminar to turbulent flow occurs at a critical value of 2300.[39] Flow disturbances can exist at Reynolds numbers well below this critical value, however.[40] Laminar flows, for example, may be steady or unsteady with respect to time and can have the appearance of being quite disturbed. Sudden changes in flow velocity, often associated with abrupt changes in geometry, can cause flow separation and the formation of eddies. Flow disturbances have been detected in the cardiovascular system in the vicinity of heart valves, for example, and distal to anastomoses where they were shown to be present at both end-to-end and end-to-side anastomoses. Stein and Sabbah[41] detected high-frequency and random velocity fluctuations, characteristic of turbulent flows, immediately downstream of normal and diseased aortic valves with a hot-film anemometer probe. The precise nature of the flow has a strong influence on the transportation of blood-borne elements to a surface and the process of cellular deposition is more complex. In cases where the streamlines do not run parallel to the vessel wall, the forces of convection increase transport toward the vessel wall and cellular material, which includes both platelets and red blood cells, is deposited at a rate which can cause abrupt occlusion of smaller vessels. While disturbed flow encourages mixing of the blood constituents, and may thus prevent local stasis of blood flow, it can increase cell–cell and cell–wall interactions to the point where cells become activated, and is generally considered to be an undesirable feature of blood-contacting devices.

C. The Response of the Vessel Wall

The development of mural thrombus subsequent to the above series of events is further complicated by the effect of local flow disturbances arising from the change in surface topology following repeated deposition of cellular matter. Furthermore, the release of ADP and potent mitogenic and chemotactic substances from degranulating platelets is now known to stimulate further platelet activity and contribute to the progressive thickening of the walls of a blood vessel, especially at the anastomoses.[8,42–46]

Intimal Hyperplasia

Intimal hyperplasia, the proliferation of smooth-muscle cells and fibroblasts in the subendothelial and medial layers, is a well-documented hypertrophic response that has been reported as early as eight days after the implantation of a vein graft.[47] The migration and subsequent proliferation of cells within the subendothelial tissues, and the build-up of an associated extracellular matrix, leads to stenosis, a narrowing of the vessel diameter that restricts the flow and eventually blocks the vessel entirely. Intimal hyperplasia has been identified as the cause of failure of endarterectomized arteries,[48] aorto-coronary bypass grafts,[49,50] and femoro-popliteal bypass grafts,[51] and following angioplasty.[52] Both vein grafts and arterial prostheses can be affected, though in the latter case the absence of sub-intimal cells within the body of the prosthesis restricts the intimal proliferation to the vicinity of the anastomoses. Smooth-muscle cell proliferation is also recognized as being an important feature of other forms of vascular disease, namely atherosclerosis.[53]

As was discussed in the previous section, the action of shear stresses on individual blood cells as they are being conveyed through a blood vessel or sheared against the surface of the blood vessel wall can alter the metabolism and morphology of these cells, often with serious consequences. It follows that opposing shear stresses are imposed on the luminal surface of a vein graft or host artery, and such stresses are now known to have a profound effect on the metabolism, structure, and function of the cells which make up the vessel wall as well (see section IID). After surgery, a decreased rate of blood flow through a bypass graft is known to increase the likelihood of graft occlusion by intimal hyperplasia, and some authors attribute this to the lower wall shear rates present in these grafts. Faulkner et al.[54] noted that intimal proliferation was absent along the length of autogenous vein grafts when these experienced the artificially-high rates of blood flow produced by arteriovenous fistulae. The intimal thickening at the anastomoses was also notably less than in the controls. In contrast, those grafts in which blood flow was reduced by an average of 44% exhibited a greater incidence of intimal thickening, with the most prominent lesions developing at the anastomoses. In a subsequent study, Rittgers et al.[55] demonstrated an inverse correlation between the thickness of intimal proliferation and the magnitude of the local average shear forces present in femoral vein grafts in dogs. These results have since been confirmed by several investigators and in other animal models.[56–58]

Inokuchi et al.[59] found a correlation between the type of flow waveform, as measured at the time of surgery, and the outcome of the bypass procedure. Waveforms were classified into five distinct categories according to their degree of pulsatility. The waveforms that resembled the normal triphasic flow waveform typical of a healthy artery (that is rapid acceleration of the flow in systole, followed by a rapid deceleration in diastole, with some reversal of flow) were most often associated with good long-term patency in these grafts. It was suggested that steps could be taken at the time of surgery to improve the measured flow waveform and thereby improve the chances of a successful outcome. These authors went on to perform an *in vitro* analysis of the velocity profiles that each of the five classified waveforms produced. The results of this study[60] showed that the desirable flow waveforms subjected the vessel wall to a large range of positive and negative (i.e., reverse flow) wall shear stress. This group has since produced experimental evidence to support their hypothesis that the *variation* in amplitude of wall shear stress, rather than simply high flow rate or velocity, is the factor influencing the development of intimal hyperplasia.[61] A more recent study by Kraiss et al.,[57] however, has generated results which cast doubt on the significance of oscillatory shear stress, and show that a modest shear stress level ≥20 dynes cm^{-2} is sufficient to inhibit intimal thickening. The authors argue that the grafts employed by Morinaga et al., in contrast to their own work, may not have been subjected to the same intraluminal pressure or tangential wall stress, two parameters that have been shown to influence the degree to intimal thickening.[62]

D. Role of Endothelial Cells in Thrombosis and Intimal Proliferation

As intimal hyperplasia is often similar in appearance to early atherosclerotic lesions, the mechanisms responsible for these two types of pathological response are thought to be related.[43,63] In 1960, Texon[64] postulated that the intima of the vessel wall may become physically detached from the media sites where there is a localized drop in pressure generated by the flow of blood through the vessel. It was suggested that the biological response to this action, which formed the basis for atherosclerosis, is essentially a reparative one that results in local proliferation of endothelial cells, fibroblasts, and underlying smooth muscle cells. Others, for example Mitchell and Schwartz,[65] argued that such proliferative lesions were the direct result of local turbulence which could either damage the vessel wall or the blood constituents, or both. Clearly, the validity of these hypotheses rests on whether the induced pressure fluctuations in turbulent flow are sufficient in magnitude to damage the cells of the vessel wall. The link between turbulence and post-stenotic dilatation in blood vessels[66] suggests that this is indeed the case, although actual denudation of the intima by fluid shear forces requires relatively high levels of shear stress (>1000 dynes cm^{-2}),[67] the magnitude of which depends on the duration that the stress is applied.[68] Fry[69] observed shear-dependent morphological changes to endothelial cells and demonstrated an increase in the influx of lipid into the vessel wall at much lower levels of shear stress (~ 400 dynes cm^{-2}) which he took to be indicative of an increase in permeability of the vessel wall under the action of fluid stress. Fry then proceeded to attempt to correlate regions of high wall shear stress with the topography of early atheromatous lesions in the canine coronary circulation.[70]

Caro et al.,[71,72] on the other hand, postulated that the distribution of early atherosclerotic lesions was determined by the mass transport of cholesterol between the arterial wall and the intraluminal blood. In contrast to the work of Fry, these authors hypothesized that the accumulation of the cholesterol in the vessel wall is dependent on the concentration gradient of lipids in the boundary layer adjacent to the vessel wall. Thus, lower rates of shear decrease the concentration gradient and, therefore, reduce the amount of cholesterol that is transported away from the wall. Conversely, higher shear rates aid the diffusion of lipid molecules from the wall owing to the greater concentration gradient, the rate at which this equilibrium process takes place being determined by the local rate of shear at the vessel wall.

Endothelial-cell Morphology and Function

The above hypotheses require modification in the light of more recent studies which have focused on the role of the vascular endothelium in mediating the biological response of the vessel wall.[63,73,74] Far from being a passive interface between the flowing blood and the structural elements in the vessel wall, endothelial cells are now thought to have a pivotal role in both hemostasis and sub-intimal cellular proliferation.

The anti-thrombogenic behavior of the natural endothelial-cell lining is the result of several factors, mainly because of the presence of heparin and fibrinolytic substances on its surface. Endothelial cells also secrete potent anticoagulant factors, such as plasminogen activator and prostacyclin, and substances that regulate vascular tone (e.g., the endothelial-cell derived relaxation factor, EDRF) and hence vessel diameter and resistance to flow.[75] On the other hand, endothelial cells are also active participants in hemostasis through the synthesis of pro-coagulants such as tissue factor (thromboplastin), plasminogen inhibitor, and factor VIII. The two antagonistic systems, both pro- and anti-coagulation, are in a state of dynamic equilibrium, with the net effect being in favor of the fibrinolytic system in the case of the normal healthy endothelium.[76,77]

In addition to their active role in hemostasis, endothelial cells are likewise far from passive in terms of smooth muscle cell (SMC) proliferation. A number of possible explanations for the intimal proliferation associated with vascular grafts have been presented in the literature, including distension and ischaemia during graft preparation and operative trauma. Each of these procedures causes damage to the endothelium. If the injury is extensive, then smooth muscle cells are seen to migrate from the media to the intima where they subsequently proliferate and form a neo-intima of longitudinally orientated smooth muscle cells. There is a concomitant change in smooth-muscle phenotype from one that is contractile to a synthetic form.[78] It has been shown that endothelial cells are capable of producing a potent mitogenic substance on the one hand (a polypeptide known as endothelial derived growth factor—EDGF),[79] and heparin-like inhibitors of SMC proliferation on the other. The latter heparin-like substances maintain smooth muscle cells in their contractile phenotype, a state in which these cells do not respond to serum mitogens, and also inhibit the proliferation of activated (synthetic state) smooth muscle cells.[80] Although the luminal smooth muscle cells that establish themselves following extensive endothelial injury are, like endothelial cells, substantially non-thrombogenic (owing to their capability for producing prostaglandin I_2 (PGI_2) in amounts that are comparable to endothelial cells), it appears that a confluent monolayer of endothelial cells is necessary if these smooth muscle cells are not to proliferate in an uncontrolled fashion. Campbell and Campbell[80] concluded that it is the proliferative state of the endothelium which determines whether these cells act to inhibit or promote SMC proliferation.

At this point, it is worth noting how endothelial cell morphology and function is influenced by the prevailing flow conditions. When subjected to uniform shear stresses in a rotational viscometer or parallel-plate flow chamber, for example, these cells tend to orient themselves in the direction of the flow and adopt an elongated, teardrop configuration.[81] In contrast, endothelial cells from regions of the circulation where the flow is relatively stagnant are more rounded in shape, as is the case when these cells are cultured under static conditions *in vitro*. The development of endothelial-cell patterns in the wake of a stenosis have been recorded *in vivo*[82] and these were found to be consistent with the flow regime that exists at an expansion, with its characteristic regions of recirculation and stagnation. Here, the residence time of activated blood cells may be of sufficient duration to contribute to thrombus formation and intimal hyperplasia, as described above. Any subsequent intimal thickening at this site results in a local constriction, or stenosis, giving rise to progressively greater disturbances to the flow. Severe vessel constriction can result in transient turbulent bursts (bruits) at relatively low Reynolds numbers. It has been demonstrated that endothelial cells subjected to a turbulent flow regime develop in a more rounded form,[83] and that these cells exhibit a significantly higher cell turnover rate (i.e., rate of proliferation)[84,85] with turbulent shear stresses as low as 1.5 dynes cm^{-2} having been found to promote cell division.[86] Furthermore, recent studies have reported that the synthesis of numerous substances involved in hemostasis (e.g., inhibitors such as PGI_2 and tPa, and EDRF) by endothelial cells is regulated by flow.[87,88] Ando et al.[89,90] have recently identified a possible mechanism whereby information concerning the flow environment is communicated to the interior of these cells (via intracellular Ca^{++}).

Mechanical Properties of the Vessel Wall

The hemodynamic effect of a change in compliance across the anastomosis has been identified as a possible contributing factor in the development of anastomotic hyperplasia. There are instances where endothelial cells and SMC proliferation has been attributed to the excessive mechanical stresses induced by compliance mismatch at the site of the anastomosis.[91] Such stresses are hemodynamically induced, but whereas blood flow exerts tangential shear stresses to the endothelial cells that line the vessel wall, the transmission of pulsatile energy through the arterial system induces, indirectly, time-varying normal stresses that are borne by all cells within the vessel wall, with stress concentrations occurring especially at the anastomoses.

Experiments carried out *in vitro* have shown that rabbit SMCs showed an increase in collagen and protein synthesis when subjected to repeated elongation and relaxation on elastin membranes (10% stretch at a rate of 52/min), as indicated by an uptake in thymidine or proline.[92,93] Others have shown that SMCs, like endothelial cells, orient themselves with respect to the axis of the applied strain,[94] but at an angle of 45°. If the cyclic strain continues to be applied over a period of several days, smooth muscle proliferation ensues.[95] Moreover, an abrupt change in mechanical properties around a suture line may give rise to local hemodynamic disturbances[96,97] at the anastomosis between the host vessels and a vascular prosthesis. In principle, compliance matching between a host artery and a prosthetic graft should help to reduce the dissipation of pulsatile energy across an anastomosis by this means[98] (*vide* section III). In addition to direct strain of SMCs, other possible mechanisms for the proliferative response include indirect activation via substances secreted from other cell types, specifically endothelial cells.[78,95] The resulting strain of the endothelial-cell membrane has been associated with increases in membrane permeability, prostacyclin production,[86,99] and cell proliferation.[100]

In conclusion, the explanation for intimal SMC proliferation at anastomoses may lie with the status of the endothelial cell: because these cells are in a state of perpetual unrest at these sites, owing to their disturbed hemodynamic (in the broadest sense of the word) environment, they do not have the opportunity to quiesce and thus provide a constant source of mitogenic and chemotactic substances. Such effects may be compounded by the fact that these cells have to survive on an artificial substrate at the anastomotic region of a vascular prosthesis; with vein grafts, on the other hand, it may be that the venous endothelium responds pathologically to the more severe hemodynamic environment that exists on the arterial side of the circulation.

III. INFLUENCE OF HEMODYNAMICS IN THE PERFORMANCE OF VASCULAR PROSTHESES

A. The Need to Maintain Adequate Blood Flow

Current vascular prostheses are fabricated from materials that are thrombogenic. Pathological studies on explanted prostheses show that although a neointima composed of organized fibrin forms on the prosthesis luminal surface, the coverage is usually incomplete.[101] Moreover, in the very rare cases where an endothelial layer is found it is confined to only a few millimeters from the anastomoses. Exposure of blood to a bare prosthesis surface or an incompletely healed neointima initiates a cascade of biochemical events that may ultimately result in thrombosis and graft occlusion (see Section IIC). Since these events, such as the initial protein adsorption, platelet activation, release of platelet constituents, and platelet deposition are flow-related, both the net flow and the local fluid dynamic pattern will influence this reactive cascade. When blood flow is high, the activated coagulated factors formed as blood contacts the artificial surface will be carried away rapidly. Conversely, if the blood flow is reduced, the activated coagulation factors may accumulate and lead to thrombosis. Thus, in order to reduce the risk of thrombosis, it is important to ensure that blood flow through the prosthesis is sufficiently high.

B. Methods of Ensuring Adequate Blood Flow

According to Sauvage et al.[18] the thrombotic threshold velocity or TTV is a good indicator of the thrombogenicity of graft materials. They classified the following materials in increasing order of TTV: autologous artery, non-crimped knitted Dacron, expanded PTFE, crimped knitted Dacron, human umbilical vein. If one accepts the concept of TTV, then it will be important to ensure that the blood flow velocity through the graft is kept well above the TTV for the material. For a given flow rate this may be achieved by reducing the prosthesis diameter. However, there is a limit on the diameter that can be used since prostheses of diameter less than 4–5 mm are associated with high failure rates. Another factor that should be considered is the hydraulic impedance of the prosthesis. This should be low in comparison with the effective impedance of the collateral vessels otherwise blood flow through the prosthesis would drop increasing the likelihood of thrombosis. The blood flow requirements of distal tissues may vary greatly depending on the level of activity. Where the prosthesis is required to supply a large tissue bed, the blood flow requirements during exercise may be several times greater than that at rest. Thus, the prosthesis diameter should be chosen such that the pressure drop does not become significant when blood flow increases above the basal rates.

Vascular prostheses are currently made of materials with elastic moduli much higher than that of natural arterial tissue. This mismatch in mechanical properties has implications for the propagation of blood pressure and flow waves since efficient transmission of pulsatile energy depends on the admittance of the graft and arterial segments being matched.[102,103] Pulse wave reflections will occur at anastomoses where abrupt changes in mechanical properties are present. Pulsatile energy is lost in the reflection and also when the incident and reflected waves are out of phase. The result is reduced blood flow through the prosthesis. To minimize these effects, various groups have been developing vascular prostheses with compliance that is more closely matched to that of the artery.[5,104–108]

The texture of the prosthesis surface should also be considered since a rough surface will present a higher hydraulic impedance, particularly when the flow is turbulent.[3] The surface of textile fabric are particularly uneven. Where the yarns cross over each other in woven and knitted fabric, projections of the order of the width of the yarn (0.3–0.5 mm) are formed. Moreover, the circumferential crimps that are provided to improve the lateral flexibility and buckling resistance of the prostheses produces even greater surface roughness. Local flow recirculation probably occurs in the crimp furrows,[109] causing fibrin and blood cellular elements to be deposited so that, eventually, any unevenness is smoothed out. The fact that the neointima in crimped prostheses is thicker than in non-crimped prostheses suggests that the surface roughness has an influence on healing of the prosthesis and formation of the neointima.[110,111] Therefore, in small diameter prostheses, where the neointima should be thin, it is important that the flow surface should be as smooth as possible.

Blood flow through a bypass graft is dependent not only on the characteristics of the graft, but also on the conditions of the inflow and outflow vessels. The presence of a stenosis upstream of the recipient artery may limit blood flow through the graft, particularly when there is increased demand for blood flow distally. An angiogram is usually performed to ensure that the proximal vessels are unobstructed. Any obstructions are usually treated by balloon angioplasty before insertion of the bypass graft. The outflow of the distal vessels in the lower limbs may be assessed angiographically by determining the number of patent outflow arteries.[112] The peripheral resistance or input impedance may also be determined intraoperatively.[113,114] If the peripheral resistance is high, blood flow through the prosthesis may drop to a point that the velocity may be lower than the thrombotic threshold for the material.

In such cases, a surgical technique of increasing blood flow through the bypass may be used. This involves the creation of an arteriovenous shunt at or near the distal anastomosis by suturing the recipient artery to a collateral vein.[115] The latter provides an additional low resistance outlet and, therefore, causes the blood flow through the graft to increase. Blood flow is significantly increased by this procedure and the patency results at periods of up to two years are encouraging.[116] However, the longer-term results are less satisfactory and it is likely that the creation of the shunt and augmentation of flow may result in highly disturbed flow patterns at the distal anastomoses thereby increasing the risk of anastomotic intimal hyperplasia. This suggests that it is not sufficient just to provide an adequate blood flow through the prosthesis; the local hemodynamics should also be an important consideration.

C. Vessel Geometry

The overriding factor that determines the local hemodynamic environment is local vascular geometry. In a long straight vessel with uniform circular geometry, contact between blood cellular elements and the artificial flow surface will be minimized since the flow will be laminar and undisturbed. In the presence of irregularities or discontinuities that occur at bifurcations, anastomoses, or stenoses, flow disturbances such as flow separation, vortex formation, and vortex shedding may result causing increased cell–wall interactions. In the case of peripheral arterial bypass, the tendency for intimal hyperplasia to develop at the toe and heel of the distal anastomosis and on the floor of the distal host artery has been attributed, partly, to local hemodynamic disturbances since the construction of a bypass can lead to major changes in local flow patterns at the anastomoses.[117] However, the question of which hemodynamic factors are implicated is still unresolved although low shear stress, turbulence, and oscillating shear stress have been proposed.[55,118–120] The difficulty arises because of the complex flow field and the lack of a full description of the flow patterns in realistic models of the anastomoses.

Arterial prostheses are normally available in a limited range of sizes which makes a mismatch in the diameter of the prosthesis and that of the host vessels inevitable. In the case of femorodistal bypass, the situation is compounded by the disparity in the diameter of the graft and the tibial or peroneal artery. The discontinuity in cross-sectional area at the anastomoses results in the flow from the proximal artery encountering either a constriction or an expansion. In an end-to-end anastomosis, this is known to cause flow separation and turbulence.[97] In proximal side-to-end and distal end-to-side anastomoses the effect of diameter mismatch may be expected to be more complex and as yet no detailed studies have been carried out. To minimize the diameter mismatch that occurs with femorodistal bypass a prosthesis with a tapered lumen may be used. Apart from the better match in size at both the proximal and distal anastomoses, it has been shown that the converging boundaries of tapered grafts are actually capable of attenuating any flow disturbances that arise at the prosthesis inlet.[121]

D. Implications for Vascular Prosthesis Design and Bypass Configuration

It is clear from the previous section that the risk of thrombosis can be minimized by ensuring an adequate flow of blood through the prosthesis. The prosthesis itself should not present a high resistance to flow even under high flow conditions. This may be achieved by ensuring that the prosthesis has similar mechanical properties (compliance) and diameter to that of the host artery so that the transmission of pulsatile energy is maximized.[98] A match in diameter may be difficult to achieve, especially when there is a large discrepancy in the diameter of the host arteries upstream and downstream (e.g., common femoral artery and tibial or peroneal arteries). In such cases the use of a tapered prosthesis may be an advantage. In addition to providing a better match in diameter with the host arteries, the tapered geometry can promote flow that is inherently more stable and produce higher wall shear rates owing to the blunt velocity profiles that exist in these conduits.

Blood flow through a bypass is determined not only by the dimensions and properties of the prosthesis. When the run-off is poor, the impedance of the distal bed will have a greater effect on the flow than the characteristic impedance of the prosthesis. Although it would be desirable in these cases to dilate the distal run-off vessels with balloon angioplasty, this is technically difficult. An alternative proposition is to use a prosthesis material of low thrombogenicity to allow lower blood flow velocities to be supported without thrombosing the conduit. Although attempts have been made to develop synthetic polymers with improved blood compatibility, their stability implanted in the body will have to be properly assessed before they will be acceptable.[122]

The lining of the prosthesis flow surface with endothelial cells would result in a non-thrombogenic surface. Since current vascular prostheses do not endothelialize in humans, except near the anastomosis, attempts have been made to promote the formation of an endothelium. One approach is to encourage the ingrowth of capillaries across the vessel wall in order that capillary endothelial cells may migrate to the prosthesis inner surface.[123] This transmural capillary ingrowth has been assessed in expanded PTFE grafts in a baboon model. It was found that a narrow range of graft porosity, corresponding to internodal distance of 30–60 mm, gave an almost complete endothelial cell coverage.[123] When the porosity was greater or smaller than this range, the lining was incomplete. The other approach is to seed endothelial cells obtained from an autologous vessel onto the prosthesis surface.[124,125] Various methods have been proposed. The cells can be harvested and seeded onto the prosthesis surface at the time of implantation (single stage seeding). When a large surface area of prosthesis is to be seeded, the number of cells may be increased by cell culture prior to seeding and implantation. Alternatively, the cells may be seeded on the prosthesis and allowed to form a complete lining on the surface in a cell culture before implantation. The main disadvantage of this method is the time delay of several weeks between cell harvest and implantation of the prosthesis.[126] The early clinical trials of endothelial cell seeded Dacron and expanded PTFE prostheses have not shown the expected improvement in patency compared with non-seeded control[127,128] although in some cases, extensive endothelialization has been demonstrated.[129] Longer-term patency of endothelial cell seeded prostheses has not yet been reported and it remains to be seen whether these

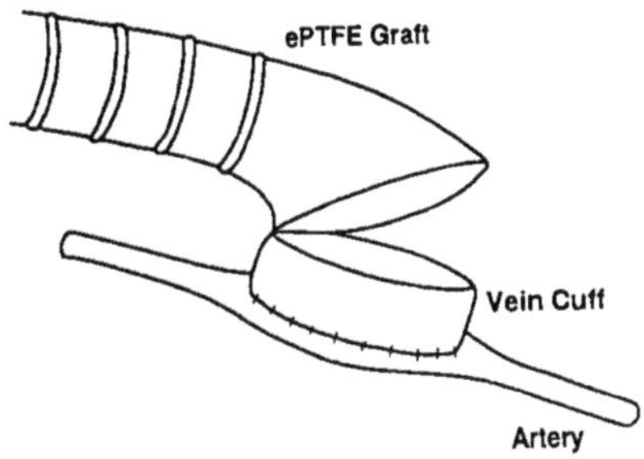

Figure 1. Diagram of a distal anastomosis with interpostion vein cuff developed by Miller et al.[118]

prostheses are less susceptible to intimal hyperplasia; it may be that such grafts are more so, depending on the status of the endothelial cell lining.[74]

When an autologous saphenous vein is unsuitable or is no longer available, the expanded PTFE is considered by many surgeons as the graft of choice for femorodistal bypass because of its excellent biostability. However, this material is far from ideal and because of its relatively incompliant wall, grafts made with this material are difficult to suture to the small tibial vessels.[130] Protrusion of the PTFE into the lumen and distortion of the recipient artery have been observed in casts of anastomoses.[131] If these obstructions occur *in vivo*, both the net blood flow rate and the local hemodynamics may be affected. The resulting low blood flow and flow disturbances may explain, at least partly, the high failure of small diameter arterial prostheses. To overcome these difficulties, new distal anastomosis configurations were developed following the introduction of the interposition vein cuff technique of Miller et al.[132] The cuff consists of a strip of vein fashioned into an elliptical tube interposed between the recipient artery and the PTFE prosthesis (Figure 1). An alternative method, known as the Taylor patch,[133] employs a vein patch that is sutured to the prosthesis and the artery (Figure 2). This results in a long anastomosis with a more gradual taper that ensures a smooth transition from the 5–6 mm diameter prosthesis to the small recipient artery. Because of the shape of the Miller

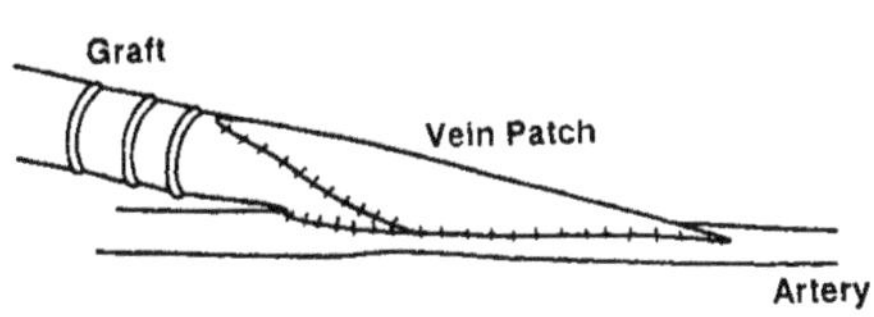

Figure 2. Diagram of a distal anastomosis with interpostion vein cuff developed by McFarland and Taylor.[119]

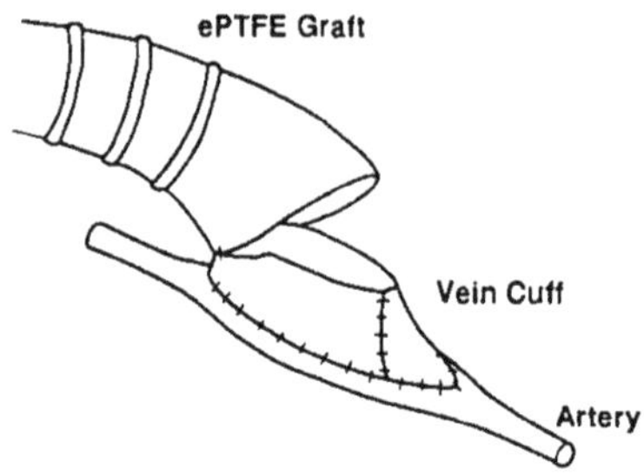

Figure 3. Diagram of a distal anastomosis with interpostion vein cuff developed by Tyrell and Wolfe.[120]

cuff, the local flow pattern may be expected to be highly disturbed[134] and a small angle of anastomosis would be difficult to achieve.[135] To overcome these limitations, a modified vein cuff was proposed by Tyrell and Wolf[136] to provide a more streamlined shape of cuff while allowing a smaller angle of anastomosis (Figure 3).

The patency rates at three years reported for expanded PTFE prostheses with an interposition vein cuff or patch for femorodistal bypass range from 52%[134] to 60%[137] compared with 14–38%[134] when the prosthesis is anastomosed directly to the recipient artery. Harris et al.[138] have investigated the performance of the vein cuff combined with an adjuvant arteriovenous fistula at the distal anastomosis and obtained a patency rate of 61% at two years. This was not significantly different from the patency results with the autologous vein grafts (64%) in a comparable group of patients.

The encouraging results obtained with the interposition vein cuffs and patches have been attributed to the reduction in the compliance mismatch that is considered to be an important factor in intimal proliferation.[139] Experimental studies have confirmed that intimal thickening was significantly greater in direct anastomoses as opposed to those with vein cuffs.[140] There is evidence that the creation of a vein cuff causes the location of intimal hyperplasia to be displaced from the small recipient artery to the cuff itself where any thickening can be more easily accommodated.[134] Another consequence of the presence of the compliant vein cuff is that the transmission of the pulsatile energy across the anastomosis may be improved thereby resulting in increased blood flow.[130] However, experimental verification for this effect is lacking.[131] Even if the net flow across the anastomosis is not changed by the presence of the vein cuff or patch, the local hemodynamics would be expected to be significantly altered and may be responsible for the redistribution of intimal hyperplasia (Figure 4). Detailed measurements of the flow field in modified anastomoses have not yet been performed. These will be required so that the different configurations may be compared and optimized.

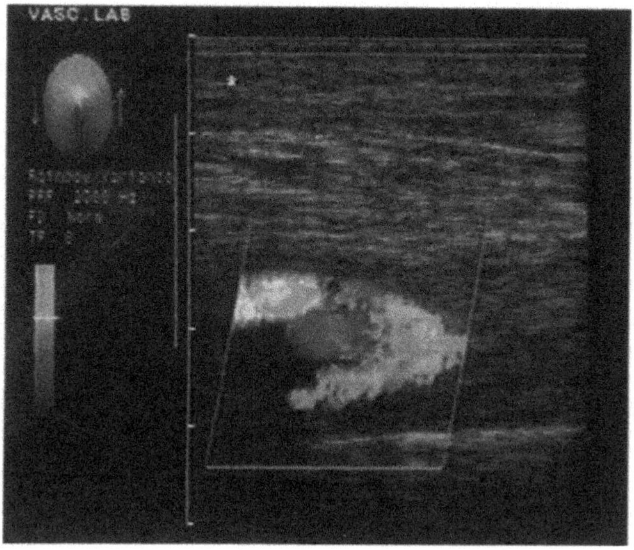

Figure 4. Color Doppler image of a distal anasomosis with a Miller cuff showing highly disturbed flow pattern.

IV. HEMODYNAMIC MEASUREMENTS IN VASCULAR PROSTHESES

Adverse flow conditions have long been associated with poor performance of natural and synthetic vascular grafts, there being an increased likelihood of graft occlusion when the diameter of the vessel is small and the rate of blood flow inadequate. Since the behavior of certain cells in the blood, and those which line the vessel wall, is thought to be influenced by local flow conditions, there have been numerous attempts to correlate hydrodynamic quantities such as flow velocity, turbulence intensity, and wall shear stress with the extent and location of thrombus deposition and intimal proliferation, the primary causes of graft failure. Nevertheless, the precise role that fluid mechanics plays in the aetiology of these disease processes remains obscure owing to the complexity of the flow within blood vessels and the difficulty in obtaining accurate, quantitative, *in vivo* measurements.

In an effort to establish a more detailed picture of the overall flow patterns and local hemodynamic conditions present within these blood vessels, our approach has been to model the cardiovascular system *in vitro* so that the flow field may be characterized under controlled physiological flow conditions by means of one or more velocity measurement techniques. When coupled with studies in which adverse biological responses are monitored under equally well controlled flow conditions, physical modeling may well reveal what improvements should be made to current designs of vascular prosthesis and indicate the most appropriate form of anastomosis configuration that should be adopted for a particular arterial bypass procedure.

A. *In Vitro* Measurements

Most theoretical analyses of blood flow have relied on numerical modeling techniques, while experimental studies have made use of flow visualization and three types of anemometry (hot-film, laser, and Doppler ultrasound) to measure those hydrodynamic quantities that are relevant to the disease process which affects the arteries, and ultimately the very bypass grafts and prostheses that divert the flow of blood past diseased or blocked blood vessels.

Flow Visualization

One of the simplest techniques used in the study of fluid flow involves visualization of the flow with an indicator dye introduced into the flow field at one or more points upstream. The source of the dye may be through gradual dissolution from dye crystals or through injection of neutrally-buoyant liquid dyes into the flow at fixed points. Rapid dispersion of the resulting dye filaments, or streaklines, in disturbed flows limits the use of this technique to laminar flow regimes, although diffusion of the dye may reduce spatial resolution even in the case of laminar flows. Nevertheless, the technique is widely used in fluid mechanics research. Numerous water-soluble dye substances are available, including potassium permanganate, carmine red, red acid fuchsin, toluidine blue, thymol blue, rhodamine, and fluorescein; the use of more than one color of dye can often assist in the interpretation of complex flow structures.[118,141,142]

Alternatives to dye injection rely on chemical reactions to induce a change in color or phase in response to an electric current or on illumination with laser light. An example of the former technique induces a stark change in color of an indicator solution at one of two electrodes. Electrolysis at the electrodes, for example, produces a change in pH that, in turn, produces a color change in an indicator solution such as thymol blue. As this chemical reaction necessarily involves the irreversible release of hydrogen and oxygen at the electrodes, direct reduction/oxidation (redox) reactions of methylene blue have been developed.[143] Alternatively, electrolysis can be actively promoted in order to generate tiny gas bubbles, usually hydrogen.[144–146] Here, the bubbles that form on a thin wire electrode placed in the flow field are carried downstream with the flow. The hydrogen-bubble technique, as it is known, has been widely used for flow visualization, however, there are problems when this technique is used for quantitative velocity measurements. Bubble rise and drag from the wire are not insignificant, and it is difficult to correct for these effects, especially in pulsatile flows.[147] As with all invasive techniques, the presence of the wire filament can interfere with the measurement in the cases where there is reverse flow is present.

The photochromic tracer technique, on the other hand, exploits the photochemical properties of a normally colorless indicator solution.[148–154] This technique, which has been applied to models of blood vessels[153] and heart valves,[155] is described in detail by Couch and Ojha in chapter 4.

Flow visualization can also be achieved by dispersing small tracer particles that follow accurately the small-scale motion of the fluid. Unlike dye visualization, the flow patterns generated by short pathlines are not contaminated by the time history of past events. Karino and Goldsmith, for example, have used cine photography to trace the motion of microspheres and blood cells *in vitro*.[156] These authors have since investigated the interaction between these particles and the vessel wall in the case of (a) tube flow, (b) flow through bi-leaflet vein valves, bifurcations, and at arterial junctions, and (c) immediately downstream of a modeled arterial stenosis.[157,158]

Flow visualization techniques yield important information regarding the overall flow field. In general, flow visualization techniques provide qualitative information about the flow field, with quantitative velocity determination having to be carried out as a separate step. Nevertheless, quantitative data can often be derived from time-lapse photographic images or video recordings of individual trajectories of neutrally-buoyant particles or dyes suspended in the flow field. The process of extracting this information can be laborious and is often very time-consuming: velocity data must be deduced from the length and orientation of individual particle trajectories. Recently, more sophisticated techniques have been developed that use state-of-the-art image processing to track the motion of many particles with time, such as particle-image velocimetry (PIV). While these techniques are very useful for characterizing the entire flow field, they are, as yet, unable to give an accurate measure of different time-varying velocity components, and hence turbulence data, with sufficient resolution.[159] A more fundamental limitation, however, is that the optical techniques are not appropriate for measurements of flow in opaque conduits such as blood-filled arterial prostheses. Invasive techniques, whereby a probe is introduced into the flow field, are used under such circumstances.

Hot-film Anemometry

The hot-film anemometer is based on the well established hot-wire anemometer used by aeronautical engineers to measure turbulence in wind tunnels.[160] Because of the small probe dimensions, both instruments are capable of measuring the small-scale, high-frequency velocity fluctuations that are characteristic of turbulent flows.[161] The flow of fluid past the heated wire filament alters the temperature, and hence resistance, of the wire. The current required to maintain a constant wire temperature is taken as a measure of the flow velocity past the wire. The analogous technique of hot-film anemometry was developed specifically for blood flow measurement.[162–166] Although the frequency response of this instrument is lower than that of the hot-wire anemometer, the probe is less fragile and has been used successfully to measure turbulent blood flow in the animal aorta.[167–171] The probe comprises a small area of platinum (typically 0.3 mm × 1.0 mm) deposited onto a quartz substrate and coated by a further layer of quartz. The probe is often conical in shape, and located at the end of a right-angled hypodermic needle so as to facilitate its insertion into the blood vessel. There are a number of inherent limitations which make the hot-film technique unsuitable for quantifying disturbed flows. Prior calibration of these instruments is essential, as the sensitivity and linearity of the flow probe are heavily dependent on its construction and the fluid being examined: when used with fluids such as blood, for example, the formation of deposits on the heated platinum surface of the probe can alter the calibration of the instrument during use; moreover, it is difficult to make reliable measurements in disturbed or pulsatile flows as there is a need for additional metal films in order to ascertain the direction of the flow. An even more fundamental objection is that the presence of the probe itself may alter the flow field adjacent to the probe and thereby invalidate the measurement.

More sophisticated instrumentation is required in order to measure one or more velocity components simultaneously. Accurate, non-invasive estimates of flow velocity can be made, however, by instruments that detect the shift in frequency of ultrasound or laser light when scattered from particles or blood cells moving within the flow field (the Doppler effect).

Ultrasonic Flowmeters

Pulsed ultrasound devices are capable of measuring flow velocity within a "sample volume" located at a fixed distance from the face of the transducer. This instrument was first described by Baker[172] and velocimeters of this kind now form part of many diagnostic ultrasound scanners. Research instruments have also been developed and commercialized by Peronneau et al.[173] and Hartley and Cole.[174] The latter system operates at a transmission frequency of 20 MHz (Crystal Biotech, Inc.), producing a sample volume that occupies approximately 1 mm^3. Pulses of ultrasound 0.4 μs in duration are transmitted at intervals of 16 μs by a frequency divider. During the remainder of the cycle, the piezoelectric crystal receives the backscattered ultrasound over an interval of time specified by an electronic "gate." The time delay set by this gate (between 1 μs and 13 μs) corresponds to a fixed distance from the face of the transducer at which echoes are received; the shape and size of the transducer and duration that the receiver gate remains open determines the sample volume dimensions. Systems of this kind have been widely used in bioengineering research for the measurement of flow velocity in physical models of blood vessels,[175,176] left-ventricular assist devices,[177] and vascular prostheses,[178–180] to name but a few applications; the latter system is described in detail elsewhere.[180] Similar systems have enabled detailed hemodynamic measurements to be made *in vivo* (e.g., refs. 174,181).

Although widely used in diagnostic medicine, the use of ultrasound for accurate velocity determination *in vitro* has certain inherent limitations. A high frequency of operation is required in order to achieve an acceptable spatial resolution,[173] however, there is an upper limit to the frequency that can be used because of increasing attenuation in acoustic energy at these high transmission frequencies: the aforementioned 20 MHz system, for example, has a range of only 10 mm in water.[180] Furthermore, the poor spatial resolution of the system can lead to significant errors in velocity measurement, especially in regions where the velocity gradient is great, as is the case in the boundary layer or when the flow is turbulent. When the sample-volume-to-vessel-diameter ratio is not small (i.e., >0.1), there will be significant distortion of the velocity profile. While correction for this distortion can be made by a process of deconvolution,[182] this technique is not entirely satisfactory when the velocity profiles are complex and contain several points of inflexion.[183]

A shortcoming of any velocity measurement that relies on the Doppler effect is the ambiguity inherent to this process—so-called "spectral broadening"—owing to transit time or geometric effects.[184,185] Laser Doppler measurements are also affected (see below),[186] but to a much lesser extent owing to the much shorter wavelength of laser light (of the order of 0.5 μm, c.f. 75 μm for ultrasound at 20 MHz). The overall effect limits the accuracy of ultrasound data, especially when these originate from disturbed flows since it is difficult to distinguish the spectral broadening arising from flow disturbances from that which is inherent in the Doppler process itself. Laser Doppler anemometry (LDA), on the other hand, offers a number of advantages in comparison with the above flow measurement techniques.[187]

Laser Doppler Anemometry

Laser anemometers measure the flow at the point where two or more laser beams intersect, known as the probe volume. LDA optics are able to define a very small, elliptical probe volume that is typically 180 μm in length by 50 μm in diameter in water. The instrument has a linear response to fluid velocity that depends only on the chosen laser wavelength and optical configuration. Apart from the requirement that the laser beams have a clear optical path, the flow may require seeding with light-scattering particles in order to provide a continuous Doppler signal and improve the signal-to-noise ratio. The instrument has a good frequency response; it can measure reverse flows, and more than one component of velocity at a time, making it particularly useful for the study of disturbed flows where the axial and radial turbulent velocity components may be different in magnitude.[188] There are a number of examples where LDA has been used to characterize the turbulent flow field downstream of a stenosis under steady and pulsatile flow conditions.[189–192] The technique has been widely used elsewhere in bioengineering research for detailed hemodynamic measurements (e.g., wall shear and turbulent shear-stress distributions) within left ventricular-assist devices,[193–195] downstream of prosthetic heart valves,[196,197] and in rigid and compliant casts of blood vessels (see below).[198,199]

Physical Modeling of Flow in Vascular Prostheses

In order to make best use of the above flow measurement techniques, considerable attention should be given to the elements of the physical model. However, desirable as it may be to carry out hemodynamic measurements in flow circuits that incorporate actual blood vessels and commercially-available vascular prostheses, such an approach is only feasible when bulk pressures and flows are the quantities being measured, or in cases where flow velocity is to be measured by means of hot-film or ultrasound measurement techniques as detailed above. Because prosthetic graft materials are porous structures that only become impervious to blood by means of the natural hemostatic process, their pores must be pre-clotted with a gelatinous substance such as gelatin or fibrin before these devices can be connected to a mock circulation. Moreover, in the case of ultrasound measurement techniques, care must be taken to ensure that there is no air trapped within the graft wall to prevent the transmission of acoustic energy. Nevertheless, such experimental procedures have enabled detailed comparisons to be made between different designs of vascular prosthesis, in terms of energy losses,[200,201] differences in velocity profile,[202] and in terms of flow stability within different geometric configurations of Dacron grafts.[203]

Often it is not feasible to make measurements through actual prosthetic graft materials and alternative engineering materials must be identified in order to benefit from the full range of flow measurement techniques that is available. There is a need for good optical characteristics for many investigations, and transparent materials, both rigid and compliant, glass and plastic, have been widely used. Glass models of aortic and carotid bifurcations have been fabricated for flow-visualization studies involving, for example, dye injection[204] and particle trajectory analysis.[205,206] Zarins et al. (1983)[206] visualized the flow patterns in their blown glass models of the carotid bifurcation by means of the hydrogen bubble technique. These authors chose to make subsequent quantitative LDA measurements of flow velocity in scale model that had been machined in clear acrylic plastic, in order to increase the effective spatial resolution of their LDA and, presumably, to reduce distortion of the laser beams caused by refraction at the solid/liquid interface. In this respect, some authors advocate refractive-index matching, especially when near-wall measurements are required for wall shear-rate determination, in order to improve the accuracy of the results.[207,208] Duncan et al. (1990),[209] for example, placed their compliant cast of the human aortic bifurcation in a bath of aqueous sodium isothiocyanate in order to minimize any refractive effects caused by non-uniformity in wall thickness.

For simple geometries, coating glass, or metallic rods in a polymeric solution may yield an adequate model. In this laboratory, an electrostatic spinning process originally developed by Annis et al.[105] for the fabrication of a microporous polyurethane vascular prosthesis, has provided a unique way of producing conduits of uniform wall thickness that have anisotropic mechanical properties.[210] The technique can also yield solid-walled conduits that have since proved suitable for studies in which the flow regimes within cylindrical and tapered geometries were compared by means of pulsed ultrasound and flow visualization techniques;[121,211] the ability to control wall thickness also provided a means of fabricating cylindrical and tapered tubes with uniform wall compliance characteristics.[212]

Sophisticated multistep modeling techniques have been developed that can reproduce detailed geometric features of prosthetic grafts and their anastomosis configurations with remarkable accuracy. For example, various techniques have been developed in order to obtain casts of major blood vessels and at anastomoses. One technique involves injecting a fast-curing resin or silicone rubber at physiological pressures; moulds are taken of the silicone cast following its excision so that an exact copy of the primary cast can be made from a less stable material such as wax. This additional step is essential if the cast is to be easily removed following its encapsulation in an acrylic block or after it has been coated with a suitable polymeric material. White et al.[213] visualized flow in scale models made of transparent silicone rubber that had been cast around clay models of an end-to-side anastomosis configuration.

The traditional means of generating reproducible pulsatile flow waveforms is to attach a piston to a piece of rotating machinery, such as a "Scotch-yoke" mechanism or rotating cam, for example; others have employed various electro-mechanical devices including vibration-generators, spool valves, or other solenoid actuators in order to modulate fluid flows. Each of these techniques is limited in terms of the complexity of the resulting flow waveforms that can be produced. Nowadays, one of the most common methods of generating arbitrary flow waveforms is to make use of servo-motor controlled positive displacement pumps (e.g., Superpump, Vivitro Systems Inc.) or valves.[213] When pulse duplicators of this kind are coupled to a flow circuit that incorporates lumped-parameter components such as compliance chambers (Windkessels) and laminar-flow resistors, it is possible to reproduce physiological pressure and flow waveforms in a given physical model with good accuracy.

Consideration must be given to the validity of the modeling process: for example, whether it is sufficient to rely on generic geometric features in the physical model rather than fabricating accurate life-size replicas of a given blood-vessel geometry, and whether one can make use of rigid conduits instead of attempting to reproduce the mechanical properties of the blood vessels and/or vascular prostheses under investigation. In the first instance, it may even be sufficient to study hemodynamic characteristics under steady-flow conditions; in most cases, however, the pulsatile nature of the flow waveform must be reproduced. In cases where the compliance of the arterial system is deemed to make a significant contribution to the quantities being measured, the modeling process must be versatile enough to allow this parameter to be controlled; and, in the case of wall shear-rate estimation, it follows that the resulting wall movement must be taken into account, thus increasing the complexity of the experiment considerably.

In addition to faithful reproduction of vascular geometry, accurate simulation of blood flow requires the use of an appropriate model fluid that shares some of the viscous properties of whole blood. The above studies were carried out with a variety of blood-analogue fluids and suspensions, including Newtonian aqueous glycerol and sucrose solutions. The studies of Thurston (*vide* Chapter 1 in this book), and those of Liepsch and others,[214–216] indicate that the detail of the flow structures, and the resulting shear-rate distribution at the vessel wall, may be influenced by the non-Newtonian flow behavior of whole blood. It remains to be shown, however, whether any discrepancy between Newtonian and non-Newtonian flow regimes that may be detected *in vitro* is significant in relation to thrombosis and intimal hyperplasia.

B. *In Vivo* Measurements

Intraoperative Measurements: Vascular Outflow Resistance

Blood flow through a vascular graft is determined by several factors including the quality of the inflow which determines the driving pressure, the resistance presented by the graft, and the resistance of the distal vascular bed. When the prosthesis is anastomosed to small distal arteries, the peripheral resistance may be the limiting factor on blood flow. Ascer et al.[217] measured the peripheral resistance intraoperatively, in patients after insertion of a femoropopliteal bypass graft. It was reported that grafts associated with peripheral resistance greater than 1.2 Hg ml^{-1} min^{-1} failed within three months while those corresponding to lower peripheral resistance remained patent for at least three months. Parvin et al.[218] obtained similar results, but drew attention to the measurement of resistance at a single flow rate. They found an inverse relationship between flow rate (and pressure) and outflow resistance: this they attributed partly to the non-Newtonian properties of blood and partly to the presence of collateral vessels which together lead to an overestimation of resistance at low flow rates and underestimation at high flow rates. These studies suggest that the measurement of peripheral resistance may be useful in identifying bypass grafts which are at risk of early occlusion. It may also prove an objective indicator for the need for increasing blood flow such as the creation of an arterio-venous shunt. Peripheral vascular resistance can be measured by injecting saline or the patient's own blood with an infusion pump that is set to known rates of infusion through the graft or the distal segment of the artery to which the graft is anastomosed. The pressure is monitored during the infusion and the resistance is determined by the ratio of pressure to flow rate.

Input impedance may provide a more physiological assessment of run-off than resistance as it relates pulsatile pressure to pulsatile flow. Few studies have been carried out since the initial reports of Cave et al.[219] and Butler and Somerville[220] found no significant difference between the impedance moduli of patent and failed femoropopliteal bypass grafts. The main reason is the difficulty in obtaining reliable and accurate blood flow measurements in small blood vessels in a confined space. However, with the availability of transit-time ultrasound flowmeters which allow *in vivo* blood flow measurements in small vessels with greater reliability than either the electromagnetic or Doppler ultrasound flowmeters, and sub-millimeter catheter-tip pressure transducers, pulsatile flow, and pressure can be now be measured reliably and rapidly.

Postoperative Measurements

The value of postoperative surveillance of femoropopliteal and femorodistal bypass with autologous vein grafts is now widely recognized and it is becoming common practice.[221] Localized stenotic lesions develop in approximately one-third of these grafts resulting in reduced blood flow and ultimately in thrombosis.[222,223] If these failing grafts are detected before they occlude, they may be treated by patch angioplasty, a short vein bypass, or balloon angioplasty.[224] Since the great majority of these stenoses develop within the first year,[225,226] measures to identify and treat them during this period can have a significant effect on the overall patency. Various methods have been proposed for early detection of stenoses. Conventional angiography is not suitable as it is too invasive for frequent graft assessment. Although the ankle/brachial pressure index has the advantage that it is a simple non-invasive technique, it performs adequately only when stenoses are hemodynamically significant. When used with a standard exercise test on a treadmill, the technique has slightly greater sensitivity in detecting lower grade stenoses.[227] Wyatt et al.[228] described a non-invasive method of assessing impedance in vein grafts using an air plethysmography cuff around the thigh and a pulse volume recorder to obtain blood pressure and a continuous-wave Doppler system to measure blood flow. They found that the mean impedance modulus in patients with graft or run-off vessel stenoses of greater than 50% were significantly higher than in the controls. Currently, the two most effective methods of assessment are intravenous digital subtraction angiography (IV DSA) where the contrast medium is injected centrally via a cannula sited in the superior vena cava under fluoroscopic control, and Duplex ultrasound (with or without Doppler color imaging).[229–231] Although IV DSA has the advantage that it provides a much greater field of view of the vascular anatomy it is demanding for both staff and patients. The ultrasonic technique is therefore preferred for screening purposes. Once a stenotic lesion has been detected by ultrasound, however, angiography is usually carried out before angioplasty or surgery is performed.

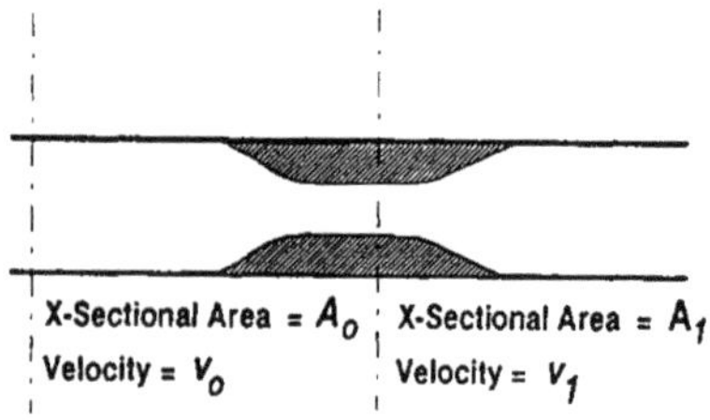

$$\% \text{ Stenosis} = 100(A_0 - A_1/A_0)$$

$$= 100\,(1 - v_0/v_1)$$

Figure 5. Diagram showing the relationship between the percentage area reduction and velocity ratio across a stenosis.

Ultrasound imaging has the advantage that it is completely non-invasive, relatively easy to perform, and provides both anatomical and hemodynamic information. With Duplex system, the graft can be imaged by B-mode scanning and the Doppler frequency spectra and velocity estimates can then be obtained at specific sites in the graft. The presence of stenoses can be detected by the narrowing of the lumen in the B-mode image. Since the reduction in cross-sectional area is accompanied by an increase in flow velocity, calculation of the degree of stenosis should be possible if it can be assumed that the stenosis has a circular cross-section (Figure 5). The increase in peak systolic velocity measured by Duplex scanning is widely used to detect and grade stenoses.[231,232] The procedure has been greatly facilitated by the introduction of color Doppler imaging as it allows a rapid evaluation of the overall flow pattern in a segment of the vessel. By using the color image as a guide, the image of the graft over its entire length can be followed. Stenoses can be localized by detecting the presence of a high velocity jet and mosaic of colors downstream relating to the velocity fluctuations in that region. The degree of stenosis may then be quantified by switching to pulsed Doppler mode ensuring the flow velocity in and upstream of the stenosis. Thus, color Doppler imaging has a distinct advantage over conventional Duplex when long graft segments are being assessed.

The color Doppler scanner has also been used for qualitative and quantitative investigation of the flow field at the distal anastomosis in an animal model.[233–235] With appropriate processing of the color Doppler data it is possible to obtain detailed three dimensional flow profiles, although the technique is limited by several factors, namely (a) the relatively poor temporal and spatial resolutions which necessitates flow measurements to be carried out in large anastomoses; (b) the inability to detect low flow velocities; (c) the measurement of only one component of velocity; and (d) the ambiguity of the Doppler velocity estimates that is caused by aliasing.[234]

V. LOCAL HEMODYNAMICS AT ANASTOMOSES

Vascular grafts are normally sutured to the recipient artery in an end-to-side configuration. Since the flow encounters a bend and an abrupt change in the cross-sectional area at the anastomosis, complex three-dimensional flow patterns, and flow disturbances may evolve. Although the flow across an end-to-end anastomosis would be much less disturbed,[175] it is rarely performed clinically because blood flow carried by collateral vessels within the bypassed segment would be lost. Energy losses across end-to-side anastomoses are generally small and are not significantly affected by the shape and angle of the anastomosis or the graft-to-artery diameter ratio.[201,236,237] On the other hand, the local hemodynamics would be expected to depend on the geometry of the anastomosis (e.g., angle, length, and shape of arteriotomy, graft-to-artery diameter ratio), as well as the fluid dynamic parameters such as the Reynolds number and the Womersley α-parameter. Since intimal hyperplasia occurs preferentially at specific sites in the distal end-to-side anastomosis (Figure 6), where flow disturbances may be present and fluid dynamic parameters such as wall shear stress may become excessive, attempts have been made to relate these parameters with the location and the degree of intimal thickening.[42,120]

The local geometry of a vascular anastomosis is dependent on the physical and mechanical properties of the graft used (this is often referred to as the handling properties by surgeons). Thus, an anastomosis constructed with an elastic vein graft will have a more rounded shape at the heel and toe than one constructed with a rigid graft. However, most of the anastomosis models used are idealized models, with straight cylindrical segments of equal diameter representing the artery and graft. These simplified models are useful since they provide basic data on the overall effects that the anastomosis geometry has on local hemodynamics and allow assessment to be carried out in a systematic way.

A. Distal Anastomosis

The majority of the studies on vascular anastomoses have been performed on the distal end-to-side anastomosis because of its greater predilection for intimal hyperplasia formation. Crawshaw et al.[118] used a dye-injection method to examine the

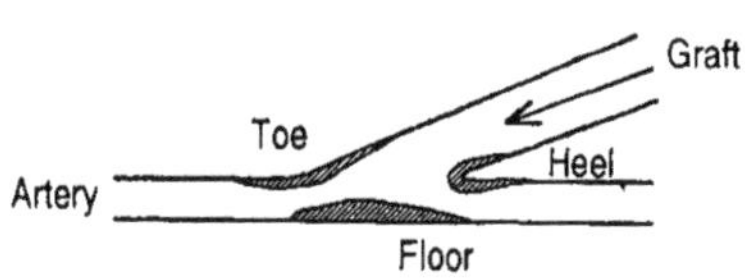

Figure 6. Distribution of intimal hyperplasia at the distal anastomosis (after Sottiurai et al.[117]).

effects of geometry and proximal outflow conditions in clear plastic models of the distal anastomosis. They observed regions of boundary layer separation near the toe of the graft and suggested that the formation of intimal hyperplasia may be reduced if the angle of anastomoses is small and the proximal artery is fully occluded. Keynton et al.[199] carried out flow visualization experiments in steady flow using the hydrogen-bubble technique in idealized models of the distal anastomosis. Velocity measurements were then performed by means of laser Doppler anemometry to quantify the effect of anastomosis angle on wall shear rates. They observed a stagnation point on the floor of the host artery and showed that the velocity profile in the distal artery was skewed toward the outer wall. They recommended that a distal anastomosis angle of 45° be used, since the minimum wall shear rate in this case was higher than the minimum values in either the 60° or 30° models. Ojha et al.[152] used a photochromic technique to estimate wall shear rate distributions in a 45° distal anastomosis model under steady and pulsatile flow conditions. Their results showed high axial velocities along the far wall of the distal artery and indicated strong secondary flow components at the near wall. In a later study, Ojha[153] observed strong fluctuations in the position of the stagnation point on the floor of the artery and noted that the sites of intimal hyperplasia correlated with regions of low shear rate at the toe and the heel of the graft and with sharp spatial and temporal variations in wall shear rate on the floor of the artery.

The flow structure may be observed in transparent models of the distal anastomoses by planar illumination of tracer particles suspended in the fluid medium with

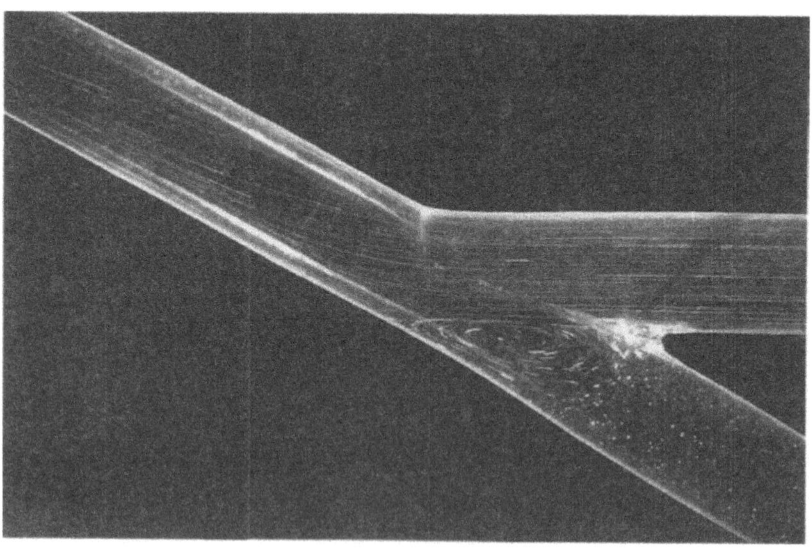

Figure 7. Steady flow pattern in a 30° distal anastomosis model at a Reynolds number of 300. The proximal outflow artery was occluded.

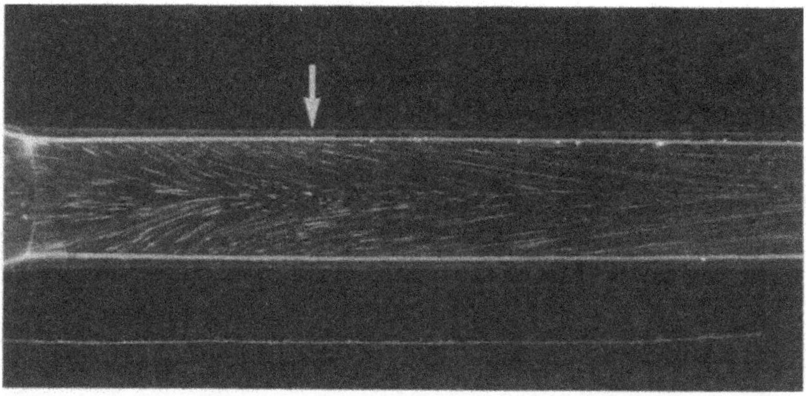

Figure 8. Steady flow pattern in the distal artery illuminated at right angle to the plane of anastomosis (30° model). Flow is from left to right.

laser light. Illumination of the flow in different planes is essential when studying complex three dimensional flows.[238] Figure 7 shows that in steady flow (Reynolds number, Re, of 300), the flow pattern is characterized by the presence of (a) a stagnation point on the floor of the anastomosis; (b) a vortex in the proximal occluded artery; and (c) helical flow in the distal artery as revealed by the outward radial component of the motion of particles from the near wall to the far wall. The double helical structure (Fig. 8) forms as the fluid from the graft impinges upon the far wall of the artery. At this point, the flow divides symmetrically along the mid-plane and the fluid passes circumferentially through both halves of the artery combining at the near wall of the artery along the mid-plane. As the Reynolds number is increased, the secondary motion becomes more evident, particularly at the near wall of the distal artery. The stagnation point on the floor of the artery moves downstream and a relatively weaker secondary vortex develops in the occluded proximal artery. A flow separation zone also occurs at the toe and a thin layer of fluid in the distal artery can be seen to move upstream toward the anastomosis toe. The formation of this separation zone is dependent on the angle of anastomosis and whether the proximal outflow artery is patent or not. Although it is not present in a 15° anastomosis even at Re in excess of 1000 when the proximal outflow artery is occluded, it forms at much lower Re even when flow in the proximal outflow artery is patent.

Figures 9a–9d show the structures under pulsatile flow at a mean Re of 500 and with the proximal outflow artery fully occluded. At the start of the sinusoidal cycle, during maximum acceleration, the flow patterns are characterized by strong axial motion in the distal artery and a weak recirculation in the occluded proximal artery (Figure 9a). The stagnation point is located almost opposite the heel of the anastomosis. At peak flow, the overall flow pattern is very similar to that in steady

(a)

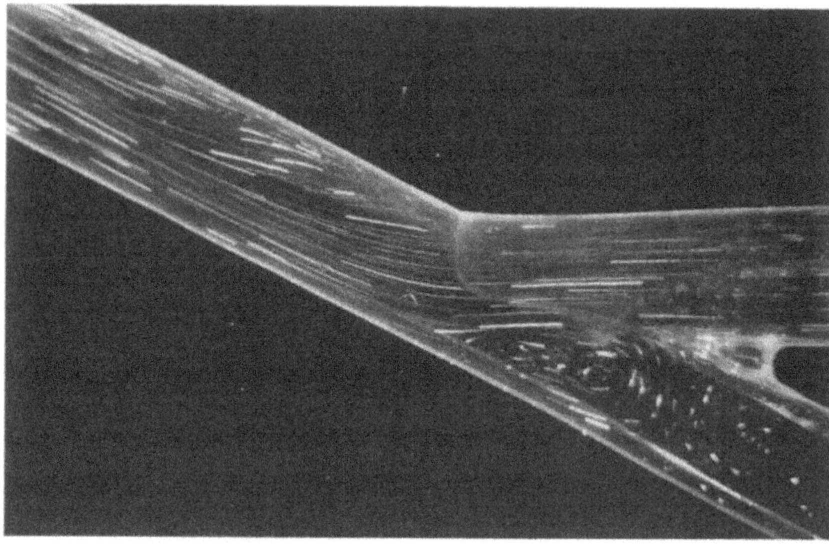

(b)

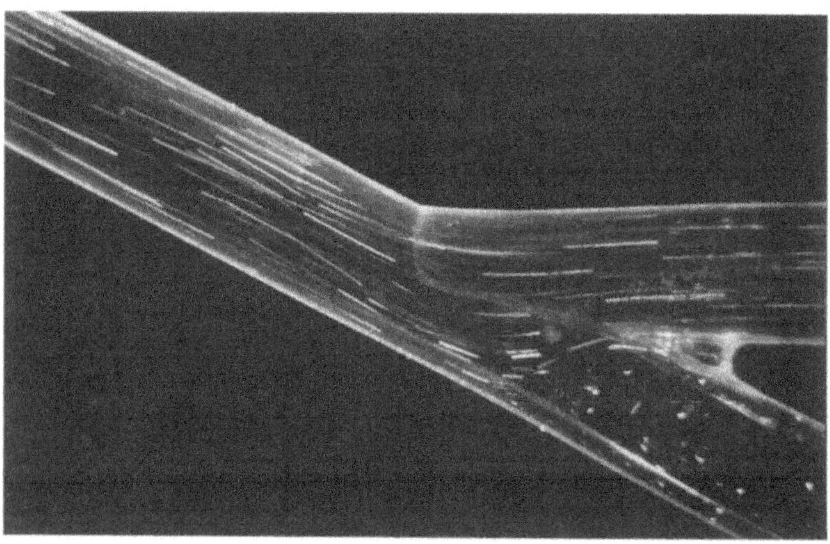

Figure 9. Pulsatile flow in a 30° distal anastomosis model at a mean Reynolds number of 500 and $\alpha = 9.7$. The proximal artery was occluded. The pulsatile flow consisted of a sinudoidal component superimposed on a steady flow. The peak-to-peak flow amplitude was twice that of the steady flow; (a) mid-acceleration phase, (b) peak flow, (c) mid deceleration phase, and (d) early acceleration phase. (Continued)

:)

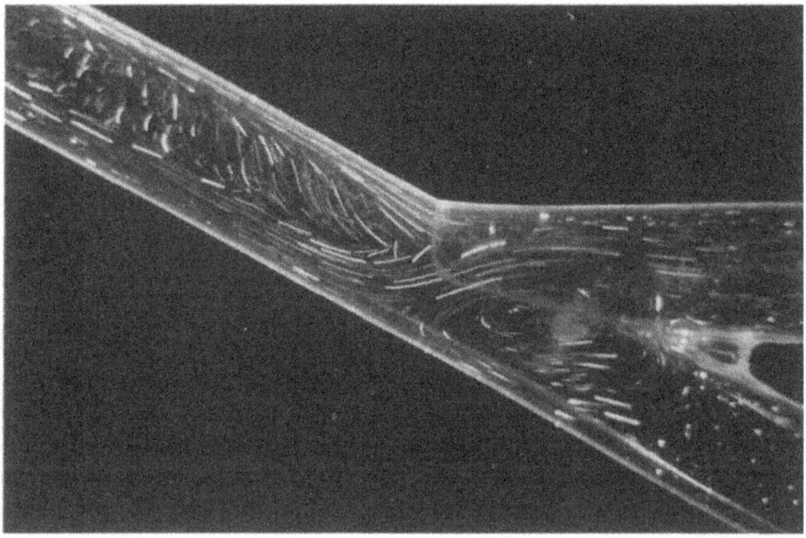

l)

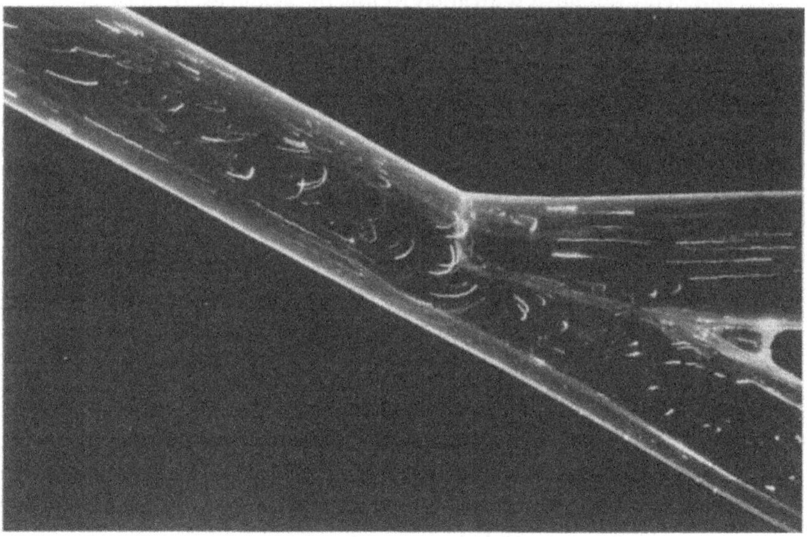

Figure 9. (Continued)

flow: secondary components are observed toward the near wall of the distal artery and the separation zone at the heel increases in size and strength while the stagnation point moves downstream (Figure 9b). During the mid-deceleration phase, the pattern in the distal artery is dominated by the secondary flow and particles at the near wall of the distal artery move upstream, particles at the far wall move axially downstream and particles in the central core move radially outward toward the far wall (Figure 9c). The flow separation region at the heel of the anastomosis moves toward the toe, and increased in size so that the stagnation point is now opposite the toe. As the flow comes to rest momentarily particles are seen to reverse in the core of the distal artery, whereas those at the far wall continued to move downstream because of their greater inertia (Figure 9d).

White et al.[213] also used particle flow visualization to study flow structures within scaled-up models of vein graft anastomoses under steady and pulsatile flow conditions. Because of the presence of a prominent sinus at the anastomosis and the extended anastomosis hood in their model, the flow patterns were more complex than those for the idealized anastomosis. The separation zone at the heel was found to contain significant helical structures with circumferential velocity components. In a parallel *in vivo* study concerning intimal thickening at the distal anastomosis, Bassiouny et al.[120] noted that the thickening around the suture line was dependent on the graft material, whereas that located on the floor of the artery was not; the site of the latter corresponded with areas of low mean and oscillatory shear rate as observed in the physical model.[213] Shu et al.[239,240] visualized the flow patterns in a model of a hemodialysis venous anastomosis and obtained detailed estimates of the pulsatile wall shear stresses with laser Doppler anemometry. They observed an oscillation of the stagnation point on the floor of the artery and the formation of separation region downstream of the anastomosis toe at the start of systole. Although the flow conditions were perhaps different from those normally obtained in illio-femoral or femoro-popliteal bypass grafts, the flow patterns still suggest that development of intimal hyperplasia is associated with low mean value or low amplitude oscillatory wall shear rates characteristic of regions of flow separation.

B. The Proximal Anastomosis

The local hemodynamics in the proximal side-to-end anastomosis models have been studied with the aid of dye injection[201,241] and photochromic tracer techniques.[242] Flow visualization studies have been carried out in this laboratory in idealized models of the proximal side-to-end anastomosis by illuminating polystyrene microspheres suspended in the fluid with a sheet of laser light.[243] The flow patterns were visualized in the symmetry plane of the model under steady flow at Re = 392. The photograph (Figure 10) reveals a separation point on the floor of the anastomosis and two counter-rotating vortices in the distal artery; note the inward radial motion in the mid-plane of the graft caused by strong secondary flow components. As the flow is increased to Re above about 420–440, there is a

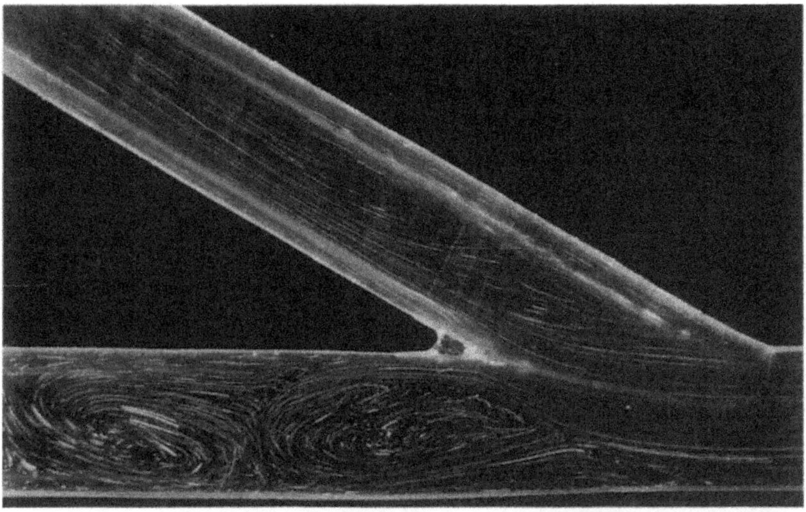

Figure 10. Steady flow pattern in a 30° distal anastomosis model at a Reynolds number of 392. The distal artery was occluded.

perceptible change in the flow pattern and time-dependent fluctuations in the primary vortex can be observed. Strong fluctuating velocity components can be observed in the primary vortex and particles are shed periodically from the occluded artery segment into the graft. As the flow rate is increased further, the vortices become stronger and the particle motions in the distal artery become even more irregular and a weak third vortex formed. The transition point can be visualized more clearly when guanine platelets (Mearlmaid, Mearl Corp) are used as tracer particles. Figures 11a and 11b were obtained just before and after the transition point (Re ≈ 440). The vortex shedding in Figure 11b is apparent from the characteristic banded structures, originating near the toe of the anastomosis, caused by the periodic shedding motion of the primary vortex in the distal occluded artery. The flow transition occurs over a narrow range of Re (425–465) and appeared to be independent of the angle of anastomosis.

The two helical vortices, one above and one below the symmetry plane of the anastomosis, are formed as the high inertia fluid entering the graft impinges on the inner wall and passes circumferentially around the upper and lower halves of the graft (Figure 12). The fluid combined along the outer wall of the graft before moving inwards along the mid-plane of the vessel.

Under pulsatile flow conditions, the flow in the graft during the peak acceleration phase is dominated by axial flow component and two vortices are present in the distal artery (Figure 13a). During the deceleration phase, secondary flow components form near the toe of the anastomosis (Figure 13b). The primary vortex

(a)

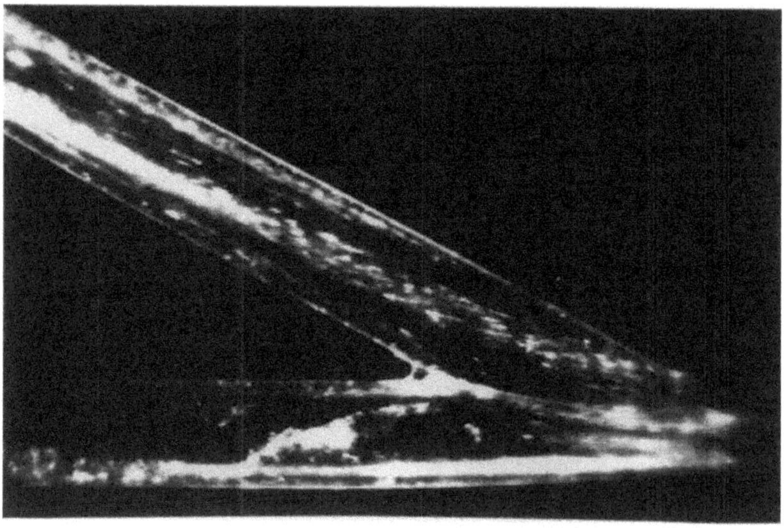

(b)

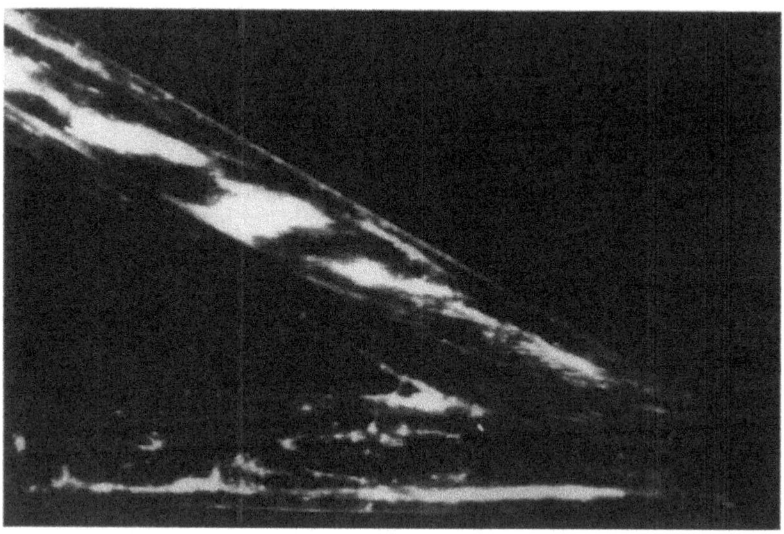

Figure 11. Flow pattern obtained in a 30° anastomosis model when a steady flow of guanine platelet particles in an aqueous suspension was used. The photographs were obtained just (a) before and (b) after the transition point (Re ≈ 440 in this case).

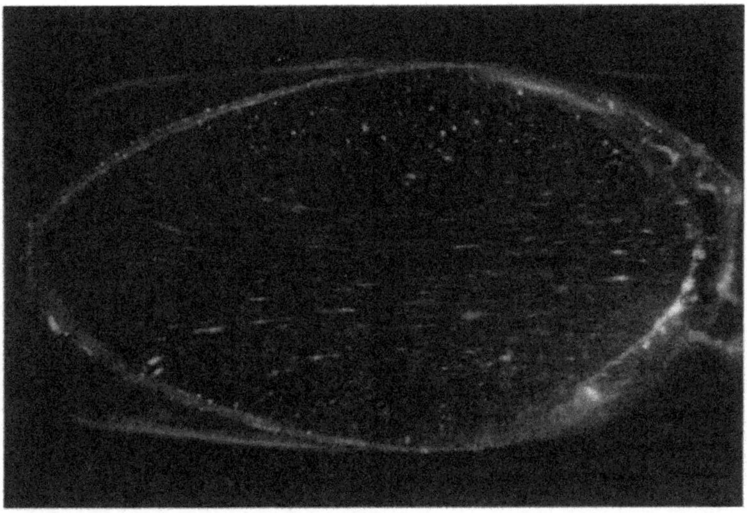

Figure 12. Flow pattern in steady flow at the junction between the artery and the graft. The 30° proximal anastomosis model was illuminated perpendicular to the plane of the anastomosis. The two helical vortices above and below the mid-plane are clearly seen. Flow is from left to right and the graft axis is normal to the page.

becomes larger and moves downstream into the distal artery. As the instantaneous flow rate approaches zero towards the end of the deceleration phase, the flow patterns in the graft are dominated by secondary components (Figure 13c). In the occluded arterial segment, the primary vortex becomes unstable and divides into two smaller vortices. As the flow begins to accelerate again, the secondary flow components in the graft are reduced and the two proximal vortices in the distal artery begin to merge (Figure 13d). The vortex at the heel of the anastomosis is shed into the graft and a small region of reverse flow is formed along the inner wall of the graft. Ojha et al.[242] found that the heel was the site where the shear stress was highest. The variation in the shear stress from cycle to cycle may be explained by the periodic nature of the vortex shedding phenomenon.

In order to explain the bias for intimal hyperplasia at the distal anastomosis, most investigators (e.g., refs. 120, 153, 213) have sought to associate local hemodynamic factors, such as the wall shear stress distribution, with this site. In cases where the proximal anastomosis has been investigated, it has been studied in isolation and the effect of graft thrombogenicity has largely been ignored.[118,201,242] The occurrence of flow disturbances at the proximal anastomosis and their propagation downstream into the graft will lead to increased interactions between platelets and the prosthetic material. Accumulation of the activated platelets in regions of high residence times at the distal anastomosis may account for the greater incidence of intimal hyperplasia at the distal anastomosis.

(a)

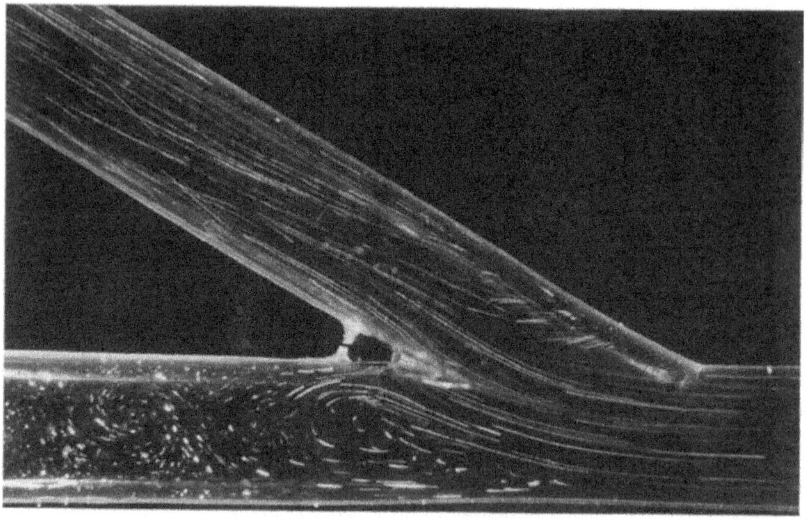

(b)

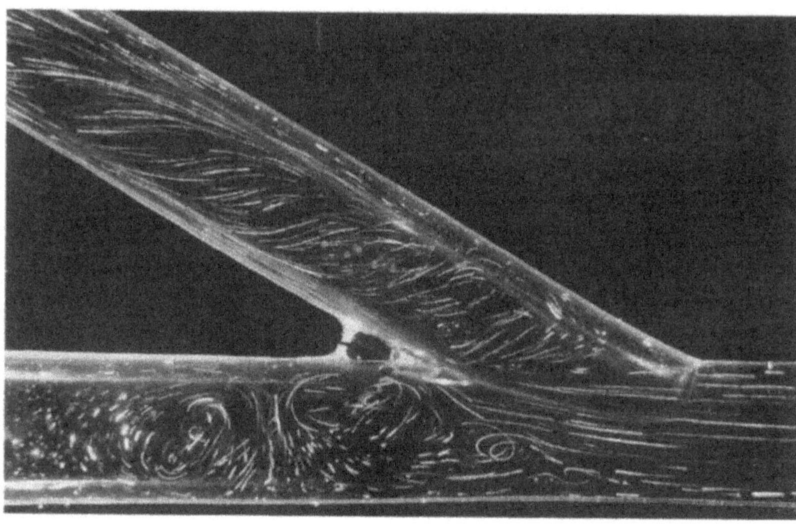

Figure 13. Pulsatile flow in a 30° proximal anastomosis at a mean Reynolds number of 500 and $\alpha = 9.7$. The distal artery was fully occluded; (a) mid-acceleration phase, (b) peak flow, (c) mid-deceleration phase, and (d) early acceleration phase. (Continued)

(c)

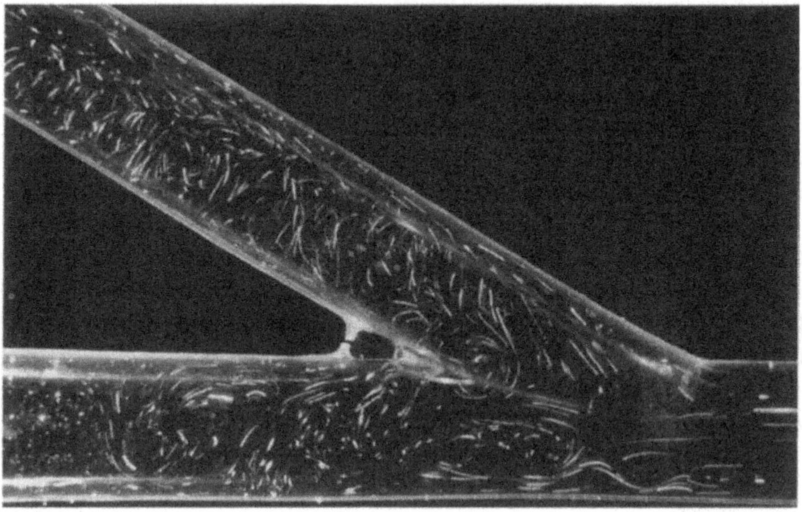

(d)

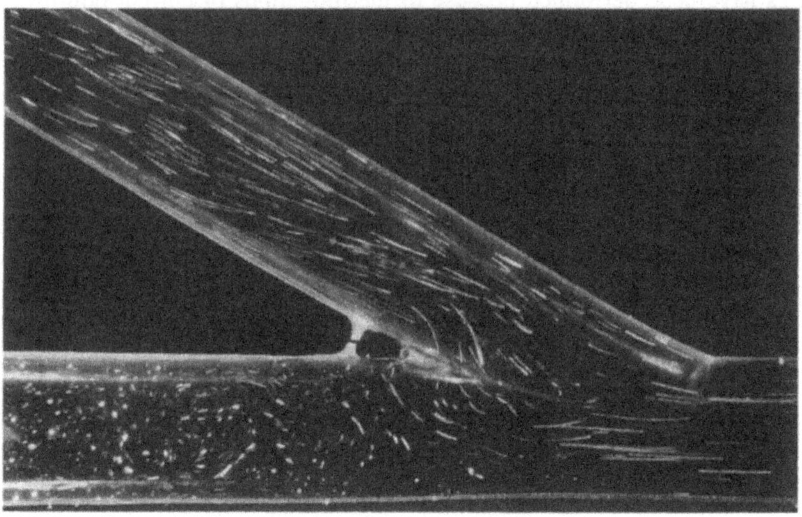

Figure 13. (Continued)

VI. CONCLUSIONS

Extensive research continues which has as its aim the development of new vascular grafts for peripheral arterial bypass in general, and novel non-thrombogenic biostable materials having good mechanical properties, in particular. Much research is also being carried out to improve the performance of existing grafts. In spite of difficulties with initial attempts at endothelial cell seeding of graft luminal surfaces, this approach is still regarded as a possible solution to the problem of poor patency rates associated with small diameter arterial grafts. The use of interposition vein cuffs and arteriovenous shunt at the distal anastomosis may provide temporary relief by minimizing the impact of intimal hyperplasia. Further work is still required since the optimum geometry of existing anastomosis configurations has yet to be established.

By definition, arterial reconstruction involves the modification of parts of the vascular system and this process may result in severe local hemodynamic disturbances. A proper understanding of the hemodynamics of arterial bypass procedures is required before bypass configurations that are associated with the least flow disturbances may be devised. The local hemodynamics in idealized models have been well characterized although experiments involving more realistic models that simulate vessel wall movement and incorporate non-Newtonian rheological behavior have yet to be performed. *In vivo* measurements are also essential but very few studies have been performed because of the limitations of current non-invasive techniques (e.g., color Doppler ultrasound imaging, magnetic resonance imaging, and digital subtraction angiography). Mathematical modeling is an important complementary technique to *in vitro* and *in vivo* studies of blood flow in vascular surgery. With the introduction of supercomputers and parallel processors, the treatment of three dimensional transient flow at vascular anastomoses is now feasible and practical. Since experimental studies can only be performed in selected cases, and under limited flow conditions, the ability to make numerical predictions in situations where direct measurements are not possible or practical is a positive advantage. Although the future is promising in this regard, until such time as numerical codes can simulate fully the three dimensional complexities of anastomotic flow patterns, there will remain a need for accurate physical models.

Finally, the advent of sophisticated imaging techniques, such as MRI and CT-scanning, when coupled with advances in computer-controlled (CNC) machining, opens the way to the fabrication of anatomically-accurate three dimensional physical models of the living cardiovascular system. This approach promises a better understanding of the hemodynamic environment in patients, both before and after surgical intervention, and its role in the progression of arterial disease.

REFERENCES

1. Whittemore, A.D.; Clowes, A.W.; Couch, N.P.; Mannick, J.A. Secondary femoropopliteal reconstruction. *Ann. Surg.* **1981**, *193*, 35–42.

2. Veith, F.J.; Gupta, S.; Daly, V. Management of early and late thrombosis of expanded polytetrafluoroethylene (PTFE) femoropopliteal bypass grafts: favourable prognosis with appropriate reoperation. *Surgery* **1980**, *87*, 581–587.
3. How, T.V.; Guidoin, R.; Young, S.K. Engineering design of vascular prostheses. *Proc. IMechE, Part H, J. Eng. Med.* **1992**, *206*, 61–71.
4. Szycher, M.; Reed, A.M.; Sciliano, A.A. *In vivo* testing of a biostable polyurethane. *J. Biomat. Appl.* **1991**, *6*, 110–130.
5. Wilson, G.J.; MacGregor, D.C.; Klement, P.; Weber, B.A.; Binnington, A.G.; Pinchuk, L. A compliant Corethane/Dacron composite vascular prosthesis. Comparison with 4-mm ePTFE grafts in a canine model. *J.A.S.A.I.O.* **1993**, *39*, M526–M531.
6. Esquivel, C.O.; Blaisdell, F.W. Why small caliber vascular grafts fail: A review of clinical and experimental experience and the significance of the interaction of blood at the interface. *J. Surg. Res.* **1986**, *41*, 1–15.
7. Towne, J.B. Role of fibrointimal hyperplasia in vein graft failure. *J. Vasc. Surg.* **1989**, *10*, 583–585.
8. Imparato, A.M.; Bracco, A.; Kim, G.E.; Zeff, R. Intimal and neointimal fibrous proliferation causing failures of arterial reconstructions. *Surgery* **1972**, *72*, 1007–1017.
9. Rodgers, V.G.J.; Teodori, M.F.; Brandt, A.M.; Borovetz, H.S. Characterization *in vitro* of the biomechanical properties of anastomosed host artery-graft combinations. *J. Vasc. Surg.* **1986**, *4*, 396–402.
10. Klein, S.R.; Goldberg, L.; Miranda, R.M.; Bosco, P.; Nelson, R.J.; White, R.A. Effect of suture technique on arterial anastomotic compliance. *Arch. Surg.* **1982**, *117*, 45–47.
11. Hasson, J.E.; Megerman, J.; Abbott, W.M. Suture technique and para-anastomotic compliance. *J. Vasc. Surg.* **1986**, *3*, 591–598.
12. Clowes, A.W.; Reidy, M.A. Mechanism of arterial graft failure: The role of cellular proliferation. In: *Blood in Contact with Natural and Artificial Surfaces* (Leonard, E.F.; Turito, V.T.; Vroman, L., Eds.). New York Academy of Sciences, New York, 1987, pp. 673–678.
13. Kaufman, B.R.; DeLuca, D.J.; Folsom, D.L.; Mansell, S.; Gorman, M.L.; Fox, P.L.; Graham, L.M. Elevated platelet-derived growth factor production by aortic grafts implanted on a long-term basis in a canine model. *J. Vasc. Surg.* **1992**, *15*, 806–816.
14. Zucker, M.B. The functioning of blood platelets. *Sci. Am.* **1980**, *242*, 86.
15. Grette, K. Studies on the mechanism of the thrombin-catalyzed hemostatic reactions in blood platelets. *Acta Physiol. Scand.* **1962**, *56*, 195–.
16. Holmsen, H. Prostaglandin endoperoxide-thromboxane synthesis and dense granule secretion as positive feedback loops in the propagation of platelet responses during the "basic platelet reaction." *Thromb. Haemostasis* **1977**, *38*, 1030–1041.
17. Salzman, E.W.; Lindon, F.; McManama, G.; Ware, J.A. Role of fibrinogen in activation of platelets by artificial surfaces. In: *Blood in Contact with Natural and Artificial Surfaces* (Leonard, E.F.; Turitto, V.T.; Vroman, L., Eds.). New York Academy of Sciences, New York, 1987, pp. 184–195.
18. Sauvage, L.R.; Walker, M.W.; Berger, K.; Robel, S.B.; Lishko, M.M.; Yates, S.G.; Logan, G.A. Current arterial prostheses. Experimental evaluation by implantation in the carotid and circumflex and coronary arteries of the dog. *Arch. Surg.* **1979**, *114*, 687–691.
19. Hashimoto, S.; Maeda, H.; Sasada, T. Effect of shear rate on clot growth at foreign surfaces. *Artif. Org.* **1985**, *9*, 345–350.
20. Chien, S. Shear dependence of effective cell volume as a determinant of blood viscosity. *Science* **1970**, *168*, 977–979.
21. Goldsmith, H.L.; Turitto, V.T. Rheological aspects of thrombosis and haemostasis: Basic principles and applications. *Thromb. Haemosta.* **1986**, *55*, 415–435.
22. Turitto, V.T.; Benis, A.M.; Leonard, E.F. Platelet diffusion in flowing blood. *Ind. Eng. Chem. Fund.* **1972**, *11*, 216–223.
23. Turitto, V.T.; Weiss, H.J. Red blood cells: Their dual role in thrombus formation. *Science* **1980**, *207*, 541–543.

24. Goldsmith, H.L. Intercellular collisions and their effect on microcirculatory transport. In: *Advances in Cardiovascular Engineering* (Hwang, N.H.C.; Turitto, V.T.; Yen, M.R.T., Eds.). Plenum Press, New York, 1992, 42–63.
25. Goldsmith, H.L.; Karino, T. Flow and vascular geometry. In: *Advances in Cardiovascular Engineering* (Hwang, N.H.C.; Turitto, V.T.; Yen, M.R.T., Eds.). Plenum Press, New York, 1992, 127–150.
26. Aarts, P.A.M.M.; Bolhuis, P.A.; Sakariassen, K.S.; Heethar, R.M.; Sixma, J.J. Red blood cell size is important for adhesion of blood platelets to artery subendothelium. *Blood* **1983**, *62*, 214–217.
27. Aarts, P.A.M.M.; van den Broek, S.A.T.; Prins, G.W.; Kuiken, G.D.C.; Sixma, J.J.; Heethar, R.M. Blood platelets are concentrated near the wall and red cells in the center in flowing blood. *Arteriosclerosis* **1988**, *8*, 819–824.
28. Alkhamis, T.M.; Bessinger, R.L.; Chediak, J.R. Red blood cell effect on platelet adhesion and aggregation in low-stress shear flow. *ASAIO Trans.* **1988**, *34*, 868–873.
29. Coller, B.S. Blood elements at surfaces: Platelets. In: *Blood in Contact with Natural and Artificial Surfaces* (Leonard, E.F.; Turitto, V.T.; Vroman, L., Eds.). New York Academy of Sciences, New York, 1987, pp. 362–379.
30. Savage, B.; Ruggeri, Z.M. Selective recognition of adhesive sites in surface-bound fibrinogen by GPIIb-IIIa on nonactivated platelets. *J. Biol. Chem.* **1991**, *266*, 11227–11233.
31. Chow, T.W.; Hellums, J.D.; Moake, J.L.; Kroll, M.H. Shear stress-induced von Willebrand factor binding to platelet glycoprotein Ib initiates calcium influx associated with aggregation. *Blood* **1992**, *80*, 113–120.
32. Ikeda, Y.; Handa, M.; Kamata, T.; Kawano, K.; Kawai, Y.; Watanabe, K.; Sakai, K.; Mayumi, F.; Itagaki, I.; Yoshioka, A.; Ruggeri, Z.M. Transmembrane calcium influx associated with von Willebrand factor binding to GPIb in the initiation of shear-induced platelet aggregation. *Thromb. Haemostas.* **1993**, *69*, 496–502.
33. Weiss, H.G.; Turitto, V.T.; Baumgartner, H.R. Effect of shear rate on platelet interaction with subendothelium in citrated and native blood. I. Shear rate-dependent decrease of adhesion in von Willebrand's disease and the Bernard-Soulier syndrome. *J. Lab. Clin. Med.* **1978**, *92*, 750–764.
34. Ikeda, Y.; Handa, M.; Kawano, K.; Kamata, T.; Murata, M.; Araki, Y.; Anbo, H.; Kawai, Y.; Watanabe, K.; Itagaki, I.; Sakai, K.; Ruggeri, Z.M. The role of von Willebrand factor and fibrinogen in platelet aggregation under varying shear stress. *J. Clin. Invest.* **1991**, *87*, 1234–1240.
35. Alevriadou, B.R.; Moake, J.L.; Turner, N.A.; Ruggeri, Z.M.; Folie, B.J.; Phillips, M.D.; Schreiber, A.B.; Hrinda, M.E.; McIntire, L.V. Real-time analysis of shear-dependent thrombus formation and its blockade by inhibitors of von Willebrand factor binding to platelets. *Blood* **1993**, *81*, 1263–1276.
36. Slack, S.M.; Cui, W.; Turitto, V.T. Fluid dynamics and thrombosis. In: *Advances in Cardiovascular Engineering* (Hwang, N.H.C.; Turitto, V.T.; Yen, M.R.T., Eds.). Plenum Press, New York, 1992, 91–102.
37. Ruggeri, Z. Mechanisms of shear-induced platelet adhesion and aggregation. *Thromb. Haemosta.* **1993**, *70*, 119–123.
38. Hanson, S.; Harker, L.; Ruggeri, Z.; Zimmerman, T. Antiplatelet GPIIb/IIa antibody (LJ-CP8) prevents thrombus formation and occlusion of small caliber vascular grafts (abstract). *Circulation* **1986**, *Suppl. II*, II–236.
39. Schlichting, H. *Boundary Layer Theory.* McGraw-Hill, New York, 1979.
40. Yongchareon, W.; Young, D.F. Initiation of turbulence in models of arterial stenoses. *J. Biomechanics* **1979**, *12*, 185–196.
41. Stein, P.D.; Sabbah, H. Turbulent blood flow in the ascending aorta of humans with normal and diseased aortic valves. *Cir. Res.* **1976**, *39*, 58–65.
42. LoGerfo, F.W.; Quist, W.C.; Nowak, M.D.; Crawshaw, H.M.; Haudenschild, C.C. Downstream anastomotic hyperplasia: A mechanism of failure in Dacron arterial grafts. *Ann. Surg.* **1983**, *197*, 479–483.

43. Ross, R.; Glomset, J.; Kariha, B.; Harker, L. A platelet-dependent serum factor that stimulates the proliferation of arterial smooth muscle cells *in vitro*. *Proc. Natl. Acad. Sci.* **1974**, *71*, 1207–1210.
44. Harker, L.A. Platelet mechanisms in the genesis and prevention of graft related vascular injury reactions and thromboembolism. In: *Vascular grafts* (Sawyer, P.N.; Kaplitt, M.J., Eds.). Appleton-Century-Crofts, New York, 1978, pp. 153–159.
45. Ross, R. The pathogenesis of atherosclerosis: An update. *N. Engl. J. Med.* **1986**, *314*, 488–500.
46. Jawien, A.; Bowen-Pope, D.F.; Lindner, V.; Schwartz, S.M.; Clowes, A.W. Platelet-derived growth factor promotes smooth muscle cell migration and intimal thickening in a rat model of balloon angioplasty. *J. Clin. Invest.* **1992**, *89*, 507–511.
47. McCabe, M.; Cunningham, G.J.; Wyatt, A.P.; Rothrie, N.G.; Taylor, G.W. A histological and histochemical examination of autogenous vein grafts. *Br. J. Surg.* **1967**, *54*, 147–155.
48. Imparato, A.M.; Baumann, F.G.; Pearson, J.; Kim, G.E.; Davidson, T.; Ibrahim, I.; Nathan, I. Electron microscopic studies of experimentally produced fibromuscular arterial lesions. *Surg. Gynecol. Obstet.* **1974**, *139*, 497–504.
49. Campeau, L.; Crochet, D.; Lesperance, J.; Bourassa, M.G.; Grondin, C.M. Postoperative changes in aortocoronary saphenous vein grafts revisited. *Circulation* **1975**, *52*, 369–377.
50. Curtis, J.J.; Stoney, W.S.; Alford, W.C.; Burrus, G.R.; Thomas, C.S. Intimal hyperplasia. A cause of radical artery aortocoronary bypass graft failure. *Ann. Thorac. Surg.* **1975**, *20*, 628–635.
51. Szilagyi, D.E.; Elliot, J.P.; Hageman, J.H.; Smith, R.F.; Dall'Olmo, C.A. Biologic fate of autogenous vein implants as arterial substitutes. *Ann. Surg.* **1973**, *178*, 232–246.
52. Austin, G.E.; Ratliff, N.B.; Hollman, J.; Tabei, S.; Phillips, D.F. Intimal proliferation of smooth muscle cells as an explanation for recurrent coronary artery stenosis after percutaneous transluminal coronary angioplasty. *J. Am. Coll. Cardiol.* **1985**, *6*, 369–375.
53. Schwartz, S.M.; Campbell, G.R.; Campbell, J.H. Replication of smooth muscle cells in vascular disease. *Circ. Res.* **1986**, *58*, 427–444.
54. Faulkner, S.L.; Fisher, R.D.; Conkle, D.M.; Page, D.L.; Bender, H.W. Effect of blood flow rate on subendothelial proliferation in venous autografts used as arterial substitutes. *Circulation* **1975**, *51/52 (Suppl. I)*, 163–172.
55. Rittgers, S.E.; Karayannocos, P.E.; Guy, J.F.; Nerem, R.M.; Shaw, G.M.; Hostetler, J.R.; Vasko, J.S. Velocity distribution and intimal proliferation in autologous vein grafts in dogs. *Cir. Res.* **1978**, *42*, 792–801.
56. Berguer, R.; Higgins, R.F.; Reddy, D.J. Intimal hyperplasia. *Arch. Surg.* **1980**, *115*, 332–335.
57. Kraiss, L.W.; Kirkman, T.R.; Kohler, T.R.; Zierler, B.; Clowes, A.W. Shear stress regulates smooth muscle proliferation and neointimal thickening in porous polytetrafluoroethylene grafts. *Arterioscler. Thromb.* **1991**, *11*, 1844–1852.
58. Kohler, T.R.; Jawien, A. Flow affects development of intimal hyperplasia after arterial injury in rats. *Arterioscler. Thromb.* **1992**, *12*, 963–971.
59. Inokuchi, K.; Kusabe, A.; Kiyose, T. Flow wave form analysis in vascular surgery. *Am. J. Surg.* **1979**, *138*, 219–223.
60. Inokuchi, K.; Kusabe, A.; Kamori, M.; Kina, M.; Okadome, K. Intraluminal velocity profile analyzed from flow waveforms. *Surgery* **1982**, *92*, 1006–1015.
61. Morinaga, K.; Okadome, K.; Kuroki, M.; Miyazaki, T.; Muto, Y.; Inokuchi, K. Effect of wall shear stress on intimal thickening of arterially transplanted autogenous veins in dogs. *J. Vasc. Surg.* **1985**, *2*, 430–433.
62. Zwolack, R.M.; Adams, M.C.; Clowes, A.W. Kinetics of vein graft hyperplasia: Association with tangential stress. *J. Vasc. Surg.* **1987**, *5*, 126–136.
63. Schettler, G.; Nerem, R.M.; Schmidt-Schonbein, H.; Morle, H.; Diehm, C. *Fluid Mechanics as a Localizing Factor for Atherosclerosis*. Springer Verlag, Berlin, 1983.
64. Texon, M. The hemodynamic concept of atherosclerosis. *Am. J. Cardiol.* **1960**, *5*, 291–294.
65. Mitchell, J.R.A.; Schwartz, C.J. *Arterial Disease*. Blackwell, Oxford, 1965.

66. Roach, M.R. Changes in arterial distensibility as a cause of poststenotic dilatation. *Am. J. Cardiol.* **1963**, *12*, 802–815.
67. Fry, D.L. Acute vascular endothelial changes associated with increased blood velocity gradients. *Cir. Res.* **1968**, *22*, 165–197.
68. Vaishnav, R.M.; Patel, D.J.; Atabek, H.B.; Deshpande, M.D.; Plowman, F.; Vossoughi, J. Determination of the local erosion stress of the canine endothelium using a jet impingement method. *J. Biomech. Eng.* **1983**, *105*, 77–83.
69. Fry, D.L. Certain histological and chemical responses of the vascular interface to acutely induced mechanical stress in the aorta of the dog. *Cir. Res.* **1969**, *24*, 93–109.
70. Fry, D.L. Certain chemorheologic considerations regarding the blood vascular interface with particular reference to coronary artery disease. *Circulation 1969, 39/40 (Suppl. IV)*, 38–57.
71. Caro, C.G.; Fitz-Gerald, J.M.; Schroter, R.C. Arterial wall shear and distribution of early atheroma in man. *Nature* **1969**, *223*, 1159–1161.
72. Caro, C.G.; Fitz-Gerald, J.M.; Schroter, R.C. Atheroma and arterial wall shear—Observation, correlation, and proposal of a shear dependent mass transfer mechanism for atherogenesis. *Proc. Roy. Soc. Lond. B.* **1971**, *177*, 109–159.
73. Nerem, R.M.; Levesque, M.J. Fluid mechanics in atherosclerosis. In: *Handbook of Bioengineering* (Skalak, R.; Chien, S., Eds.). McGraw-Hill, New York, NY, 1987, Chapter 21.
74. Nerem, R.M. Vascular fluid mechanics, the arterial wall, and atherosclerosis. *ASME J. Biomech. Eng.* **1992**, *114*, 274–282.
75. Vanhoutte, P.M.; Rubanyi, G.M.; Miller, V.M.; Houston, D.S. Modulation of vascular smooth muscle contraction by the endothelium. *Ann. Rev. Physiol.* **1986**, *48*, 307–320.
76. Simionescu, M.; Simionescu, N. Functions of the endothelial cell surface. *Ann. Rev. Physiol.* **1986**, *48*, 279–93.
77. Gimbrone, M.A. Nature's blood-compatible container. In: *Blood in Contact with Natural and Artificial Surfaces* (Leonard, E.F.; Turitto, V.T.; Vroman, L., Eds.). New York Academy of Sciences, New York, 1987, pp. 5–11.
78. Chamley-Campbell, J.H.; Campbell, G.R.; Ross, R. Phenotype-dependent response of cultured aortic smooth muscle to serum mitogens. *J. Cell. Biol.* **1981**, *89*, 379–383.
79. Gajdusek, C.; DiCorlett, P.; Ross, R.; Schwartz, S.M. An endothelial cell-derived growth factor. *J. Cell. Biol.* **1980**, *85*, 467–472.
80. Campbell, J.H.; Campbell, G.R. Endothelial cell influences on vascular smooth muscle phenotype. *Ann. Rev. Physiol.* **1986**, *48*, 295–306.
81. Levesque, M.J.; Liepsch, D.; Moravec, S.; Nerem, R.M. Correlation of endothelial cell shape and wall shear stress in a stenosed dog aorta. *Arteriosclerosis* **1986**, *6*, 220–229.
82. Levesque, M.J.; Nerem, R.M. The study of rheological effects on vascular endothelial cells in culture. *Biorheology* **1989**, *26*, 345–357.
83. Helmlinger, G.; Geiger, R.V.; Schreck, S.; Nerem, R.M. Effects of pulsatile flow on cultured vascular endothelial cell morphology. *J. Biomech. Eng.* **1991**, *113*, 123–131.
84. Levesque, M.J.; Nerem, R.M.; Sprague, E.A. Vascular endothelial cell proliferation in culture and the influence of flow. *Biomaterials* **1990**, *11*, 702–707.
85. Schwartz, C.J.; Sprague, E.A.; Fowler, S.R.; Kelley, J.L. Cellular participation in atherogenesis: Selected facets of endothelium, smooth muscle, and peripheral blood monocyte. In: *Fluid Dynamics as a Localizing Factor for Atherosclerosis* (Schettler, G.; Nerem, R.M.; Schmid-Schonbien, H.; Morle, H.; Diehm, D., Eds.). Springer-Verlag, Berlin/Heidelberg, 1983, pp. 200–207.
86. Davies, P.F.; Dewey, C.F.; Bussolari, S.R.; Gordon, E.J.; Gimbrone, M.A. Influence of hemodynamic forces on vascular endothelial function: *In vitro* studies of shear stress and pinocytosis in bovine aortic endothelial cells. *J. Clin. Invest.* **1983**, *73*, 1121–1129.
87. Frangos, J.A.; Eskin, S.G.; McIntire, L.V.; Ives, C.L. Flow effects on prostacyclin production by cultured human endothelial cells. *Science* **1985**, *227*, 1477–1479.

88. Diamond, S.L.; Eskin, S.G.; McIntire, L.V. Fluid flow stimulates tissue plasminogen activator secretion by cultured human endothelial cells. *Science* **1989**, *243*, 1483–1485.
89. Ando, J.; Komatsuda, T.; Ishikawa, C.; Kamiya, A. Fluid shear stress enhanced DNA synthesis in cultured endothelial cells during repair of mechanical denudation. *Biorheology* **1990**, *27*, 675–684.
90. Ando, J.; Ohtsuka, A.; Katayama, Y.; Korenaga, R.; Ishikawa, C.; Kamiya, A. Intracellular calcium response to directly applied mechanical shearing force in cultured vascular endothelial cells. *Biorheology* **1994**, *31*, 57–68.
91. Glagov, S.; Zarins, C.; Giddens, D.P.; Ku, D.N. Hemodynamics and atherosclerosis. *Arch. Pathol. Lab. Med.* **1988**, *112*, 1018–1031.
92. Leung, D.; Glagov, S.; Mathews, M. Cyclic stretching stimulates synthesis of matrix components of matrix components by arterial smooth muscle cells *in vitro*. *Science* **1976**, *191*, 475–477.
93. Hume, W.R. Proline and thymidine uptake in rabbit ear artery segments *in vitro* increased by chronic tangential load. *Hypertension* **1980**, *2*, 738–743.
94. Buck, R.C. Behavior of vascular smooth muscle cells during repeated stretching of the substratum *in vitro*. *Atherosclerosis* **1983**, *46*, 217–223.
95. Seidel, C.L.; Schildmeyer, L.A. Vascular smooth muscle adaptation to increased load. *Ann. Rev. Physiol.* **1987**, *49*, 489–499.
96. Kinley, C.E.; Marble, A.E. Compliance: A continuing problem with vascular grafts. *J. Cardiovasc. Surg.* **1980**, *21*, 163–170.
97. Miyawaki, F.; How, T.V.; Annis, D. Effect of compliance mismatch on flow disturbances in a model of an arterial graft replacement. *Med. & Biol. Eng. & Comput.* **1990**, *28*, 457–464.
98. Baird, R.N.; Abbott, W.M. Pulsatile blood flow in arterial grafts. *Lancet* **1976**, *2*, 948–950.
99. Sumpio, B.E.; Banes, A.J. Prostacyclin synthetic activity in cultured aortic endothelial cells undergoing cyclic stretch. *Surgery* **1988**, *104*, 383–389.
100. Sumpio, B.E.; Banes, A.J.; Levin, L.G.; Johnson, G. Mechanical stress stimulates aortic endothelial cells to proliferate. *J. Vasc. Surg.* **1987**, *6*, 252–256.
101. Guidoin, R.; Chafke, N.; Maurel, S.; How, T.; Batt, M.; Marois, M.; Gosselin, C. Expanded polytetrafluoroethylene arterial prostheses in humans: Histopathological study of 298 surgically excised grafts. *Biomaterials* **1993**, *14*, 678–693.
102. Nahon, D.; Lee, J.M.; Wilson, G.J. A two-dimensional incremental study of the static mechanical properties of vascular grafts. *Clinical Materials* **1986**, *1*, 177–197.
103. How, T.V.; Annis, D. Viscoelastic behaviour of polyurethane vascular prostheses. *J. Biomed. Mater. Res.* **1987**, *21*, 1093–1108.
104. Lyman, D.J.; Albo, D.; Jackson, R.; Knutson, K. Development of small diameter vascular prostheses. *Trans. Am. Soc. Artif. Intern. Organs* **1977**, *23*, 253–261.
105. Annis, D.; Bornat, A.; Edwards, R.O.; Higham, A.; Loveday, B.; Wilson, J. An elastomeric vascular prosthesis. *Trans. Am. Soc. Artif. Intern. Organs* **1978**, *24*, 209–214.
106. Leidner, J.; Wong, E.W.C.; MacGregor, D.C.; Wilson, G.J. A novel process for the manufacturing of porous grafts: Process description and product evaluation. *J. Biomed. Mater. Res.* **1983**, *17*, 229–247.
107. Charlesworth, D.; White, E.T.; Kent, S. A synthetic tubular structure. UK Patent, GB 2,130,521B, 11 December 1985.
108. Kowligi, R.R.; Calcotte, R.W. Vascular prostheses from polyurethanes: Methods for fabrication and evaluation. In: *High Performance Biomaterials* (Szycher M., Ed.). Technomic Press, Lancaster, PA, 1991, pp. 425–442.
109. Savvides, C.N.; Gerrard, J.H. Numerical analysis of the flow through a corrugated tube with application to arterial prostheses. *J. Fluid Mech.* **1984**, *138*, 129–160.
110. Takebayashi, J.; Kamatani, M.; Katagami, Y.; Hayashi, K.; Yanagi, Z.; Nakaue, T.; Kise, K.; Tokura, K.; Asada, K. A comparative study on the patency of crimped and noncrimped vascular prostheses, with emphasis on the earliest morphological changes. *J. Surg. Res.* **1975**, *19*, 209–218.

111. Herring, M.; Rajah, F.M.; Crow, M.J.; Kester, R.C. Evaluation of the thrombogenic potential of three types of arterial grafts studied in an artificial circulation. *Br. J. Surg.* **1978**, *65*, 272–276.
112. Klimach, O.; Underwood, C.J.; Charlesworth, D. Femoropopliteal bypass with Gore-Tex: A long term follow-up. *Br. J. Surg.* **1984**, *B71*, 821–824.
113. Acer, E.; Veith, F.J.; Morin, L.; Lesser, M.; Gupta, S.K.; Samson, R.H.; Scher, L.A.; White-Flores, S.A. Components of outflow resistance and their correlation with graft patency in lower extremity arterial reconstructions. *J. Vasc. Surg.* **1984**, *1*, 817–828.
114. Cave, F.D.; Walker, L.; Naylor, G.P.; Charlesworth, D. The hydraulic impedance of the lower limb: Its relevance to the success of bypass operations for occlusions of the superficial femoral artery. *Br. J. Surg.* **1976**, *63*, 408–412.
115. Ibrahim, I.M.; Sussman, B.; Dardik, I.; Kahn, M.; Israel, M.; Kenny, M.; Dardik, H. Adjunctive arteriovenous fistula and tibial and peroneal reconstruction for limb salvage. *Am. J. Surg.* **1980**, *140*, 246–251.
116. Dardik, H.; Berry, S.M.; Dardik, A.; Wolodiger, F.; Pecorato, J.; Ibrahim, I.M.; Sussman, B. Infrapopliteal prosthetic graft patency by use of the distal adjunctive arteriovenous fistula. *J. Vasc. Surg.* **1991**, *13*, 685–691.
117. Sottiurai, V.S.; Yao, J.S.T.; Batson, R.C.; Lim Sue, S.; Jones, R.; Nakamura, Y.A. Distal anastomotic intimal hyperplasia: Histopathologic character and biogenesis. *Ann. Vasc. Surg.* **1989**, *3*, 26–33.
118. Crawshaw, H.M.; Quist, W.C.; Serralach, E.; Valeri, R.; LoGerfo, F.W. Flow disturbance at the distal end-to-side anastomosis. *Arch. Surg.* **1980**, *115*, 1280–1284.
119. Ojha, M.; Ethier, C.R.; Johnston, K.W.; Cobbold, R.S.C. Steady and pulsatile flow fields in an end-to-side arterial anastomosis model. *J. Vasc. Surg.* **1990**, *12*, 747–753.
120. Bassiouny, H.S.; White, S.; Glagov, S.; Choi, E.; Giddens, D.P.; Zarins, C.K. Anastomotic intimal hyperplasia: Mechanical injury or flow induced. *J. Vasc. Surg.* **1992**, *15*, 708–717.
121. Black, R.A.; How, T.V. Attenuation of flow disturbances in tapered vascular grafts. *J. Biomech. Eng.* **1989**, *111*, 303–310.
122. Bamford, C.H.; Al-Lamee, K.G. Chemical methods for improving the haemocompatibility of synthetic polymers. *Clinical Materials* **1992**, *10*, 243–261.
123. Clowes, A.W.; Kirkman, T.R.; Reidy, M.A. Mechanisms of arterial graft healing: Rapid transmural capillary ingrowth provides a source of intimal endothelium and smooth muscle in porous PTFE prostheses. *Am. J. Path.* **1986**, *123*, 220–230.
124. Herring, M.; Gardner, A.; Glover, J. A single-staged technique for seeding vascular grafts with autogenous endothelium. *Surgery* **1978**, *84*, 498–504.
125. Graham, L.M.; Burkel, W.E.; Ford, J.W. Expanded polytetrafluoroethylene vascular prostheses seeded with enzymatically derived and cultured canine endothelial cells. *J. Surg. Res.* **1982**, *30*, 305–324.
126. Zilla, P. Endothelialization of vascular grafts. *Curr. Opin. Cardiol.* **1991**, *6*, 877–886.
127. Herring, M.B.; Gardner, A.; Glover, J. Seeding human arterial prostheses with mechanically derived endothelium: The detrimental effect of smoking. *J. Vasc. Surg.* **1984**, *1*, 279–289.
128. Fasol, R.; Zilla, P.; Deutsch, M.; Grimm, M.; Fischlein, T.; Laufer, G. Human endothelial cell seeding: Evaluation of its effectiveness by platelet parameters after one year. *J. Vasc. Surg.* **1989**, *9*, 432–436.
129. Herring, M.B.; Baughman, M.S.; Glover, J. Endothelium develops on seeded human arterial prostheses: A brief clinical note. *J. Vasc. Surg.* **1985**, *2*, 727–730.
130. Beard, J.D.; Beneviste, G.L.; Miller, J.H.; Baird, R.N.; Horrocks, M. Hemodynamics of the interposition vein cuff. *Br. J. Surg.* **1986**, *73*, 823–825.
131. Tyrell, M.R.; Chester, J.F.; Vipond, M.N.; Clarke, G.H.; Taylor, R.S.; Wolfe, J.H.N. Experimental evidence to support the use of the interposition vein collars/patches in distal PTFE anastomoses. *Eur. J. Vas. Surg.* **1990**, *4*, 95–101.

132. Miller, J.H.; Foreman, R.K.; Ferguson, L.; Faris, I. Interposition vein cuff for anastomosis of prosthesis to small artery. *Aust. NZ J. Surg.* **1984**, *54*, 283–285.
133. McFarland, R.J.; Taylor, R.S. An improved technique of anastomosis of femero-distal arterial prostheses. [*Une amélioration technique d'anastomose des prothèse arterielles fémoro-distales.*] *Phlébologie* **1988**, *41*, 229–233.
134. Wolfe, J.H.N.; Tyrell, M.R. Justifying arterial reconstruction to crural vessels—Even with a prosthetic graft. *Br. J. Surg.* **1991**, *78*, 897–899.
135. Taylor, R.S.; Loh, A.; McFarland, R.J.; Cox, M.; Chester, J.F. Improved technique for polytetrafluoroethylene bypass grafting: Long-term results using anastomotic vein patches. *Br. J. Surg.* **1992**, *79*, 348–354.
136. Tyrell, M.R.; Wolfe, J.H.N. New prosthetic venous collar anastomotic technique: Combining the best of other procedures. *Br. J. Surg.* **1991**, *78*, 1016–1017.
137. Taylor, R.S.; McFarland, R.J.; Cox, M.L. An investigation into the causes of failure of PTFE grafts. *Eur. J. Vas. Surg.* **1987**, *1*, 335–343.
138. Harris, P.L.; Bakran, A.; Enabi, L.; Nott, D.M. ePTFE grafts for femoro-crural bypass—Improved results with combined adjuvant venous cuff and arteriovenous fistula? *Eur. J. Vas. Surg.* **1993**, *7*, 528–537.
139. Tyrell, M.R.; Clarke, G.H.; Wolfe, J.H.N. Consideration of the mechanical properties at the distal anastomosis of PTFE grafts may improve patency rates. *J. Cardiovasc. Surg.* **1990**, *30(Suppl)*, 91.
140. Suggs, W.D.; Henriques, H.F.; DePalma, R.G. Vein cuff interposition prevents juxta-anastomotic neointimal hyperplasia. *Ann. Surg.* **1988**, *207*, 717–723.
141. Singh, M.; Lucas, C.L.; Henry, G.W.; Ferreiro, J.I.; Wilcox, B.R. Multiangle visualization of flow patterns in arterial bifurcation models. *Biorheology* **1990**, *27*, 963–970.
142. Singh, M.; Lucas, C.L.; Henry, G.W.; Ferreiro, J.I.; Wilcox, B.R. Multiangle visualization of flow patterns in saccular aneurysms. *Biorheology* **1991**, *28*, 333–339.
143. Lister, C.R.B. Flow visualization by redox-reaction dyes. *Experiments in Fluids* **1992**, *13*, 224–228.
144. Clutter, D.W.; Smith, A.M.O. Flow visualization by electrolysis of water. *Aero. Eng.* **1961**, *20*, 24–27, 74–77.
145. Schraub, F.A.; Kline, S.J.; Henry, J.; Runstadler, P.W.; Littell, A. Use of hydrogen bubbles for quantitative determination of time-dependent velocity fields in low-speed water flows. *ASME J. Basic Eng.* **1965**, *87*, 429–444.
146. Davis, W.; Fox, R.W. An evaluation of the hydrogen bubble technique for the quantitative determination of fluid velocities within clear tubes. *ASME J. Basic Eng.* **1967**, *89*, 771–781.
147. Hughes, P.E.H. The hemodynamics of arterial prostheses. Ph.D. Thesis, University of Liverpool, U.K., 1993.
148. Popovich, A.T.; Hummel, R.L. A new method for non-disturbing turbulent flow measurements very close to the wall. *Chem. Eng. Sci.* **1967**, *22*, 21–25.
149. Poots, K.; Cobbold, R.S.C.; Johnston, K.W.; Appugliese, R.; Kassam, M.; Zuech, P.E.; Hummel, R.L. A new pulsatile flow visualization method using a photochromic dye with application to Doppler ultrasound. *Ann. Biomed. Eng.* **1986**, *14*, 203–218.
150. Ojha, M.; Hummel, R.L.; Cobbold, R.S.C.; Johnston, K.W. Development and evaluation of a high resolution photochromic dye method for pulsatile flow studies. *J. Phys. E: Sci. Instrum.* **1988**, *21*, 998–1004.
151. Ojha, M.; Cobbold, R.S.C.; Johnston, K.W.; Hummel, R.L. Pulsatile flow through constricted tubes: An experimental investigation using photochromic tracer methods. *J. Fluid Mech.* **1989**, *203*, 173–197.
152. Ojha, M.; Ethier, C.R.; Johnstone, K.W.; Cobbold, R.S.C. Steady and pulsatile flow fields in an end-to-side arterial anastomosis model. *J. Vasc. Surg.* **1990**, *12*, 747–753.
153. Ojha, M. Spatial and temporal variations of wall shear stress within an end-to-side arterial anastomosis model. *J. Biomechanics* **1993**, *26*, 1377–1388.

154. Yurechko, V.N.; Ryazantsev, Y.S. Fluid motion investigation by the photochromic flow visualization technique. *Exp. Therm. Fluid Sci.* **1991**, *4*, 273–288.
155. Yurechko, V.N. Photochromic flow visualization for the investigation of artificial heart valves. *Int. J. Artif. Organs* **1993**, *16*, 146–150.
156. Karino, T.; Goldsmith, H.L. Role of cell-wall interactions in thrombogenesis and atherogenesis. A microrheological study. *Biorheology* **1984**, *21*, 587–601.
157. Karino, T.; Goldsmith, H.L.; Motomiya, M.; Mabuchi, S.; Sohara, Y. Flow patterns in vessels of simple and complex geometries. In: *Blood in Contact with Natural and Artificial Surfaces* (Leonard, E.F.; Turitto, V.T.; Vroman, L., Eds.). New York Academy of Sciences, New York, 1987, pp. 422–441.
158. Goldsmith, H.L.; Karino, T. Flow and vascular geometry. In: *Advances in Cardiovascular Engineering* (Hwang, N.H.C.; Turitto, V.T.; Yen, M.R.T., Eds.). Plenum Press, New York, 1992, pp. 127–150.
159. Hesselink, L.; Helman, J.; Ning, P. Quantitative image processing in fluid mechanics. *Exp. Therm. Fluid. Sci.* **1992**, *5*, 605–616.
160. Reynolds, A.J. *Turbulent Flows in Engineering*. Wiley-Interscience, London, 1974.
161. Khalifa, A.M.A.; Giddens, D.P. Analysis of disorder in pulsatile flows with application to poststenotic blood velocity measurement in dogs. *J. Biomechanics* **1978**, *11*, 129–141.
162. Ling, S.C.; Atabek, H.B. Measurement of aortic blood flow in dogs by the hot film technique. *Proc. 19th Ann. Conf. on Engng. in Med. & Biol.* **1966**, *8*, 113.
163. Ling, S.C.; Atabek, H.B.; Fry, D.L.; Patel, D.J.; Janicki, J.S. Application of heated-film velocity and shear probes to hemodynamic studies. *Cir. Res.* **1968**, *23*, 789–801.
164. Bellhouse, B.J.; Bellhouse, F.H. Thin film gauges for the measurement of velocity or skin friction in air, water or blood. *J. Sci. Instrum.* **1968**, *1*, 1211–1213.
165. Schultz, D.L.; Tunstall-Pedoe, D.S.; Lee, G.D.; Gunning, A.J.; Bellhouse, B.J. Velocity distribution and transition in the arterial system. In: *Ciba Foundation Symposium on Circulatory and Respiratory Mass Transport* (Wolstenholme, G.E.W.; Knight, J., Eds.). Churchill, London, 1969.
166. Seed, W.A.; Wood, N.B. Use of a hot-film velocity probe for cardiovascular studies. *J. Phys. E: Sci. Instrum.* **1970**, *3*, 377–384.
167. Seed, W.A.; Wood, N.B. Development and evaluation of a hot-film velocity probe for cardiovascular studies. *Cardiovasc. Res.* **1970**, *4*, 253–263.
168. Seed, W.A.; Wood, N.B. Velocity patterns in the aorta. *Cardiovasc. Res.* **1971**, *5*, 319–330.
169. Seed, W.A.; Wood, N.B. Application of constant temperature anemometry in measurement of intra-arterial blood flow velocity. *Int. J. Engng. Sci.* **1972**, *10*, 1009–1021.
170. Nerem, R.M.; Seed, W.A. An *in vivo* study of the nature of aortic flow disturbances. *Cardiovasc. Res.* **1972**, *6*, 1–14.
171. Paulsen, P.K. Use of a hot-film anemometer system for cardiovascular studies, with special reference to the ascending aorta. *Dan. Med. Bull.* **1989**, *36*, 430–443.
172. Baker, D.W. Pulsed ultrasonic flowmeter. In: *Methods in Medical Research* (Rushmer, R.F., Ed.). Year Book Medical Publishers Inc., Chicago, IL, 1966.
173. Perroneau, P.A.; Hinglais, J.; Pellet, M.; Leger, F. Pulsed-wave Doppler ultrasound blood velocimeter. A description of the device—Results. [*Velocimetre sanguin par effet Doppler a emission ultra-sonore pulsee. A. Description de l'appareil—Resultats.*] *L'onde Electrique* **1970**, *50*, 3–18.
174. Hartley, C.J.; Cole, J.S. An ultrasonic pulsed Doppler system for measuring blood flow in small vessels. *J. Appl. Physiol.* **1974**, *37*, 626–629.
175. Bandyk, D.F.; Zierler, R.E.; Berni, G.A.; Thiele, B.L. Pulsed Doppler velocity patterns produced by arterial anastomoses. *Ultrasound in Med. & Biol.* **1983**, *9*, 79–87.
176. Giddens, D.P.; Khalifa, A.M.A. Turbulence measurements with pulsed Doppler ultrasound employing a frequency tracking method. *Ultrasound in Med. & Biol.* **1982**, *8*, 427–437.

177. Tarbell, J.M.; Gunshinan, J.P.; Geselowitz, D.B.; Rosenberg, G.; Shung, K.K.; Pierce, W.S. Pulsed ultrasonic Doppler velocity measurements inside a left ventricular assist device. *J. Biomech. Eng.* **1986**, *108*, 232–238.
178. Hughes, P.E.; How, T.V. Pulsatile velocity distribution and wall shear rate measurement using pulsed Doppler ultrasound. *J. Biomechanics* **1994**, *42*, 103–110.
179. Black, R.A. The hydrodynamics of tapered arterial prostheses. Ph.D. Thesis, University of Liverpool, 1989.
180. Black, R.A.; How, T.V. A pulsed ultrasound system for the measurement of velocity distribution and flow disturbances in arterial prostheses. *J. Biomed. Eng.* **1989**, *11*, 35–42.
181. Nygaard, H.; Hasenkam, J.M.; Pedersen, E.M.; Kim, W.Y.; Paulsen, P.K. A new perivascular multi-element pulsed Doppler ultrasound system for *in vivo* studies of velocity fields and turbulent shear stresses in large vessels. *Med. & Biol. Eng. & Comput.* **1994**, *32*, 55–62.
182. Jorgensen, J.E.; Campau, D.N.; Baker, D.W. Physical characteristics and mathematical modelling of the pulsed ultrasonic flowmeter. *Med. Biol. Eng.* **1973**, *11*, 404–420.
183. Jorgensen, J.E.; Garbini, J.L. An analytical procedure of calibration for the pulsed ultrasonic Doppler flowmeter. *J. Fluids Eng.* **1974**, *96*, 158–167.
184. Garbini, J.L.; Forster, F.K.; Jorgensen, J.E. Measurement of fluid turbulence based on pulsed ultrasound techniques, part 1. *J. Fluid Mech.* **1982**, *118*, 445–470.
185. Garbini, J.L.; Forster, F.K.; Jorgensen, J.E. Measurement of fluid turbulence based on pulsed ultrasound techniques, part 2. *J. Fluid Mech.* **1982**, *118*, 471–505.
186. George, W.K.; Lumley, J.L. The laser Doppler velocimeter and its application to the measurement of turbulence. *J. Fluid Mech.* **1973**, *60*, 321–362.
187. Durst, F.; Melling, A.; Whitelaw, J.H. *Principles and Practice of Laser-Doppler Anemometry. Academic Press, London, 1981.*
188. Deshpande, M.D.; Giddens, D.P. Turbulence measurements in a constricted tube. *J. Fluid Mech.* **1980**, *97*, 65–89.
189. Ahmed, S.A.; Giddens, D.P. Flow disturbance measurements through a constricted tube at moderate Reynolds numbers. *J. Biomechanics* **1983**, *16*, 955–963.
190. Ahmed, S.; Giddens, D.P. Pulsatile poststenotic flow studies with laser Doppler anemometry. *J. Biomechanics* **1984**, *17*, 695–705.
191. Khalifa, A.M.A.; Giddens, D.P. Characterization and evolution of poststenotic flow disturbances. *J. Biomechanics* **1981**, *14*, 279–296.
192. Lieber, B.B.; Giddens, D.P. Post-stenotic core flow behavior in pulsatile flow and its effects on wall shear stress. *J. Biomechanics* **1990**, *23*, 597–605.
193. Baldwin, J.T.; Deutsch, S.; Geselowitz, D.B.; Tarbell, J.M. LDA measurements of mean velocity and Reynolds stress fields within an artificial heart ventricle. *J. Biomech. Eng.* **1994**, *116*, 190–200.
194. Baldwin, J.T.; Tarbell, J.M.; Deutsch, S.; Geselowitz, D.B. Mean flow velocity patterns within a ventricular assist device. *ASAIO-Trans.* **1989**, *35*, 429–433.
195. Baldwin, J.T.; Deutsch, S.; Geselowitz, D.B.; Tarbell, J.M. LDA Estimation of Reynolds stresses within the Penn State left ventricular assist device. *ASAIO-Trans.* **1990**, *36*, M274–278.
196. Baldwin, J.T.; Tarbell, J.M.; Deutsch, S.; Geselowitz, D.B. Mean velocities and Reynolds stresses within regurgitant jets produced by tilting disc valves. *ASAIO-Trans.* **1991**, *37*, M348–349.
197. Yoganathan, A.P.; Woo, Y.R.; Sung, H.W. Turbulent shear stress measurements in the vicinity of aortic heart valve prostheses. *J. Biomechanics* **1986**, *19*, 433–443.
198. Friedman, M.H.; Bargeron, C.B.; Duncan, D.D.; Hutchins, G.M.; Mark, F.F. Effects of arterial compliance and non-Newtonian blood rheology on correlations between intimal thickness and wall shear. *J. Biomech. Eng.* **1992**, *114*, 317–320.
199. Keynton, R.S.; Rittgers, S.E. The effect of angle and flow rate upon hemodynamics in distal vascular graft anastomoses: An *in vitro* model study. *J. Biomech. Eng.* **1991**, *113*, 458–463.
200. Schultz, R.D.; Hokanson, D.E.; Strandness, D.E. Pressure-flow relations of the end-to-side anastomosis. *Surgery* **1967**, *62*, 319–324.

201. LoGerfo, F.W.; Soncrant, T.; Teel, T.; Dewey, C.F. Boundary layer separation in models of side-to-end arterial anastomoses. *Arch. Surg.* **1979**, *114*, 1369–1373.
202. Rieu, R.; Pelissier, R. *In vitro* study of a physiological type flow in a bifurcated vascular prosthesis. *J. Biomech.* **1991**, *24*, 923–933.
203. Shombert, D.G. Measurement of steady-flow instability and turbulence levels in Dacron vascular grafts. *J. Biomech. Eng.* **1992**, *114*, 521–526.
204. Fergusson, G.C.; Roach, M.R. Flow conditions at bifurcations as determined in glass models, with reference to the focal distribution of vascular lesions. In: *Cardiovascular Fluid Dynamics* (Bergel, D.H., Ed.). Academic Press, New York, 1972, 141–156.
205. Pedersen, E.M.; Yoganathan, A.P.; Lefebvre, X.P. Pulsatile flow visualization in a model of the human abdominal aorta and aortic bifurcation. *J. Biomech.* **1992**, *25*, 935–944.
206. Zarins, C.K.; Giddens, D.P.; Bharadvaj, B.K.; Sottiurai, V.S.; Mabon, R.F.; Glagov, S. Carotid bifurcation atherosclerosis: Quantitative correlation of plaque localization with flow velocity profiles and wall shear stress. *Circ. Res.* **1983**, *53*, 502–514.
207. Ku, D.N.; Liepsch, D. The effects of non-Newtonian viscoelasticity and wall elasticity on flow at a 90° bifurcation. *Biorheology* **1986**, *23*, 359–370.
208. Mark, F.F.; Bargeron, C.B.; Deters, O.J.; Friedman, M.H. Variations in geometry and shear rate distribution in casts of human aortic bifurcations. *J. Biomech.* **1989**, *22*, 577–582.
209. Duncan, D.D.; Bargeron, C.B.; Borchardt, S.E.; Deters, O.J.; Gearhart, S.A.; Mark, F.F. The effect of compliance on wall shear in casts of a human aortic bifurcation. *J. Biomech. Eng.* **1990**, *112*, 183–188.
210. How and Clarke. The elastic properties of a polyurethane arterial prosthesis. *J. Biomechanics* **1984**, *17*, 597–608.
211. How, T.V.; Marois, Y.; Guidoin, R.; Black, R.A. Effect of geometric taper on thrombosis in vascular prostheses: An *ex vivo* study. *J. Biomat. Sci. Polymer Edn.* **1994**, *6*, 111–122.
212. Black, R.A.; How, T.V. Effect of wall compliance and geometry on flow disturbances in arterial prostheses. *Clin. Hemorheol.* **1992**, *12*, 45–54.
213. White, S.S.; Zarins, C.K.; Giddens, D.P.; Bassiouny, H.; Loth, F.; Jones, S.A.; Glagov, S. Hemodynamic patterns in two models of end-to-side vascular graft anastomoses: Effects of pulsatility, flow division, Reynolds number and hood length. *J. Biomech. Eng.* **1993**, *115*, 104–111.
214. Liepsch, D.; Moravec, S. Pulsatile flow of non-Newtonian fluid in distensible models of human arteries. *Biorheology* **1984**, *21*, 571–586.
215. Liepsch, D.; Thurston, G.; Lee, M. Studies of fluids simulating blood-like rheological properties and applications in models of arterial branches. *Biorheology* **1991**, *28*, 39–52.
216. Liepsch, D. Flow studies in a rigid T-junction model with a non-Newtonian fluid using a 3-D laser-Doppler-anemometer. In: *Biofluid Dynamics* (Liepsch, D., Ed). Springer Verlag, Berlin, 1990, pp. 307–320.
217. Ascer, E.; Veith, F.; Morin, L.; White-Flores, S.A.; Scher, L.A.; Samson, R.H.; Weiser, R.K.; Rivers, S.; Gupta, S.K. Quantitative assessment of outflow resistance in lower extremity arterial reconstructions. *J. Surg. Res.* **1984**, *37*, 8–15.
218. Parvin, S.D.; Evans, D.H.; Bell, P.R.F. Peripheral resistance measurement in the assessment of severe peripheral vascular. *Br. J. Surg.* **1985**, *72*, 751–753.
219. Cave, F.D.; Walker, A.; Naylor, G.P.; Charlesworth, D. The hydraulic impedance of the lower limb: Its relevance to the success of bypass operations for occlusion of the superficial femoral artery. *Br. J. Surg.* **1976**, *63*, 408–412.
220. Butler, P.A.; Somerville, P.G. The use of hydraulic input impedance in the evaluation of obliterative arterial disease In: *Cardiovascular and Pulmonary Dynamics* (Jaffrin, M. Y., Ed.). Edition INSERM Paris, 1978, pp. 155–164.
221. Taylor, P.R.; Wolfe, J.H.N.; Tyrell, M.R.; Mansfield, A.O.; Nicolaides, A.N.; Houston, R.E. Graft stenosis: Justification for 1-year surveillance. *Br. J. Surg.* **1990**, *77*, 1125–1128.

222. Bandyk, D.F.; Seabrook, G.R.; Moldenhauer, P.; Lavin, J.; Edwards, J.; Cato, R.; Towne, J.B. Hemodynamics of vein graft stenosis. *J. Vasc. Surg.* **1988**, *8*, 688–695.
223. Brennan, J.A.; Thrush, A.J.; Evans, D.H.; Bell, P.R.F. Intensive blood flow monitoring following femoro-distal bypass: Can early outcome be reliably predicted? *Eur. J. Vas. Surg.* **1992**, *6*, 148–152.
224. Wolfe, J.H.N.; Lea Thomas, M.; Jamieson, C.W.; Browse, N.L.; Burnand, K.G.; Rutt, D.L. Early diagnosis of femorodistal graft stenoses. *Br. J. Surg.* **1987**, *74*, 268–270.
225. Harris, P.L.; How, T.V.; Jones, D.R. Prospectively randomised clinical trial to compare *in situ* and reversed vein grafts for femoro-popliteal bypass. *Br. J. Surg.* **1987**, *74*, 252–255.
226. Berkowitz, H.D.; Hobbs, C.L.; Roberts, B.; Frieman, D.; Oleaga, J.; Ring, E. Value of routine vascular laboratory studies to identify vein graft stenosis. *Surgery* **1981**, *90*, 971–979.
227. Laing, S.P.; Greenhalgh, R.M. Standard exercise test to assess peripheral arterial disease. *Br. Med. J.* **1980**, *280*, 13–16.
228. Wyatt, M.G.; Muir, R.M.; Tennant, W.G.; Scott, D.J.A.; Baird, R.N.; Horrocks, M. Impedance analysis to identify the at risk femorodistal graft. *J. Vasc. Surg.* **1991**, *13*, 284–293.
229. McShane, M.D.; Gazzard, V.M.; Clifford, P.C.; Hacking, C.N.; Fairhurst, J.J.; Humphries, K.N.; Birch, S.J.; Webster, J.H.H.; Chant, A.D.B. Duplex ultrasound assessment of femorodistal grafts: Correlation with angiography. *Eur. J. Vas. Surg.* **1987**, *1*, 409–414.
230. Moody, P.; de Cossart, L.M.; Douglas, H.M.; Harris, P.L. Asymptomatic strictures in femoro-popliteal vein grafts. *Eur. J. Vas. Surg.* **1989**, *3*, 389–392.
231. Taylor, P.R.; Tyrell, M.R.; Crofton, M.; Bassan, B.; Grigg, M.; Wolfe, J.H.N.; Mansfield, A.O.; Nicolaides, A.N. Colour flow imaging in the detection of femoro–distal graft and native artery stenosis: Improved criteria. *Eur. J. Vas. Surg.* **1992**, *6*, 232–236.
232. Grigg, M.J.; Nicolaides, A.N.; Wolfe, J.H.N. Detection and grading of femorodistal vein graft stenoses: Duplex velocity measurements compared with angiography. *J. Vasc. Surg.* **1988**, *8*, 661–666.
233. Staalsen, N-H.; Pedersen, E.M.; Ulrich, M.; Winther, J.; How, T.V.; Hasenkam, J.M. An *in vivo* model for studying local hemodynamics at vascular end-to-side anastomoses. *Eur. J. Vas. Surg.* **1995**, *9*, 152–161.
234. Staalsen, N-H.; Ulrich, M.; Kim, W.Y.; Pedersen, E.M.; How, T.V.; Hasenkam, J.M. *In vivo* analysis and three dimensional visualization of blood flow patterns at vascular end-to-side anastomoses. *Eur. J. Vas. Surg.* **1995**, *117*, 224–232.
235. Staalsen, N-H.; Ulrich, M.; Winther, J.; Pedersen, E.R.; How, T.V.; Nygaard, H. The anastomosis angle does change the local flow field at vascular end-to-side anastomosis *in vivo*. *J. Vasc. Surg.* **1995**, *10*, 168–181.
236. Szilagyi, D.E.; Whitcomb, J.G.; Schenker, W.; Waibel, P. The laws of fluid flow and arterial grafting. *Br. J. Surg.* **1960**, *47*, 55–73.
237. Schultz, R.D.; Hokanson, D.E.; Strandness, D.E. Pressure-flow relations of the end-to-side anastomosis. *Br. J. Surg.* **1967**, *62*, 319–324.
238. Hughes, P.E.; How, T.V. Effects of geometry and flow division on flow structures in models of the distal end-to-side anastomosis. *J. Biomechanics*, in press.
239. Shu, M.C.S.; Noon, G.P.; Hwang, N.H.C. Phasic flow patterns at a hemodialysis venous anastomosis. *Biorheolgy* **1987**, *24*, 711–722.
240. Shu, M.C.S.; Noon, G.P.; Hwang, N.H.C. Flow profiles and wall shear stress distribution at a hemodialysis venous anastomosis: Preliminary study. *Biorheolgy* **1987**, *24*, 723–735.
241. Nowak M.D.; LoGerfo, F.W.; Quist, W.C. Models of side-to-end anastomoses: Effects of angle and flow split. *J. Surg. Res.* **1982**, *32*, 489–498.
242. Ojha M.; Cobbold, R.S.C.; Johnston, K.W. Hemodynamics of a side-to-end proximal arterial anastomosis model. *J. Vasc. Surg.* **1993**, *17*, 646–655.
243. Hughes, P.E.; How, T.V. Flow structures at the proximal side-to-end anastomosis. Influence of geometry and flow division. *J. Biomech. Eng.*, in press.

INDEX

www.ingramcontent.com/pod-product-compliance
Ingram Content Group UK Ltd.
Pitfield, Milton Keynes, MK11 3LW, UK
UKHW020228290926
15470UKWH00001B/424

9 781559 3863